W9-CDD-420

A Review of Science and Technology During the 1976 School Year

Science Year

The World Book Science Annual

1977

Field Enterprises Educational Corporation
Chicago Frankfurt London Paris Rome Sydney Tokyo Toronto

A subsidiary of
Field Enterprises, Inc. fe

ISBN 0-7166-0577-5
Library of Congress Catalog Card Number: 65:21776
Printed in The United States of America

The publishers of *Science Year* gratefully
acknowledge the following for permission to use
copyrighted illustrations. A full listing of illustration
acknowledgments appears on pages 430 and 431.

136 L. A. Nagy, from *Science*. Copyright 1974 by the American Association for the Advancement of Science

181 © 1975 by the Association of Universities for Research in Astronomy, Inc. John Luthes, Kitt Peak National Observatory

231 © National Geographic Society
Cleveland Museum of Natural History; © National Geographic Society
David Brill; © National Geographic Society

238 Copyright 1976 by permission of *Saturday Review* and Nick Hobart

253 Andrew H. Knoll and Elso S. Barghoorn, Harvard University. Copyright 1975 by the American Association for the Advancement of Science

270 Copyright 1976 by permission of *Saturday Review* and James Estes

296 Drawing by Ross; © 1975 The New Yorker Magazine, Inc.

309 Richard Blakemore, Woods Hole Oceanographic Institution from *Science*. Copyright 1975 by the American Association for the Advancement of Science

312 Martha Constantine-Paton and Robert E. Caprancia, Cornell University from *Science*. Copyright 1975 by the American Association for the Advancement of Science

389 © Karsh, Ottawa, from Kittay Science Foundation

The Cover: The theater of the microsurgeon

Preface

Reporting on the excitement and promise of science and technology is a primary purpose of *Science Year*. To be useful, however, a science annual must also report instances where the promise is delayed or undelivered. Advances in scientific knowledge frequently bring problems that are not always easy to solve. And they raise questions about the pace and direction that are not always easy to answer.

Nowhere is this more evident than on the medical scene, where doctors and other people are wrestling with the problem of whether to turn off life-supporting equipment on comatose patients who have no chance of recovering. In his essay "The Choice to Die," Robert M. Veatch points out that the medical equipment and surgical skills that save or prolong lives can also save or prolong those that, by certain measures, are not worth prolonging. The moral, ethical, and legal aspects of decisions in such matters indicate the complexity of a problem that technology can create.

Another area where technology has been causing confusion and controversy is nuclear energy. This year the debate on the safety of nuclear plants reached the voting booth, as reported in ENVIRONMENT. On June 8, 1976, a referendum was held in California to consider a moratorium on new nuclear plants and sharply curtail the operations of the three existing ones until the state legislature could pass judgment on the safety systems. Although the proposition was defeated by a 2 to 1 margin, it indicates that a growing number of people are concerned about this technology. Interestingly, a map that accompanies ENVIRONMENT shows that a California nuclear reactor, built to withstand tremors generated by two relatively distant faults, is only 2½ miles from a newly discovered one.

Safety is only one of the problems plaguing the nuclear energy industry. Some people are looking at the increasing costs of plant construction and the steady rise in the price of uranium and declaring that this source of energy will soon become uneconomic. Certainly the debate takes on new meaning when we read in GEOLOGY that new, more accurate methods of estimating uranium ore reserves in the United States indicate that there is much less left than we previously thought. Thus, when we turn to ENERGY, we are not cheered to learn that the first demonstration model for a fast breeder reactor, which theoretically produces more fuel than it consumes, is not scheduled to begin operating until 1983. And, although there are encouraging advances in the development of fusion reactors, as explained in PLASMA PHYSICS, this technology also still has a long way to go.

Before we decide that perhaps the scientists should close up their laboratories and the technologists their factories, we should remember that the successes far outnumber the failures. Most of the articles in this edition prove that the excitement and promise of science are alive and well. What is indicated is that more people should be asking more questions, and deciding whether they need or want all of the things that science and technology seem able to give. To do this, they have to be informed. The editors of *Science Year* consider it our responsibility to help inform them.

[Arthur G. Tressler]

Contents

Staff

Editorial Director
William H. Nault

Editorial

Executive Editor
Arthur G. Tressler

Managing Editor
Michael Reed

Chief Copy Editor
Joseph P. Spohn

Senior Editors
Robert K. Johnson
Edward G. Nash
Kathryn Sederberg
Darlene Stille
Foster P. Stockwell

Assistant Copy Editor
Irene B. Keller

Editorial Assistant
Madelyn Krzak

Art

Executive Art Director
William Dobias

Art Director
James P. Weren

Senior Artists
Roberta Dimmer
Mary-Ann Lupa

Artist
Wilma Stevèns

Photography Director
Fred C. Eckhardt, Jr.

Photo Editing Director
Ann Eriksen

Senior Photographs Editors
Blanche Cohen
Leslie J. Cohn
Marilyn Gartman
John S. Marshall

Research and Services

Head, Editorial Research
Jo Ann McDonald

Head, Research Library
Vera Busselle

Head, Cartographic Services
H. George Stoll

Index Editor
Judith Deraedt

Pre-Press Services

Director
Richard A. Atwood

Manager, Production
John Babrick

Assistant Manager, Production
Lynn Iverson

Manager, Film Separations
Alfred J. Mozdzen

Assistant Manager, Film Separations
Barbara J. McDonald

Manager, Traffic
Joseph J. Stack

Assistant Manager, Traffic
Marguerite DuMais

Manufacturing

Executive Director
Philip B. Hall

Production Manager
Jerry R. Higdon

Manager, Research and Development
Henry Koval

Editorial Advisory Board

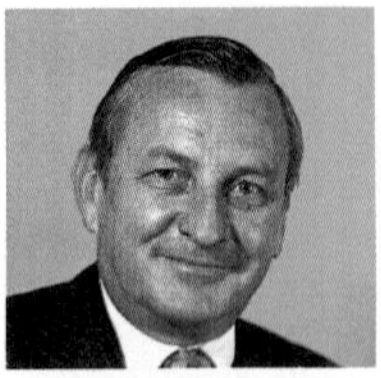

Dr. Ivan L. Bennett, Jr., is Provost, New York University Medical Center, Dean of the N.Y.U. School of Medicine, and Professor of Medicine at N.Y.U. He received the A.B. degree in 1943 and his M.D. degree in 1946 from Emory University. Dr. Bennett's activities include chairmanships of the Panel on Biological and Medical Science of the President's Science Advisory Committee, and the Health Policy Fellowship Board of the National Academy of Sciences' Institute of Medicine.

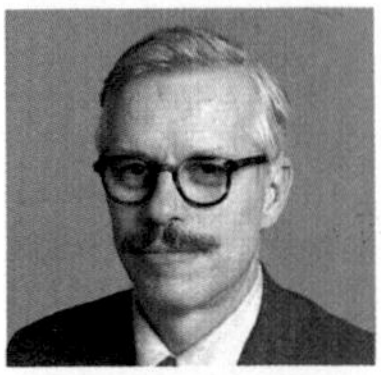

Wilbur B. Davenport, Jr., is head of the Department of Electrical Engineering and Computer Science at the Massachusetts Institute of Technology. He received the B.E.E. degree from Alabama Polytechnic Institute in 1941, and the S.M. and Sc.D. degrees from M.I.T. in 1943 and 1950, respectively. He spent several years with the Lincoln Laboratory, serving as Assistant Director from 1963 to 1965. Professor Davenport has been consultant to the Office of Science and Technology and the Communications Satellite Corporation.

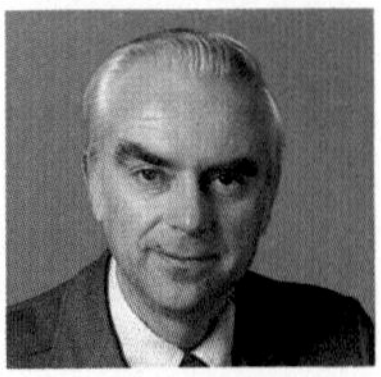

Frank D. Drake is director of the National Astronomy and Ionosphere Center, which operates the Arecibo Observatory, and Goldman Smith Professor of Astronomy at Cornell University. He received the B.E.E. degree from Cornell in 1952, and the M.A. and Ph.D. degrees in astronomy from Harvard University in 1956 and 1958, respectively. Dr. Drake is a member of the International Astronomical Union, and the U.S. Commission V of the International Scientific Radio Union.

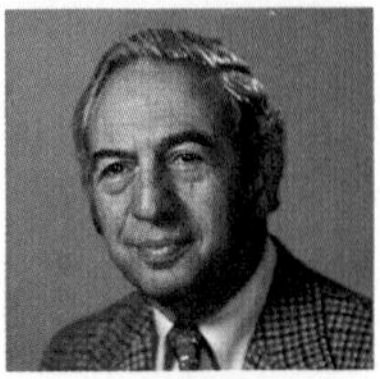

Arthur W. Galston is Eaton Professor of Biology at Yale University. He received the B.S. degree from Cornell University in 1940, the M.S. degree and the Ph.D. degree from the University of Illinois in 1942 and 1943, respectively. Professor Galston was among the first American scientists to visit the People's Republic of China in 1971 and he was invited to North Vietnam in 1975 to lecture and give instruction on plant physiology and biochemistry.

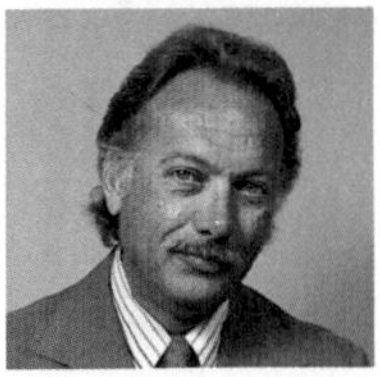

F. Clark Howell is Professor of Anthropology at the University of California, Berkeley. He received the Ph.B. degree in 1949, the M.A. degree in 1951, and the Ph.D. degree in 1953, all from the University of Chicago. He served as Chairman of the University of Chicago's Department of Anthropology from 1966 to 1969. He is a member of the Editorial Advisory Boards of *Quaternary Research* and the *American Journal of Physical Anthropology.*

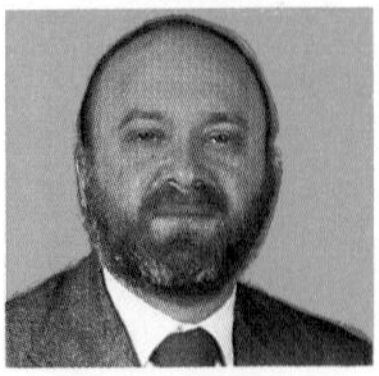

Leo P. Kadanoff is Professor of Physics at Brown University. He received the A.B. degree in 1957, the M.A. degree in 1958, and the Ph.D. degree in 1960, all from Harvard University. His work has been in the theory of solid-state and particle physics. He is a Fellow of the American Physical Society. He has also been involved in research on urban growth, and served from 1974 to 1976 as President of Rhode Island Urban Observatory.

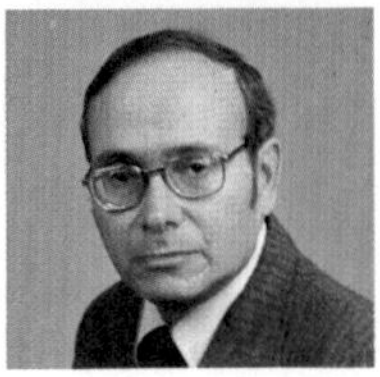

Frank Press is Chairman, Department of Earth and Planetary Sciences at the Massachusetts Institute of Technology. He received the B.S. degree from City College of New York in 1944; the M.S. degree in 1946 and the Ph.D. degree in 1949 from Columbia University. Dr. Press is President of the American Geophysical Union, Chairman of the Board of Advisors of the National Center for Earthquake Research, and Chairman of the Committee on Scholarly Communications with the People's Republic of China.

Contributors

Adelman, George, M.S.
Managing Editor & Librarian
Neurosciences Research Program
Massachusetts Institute of Technology
Neurology

Alderman, Michael H., M.D.
Assistant Professor of Public Health
Cornell University Medical College
Medicine, Internal
Public Health

Araujo, Paul E., Ph.D.
Assistant Professor
Department of Food Science
University of Florida
Nutrition

Auerbach, Stanley I., Ph.D.
Director, Environmental Sciences Division
Oak Ridge National Laboratory
Ecology

Bakker, Robert T., Ph.D.
Assistant Professor
Johns Hopkins University
Department of Vertebrate Paleontology
A New Image for
Dinosaurs

Bell, William J., Ph.D.
Associate Professor of Entomology
University of Kansas
Zoology

Belton, Michael J. S., Ph.D.
Astronomer
Kitt Peak National Observatory
Unveiling Venus
Astronomy, Planetary

Bromley, D. Allan, Ph.D.
Henry Ford II Professor and Chairman
Department of Physics
Yale University
Physics, Nuclear

Bylinsky, Gene, B.A.
Associate Editor
Fortune Magazine
Invasion of the Microcomputers

Chiller, Jacques M., Ph.D.
Associate Professor
National Jewish Hospital
Immunology

Copeland, John A., Ph.D.
Senior Member
Institute of Electronics and
Electrical Engineers
Communications

Cromie, William J., B.S.
Executive Director
Council for the Advancement of
Science Writing
The International Science and
Engineering Fair

Davies, Julian, Ph.D.
Professor of Biochemistry
University of Wisconsin
Biochemistry

Dawe, Albert R., Ph.D.
Deputy Director and Chief Scientist
Office of Naval Research Branch Office
The Secrets of Winter Sleep

Deffeyes, Kenneth S., Ph.D.
Professor of Geology
Princeton University
Geoscience, Geology

Dorozynski, Alexander
Science Writer
Georges Mathé

Dowler, Michael J., Ph.D.
Associate Professor
San Diego State University
Our Chemical Ancestors

Drake, Charles L., Ph.D.
Professor of Earth Sciences
Dartmouth College
Geoscience, Geophysics

Eckholm, Erik P., M.A.
Senior Researcher
Worldwatch Institute
Close-Up, Energy

Edson, Lee
Science Writer
Airships Make a Comeback

Eberhart, Jonathan
Space Sciences Editor
Science News
Space Exploration

Ensign, Jerald C., Ph.D.
Professor of Bacteriology
University of Wisconsin
Microbiology

Giacconi, Riccardo, Ph.D.
Professor of Astronomy
Harvard University and
Associate Director
High Energy Astrophysics Division
Center for Astrophysics
Astronomy, High Energy

Goldhaber, Paul, D.D.S.
Dean and Professor of Periodontology
Harvard School of Dental Medicine
Medicine, Dentistry

Gray, Harry B., Ph.D.
W. R. Kenan, Jr. Professor
California Institute of Technology
Chemistry

Griffin, James B., Ph.D.
Curator, Museum of Anthropology and
Professor of Anthropology
University of Michigan
Archaeology, New World

Gump, Frank E., M.D.
Professor of Surgery
College of Physicians and Surgeons
Columbia University
Medicine, Surgery

Hamburger, Robert N., M.D.
Professor and Head
Pediatric Immunology and Allergy Division
University of California, San Diego
Attack on Allergy

Hartl, Daniel L., Ph.D.
Associate Professor
Department of Biological Sciences
Purdue University
Genetics

Hayes, Arthur H., Jr., M.D.
Chief, Division of Clinical Pharmacology
Milton S. Hershey Medical Center
Pennsylvania State University
Drugs

Henahan, John F., B.S.
Science Writer
Surgery in Miniature
Plants: The Renewable Resource

Hoffman, Ruth I., Ed.D.
Professor of Mathematics
Director, Math Laboratory
University of Denver
Close-Up, Electronics

Irwin, Howard S., Ph.D.
President
New York Botanical Garden
Botany

Jennings, Feenan D., B.S.
Head,
International Decade of Ocean Exploration
National Science Foundation
Oceanography

Jernigan, Sherry L., M.A.
Research Director
Garbage Project
Department of Anthropology
University of Arizona
Close-Up, Archaeology

Jones, David L., Ph.D.
Professor of Physical Geography
Southern Illinois University
Meteorology

Kessler, Karl G., Ph.D.
Chief, Optical Physics Division
National Bureau of Standards
Physics, Atomic and Molecular

Ketterson, John B., Ph.D.
Professor of Physics
Northwestern University
Physics, Solid State

Kritchevsky, David, Ph.D.
Associate Director
The Wistar Institute
Close-Up, Nutrition

Maglio, Vincent J., Ph.D.
Research Associate
Department of Geology
Princeton University
Geoscience, Paleontology

Maran, Stephen P., Ph.D.
Head, Advanced Systems and
Ground Observations Branch
Goddard Space Flight Center
Astronomy, Stellar

March, Robert H., Ph.D.
Professor of Physics
University of Wisconsin
Disabling the White Death
Physics, Elementary Particles

Maugh, Thomas H. II, Ph.D.
Science Writer
Science Magazine
Close-Up, Medicine

McBride, Gail, M.S.
Associate Editor
JAMA Medical News
American Medical Association
Close-Up, Public Health

Meade, Dale M., Ph.D.
Research Physicist
Princeton University
Physics, Plasma

Merbs, Charles F., Ph.D.
Professor and Chairman
Department of Anthropology
Arizona State University
Anthropology

Nelson-Rees, Walter A., Ph.D.
Research Geneticist
Cell Culture Laboratory
University of California
School of Public Health
Close-Up, Microbiology

Norman, Colin, B.Sc.
Washington Correspondent
Nature
Science Policy

Novick, Sheldon
Editor, *Environment*
Environment

Peebles, P. James E., Ph.D.
Professor of Physics
Princeton University
The Fate of The Cosmos

Price, Frederick C., B.S.
Managing Editor
Chemical Engineering
Chemical Technology

Randal, Judith E., B.A.
Science Correspondent
New York *Daily News*
Stress: The Ticking Bomb

Reyman, Theodore A., M.D.
Director of Laboratories
Mount Carmel Mercy Hospital, Detroit
Dead Men Do Tell Tales

Richardson, Herbert H., Sc.D.
Professor and Head of
Mechanical Engineering Department
Massachusetts Institute of Technology
Transportation

Rodden, Judith, M.Litt.
Research Archaeologist
Archaeology, Old World

Savit, Carl H., M.S.
Senior Vice President, Technology
Western Geophysical Company of America
Refining the Search for Oil

Shank, Russell, D.L.S.
Director of Libraries
Smithsonian Institution
Books of Science

Silk, Joseph, Ph.D.
Associate Professor of Astronomy
University of California, Berkeley
Astronomy, Cosmology

Sleeper, Barbara L., B.A.
Education Director
Center for Short-Lived Phenomena
Fingers on the Planet's Pulse

Sperling, Sally E., Ph.D.
Professor of Psychology
University of California, Riverside
Psychology

Tilton, George R., Ph.D.
Professor of Geochemistry
University of California, Santa Barbara
Geoscience, Geochemistry

Veatch, Robert M., Ph.D.
Staff Director,
Research Group on Death and Dying
Institute of Society, Ethics,
and the Life Sciences
The Choice to Die

Vietmeyer, Noel D., Ph.D.
Professional Associate
National Academy of Sciences
Prospecting for Green Gold

Weber, Samuel, B.S.E.E.
Executive Editor
Electronics Magazine
Electronics

Wittwer, Sylvan H., Ph.D.
Director, Agricultural Experimental Station
Michigan State University
Agriculture

Contributors not listed on
these pages are members of the
Science Year editorial staff.

C
O
O
O
O
C
N
C

Special Reports

The Special Reports and an exclusive *Science Year* Special Feature give in-depth treatment to the major advances in science. The subjects were chosen for their current importance and lasting interest.

Refining the Search for Oil

By Carl H. Savit

Petroleum scientists are now turning to high-speed computers for help in decoding seismic data that tell them where to drill

As dawn breaks on the Colorado plains, trucks equipped with special vibrators to "shake" the ground move out in a never-ending search for underground oil fields.

Scientists and engineers, spurred on by the 1973 oil crisis, are studying alternative sources of energy–solar, geothermal, and wind power. But, even with the best technological expertise, studies show clearly that it will take several generations of scientific development before alternative energy sources can fully meet our power needs. Even then, we will still need petroleum for lubricants and for the cloth, plastics, and fertilizers that are produced from it.

Meanwhile, the discovery of new oil wells is slipping far behind anticipated needs, and some experts predict that we will have to ration our supplies within the next decade. Fortunately, scientists have recently developed new techniques and vastly improved old ones for oil exploration. They have been aided in this by the sciences of electronics, geology, mathematics, and physics. Using massive computers to process millions of complex records of seismic waves, they can now get detailed views of sections of the earth's crust where oil is buried. But the amount of data that must be processed is so great that geophysicists have even had to turn to a show-business tool–the magnetic-tape systems used for high-fidelity recording of color-television programs. These magnetic-tape systems enable scientists to record much more data for the computers. Eventually, the new techniques should lower the traditional average of nine holes that oilmen must drill to find one commercially producing well.

As yet, the new surveying techniques cannot çompletely guarantee that oil lies at a given spot in the earth. But they give us a much better idea of where to look–and where not to look–for it. However, the presence or absence of oil, and the natural gas that often accompanies it in the earth, can still only be confirmed by actual drilling.

Almost all the oil and gas we use comes from sedimentary rocks that were laid down on the floor of ancient, shallow seas. These rocks began as great layers of mud, silt, sand, and organic–or plant and animal–matter. As more and more of this sediment piled up, the older layers on the bottom were compressed and heated by the pressure. The pressure also squeezed most of the liquid out of the old muds and silts and transformed the solid materials that remained into the fine-grained, impervious sedimentary rocks we call shale and siltstone. The soup of salt water and dissolved organic matter squeezed out of the muds and silts found its way into the sandy sediments that ultimately congealed into the relatively porous sedimentary rocks called sandstones and conglomerates.

The author:
Carl H. Savit is senior vice-president of technology for the Western Geophysical Company in Houston.

In the absence of oxygen and under the influence of both heat and pressure, the organic matter slowly changed into one or more of the hydrocarbon mixtures that make up oil and natural gas. Plate tectonics then acted as the mechanism for concentrating the oil and gas in structural traps, or localized deposits, where they are now found in sufficient quantities to be economically extracted by man.

Plate tectonics is the slow geological process in which plates, or sections of the earth's crust, tear apart, scrape past each other, and collide with tremendous forces that create mountains and rifts in the earth's surface. Bends and folds created in the rock layers become the traps in which hydrocarbons can collect. These traps that we seek with sophisticated geophysical techniques typically consist of a layer of sandstone or other porous rock between two layers of impervious rock such as shale. The tectonic forces press the layers into a shape that is concave downward, like a bent sponge between two upside-down soup plates. The oil, gas, and water in the porous rock separate according to density. The gas, which is least dense, collects at the top; the oil, which is denser than gas but not as dense as water, collects below the gas; and the water, which is densest, settles at the bottom.

Structural traps are so called because they depend on the shape, or structure, of the rock layers produced by the tectonic forces. Oil is also found in traps formed in other ways. One of these is called a stratigraphic trap because it depends on changes in the composition of rock strata rather than on its shape. A common example of a stratigraphic trap is an ancient sandy beach buried between layers of shale.

Almost any geologic process that results in porous rock being covered by nonporous rock can form a trap for commercially recoverable amounts of oil or gas. But, until recently, geophysicists could only find structural traps and a few special types of stratigraphic traps.

Once geophysicists have found a trap, their next step is to drill a hole into it to find out if oil or gas is really there. Geophysicists have been trying to improve their methods so that they could find out if oil or gas is in the trap without drilling. The method that now yields the most useful information of this sort is seismic reflection.

Seismic-reflection mapping is based on the concept of thumping on the ground, then listening to the echoes that come back from beneath the surface. The basic idea was patented in 1918 by physicist Reginald A. Fessenden. Other scientists published a paper that "proved" that reflected seismic waves would be too weak to be detected, but identifiable reflections of man-made seismic waves from subsurface rock layers were recorded in 1926. At that time, explosive charges were detonated in shallow holes to produce the thumping. The detection equipment was heavy and crude, but it worked.

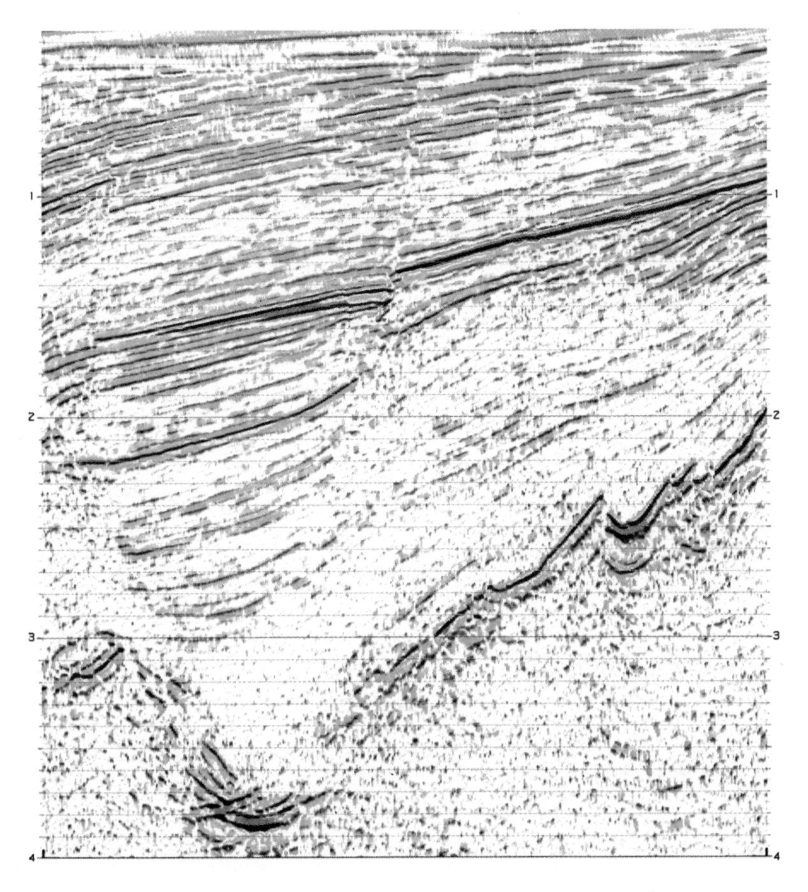

Fuchsia-colored areas are "bright spots" that show oil deposits in color-coded reflection profile of an offshore oil field near Alaska.

Seismic boat in the Gulf of Mexico is relatively small, but it bristles with navigation and communication antennas, *above left.* Spacious bridge of the geophysical "supership" *Anne Bravo, above right,* is equipped with navigation instruments that can locate the ship's position exactly by satellite tracking. Its Aquapulse gun, *below,* produces seismic waves that are reflected to hydrophones attached to the ship's rear cable.

With the development of more sensitive detectors and modern electronic computers, seismic-wave information has become our most important tool. In its most elementary form, modern seismic-reflection mapping uses a "thumping" source, such as dynamite, to produce seismic waves at or near the earth's surface and a nearby sensitive detector. By measuring the elapsed time between setting off the shot and receiving the returned reflections, geophysicists can estimate the depth of the rock layers from which the reflections came. They record all reflections that the detector registers from the time the charge is set off until the last reflection reaches the detector. Usually, this is only a few seconds, rarely more than six or eight.

In exploration at sea, the scientists working on a typical research ship produce seismic waves that go through the water and into the ocean floor. They tow a 3-kilometer (2-mile) -long plastic tube filled with oil from the back of the vessel. The tube contains about 2,000

Undersea Exploration
Seismic waves produced by four Aquapulse guns hanging from cables at the sides of the ship penetrate rocks below and are reflected back to hydrophones in the 3-kilometer-long cable at the rear of the ship. (For clarity, angles have been exaggerated.)

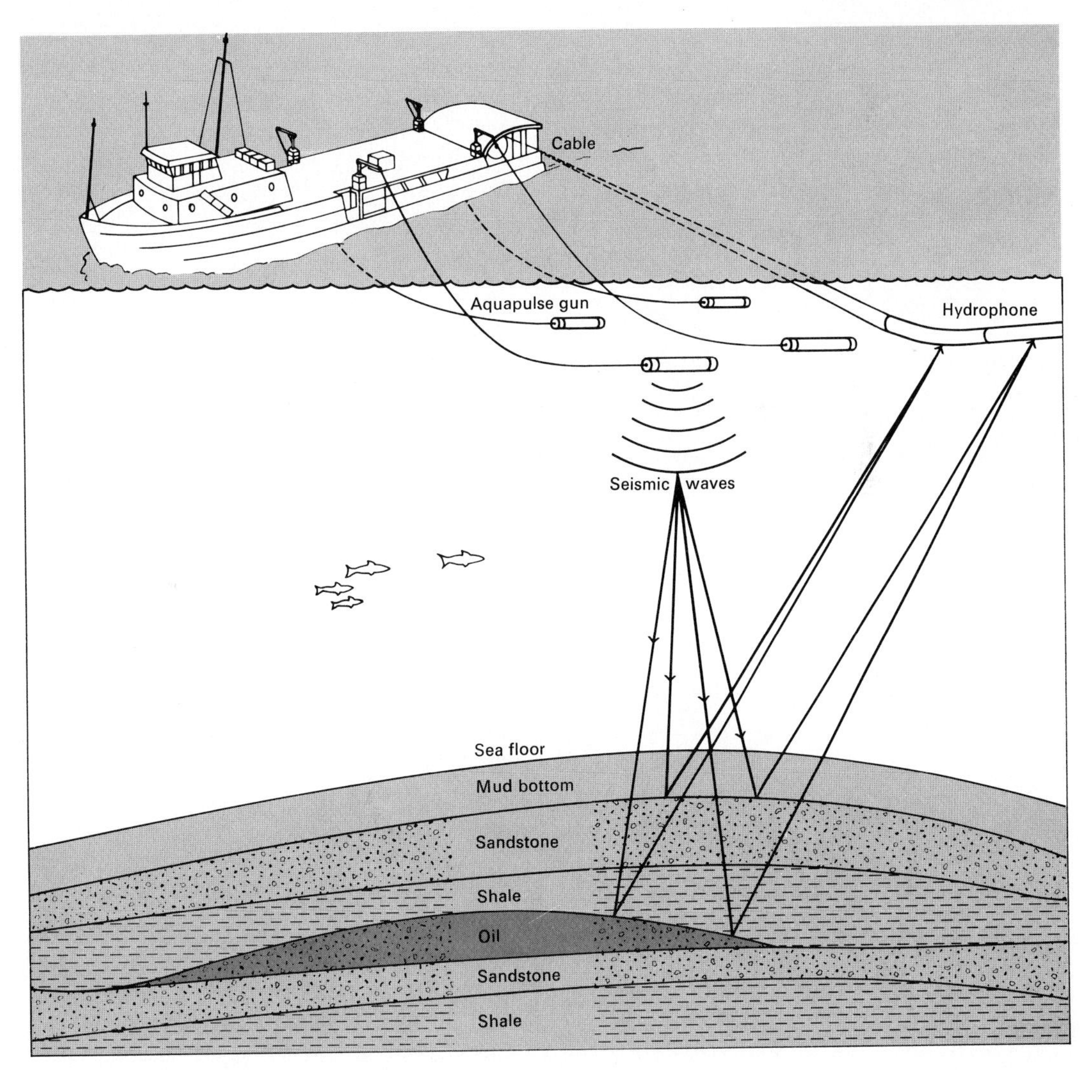

detectors and has various depth-control devices and other instruments attached to it. The ship also tows four rubber-walled Aquapulse guns, two on each side of the ship. Shock waves are produced by exploding a mixture of propane and oxygen inside the Aquapulse guns. The research ship uses highly accurate navigation equipment, including atomic clocks, Doppler sonar devices, inertial navigation systems, satellite-tracking equipment, and radio signals, to ensure that the cable follows exactly a preset course above the sea floor.

On land, a ground crew can insert explosive devices at precise points along a surveyed line. Engineers have also developed large trucks that can follow the surveyed line, stop at any point, place big metal disks on the ground, and vibrate the surface.

Land-exploration crews usually conduct their research along a straight-line track, setting off the seismic explosions at closely spaced points along that line. Several detectors are connected to the same

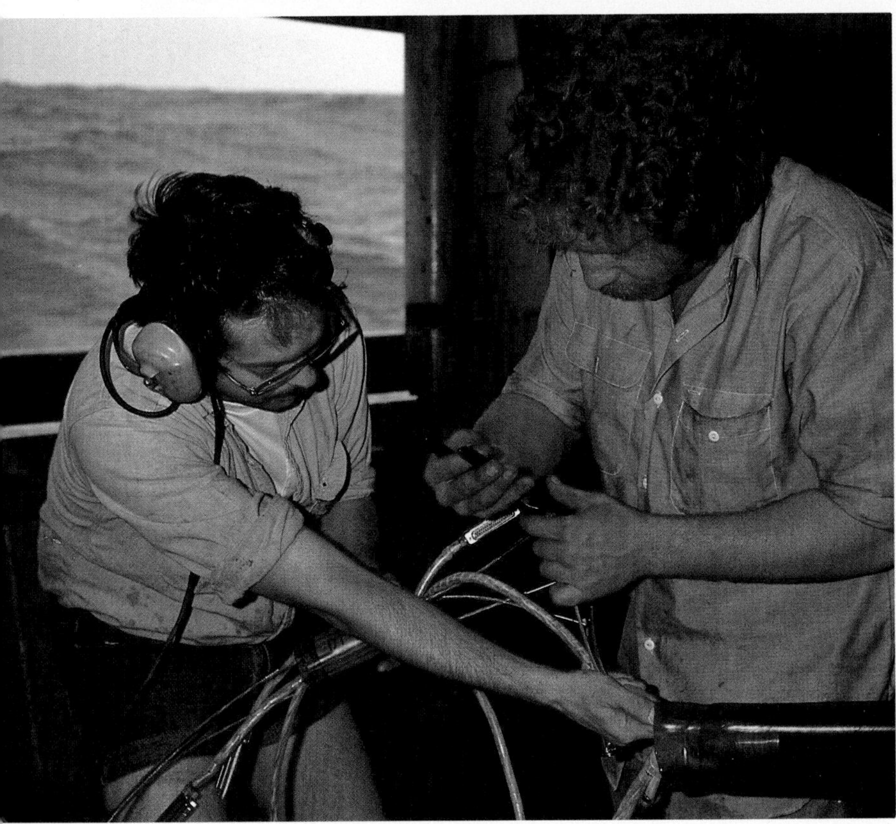

Huge hydraulic reel is almost empty, *opposite page,* after nearly three kilometers of detector cable have been let out behind a ship. The cable emerges from the ship's stern between rollers, *above.* Crew members must repair the cable breaks quickly, *left.*

A surveyor, *left,* is the first of a land exploration team into the field. He lays out positions for the instrument cables for recording seismic waves. A technician, *below,* monitors the explosions that produce the waves. When trucks are used, *bottom,* they have vibrators that mechanically pound on the ground to produce waves. They also record data.

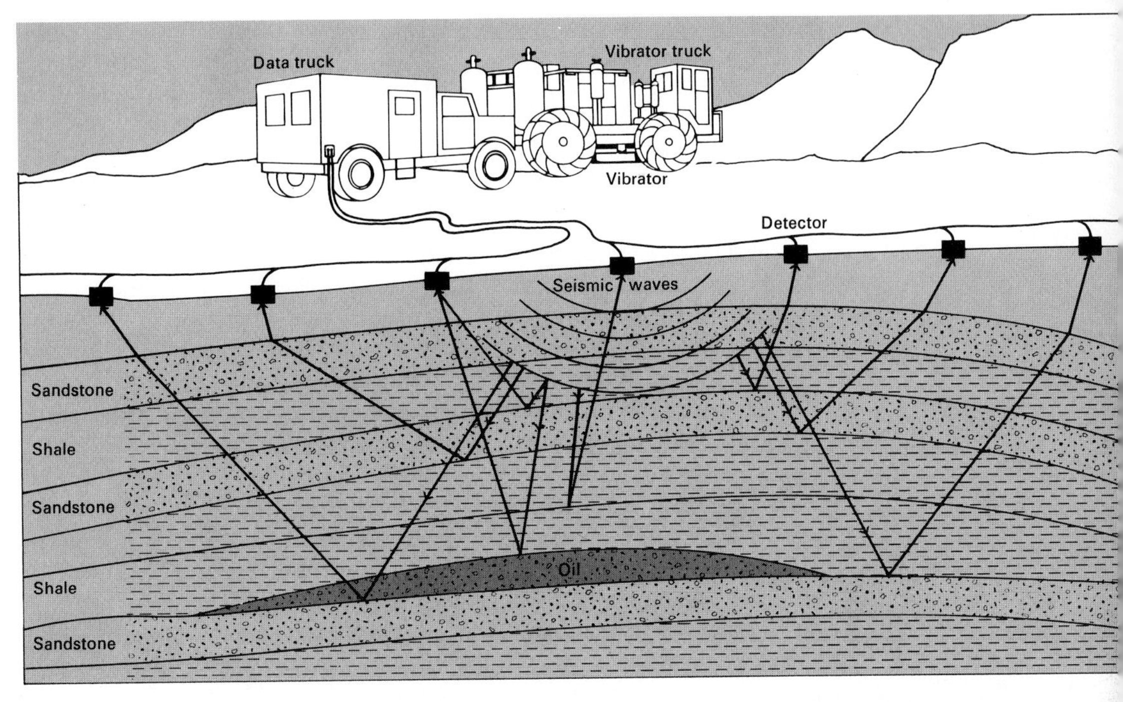

Land Exploration
Seismic waves produced by truck vibrators enter the layers of rock and earth below and are reflected back to the many sensors along the ground. These sensors relay the seismic wave information to a truck where it is recorded. (For clarity, angles have been exaggerated.)

receiver circuit in order to average the received seismic signals over a few hundred meters of surface. Averaging the results helps to overcome the noise factor produced by extraneous waves. Fifty to 100 such groups of receivers are spread out in a line 3 kilometers or more long and connected by a cable to elaborate recording instruments. Shots are fired every few hundred meters along the survey line, and the whole array of detector groups is moved forward to keep up with the shots. All the reflections resulting from each shot are separately recorded by each of the detector groups. The records from sea and land explorations are stored on magnetic tape, and later processed at a central location by big computers.

Until the late 1960s, only the time it took the reflections to reach the detectors was considered important. Careful analysis of these times yielded a picture of underground rock structures from which geologists could pick the likely oil and gas traps. The available recording instruments could not record the full range of variations in the strength of the reflected signals. Even after systems were developed that could record these variations, they were systematically eliminated in the computerized processing of the data. We were discarding valuable information in ignoring this data.

We are now using it. New data-processing techniques followed the introduction in the late 1960s of computer equipment that recorded the reflected seismic signals as a series of numbers called floating point numbers. These numbers can indicate not only that a reflected signal has arrived at the detector, but also indicate the differences in strength. Floating point numbers are those in which the position of the decimal point is recorded separately from the number itself so that any size figure can be written with just a few digits. For example, a number such as 0.0000053 can be written in "floating point" as 53−7. The −7 means that the decimal point is seven places to the left of its natural position after the 53.

Floating point numbers are recorded at a rate of 500 per second by each detector group along the survey line. A typical six-second seismic recording from 100 detector groups contains 300,000 floating point numbers. A marine seismic-reflection crew can easily produce 75,000 such recordings, or 22.5 billion floating point numbers each month. Only the largest electronic computers available today can perform the intricate computations needed to sort out and assemble all this reflection data in a form that can be used by geologists.

The reflections detected by the new systems seemed to show that porous rock in which natural gas is trapped reflects a much stronger seismic echo than does porous rock containing water. The profile produced by the computer shows this contrast as a "bright spot," an intense dark section, on the plotted picture of the amplitude of the reflected seismic signals. Such bright spots show up in computer records from surveys off Louisiana's Gulf Coast, on Africa's West Coast, in the North Sea and the Mediterranean, as well as in Indonesia, South America, and several other places. Many of them proved to be traps containing gas and sometimes oil as well. Best of all, many of the traps found by the bright spots were the stratigraphic kind and had been overlooked by the older techniques.

Unfortunately, wells were also drilled into bright spot areas that produced no gas and oil. After the initial excitement over the discovery of bright spots died down, it became apparent that geologic phenomena other than gas also produced bright spots. It was also found that some gas accumulations do not give especially strong reflections, and that the strength of the reflection is determined by many factors, including the pressure of the gas and the nature of the rock it is in.

The reason for such irregularities is complex. Part of the seismic wave is reflected each time it strikes an interface between different types of rock with different densities. The strength of the reflected signal is related to the number of partial reflections as the wave travels through the rock layers. The wave travels rapidly through incompressible materials such as limestone. It travels less rapidly through more compressible material such as shale.

The presence of minute amounts of gas interspersed in a layer of rock can greatly increase the rock's compressibility, and the velocity of the seismic wave may diminish by as much as 50 per cent, when compared with the same kind of rock containing minute amounts of water. The velocity of the wave thus becomes a clue as to the presence of gas. Rocks that make up the earth's outer layers in a given locality do not usually vary drastically in composition. As a result, contrasts

All the equipment and supplies needed for any seismic survey in the jungles of Ecuador must be carried in on the surveyors' backs since there are few roads.

between layers are not great and the reflection coefficients (the ratio of the strength of the reflected signal to that of the incident signal) are small. A reflection coefficient of 2 per cent means that 2 per cent of the wave is reflected back to the detector and 98 per cent goes on through the rock to be partially reflected, with a different reflection coefficient, at the next interface. Because most of the wave goes on through, the seismologist can get reflections from many layers. Where the sequences of rock strata produce several reflection coefficients of 10 per cent or more, seismologists can determine little about the deeper rocks.

Now we can understand why a gas-filled sandstone may sometimes return an unusually strong reflection. If the grains of sand making up the sandstone are not tightly packed together and gas under high pressure lies between them, sandstone will appear compressible like a gas, not incompressible like a rock. As a result, the seismic velocity in that layer will be low, contrasting greatly with the velocity in the impervious shale layer directly above. But if the sand grains are firmly packed together, any gas in the pores will have little or no effect on the compressibility of the rock.

Gas accumulations in sandstone can produce reflection coefficients ranging from as high as 50 per cent down to almost zero. To compli-

In one of the largest computer rooms in the world, thousands of tape reels containing seismic recordings from many ship and land crews are processed each month.

cate the matter still further, some shales under great pressure from surrounding rock layers have been found to mimic the high reflectivity of gas-filled sandstone.

Floating point computer profiles have two indicators that help reflection seismologists determine which bright spots are associated with gas and which are not. They also help in identifying types of gas and oil accumulations that do not generate strong reflections.

The first of these two indicators is phase, a property that distinguishes one seismic wave from another. Like sound waves, seismic waves are transmitted through the earth as a sequence of compressions and rarefactions. When a wave reflects from the surface of a material that is less compressible than the one in which it has been traveling, the reflected wave is in the same phase just after reflection as it was before reflection. In other words, a compression is reflected as a compression and a rarefaction as a rarefaction. But if the wave is reflected from a surface that is more compressible, the phase is reversed, a compression returning as a rarefaction and vice versa.

Phase reversal of seismic waves reflected from rock that is more compressible, such as gas-filled sandstone, can confirm the presence of gas that was indicated by a bright spot. It has therefore proved to be a relatively good additional indicator for reflection seismologists.

The second indicator is a reflecting interface that is perfectly horizontal. A unique characteristic of gas and oil accumulations is that their bottom surfaces are horizontal. Gas floats on either oil or water and oil floats on water. A flotation surface between any two fluids is formed by the force of gravity and is necessarily horizontal.

Most reflections from interfaces between two liquids in porous rock are quite weak and hard to find in the welter of broad, booming reflections, but they are always sharply defined because of their horizontal nature. These horizontal reflections are positive proof that a given trap holds hydrocarbons, and they confirm the indications of gas given by bright spots with reversed phase.

Finding and confirming phase reversals and thin horizontal reflectors strains the resolving power of today's seismic-reflection technology. The deeper the rocks, the thicker they must be if computers are to map them with any precision. Relatively thin rock layers also present a problem no matter how close to the surface they are. If geophysicists could double or treble their resolution capabilities and detect and map rock layers one-half or one-third as thick as they now can, a whole new vista of oil exploration would open up. It would then be possible to record many of the horizontal surfaces that prove the presence of hydrocarbons. We would then be finding direct proof of the presence of oil or gas, not just traps in which oil and gas is only a possibility.

However, to achieve this greater resolution, we will have to record many more floating point numbers from the seismic reflections after each shot. Seismic-exploration crews already are the world's principal users of digital recording tape, surpassed only by the U.S. government. To get more data, a seismic-reflection crew will probably have to make at least twice as many shots and record 10 times as many floating point numbers as they do now.

The technology developed for high-fidelity recording of color TV programs was adapted to record numerical information for oil exploration in 1976. Now, with seismic exploration on the verge of a 10- to 20-fold increase in data gathered per crew, the color-TV technology will make it possible to record data more than 20 times faster and to put nearly 200 times more of it on one reel of tape.

A variety of special instruments and special-purpose computers were also tested in 1976. Microcircuit and microcomputer technologies developed for space flights, digital watches, and hand calculators are being stretched to the limit to process the expected avalanche of geophysical data. See INVASION OF THE MICROCOMPUTERS.

A new generation of computers with 100 to 1,000 times the capacity of present models is expected to be in use in a few years. With these computers to handle the required increase in seismic data, scientists should also be able to produce new kinds of information. For example, we might be able to determine the differences in the time it takes one seismic signal to travel different paths to the same subterranean point accurately enough to determine the velocity of seismic waves in a single layer of rock. Accurate information about individual layers would further our ability to find oil and gas.

There is still a great deal of petroleum in the earth, enough to meet all our projected needs. The technology is available to economically locate the oil and extract it. This is the primary goal of the new seismic-reflection systems. They will describe the shapes and locations of underground layers in much more detail than before. And the additional data about the kinds of rock making up those layers could eventually eliminate drilling dry wells in the search for gas and oil.

Stress: The Ticking Bomb

By Judith E. Randal

Prolonged stress, brought about by the trials, triumphs, and disappointments in our lives, can wear down our body's defenses against disease

One by one, the new soldiers finished running the Fort Dix obstacle course and lined up at attention. The drill sergeant strode slowly along the line of perspiring men, his cold gaze moving from one face to the next. Then the sergeant stopped in front of a young recruit, fixed him with a contemptuous glare, and roared at the top of his lungs: "You have disgraced this entire platoon!"

The young recruit stared straight ahead; his face betrayed no emotion as the sergeant berated and humiliated him in front of the whole platoon. But three days later, that recruit came down with a particularly severe case of influenza.

John W. Mason, who witnessed the incident, was not surprised. As a physician, he had often noted that people under severe emotional strain were likely to come down with ailments ranging from peptic ulcers in New York City taxi drivers to influenza in Army recruits. As a researcher at the Walter Reed Army Institute of Research in Washington, D.C., Mason had gone to Fort Dix in New Jersey to study the apparent association between stress and vulnerability to disease.

Part of Mason's study included analyzing blood and urine samples taken periodically from men in the young recruit's platoon. Mason noted that the level of certain hormones in these samples changed whenever one of the recruits suffered an emotional blow. Those who were felled by influenza had almost always undergone these hormonal changes first. So Mason was convinced that most of the cases of influenza among the recruits had every bit as much to do with emotional factors as they did with a virus.

Many other investigators share Mason's interest in the connection between stress and disease. Some researchers have linked personality type to susceptibility to heart attack, cancer, or other diseases. Other scientists are investigating the possible stress-relieving effects of meditation, biofeedback, and various relaxation techniques. In late 1975 and early 1976, there came a flood of books on how to relieve stress.

Despite this sudden burst of interest, the idea that stress can affect our health is a very ancient one. Some methods of treatment–acupuncture, for example–are based on the philosophy that bodily imbalances result from physical and emotional pressures, thus causing disease. And we have all noticed, for instance, that a young child may come down with a case of the sniffles after the excitement of a birthday party or that the death of one spouse is often followed by the serious illness or death of the other.

People often say "I'm under stress," or talk about the "stresses of living." But theirs is a limited concept of stress. The word is immensely difficult to define. It means different things to different people, and even scientists disagree on how to define it. Actually, to learn something about stress we do not need to think of it in terms of what it is, but rather in terms of what it does.

High levels of prolonged stress can be dangerous to our health. But a certain amount of stress is necessary to keep us alive. If no demands were placed on our bodies to make them exert energy, our hearts would pump poorly, our digestive tracts would become sluggish, our kidneys would falter, and our mental powers would decline. Some stress, then, is essential. Without its more or less constant prodding, our bodies would lose their capacity to adapt to the ever-changing conditions of our environment.

The author:
Judith E. Randal, a New York *Daily News* science correspondent, is based in Washington, D.C.

In the early 1900s, physiologist Walter B. Cannon of Harvard University became interested in how the body adapts to meet demands. In a series of animal experiments, he demonstrated that fear, rage, and other intense emotional responses put the body on a war footing and prepare it for "fight or flight."

In this fight-or-flight response, scientists now know that a part of the brain called the hypothalamus sends a signal through the autonomic nervous system to the adrenal medulla, or inner part, of the adrenal gland. The adrenal glands, located above the kidneys, release an additional, emergency supply of hormones, one of which is adrenalin, into the bloodstream. Adrenalin and the autonomic nervous system are the major factors in the body's response. They work together to increase blood pressure, blood sugar, and respiration; slow digestion; and divert blood to muscles in the arms and legs.

At the same time, but playing a lesser role, the hypothalamus chemically signals the pituitary, a gland below the brain, to secrete hormones, including one known as ACTH. The ACTH stimulates the adrenal cortex, the outer part of the adrenal gland, to produce various steroid hormones that are called corticoids. Scientists believe the corticoids cause metabolic changes that help the body sustain a ready supply of energy during the emergency.

After the danger has passed, our hearts may pound and our hands may tremble, but our bodies usually get back to normal fairly quickly. Cannon observed this and wanted to know how the body could readjust so swiftly.

Through experiments, he demonstrated that such factors as the chemical content of the blood, body temperature, blood pressure, respiration, and many others can vary, within limits. But these factors tend to adjust to each other; sensitive regulatory mechanisms keep them on a fairly even keel. For example, if the concentration of salt in a person's bloodstream rises above normal, an adrenal hormone will influence the kidney to adjust this by eliminating the salt through more wastewater. Cannon called this delicate balancing and adjusting of the body homeostasis. As long as homeostasis is operating smoothly, we have a good chance of staying well.

In 1926, the year in which Cannon coined the term homeostasis, Hans Selye, then a second-year medical student at the University of Prague in Czechoslovakia, began to wonder why many of the patients he saw had certain symptoms in common no matter what their illness. "Whether a man suffers from severe loss of blood, an infectious disease, or advanced cancer," Selye noted, "he loses his appetite, his muscular strength, and his ambition to accomplish anything...." Selye called this the "syndrome of just being sick," and he wondered whether there were certain common factors in the way a person reacts to any injury or illness.

He dismissed the question from his mind until about 10 years later, when he had moved to Canada and was doing research on sex hormones. In an experiment, he injected a hormonal extract from cattle ovaries into rats. Within a few days, the rats' adrenal glands enlarged, their thymus and lymph glands shrank, and they developed deep bleeding ulcers in their stomachs and intestines. At first, Selye thought the ovary extract caused these symptoms. But he soon discovered that the same symptoms appeared after he injected any toxic substance into the rats. He then found he could produce these three symptoms by subjecting the rats to extreme cold or heat or even by making them afraid and nervous for long periods of time.

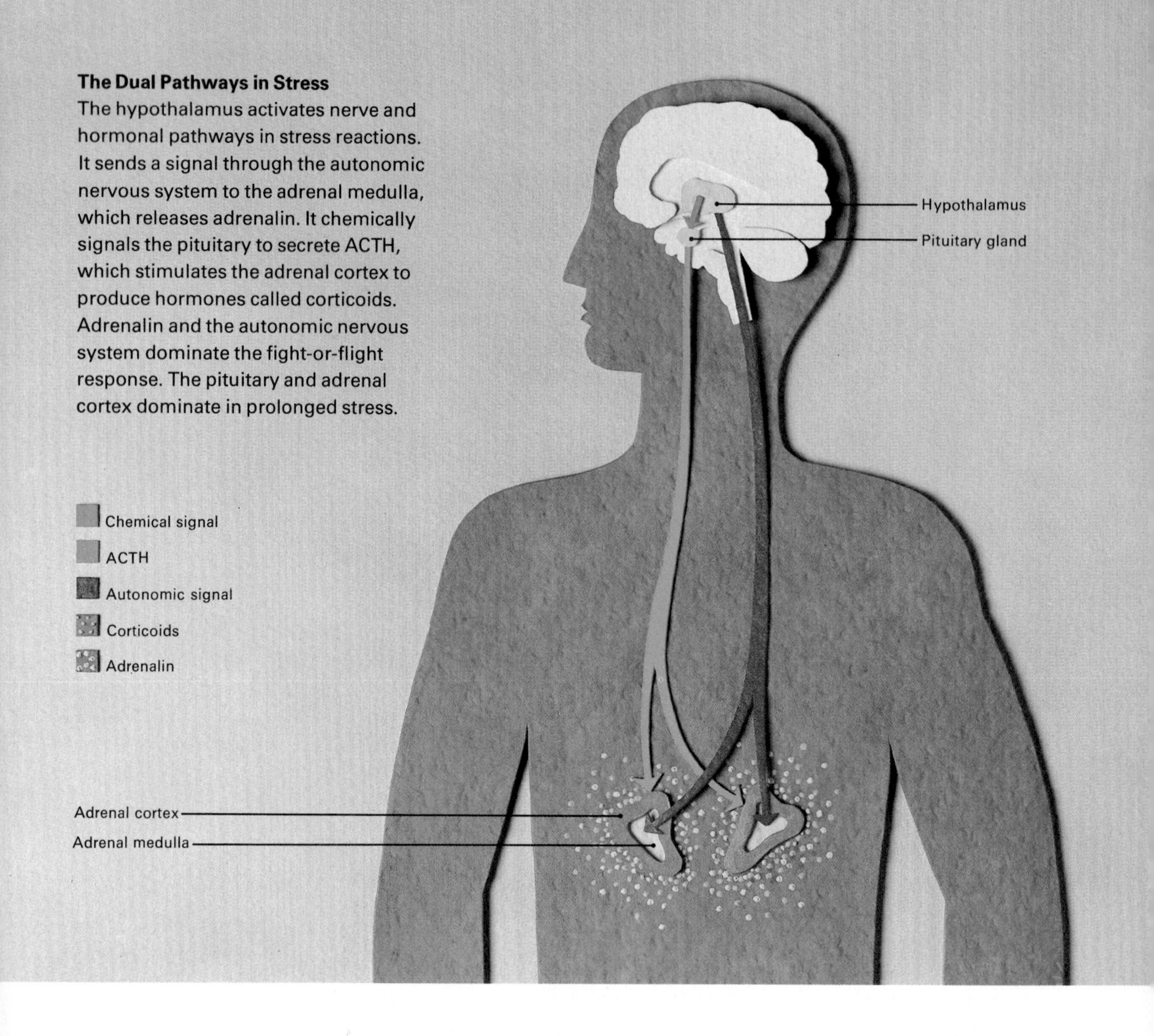

Selye recalled the observations he had made of patients during his medical-student days in Prague. What did the rats have in common that caused the same symptoms no matter what kind of shock or strain they encountered? He theorized that it was a general body response to stress-causing situations.

Selye, now director of the Institute of Experimental Medicine and Surgery at the University of Montreal, came to define stress as "the nonspecific response of the body to any demand." He called any demand that causes wear and tear on the body a stressor. A stressor may be physical or emotional, pleasant or unpleasant–a toothache, a kiss, a vacation–anything out of the ordinary requiring that the body readjust, thus calling the mechanisms of homeostasis into play.

Just as Cannon investigated how the body responds when confronted with an emergency demand, Selye began to study how it responds when faced with stressors for long periods of time. He concluded that the body responds to continued stress in three stages, which together he called the general adaptation syndrome: the alarm reaction, in which the person or animal becomes aware of the stressor; the stage of resistance, in which the body adapts to the stressor; and the stage of exhaustion, in which the body loses its ability to adapt. If the stress continues beyond this point, homeostasis cannot be maintained and illness will result. Selye calls the illnesses that result after the body can no longer adjust diseases of adaptation.

Hormones play a complex and crucial role in the stress response. Scientists understand how some hormones act during the stress response, but they do not yet know all of the hormones that are involved or how these hormones react with other chemicals in the body. Unlike the minor role they play during the fight-or-flight response, hormones produced by the pituitary gland and the adrenal cortex dominate during prolonged stress.

Chronically high levels of corticoids, the adrenal cortex hormones, can be destructive to the body. For example, in addition to raising blood pressure and blood sugar, one group of corticoids can slow down the production of antibodies, the guardians against viruses and other invaders. Also, as Selye's animal experiments showed, these stress hormones can eventually damage the thymus and lymph glands–the core of the body's immune system. This could leave the body vulnerable to diseases ranging from flu to cancer.

Not all persons or animals under stress immediately become mortally ill. The type of illness and when it strikes depend on a complex set of internal and external conditions. Internal conditions include age, sex, genetic predisposition to certain diseases, and general health. External conditions include diet, exercise, and the atmosphere in the home or at work. Under continuous stress, the weakest link in the body will break first, resulting in disorders such as heart disease, ulcers, or mental illness.

There is general agreement among scientists that the physical and emotional demands placed upon a person and the way his body handles them can influence whether he becomes ill. But there are two schools of thought about stressors.

One of these schools is led by Selye. He contends that any stimulus, physical or emotional, will evoke the same response from the body–a nonspecific stress response, characterized by the general adaptation syndrome. Selye maintains that a physical demand–cold, for example –evokes two responses. One is specific, such as shivering, the other is the nonspecific stress response.

The other school, led by Mason, argues that physical strain alone is usually not enough to cause illness. Mason claims that the stress response comes into play only when physical strain becomes extreme enough to bring out an emotion such as fear, anger, frustration, or excitement. He points out that emotions are the greatest known activators of hormones controlled by the pituitary and adrenal cortex.

Mason cites experiments to support his theory that changes in the body caused by stress are mainly triggered by emotional factors. In one study, he used heat as a physical stressor to test the reactions of monkeys. He gradually accustomed the monkeys to the laboratory setting so that they would feel comfortable and at home. Then he began to raise the temperature in their cages. Mason found that the monkeys' hormonal profiles showed a stress reaction only when the temperature became quite high and the animals became clearly anxious as well as physically uncomfortable.

In a similar experiment, Mason used monkeys to test fasting as a physical stressor. By Selye's definition, depriving the monkeys of nutrition should produce a nonspecific stress response. Mason deprived the monkeys of food but provided them with every physical and psychological comfort, including nonnutritious pellets to eat. The hormonal levels of the fasting monkeys showed no stress response. However, other monkeys that were simply deprived of food became anxious and registered the hormonal changes. Mason believes this might mean that psychological factors play a greater role in stress than scientists had previously imagined. Therefore, the stress response may be more specific than Selye's concept indicates.

Another experiment implicating emotions as the main factor in stress-related illness was reported in 1975 by biologist Vernon T. Riley of the Pacific Northwest Research Foundation and the Fred Hutchinson Cancer Research Center in Seattle. Riley was able to delay the appearance of breast tumors in cancer-prone strains of female mice by providing them with a minimum-stress environment. They were handled as little as possible and housed in cages where they were neither isolated nor overcrowded.

Another key factor in stress-caused disease, some researchers suspect, may be feelings of helplessness and despair. For example, in one experiment led by psychologist Jay Weiss in 1972 at Rockefeller University in New York City, researchers restrained three rats in confining cages. They subjected two of them to electric shocks after sounding a buzzer. One rat could prevent the shocks by pressing a lever when he heard the buzzer; the second could not. The third rat received no shocks. Only the rat that had no control over the shocks developed severe stomach ulcers.

At least one scientist believes that helplessness, despair, and uncertainty make people more vulnerable to illness. M. Harvey Brenner, a medical sociologist at the Johns Hopkins University School of Hygiene and Public Health in Baltimore, has studied what happens to the death rate in the United States when unemployment rises or when prices rise while personal buying power falls.

He has charted the death statistics for kidney disease, heart and blood vessel disorders, cirrhosis of the liver, alcoholism, and suicide after economic recessions that occurred between 1902 and 1970. Two to four years after an economic downturn, the statistics show higher

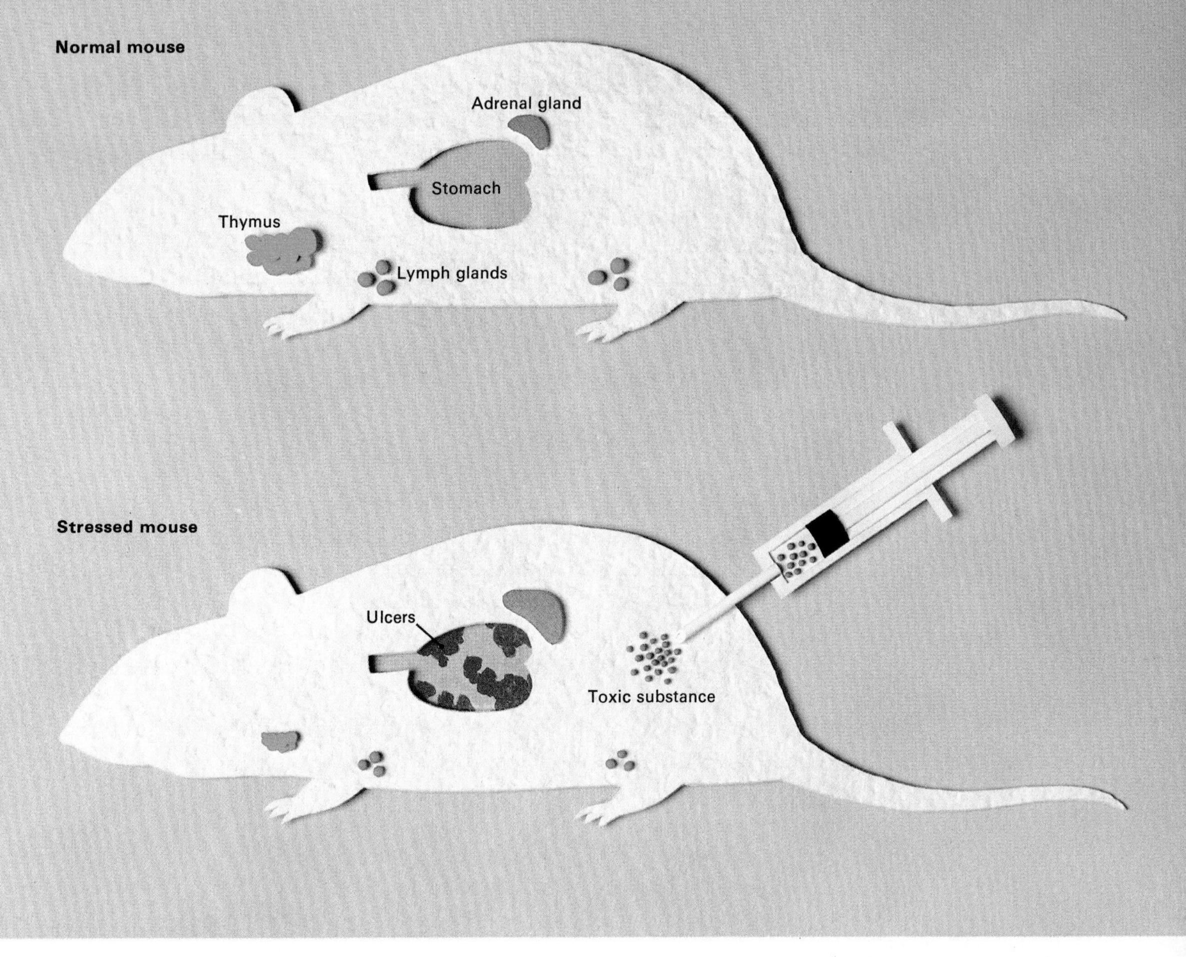

Mice subjected to any kind of stressor, such as a toxic substance, always developed the same three symptoms in experiments by Hans Selye: Their adrenal glands enlarged, their thymus and lymph glands shrank, and they also developed ulcers. This led to Selye's theory of nonspecific response.

death rates. According to Brenner, this reflects increased stress among the population. The impact is very pronounced, regardless of race or sex. As a matter of fact, even the infant-mortality rate climbs. Brenner believes the rise in infant deaths may be partly because parents cannot afford proper care. But stress on the mother might also weaken the fetus or cause premature birth.

Climbing mortality rates following the recession of the mid-1970s, Brenner believes, will lend further confirmation to his theory–that there has been a historical change in the nature of stress in industrial U.S. society. Some of the human wreckage, says Brenner, can be traced to simple loss of income–lack of funds to see a doctor, for example, or not enough money to buy medicine or proper food. But this is far from the whole story.

"As recently as 1900," explains Brenner, "stress was primarily a matter of things having to do with the physical environment–heat, cold, shelter, disease germs, and getting enough food to eat. But man's growing control over such factors, which began with the technology of the 1920s, has seen this kind of stress largely give way to be replaced by social pressures and strains."

This shift has brought more complex life styles, rising expectations, and greater inner turmoil. Brenner believes it is only logical that the psychological component in many illnesses has grown accordingly. And this psychological factor in triggering illness is never more evident than when times are bad.

There is also a growing body of evidence indicating that personality factors have something to do with which disease strikes whom and when. Doctors have noted for many years that individuals with different personality traits are sensitive in varying degrees to stressful situations. For example, a financial setback might be devastating to one person, while another might take such an event in stride. Furthermore, there appear to be two ways in which psychological factors might be involved in disease. In the first case, long-term stress makes a person vulnerable to any number of diseases. In the second, certain personality traits and behavior patterns may be linked somehow to very specific diseases.

Duodenal ulcers, for example, commonly afflict those who bottle up their anger. The ulcer patient–male or female–usually has a need to be nurtured and a deep-seated fear of being disliked. When he feels that others are taking advantage of him or have let him down, he dares not protest outwardly. So his gastric juices protest for him by eating a hole into his stomach or small intestine wall.

In the 1930s, H. Flanders Dunbar, a New York City psychoanalyst, observed that hard-driving, competitive people who derive little satisfaction from their achievements are especially prone to premature heart attacks. More recently, San Francisco cardiologists Meyer Friedman and Ray H. Rosenman became interested in the same idea. In their book, *Type A Behavior and Your Heart* (1974), they claim to have identified a heart attack personality. They examined many men who had suffered heart attacks and found that they tended to have common personality traits. Friedman and Rosenman called this grouping of traits the Type A personality.

Type A people are always in a race against time. They are fast talkers and poor listeners. They also find it hard to relax and constantly try to do more than one thing at a time. For example, a Type A man might simultaneously listen to the radio, read a newspaper, and try to eat breakfast–and meanwhile be planning in his mind all the things he must do that day. Type A personalities are also easily provoked to anger or rage.

As a counterpart to the Type A personality, Friedman and Rosenman claim to have identified a more relaxed and easygoing Type B individual who, if he ever has a heart attack, rarely has one until he is well past middle age.

One of the more intriguing aspects of the Type A personality is that it is apparently no respecter of sex. Historically, few women have had heart attacks before menopause. Consequently, doctors suspected the female hormones released periodically during childbearing years had some kind of protective effect. However, the heart attack rate among younger women is now climbing steadily. Some experts believe this is because more women are smoking. Others think it may also be because more women have entered the business world, and their careers subject them to the same stressful conditions that men face. Besides, they point out, cigarette smoking is a type of behavior that often indicates stress in an individual.

Some of the most impressive recent findings on personality and disease involve cancer. Evidence is accumulating that certain personality traits are common in cancer victims. The emotionally contained, self-effacing person who was starved for affection in childhood and has known great loneliness appears to face the greatest cancer risk.

One of several researchers whose findings support this impression is Caroline B. Thomas of Johns Hopkins University School of Medicine. Beginning with the class that graduated in 1948, Thomas studied 17 successive classes of medical students and recorded everything about them that might have some bearing on their future health. The students' profiles included information about their families, what their relatives had died of, what diseases the students had contracted, their general physical condition, their drinking and smoking habits, and their reactions to various physical stresses such as a salt-free diet, cold, and exercise. Thomas also recorded various psychological factors such as their dispositions, how they felt about their family lives as children, their schooling, hobbies, and goals. The students also took various psychological tests to measure their aggression, passivity, anxiety, depression, anger, and other personality traits.

All but 6 of the 1,337 men and women involved in the study have kept in touch with Thomas. As they have grown older, Thomas has learned who among them has been afflicted with mental illness, developed high blood pressure, suffered a heart attack, committed suicide, or become a cancer victim.

As the data unfolded, it became increasingly clear that each type of disorder grew in a soil containing a distinctive mix of specific psychological and physical nutrients. But what Thomas has found truly astonishing is that certain aspects of the psychological profiles of the 48 cancer victims closely resemble the profiles of the 17 suicides. Both the cancer victims and the suicides felt cheated of warm childhood relationships with their parents. Neither the cancer victims nor the suicides had shown obvious signs of mental illness, but the individuals in both groups had been unusually sensitive and often inwardly depressed. They had concealed their anger and disappointment behind a cheerful facade. In contrast, those who became heart attack victims had vented their anxieties and frustrations openly.

Other researchers had found the same personality pattern linked to cancer before Thomas, but they had studied people already afflicted with the disease. Thomas' study is the first in which the danger signals were documented while the victims were still healthy. The results have led her to pose a fascinating question: Are unconscious dreads and morbid fears in some individuals "everpresent stress undermining the biological guardians of general resistance"?

If stress is indeed turning out to be a modern plague, what can be done about it? Can personality traits and behavior slanted toward the production of disease be redirected toward health? Can we learn to prevent illness by coping better with the stress created by modern life?

Researchers are trying to answer these questions by learning more about the physical and psychological mechanisms of stress and how it affects our bodies. Mason is trying to understand the hormonal links in

Herbert Benson, in his research on ways to alleviate prolonged stress, measures the heart, breathing, and metabolism rates of a woman meditating.

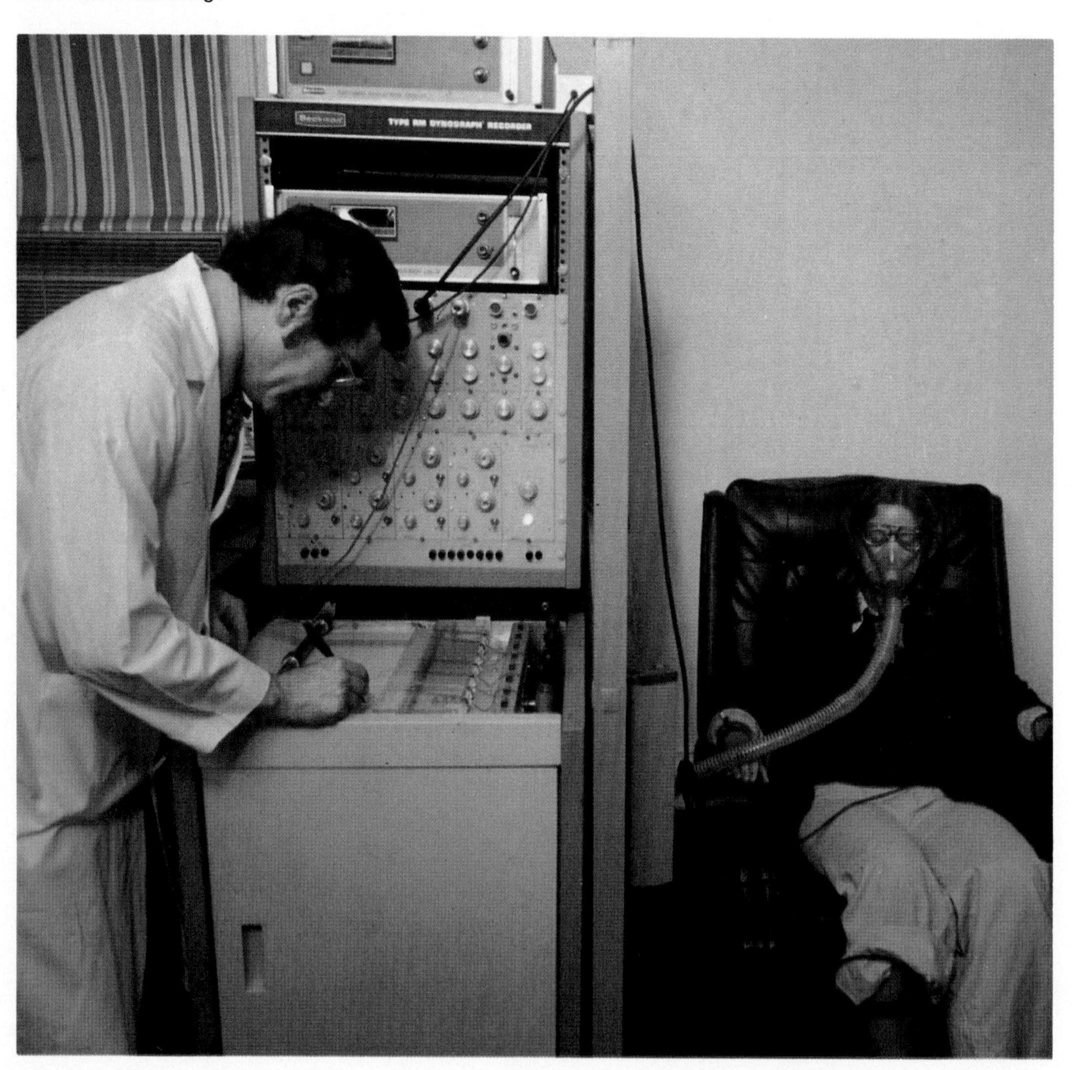

Biofeedback pioneer Elmer Green explains how the electronic equipment will help the patient learn to relax his muscles. Sensors on his arms detect muscle tension, and tones transmitted through the headset tell him if he is relaxing.

the mind-body chain. "We must go beyond clinical observations that psychological factors are related to disease and get into the bodily mechanisms," he says. "We have to study the many interdependent hormones–perhaps 15 or 20 at once–and ask how it all works. How is it all coordinated? But," he continues, "we're just in the early stages of this work. Considering the complexities of biomedical research, it will be many years before we can answer this question."

In Selye's view, evolution may play a role in the problems that human beings are having with stress and disease. The fight-or-flight response was appropriate for early man in his struggle to survive in a primitive world populated with wild beasts and other physical dangers. But in many modern situations–receiving a supervisor's angry memo, failing a course, hearing a loud noise in the street–the fight-or-flight response is inappropriate. We can neither fight nor run. We can only suffer the wear and tear that stress inflicts on our bodies.

Selye likens this to what occurs in allergy victims. The body's immune response to a foreign invader, such as ragweed pollen, is what causes the allergy. Hay fever sufferers would be better off if their immune systems ignored the harmless pollen. "Evolution has not yet come to its end," he says, "and meanwhile the body does not know how to recognize when the results of preparing to fight are worse than passive tolerance." Selye recommends that each of us find the level of stress with which we are comfortable and try to avoid situations that will cause us to exceed it.

Test Your Stress Level

Place a number from 0 to 5 to the left of each event, indicating how many times it has occurred during the past 12 months. Multiply the number of times each event occurred by the value in the right-hand column.

	Desirable Personal Events	Score
___	The end of a serious problem with your parents	x 52 = ___
___	*Going on the first date of your life	x 46 = ___
___	Being told you are attractive by someone your age	x 26 = ___
___	Finding a boyfriend or girlfriend	x 66 = ___
___	*Beginning the first year of senior high school	x 30 = ___
___	Increase in acceptance by those your age	x 40 = ___
___	Recognition for doing well in a sport or other activity	x 36 = ___
___	Becoming a member of a church	x 29 = ___
___	Being invited to join a social organization	x 23 = ___
___	Getting a summer or part-time job	x 50 = ___
___	*Getting your first driver's license	x 60 = ___
___	Quitting the use of drugs or alcohol after using them regularly	x 41 = ___
___	Being accepted at the college of your choice	x 68 = ___
___	*Graduating from high school	x 75 = ___
___	*Getting your first permanent job	x 65 = ___
___	*Deciding to leave home and doing so	x 95 = ___
___	*Boys—getting married	x 101 = ___
___	*Girls—getting married	x 158 = ___
___	Outstanding personal achievement (receiving special award or prize)	x 68 = ___
	Add the right-hand column to get your Desirable-Events Score.	___

	Undesirable Personal Events	Score
___	Development of a serious problem between you and your parents	x 65 = ___
___	Breaking up with a boyfriend or girlfriend	x 41 = ___
___	Being told to break up with a boyfriend or girlfriend	x 52 = ___
___	Decrease in acceptance by those your age	x 44 = ___
___	Not making an extracurricular activity in which you wanted to be involved	x 44 = ___
___	Being involved in an automobile accident	x 46 = ___
___	Serious illness requiring your hospitalization	x 77 = ___
___	Death of a close friend	x 105 = ___
___	Starting to use drugs or alcohol	x 46 = ___
___	Being asked by friends to do something illegal	x 33 = ___
___	Appearing in juvenile court	x 56 = ___
___	Moving to a new school district	x 56 = ___
___	*Failing a grade in school	x 62 = ___
___	Being suspended from school	x 23 = ___
___	Being sent away from home	x 103 = ___
___	For boys—fathering an unwed pregnancy	x 72 = ___
___	*For girls—becoming pregnant while unmarried	x 127 = ___
	Add the right-hand column to get your Undesirable-Events Score.	___
	Subtract the Undesirable-Events Score from the Desirable-Events Score.	___

The total should fall between 0 and 450. The average is about 200.

	Family Events	Score
___	Change in your father's occupation so that he spends a lot less time with you	x 40 = ___
___	Very large improvement in your parents' financial status	x 55 = ___
___	Decrease in your parents' financial status	x 60 = ___
___	Loss of a job by a parent	x 82 = ___
___	Mother beginning to work outside the home	x 27 = ___
___	Development of a serious personal problem between your parents	x 60 = ___
___	End of a serious personal problem between your parents	x 43 = ___
___	Separation of your parents	x 98 = ___
___	*Divorce of your parents	x 106 = ___
___	*Marriage of a parent to a stepparent	x 84 = ___
___	Addition of a third adult to the home (such as a grandmother)	x 38 = ___
___	Death of a parent	x 243 = ___
___	Death of a grandparent	x 113 = ___
___	Death of a brother or sister	x 192 = ___
___	Serious illness of a parent requiring hospitalization	x 114 = ___
___	Serious illness of a brother or sister requiring hospitalization	x 99 = ___
	Add the right-hand column to get your Family-Events Score	___
	Subtract your Family-Events Score from your Desirable-Events minus Undesirable-Events Score to get your total Environmental Score.	___

Most Family-Events Scores fall between 0 and 400, with the average about 200. The Environmental Score may vary between −300 and +400; the average is less than 100.

*These events call for either 0 or 1, because they are unlikely to have occurred more than once in 12 months.

A test for measuring stress in persons age 12 to 19, *opposite page,* was developed by R. Dean Coddington, a child psychiatrist at the Louisiana State University School of Medicine. You can determine whether your stress level is above or below average by completing the test. Coddington established the averages from a relatively small number of teen-agers, and he wants to study the scores of many more. If you or your class would like to help with this research, send a copy of your scores to Dr. Coddington at the LSU School of Medicine, 1542 Tulane Ave., New Orleans, La. 70112. Also list your age, sex, state in which you live, father's and/or mother's job, and how much schooling they had. You need not give your name.

Several investigators are searching for ways to reduce or reverse the damaging effects of continued stress. For example, Herbert Benson of the Harvard Medical School in Boston has studied the effects of meditation on patients suffering from high blood pressure, one indicator of susceptibility to heart disease. He found that not only did blood pressure drop during meditation, but the body's metabolism slowed down. Benson believes this might indicate the existence of an opposite reaction to the fight-or-flight response, a reaction that is set in motion by the hypothalamus to repair naturally the wear and tear on the body caused by prolonged stress. He calls this reaction the relaxation response and claims we can learn to evoke it by performing a simple exercise that consists mainly of sitting quietly for 20 minutes twice a day, being aware of our breathing, and passively accepting–but not concentrating on–any thoughts that come into our mind. Benson says these findings have been confirmed by other investigators, and researchers are now looking into ways in which the relaxation response may be used to prevent and treat other stress-related diseases.

Scientists are also investigating the stress-relieving potential of biofeedback. Biofeedback machines attached to patients' bodies allow them to monitor otherwise involuntary bodily processes. Theoretically, if they are aware of these processes, they can learn to control them.

"You visualize what you want to happen and let the body do it," explains Elmer Green, director of the Menninger Foundation's Voluntary Control Program in Topeka, Kans. Green has used biofeedback to help patients control epileptic seizures, high blood pressure, and migraine headaches. Biofeedback is also used to help individuals learn to relax by generating a certain type of brain wave associated with deep relaxation.

Ten years ago, most doctors and the general public were so deeply involved in their love affair with the health-care technology of wonder drugs, artificial kidneys, heart transplants, and other complicated treatments that questions about controlling disease by coping with stress rarely came up. Today, an entirely different attitude is emerging. The philosophy that the body should be treated as more than just the sum of its physical parts is growing in popularity. And along with this philosophy goes a strong dose of the old adage that an ounce of prevention is worth a pound of cure.

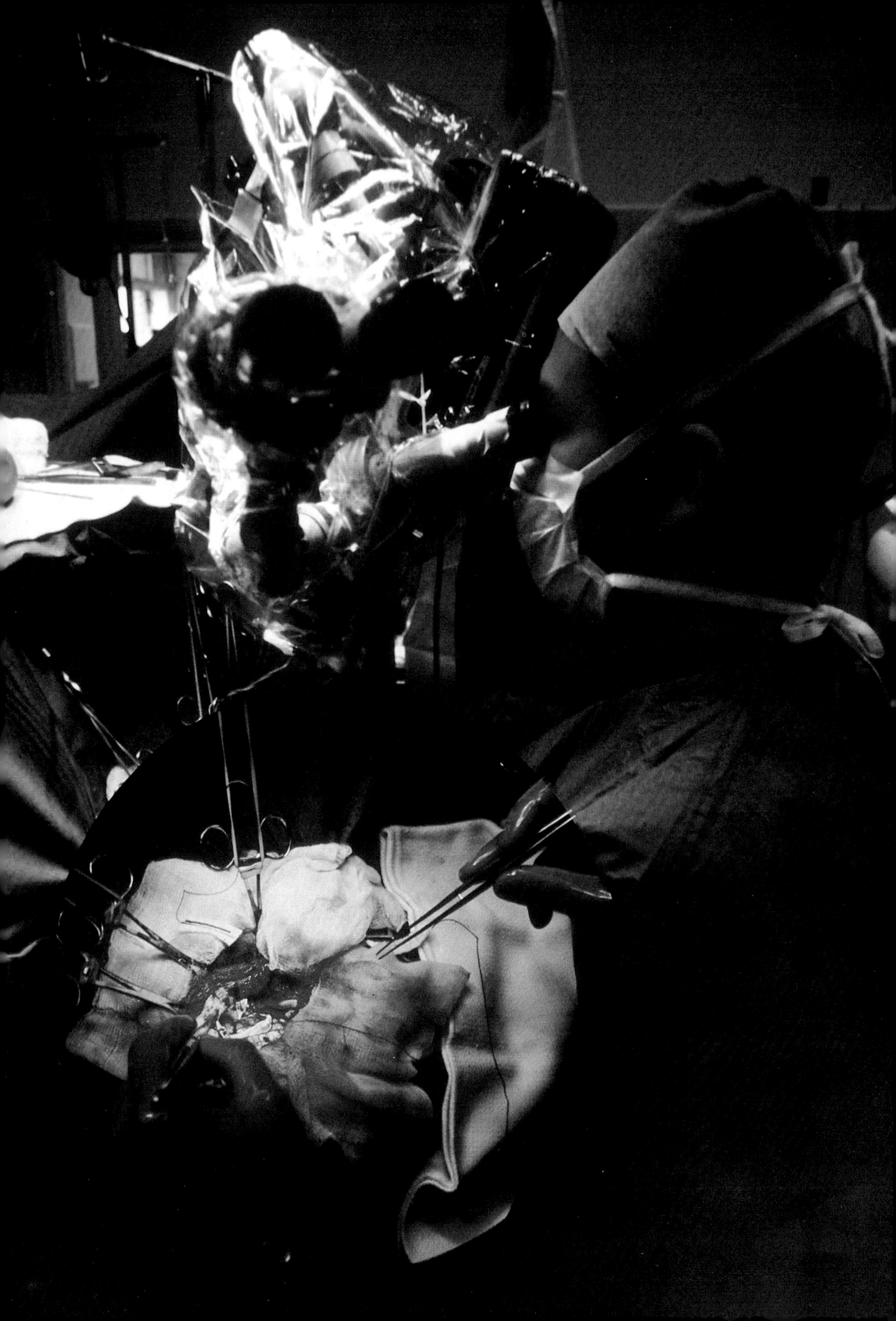

Surgery in Miniature

By John F. Henahan

Tiny instruments and special microscopes open a new life- and limb-saving era in modern surgery

It happened on Jan. 19, 1976, during the night shift at a plastics plant in Santa Clara, Calif. Martha Carpenter, a 25-year-old working mother, was cutting out plastic cookie molds when her cutting machine jammed. She reached in with her left hand to clear the machine, but could not pull her hand out quickly enough. The sharp blade came down and neatly sliced off all four fingers.

Through her shock and pain came a horrifying thought, "My fingers are gone forever." But she was wrong. Luckily, the foreman remembered reading that doctors had developed ways of reattaching severed fingers. So he picked them up and packed them in ice. Within minutes, the local rescue squad arrived and took Martha Carpenter and her fingers to the Ralph K. Davies Medical Center in San Francisco. Two hours later, a team of three surgeons began an incredibly intricate operation. Guiding tiny instruments beneath the lens of a microscope, sewing together nerves and blood vessels no thicker than a human hair, the surgeons worked for 15 hours. And when the operation was completed, the nerves, muscles, and blood vessels of all four fingers were back in place with circulation restored. It might take a year, but Carpenter will probably regain almost full use of her hand.

Martha Carpenter is one of the thousands of patients whose limbs and even lives have been salvaged by microsurgery. Before the development of these meticulously detailed operations performed under a microscope, doctors could not reattach amputated fingers because of the difficulties in reconnecting tiny blood vessels and nerve fibers. Now, using an impressive array of recently developed miniaturized needles, forceps, scissors, sutures, and clamps, microsurgeons neatly sew those blood vessels and nerves together again. The operations have become almost routine in a number of hospitals around the world. In addition to reattaching fingers, microsurgeons can, in some cases, reverse the effects of vasectomies, perform previously impossible brain operations and feats of plastic surgery, and operate on difficult-to-reach and delicate parts of the spinal cord, eye, and inner ear.

Today, surgical supply companies manufacture the sophisticated microscopes and instruments that are required for microsurgery, but pioneers in the field had to invent their own instruments. One of the pioneers in finger surgery, Harry J. Buncke, Jr., a plastic surgeon in San Mateo, Calif., and one of the team that operated on Martha Carpenter, began laying the groundwork in the 1950s. One of the major problems he faced was devising needles and thread fine enough to repair blood vessels that are less than 1 millimeter in diameter.

Buncke found that the single-stranded filament of a silkworm's cocoon made excellent suture material. But making a needle that was small enough and sharp enough to sew the ends of extremely small blood vessels back together without damaging them was a more difficult problem. Buncke and Werner P. Schulz, a German-born physicist and microelectronics expert at Stanford University, eventually made the needles by electroplating a small amount of chromium metal on a single nylon thread, thinner than a human hair. Other early microsurgeons made small scalpels from tiny pieces of razor blade. They also found that jeweler's forceps with their needle-sharp tips were perfect for operating under a microscope.

The author:
John F. Henahan, a science writer, also wrote "Plants: The Renewable Resource," which appears in this edition of *Science Year.*

Commercial manufacturers now make microsutures from single-stranded nylon fibers so thin they can only be seen when held up to light. They produce metal needles with a hole small enough to accommodate the finest threads. They also offer instruments such as fine-tipped forceps and scissors with blades about 1.3 centimeters long attached to spring-steel handles that allow precision, fingertip control for the most delicate surgical maneuvers.

The first microsurgeons used fairly crude operating microscopes. These microscopes, attached to a floor stand, had been used since the 1920s for a limited number of operations, mainly on the eye and ear. By 1976, several optical companies had manufactured thousands of

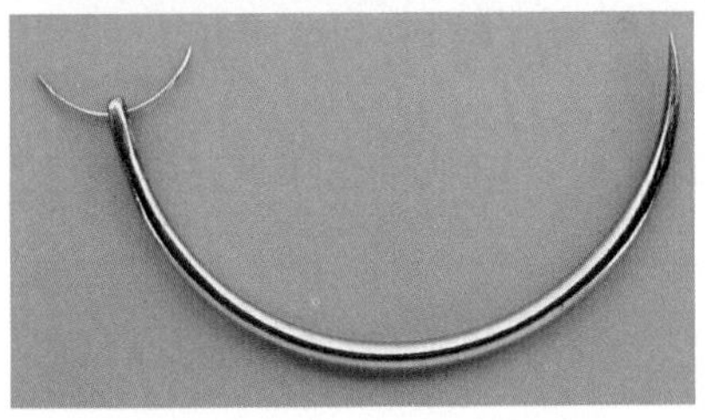

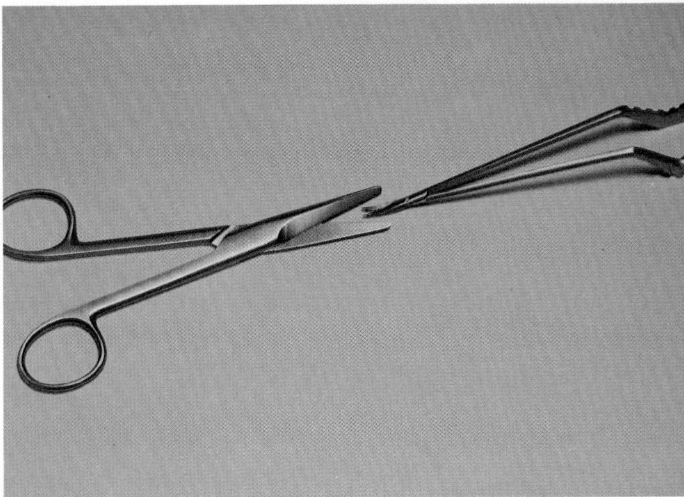

The microsurgeon uses an operating microscope, *top,* sheathed in sterile plastic, to view tiny nerves and blood vessels. A microneedle can pass through the eye of its conventional counterpart, *above,* and microscissors, *right,* are swallowed by standard surgical scissors.

highly refined operating microscopes. Some are equipped with two or more viewers; pedal controls to adjust the height of the microscope; zoom lenses to focus in on a particular area; and bright lamps to provide the vivid illumination needed for seeing clearly in a tangle of threadlike vessels or nerves. Many of these microscopes can be fitted with still- and motion-picture cameras, and some use a television camera to transmit views of the operating field to students and other observers through closed-circuit television.

Before the early microsurgeons could dare to operate on human beings, they had to develop their techniques by experimenting on animals. Buncke, for example, began replacing monkey fingers and amputated rabbit ears in the late 1950s. He chose rabbit ears because their key blood vessels are about the same diameter as the smallest human blood vessels. In 1971, Buncke's skills were put to a most unusual test. A young fireman had cut off his thumb with a power

New Lifeline for the Brain

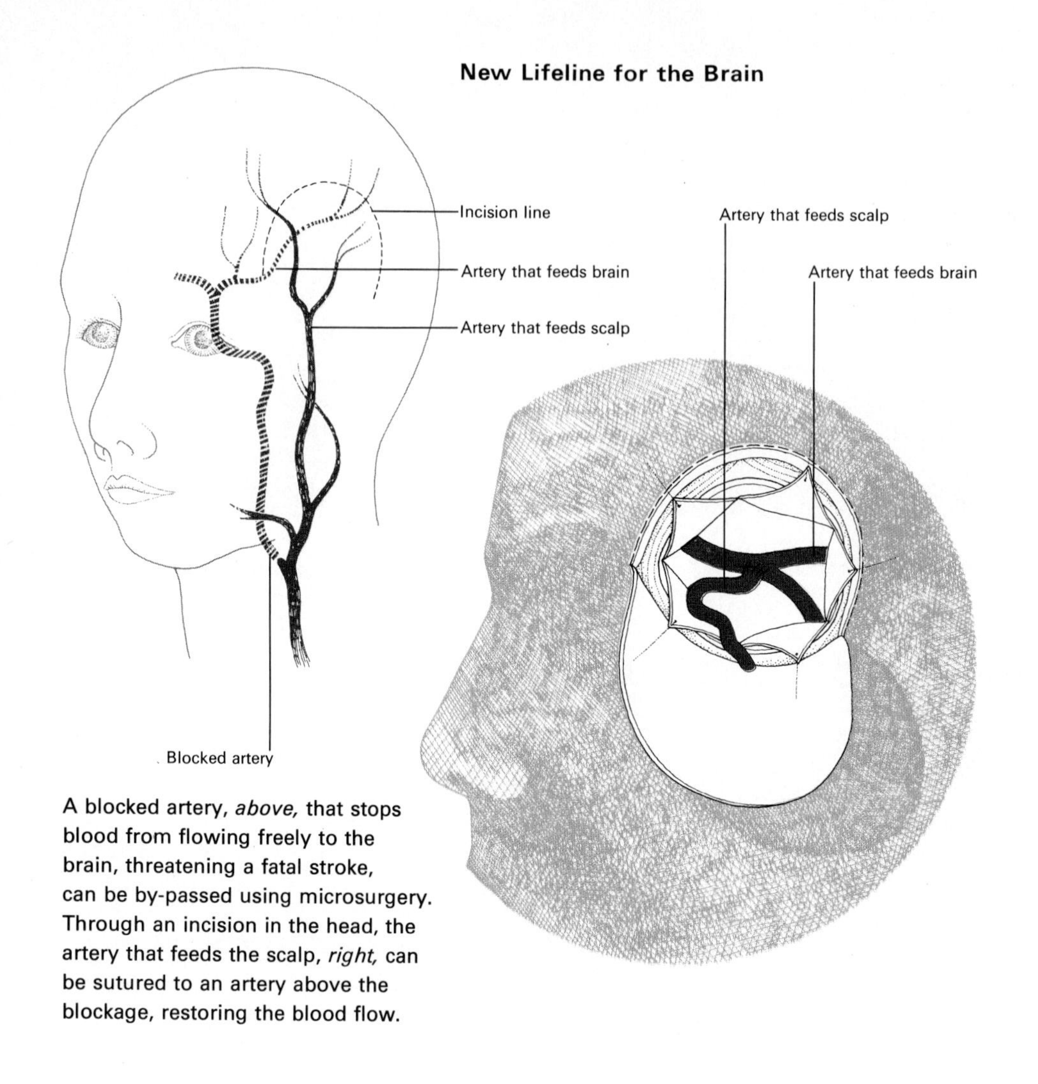

A blocked artery, *above,* that stops blood from flowing freely to the brain, threatening a fatal stroke, can be by-passed using microsurgery. Through an incision in the head, the artery that feeds the scalp, *right,* can be sutured to an artery above the blockage, restoring the blood flow.

saw. This accident could have put an end to his career, because he could not grasp tools without a thumb. Buncke devised an ingenious solution to the problem. He replaced the lost thumb with one of the fireman's large toes. Today, the fireman can handle tools almost as well as he could before the accident.

Another pioneer in the late 1950s, Theodore Kurze, head of the Department of Neurosurgery at the University of Southern California, became one of the first doctors to use microsurgery for removing almost inaccessible brain tumors without damaging or destroying sensitive nerve fibers. Before Kurze's pioneering work, from 70 to 90 per cent of the patients were afflicted with facial paralysis after operations with traditional tools. Now about 80 per cent of these patients recover without any paralysis.

Undoubtedly, the largest number of successful finger and limb reattachments have been achieved in the People's Republic of China.

Chinese microsurgeons in 1964 were the first to reattach an amputated finger successfully. They were also the first to reattach as many as four amputated fingers on one hand. In another spectacular operation, they used microsurgical techniques to replace the mangled right foot of a train-wreck victim with the uninjured foot from his left leg, which had to be amputated. He can now walk with crutches. In China, many people have had fingers reattached, yet can manipulate chopsticks, play table tennis, operate machinery, and even play the accordion. "Their [the surgeons'] expertise far exceeds anything we have seen in the United States," says Malcolm C. Todd, former president of the American Medical Association, who recently visited China.

One reason for their skill may be that the Chinese have set up microsurgical centers throughout their country, each serving hundreds of millions of people. At the world's largest such facility, the Sixth People's Hospital in Shanghai, Ch'en Chung-wei, one of China's foremost microsurgeons, has reattached more than 300 amputated fingers since 1966. In about 50 per cent of these cases, the finger reattachments were successful.

The Loma Linda (Calif.) School of Medicine has one of the busiest U.S. centers for microsurgery on hands and fingers. One or two reattachments are performed there every month.

"Normally, a patient can get along without one finger and could be back on the job in a few days after it's amputated," says Gary K. Frykman, assistant professor of orthopedic surgery at Loma Linda. "However, if more than one finger has been lost, or a thumb, the patient may feel that he needs to have them reattached to carry out his job, or even for cosmetic reasons.

"Under those circumstances, we tell the patient there is a 50-50 chance that we can reattach the fingers or thumb successfully, but we warn him that it may be several months before he will be able to get anything like full use out of them."

Once the decision for reattachment is made, the patient and the surgical team must brace themselves for an operation that can take anywhere from three to six hours for each finger. Most surgeons prepare themselves for the ordeal by getting plenty of sleep. Some abstain from coffee for several days, and others give it up altogether, because caffeine can cause fine tremors in their fingers.

In a typical finger reattachment, the surgeon must carefully clean the bone on both ends of the amputated finger. He then fastens the bone together with small hooks and wires. Because even a single drop of blood can completely block the microsurgeon's view of the tiny operating area, he must temporarily stop all flow of blood. He may use a tourniquet, small clamps, or the bipolar coagulator, used in brain surgery since the 1940s but adapted for microsurgery by Leonard I. Malis, a neurosurgeon at Mount Sinai School of Medicine in New York City. The bipolar coagulator passes a weak electric current into the tips of a jeweler's forceps. The surgeon touches the walls of the

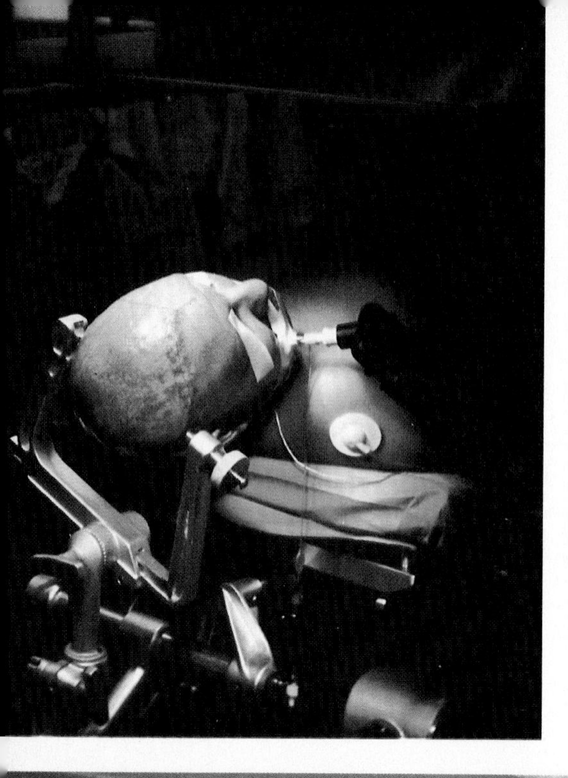

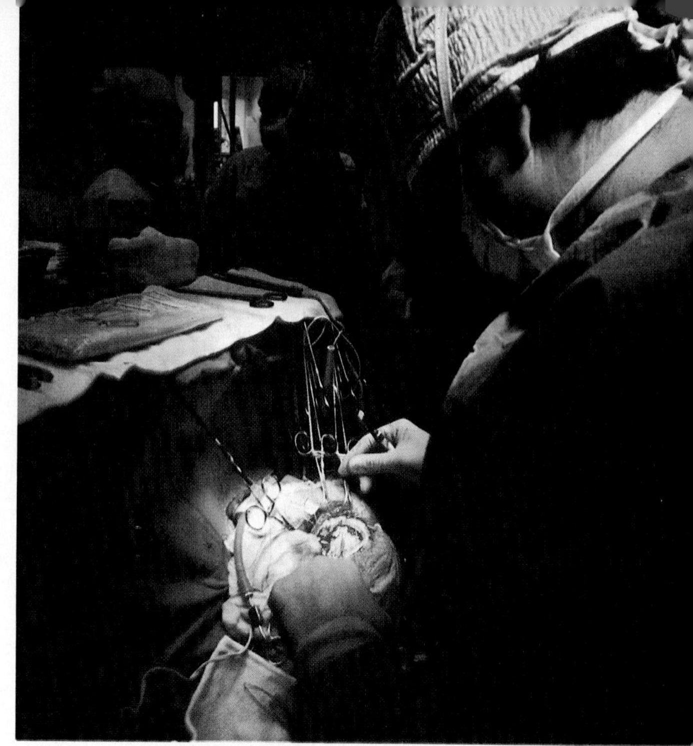

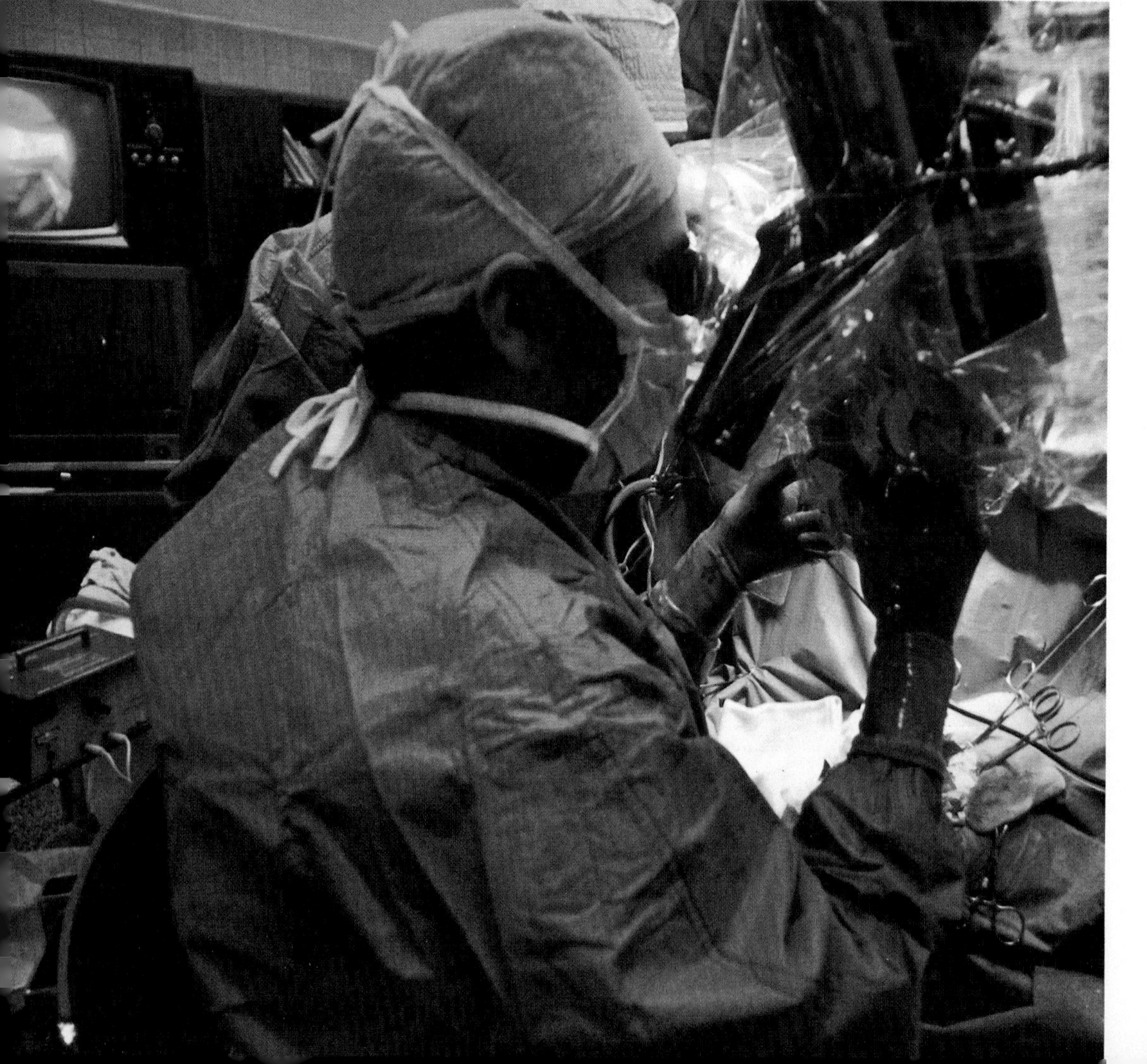

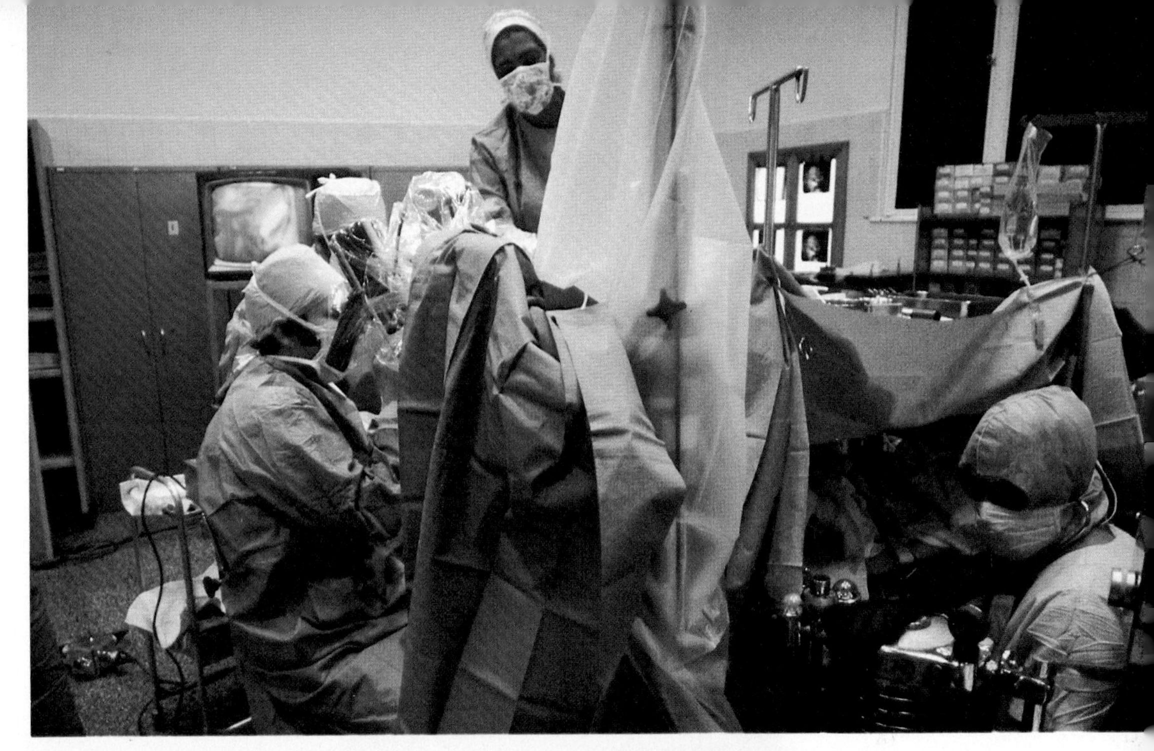

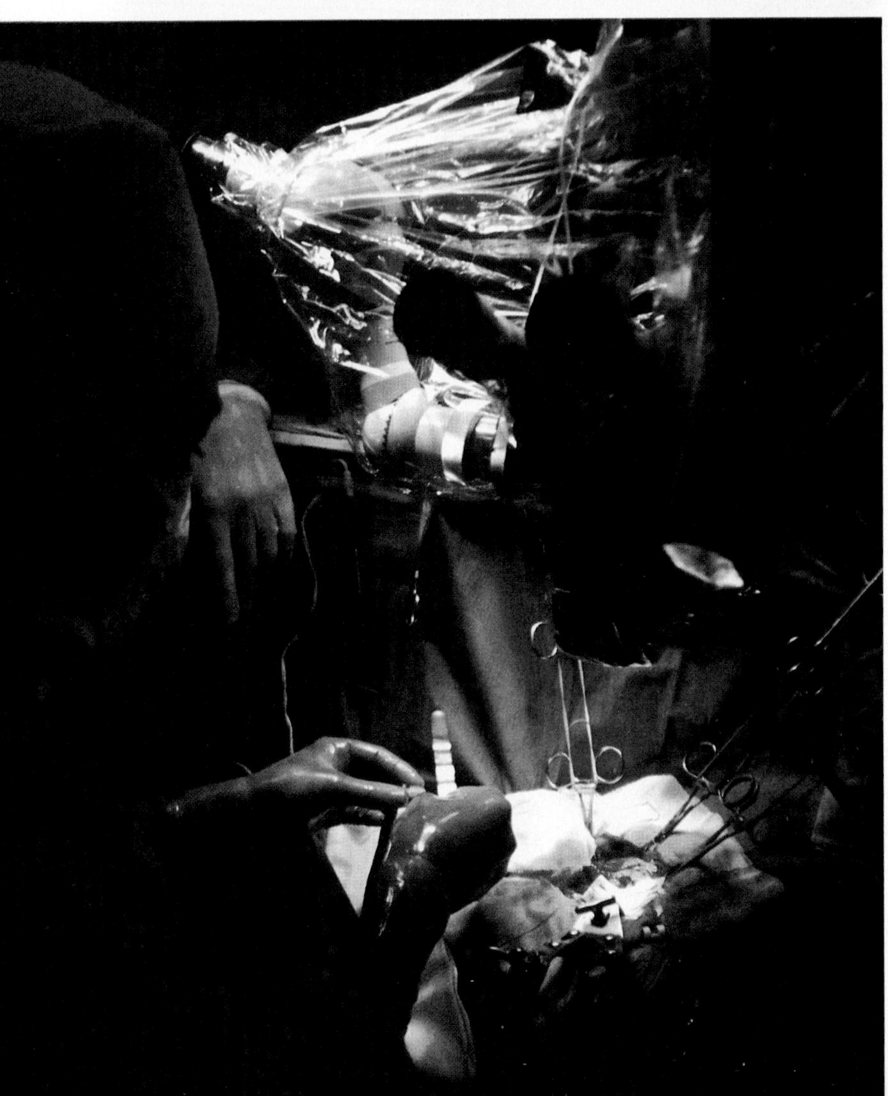

In a stroke-prevention operation, *opposite page, clockwise,* the patient's shaved head is clamped before a surgeon removes a section of scalp bone. Then the microsurgeon brings the operating microscope into position, and takes up the fine-tipped instruments, *left,* that enable him to work on the tiny arteries. A TV camera attached to the microscope projects a view of the operating field on a monitor, *above,* so students and other observers can see the surgeon's technique.

After the dura mater, the brain's protective membrane, is pulled back, a rubber shield is slipped under the operating area, *top,* to protect the brain. The brain artery is clamped above and below the incision site. The surgeon then sutures the scalp artery to the incision in the brain artery, *middle and bottom.*

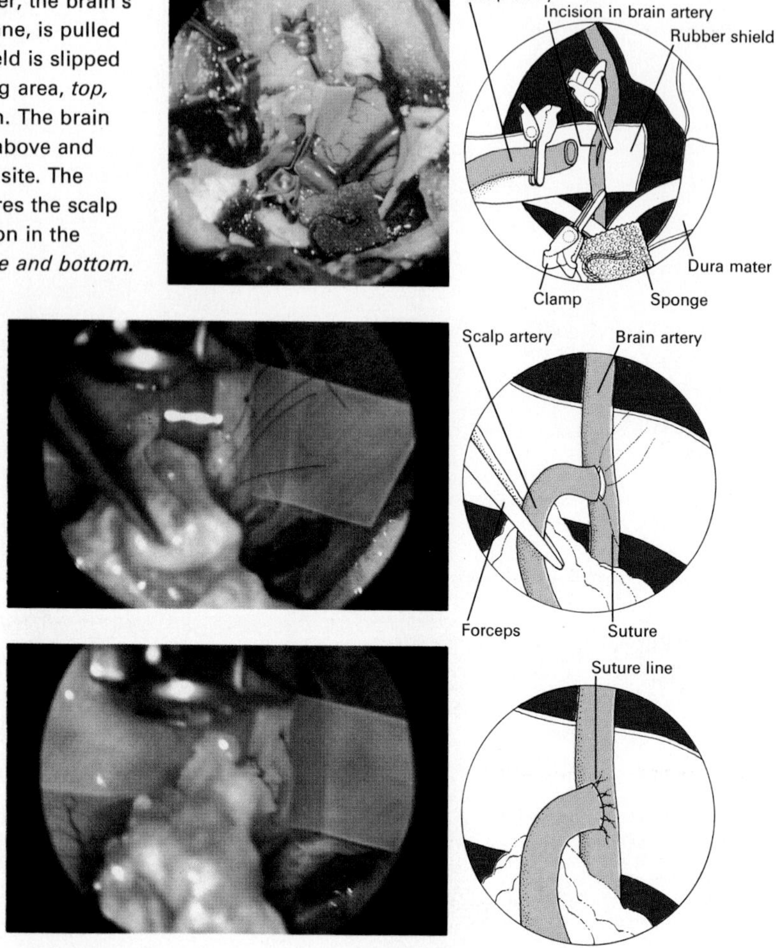

blood vessel with the charged tips of the forceps, causing the blood to coagulate temporarily.

Then the microsurgeon sews together the ends of tendons, arteries, nerves, and the extremely thin-walled veins. His first assistant ties up and clips off the ends of the sutures. During and after the operation, the patient is given drugs to prevent his blood from clotting. Blood must be able to flow freely through the blood vessels as soon as they are reattached for the operation to succeed.

Several thousand surgeons throughout the world are now using microsurgery to break through to new frontiers. They have, for example, greatly improved techniques for repairing cerebral aneurysms, weakened spots in the wall of an artery in the head. Previously, only about 50 per cent of these operations were successful; now the success rate is about 95 per cent. Microsurgery is also promising new hope for potential victims of strokes, a leading cause of death in the United States. Doctors have perfected a technique for cutting through, cleaning out, and reattaching small arteries that carry blood to the brain. Left partially clogged, these arteries almost certainly would cause strokes. About 85 per cent of the patients who undergo this operation have excellent chances for long-time survival.

However, this technique cannot be used when the patient's arteries are completely blocked. Instead, a method developed by microsurgeons at the University of Vermont during the mid-1960s by-passes the blocked arteries by attaching the artery that feeds the scalp directly to an artery that feeds the brain.

Neurosurgeon Jose Luis Salazar of the University of Illinois Medical School in Chicago is a leading expert in this type of operation. He first cuts a hole about 4 centimeters in diameter in the patient's skull to expose the brain. Then, using microsurgical techniques, he clamps off and cuts a hole in one of the brain's three main arteries and inserts the scalp artery. His hand movements are so slight that they can barely be detected unless viewed through the microscope. Salazar deftly stitches the scalp artery in place with a hair-thin needle and thread that is almost invisible. Then, releasing the clamps, he opens a new blood channel to the brain. Thanks to the precision of microsurgery, Salazar's instruments barely touch the surrounding delicate brain tissue.

Buncke and other plastic surgeons are now exploring another promising area of microsurgery. They are developing a technique for transplanting free skin flaps–chunks of skin, complete with the underlying flesh and blood vessels–from hidden areas, such as the groin, to severely damaged visible areas such as the face.

In the conventional procedure for making such skin grafts, the surgeon cuts a flap of skin from the patient's arm. But he leaves the skin partially attached so it will be kept supplied with blood by vessels in the arm. The surgeon then sews the flap, and with it the patient's arm, to the damaged facial area. The patient must remain in this awkward position for several weeks, until a new system of blood vessels grows naturally from flap to face. With the new skin-flap technique, however, microsurgeons can hook up the blood vessels immediately. As more skilled microsurgeons enter the field, this and other exotic-sounding operations will become almost commonplace.

But microsurgery is not the kind of field in which every surgeon could or would want to be involved. A surgeon must go through at least two years of rigorous discipline and training–from practicing on thin plastic tubes to operating on rat brains and monkey fingers–before he can work on a human brain or reattach amputated limbs.

Yet, microsurgeons who have run this difficult course feel it is worth all their trouble. Undoubtedly, there are many people who now do or someday will owe their lives to microsurgery–people otherwise doomed to suffer paralyzing strokes or fatal brain hemorrhages. Many others, like Martha Carpenter and the young fireman, will be saved from the crippling loss of fingers and thumbs by the skillful surgeons who operate through the microscope.

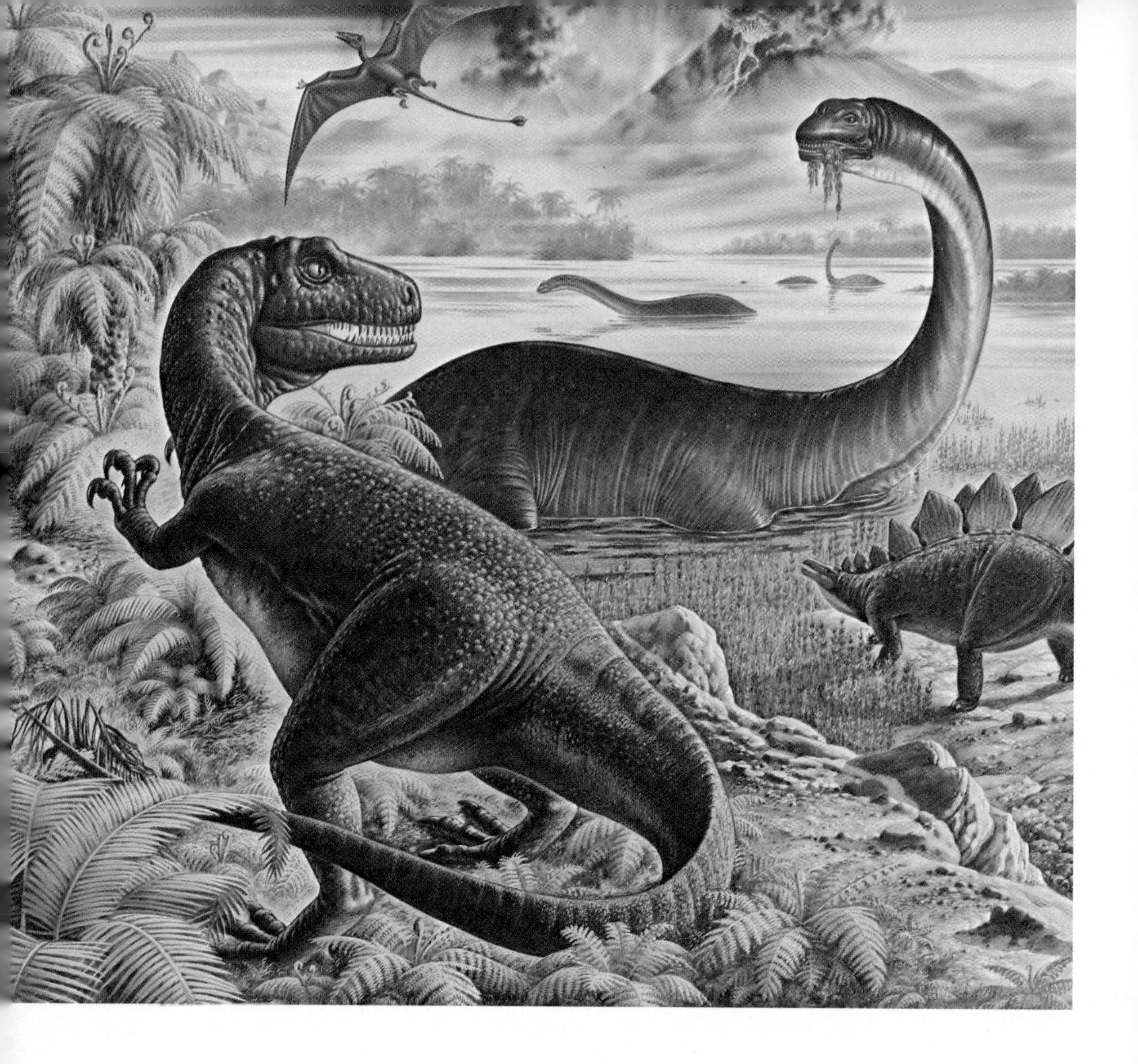

A New Image For Dinosaurs

By Robert T. Bakker

Paleontologists are taking a new look at old evidence about dinosaurs and concluding that they were warm-blooded ancestors of the birds

Old view of dinosaurs and their environment during the Jurassic Period, *opposite page,* is in marked contrast to the picture painted by latest interpretations of fossil information about these animals.

If you were asked to describe the dinosaurs, which roamed the earth 225 to 65 million years ago, you would probably say that they were huge, sluggish, solitary creatures–cold-blooded relatives of today's lizards and crocodiles–who spent most of their time in tropical swamps or lakes. And you would probably say that all the dinosaurs have long been extinct.

You would be wrong on all counts. Some paleontologists are now convinced that dinosaurs were warm-blooded and lived in a wide range of environments. We also now know that some of them traveled in herds, that they could run with some speed, and that they are not completely extinct. One group of direct descendants of the dinosaurs still thrives today–the birds.

Studies of dinosaurs in the last few years–some of them based on new views of old evidence–have overturned many traditional theories about these animals. The studies have also revived scientific interest in these creatures and their role in the evolution of modern vertebrates, or animals with backbones.

Fittingly, the results of this re-evaluation are being revealed almost exactly 100 years after the word dinosaur became familiar to the general public. Early in 1877, amateur naturalists discovered three huge beds of the bones of dinosaurs that lived in the late Jurassic Period (about 140 million years ago) in what is now Wyoming and Colorado. The naturalists notified two competing American paleontologists, Yale University's O. Charles Marsh and Edward D. Cope of the Philadelphia Academy of Natural Sciences. Marsh and Cope hired teams of fossil hunters, and within a few months, railroad cars were carrying huge loads of fossil dinosaur bones east to their laboratories. When the bones were assembled, they included complete skeletons of more than a dozen species. Some of the animals were 30 meters (100 feet) long and weighed an estimated 27 to 36 metric tons (30 to 40 short tons) when alive.

A few years later, paleontologists found other rich dinosaur bone beds in the rocks of Montana, South Dakota, and Wyoming. These bones were from the late Cretaceous Period (between 65 and 75 million years ago). Newspaper and magazine accounts described the brilliant successes and the frontier hardships of the men who hunted for and gathered these bones. From Marsh and Cope came scientific and popular descriptions of animals whose names soon became household words, such as *Brontosaurus* (thunder lizard); *Stegosaurus* (plated lizard); and *Triceratops* (three-horned face).

But after this initial flush of excitement, most scientists studying vertebrate fossils lost interest in dinosaurs, which had seemingly become extinct. They began to concentrate on other groups of animals, particularly fossil mammals. These were the ancestors of several important modern groups, such as bears, cats, dogs, elephants, horses, and, most important of all, man.

The author:
Robert T. Bakker is an assistant professor of vertebrate paleontology and geology at Johns Hopkins University.

Many scientists considered dinosaurs an evolutionary novelty, not worth much serious study because they did not appear to be related to any living species. The huge, long-necked sauropod dinosaurs, the largest land animals that ever lived, were portrayed as dimwitted brutes slowly plodding through steamy swamps where the water helped support their weight and protected them from such predators as the more mobile *Allosaurus*. When pictures of these sauropods were painted, the artists frequently placed tiny mammals in the scene, mammals that seemed to scurry actively through the forest undergrowth, waiting for the time when the dinosaurs would become extinct so that they might evolve as the dominant vertebrates.

Two bits of evidence were cited as proof of the picture of sauropod dinosaurs–the position of their nostrils and the texture of the bones at the joints. In sauropods, the external nares (openings in the skull for the nostrils) were set far back from the snout, just in front of the eyes in some species and between the eyes in others. Today, some swimming vertebrates, such as whales, have nostrils set back this far. The structure is also found in fossils of icthyosaurs, or extinct sea reptiles. It was assumed that the position of their nostrils enabled sauropods to breathe with only the top of the head exposed above the water of a lake or swamp pond. But this conclusion is not as logical as it appears. Some fully terrestrial vertebrates, such as savanna monitor lizards and elephants, also have nostrils set far back, and so did some South American land mammals that are now extinct.

The limb-joint bone surfaces in most adult mammals and birds are smooth and covered with a thin layer of cartilage. Sauropod limb-joint bone surfaces were not smooth; they had many grooves and pits that must have been filled in life with cartilage. It was argued that sauropod limb joints were so thick with cartilage, a weaker substance than bone, that they could not have taken the strain of fast movement on firm, dry ground.

That conclusion was incorrect. The general shape of sauropod limb joints closely resembled those of present-day land mammals, especially elephants. The sauropod joint cartilage would have been a few centimeters thick in the pits and grooves, but quite thin on the surfaces between the pits and grooves. Most important of all, only adult birds and mammals have smooth joint surfaces on their bones. Rapidly growing juveniles do not, yet juveniles often are faster and more active than adults. The sauropods continued to grow throughout life–as do some living reptiles–and the cartilage-filled pits were growing areas where new bone was added to the limbs. The sauropod joints thus reflect growth patterns, not aquatic habits.

Many other early conclusions about dinosaurs were also based on assumptions that do not stand up under intensive investigation. If we change these assumptions and look again at the fossil evidence, we get a totally different picture of dinosaurs.

Consider the environment in which these animals lived. Until the early 1900s, the Lake Basin theory dominated interpretations of the fossil-bearing rocks in the Western United States, where the Jurassic dinosaurs and many mammal fossils have been found. According to that theory, the sedimentary rocks containing the fossils were formed by deposition in huge lakes. The Morrison Formation, which extends through parts of Wyoming, Colorado, and Utah and yielded many sauropod fossils, was supposedly the product of one of these great lake basins. But intensive research over the last 100 years has shown that few of the rocks in the Morrison Formation were formed by lake deposits. Geologist John B. Hatcher of Princeton University and the Carnegie Institute's Museum of Natural History in Pittsburgh, a combination of fossil hunter, anatomist, and frontier poker player, was the first to point this out. He explored the Western frontier bone beds, including the Morrison, intensively between 1880 and 1903, and became convinced that the rocks were formed mostly by sedimentary deposits in river channels and on river flood plains, not in lakes. Al-

Study of the geology of the Morrison Formation in Utah reveals the environment (as shown in cutaway section) when the dinosaurs lived.

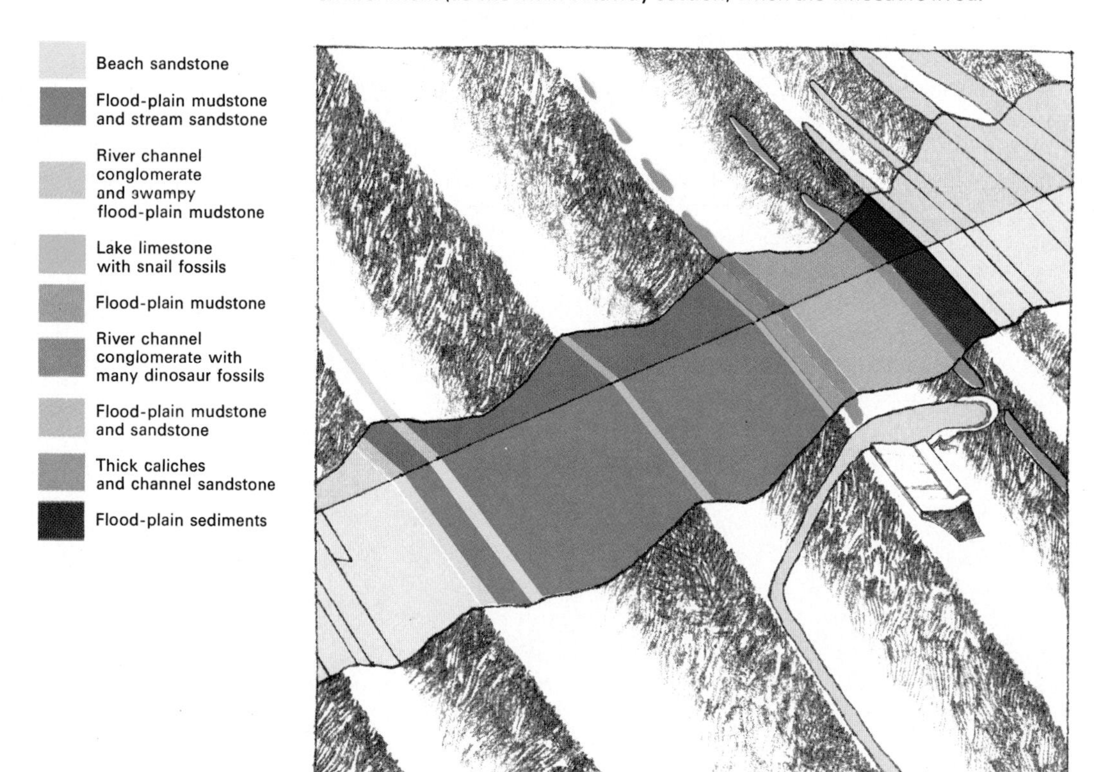

though other geologists accepted his conclusions regarding the rocks in which mammal fossils were found, they clung to the Lake Basin theory in regard to the Morrison deposits.

Anatomical studies of dinosaur bones convinced me that sauropods were basically land-living animals, and I was eager to analyze the Morrison rocks to confirm Hatcher's conclusions. So, in the spring of 1974, the Morrison Dinosaur Ecology Study Group was formed. This group included paleontologists Kay Behrensmeyer of the University of California, Santa Cruz, who had intensive experience studying sediments in the arid East African rift valleys; Peter Dodson of the University of Pennsylvania, who had analyzed dinosaur beds from the Cretaceous Period in Canada; theoretical physicist John S. McIntosh of Wesleyan University in Middletown, Conn., a leading authority on sauropod dinosaurs and the history of the early Morrison exploration; and myself. McIntosh's knowledge of the correspondence between the early bone collectors and their employers in the Eastern museums and universities enabled us to locate the quarries where most of the Morrison Formation dinosaur specimens were found.

Our studies of the sedimentary deposits in those quarries sharply contradict the view of the Morrison environment as steamy tropical swamps and lakes. We confirmed Hatcher's conclusion that most of the deposits were flood-plain and channel sediments. And the deposits showed clearly that these dinosaurs lived in a dry climate, like that of East Africa today. There were lakes in early Morrison times, but the evidence we found showed that they were shallow and short-lived.

The best proof of a seasonal climate found by the Morrison Dinosaur Ecology Study Group was the discovery of a massive caliche zone in the later Morrison deposits in what is now Dinosaur National Monument at Jensen, Utah. Caliches are zones of hard, calcium carbonate nodules that form in or on the soil. They occur only in seasonally dry climates, such as the Southwestern United States, and are especially well developed in the East African rift valleys. Caliches form when evaporation at the surface during the dry season causes ground water to rise through the soil. The evaporating water leaves behind the minerals that it carried upward.

Fossil dinosaur bones, which are embedded in rocks of the Morrison Formation, must first be carefully removed and cleaned before they are assembled for study.

But even though the climate was seasonally arid, sauropods could have been aquatic animals, restricting themselves to a few large channels and lakes much as hippopotamuses do in East Africa today. So we made a taphonomic analysis to determine if they were aquatic or land animals. Taphonomy is the study of the natural movement of an animal's carcass from the point of death to its final fossilized resting place. The main agents of carcass movement are flowing water and predators that scatter parts of an animal.

Behrensmeyer's taphonomic research in East Africa had shown that the highest relative abundance of hippo, crocodile, and swimming turtle bones were usually found in river-channel deposits and lake-shore beds. Markedly fewer were found in flood-plain sediments. The bones of such land mammals as antelopes, rhinoceroses, and elephants were found in a wide range of deposits–lake margins, channels, and flood plains. These animals travel across many types of terrain–lakeshores, plains, and riverbanks. They die and leave their bones in all these locales. Thus the taphonomic test of whether a fossil vertebrate was aquatic or not is whether its relative frequency decreases rapidly going from lake and channel to flood-plain sediments.

The only truly aquatic vertebrates found in the Morrison deposits were crocodiles and lungfish. The latter were rare, but hundreds of crocodile teeth and bone fragments were found in channel beds. In flood-plain beds, crocodiles are rare. However, the sauropod dinosaurs were common in all the environments. The wide distribution of their bones matches the pattern of fossil elephants and antelopes, not fossil hippos. Clearly these giant dinosaurs roamed freely over the landscape, crossing streams, ponds, and great arid plains.

Furthermore, many of them probably did not roam alone. The spectacular sauropod graveyards discovered in 1877 are typical of the Morrison dinosaur deposits. Most of these contained dozens of skeletons of three, four, or even five species of sauropod. So we can conclude that these huge animals traveled in great mixed herds, much as do zebras, wildebeests, and other antelopes today on the African savannas. The sauropods moved from one locale to another, lingering to feed, mate, and some, to die, leaving their skeletons to be buried by mud, silt, and sand.

As they roamed over the land, the dinosaurs browsed on conifers, cone-bearing trees and shrubs. Their long necks, once thought to be adaptations for deep water, were adaptations for reaching high into the foliage of the tall trees. Some sauropods, such as the *Brachiosaurus*, had a relatively small tail, long forelegs, and an exceptionally long neck. Other sauropods, such as the *Brontosaurus* (also known as *Apatosaurus*) and *Diplodocus*, had a short body and forelimbs, but their hind legs and tail were enormously powerful. The tail had sledlike devices on the bottom for supporting its weight, and these sauropods could rear up on their hind legs and tail to reach into the trees for feeding.

The first predators that could kill prey as large as the crocodile-sized *Eryops* (gray) and the *Ophiacodons* (tan) were these fin-backed *Dimetrodons* that lived about 275 million years ago. The cold-blooded animals were ancestors of the warm-blooded dinosaurs and the mammallike reptiles.

Sauropods divided the conifer resources among themselves partly by feeding at different levels. For example, a big *Camarasaurus*, with only a moderately long neck, could not stand on its hind legs and had to feed at lower levels than the *Diplodocus*, which could stand on its hind legs and tail to reach higher. Those sauropods that fed at the same level could divide the resources by taking different parts of the conifers for their food. A young *Diplodocus* might feed at the same level as an adult *Camarasaurus*, but the *Camarasaurus*, with strong, chisellike teeth, could tear off tough leaves and branchlets while the *Diplodocus*, with thin, delicate, pencillike teeth, could pluck out the smaller and difficult-to-separate plant parts.

Eating conifers is a prickly business–needles, cones, and branchlets are tough and rough in texture. This may explain the peculiar position

Support for Warm-Blooded Dinosaurs

The ratio of predators to prey that is found in the dinosaur fossil record resembles that of modern mammals and differs greatly from that of both ancient and modern reptiles. Most dinosaurs preyed on other species, but occasionally they killed their own kind.

Dinosaurs

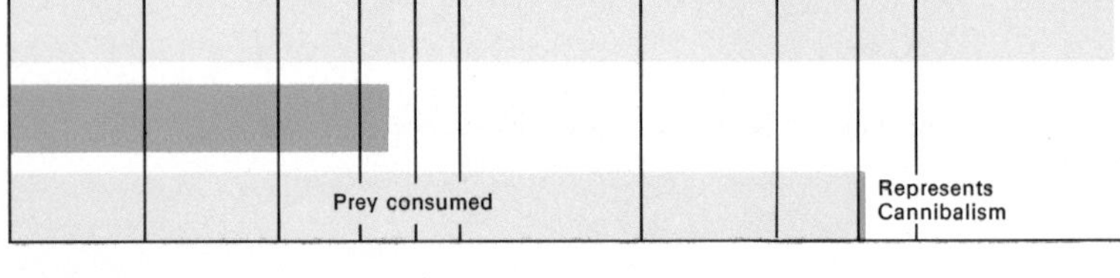

Reptiles

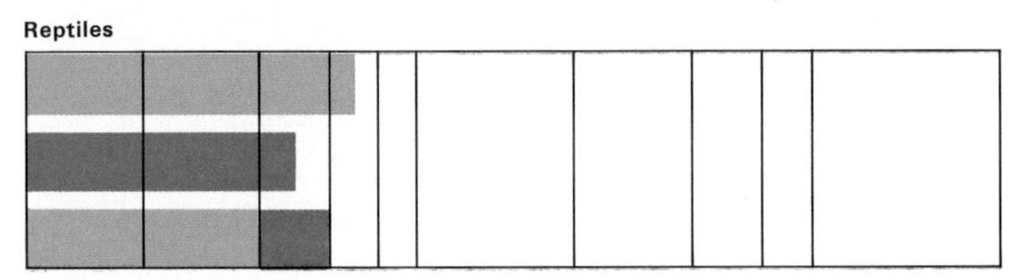

Mammals

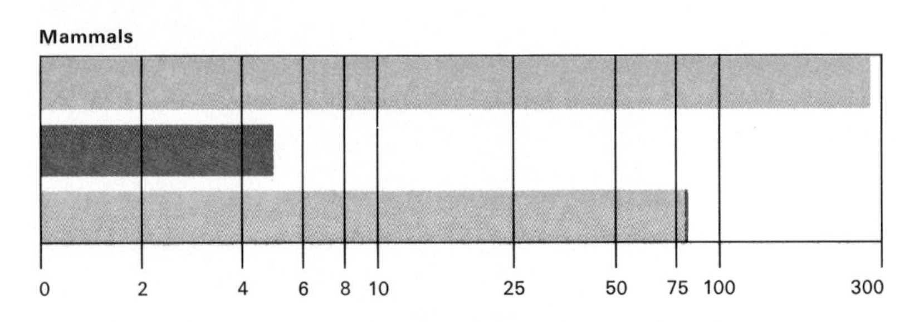

of the sauropod's nostrils. If they were at the tip of the snout, as they are in most land vertebrates, the soft nostril tissue would be constantly jabbed by the needles and other sharp bits of foliage. Natural selection would therefore favor the posterior position for the nostrils.

A comparison of the *Camarasaurus* and *Diplodocus* snouts seems to confirm this explanation. The short snout of *Camarasaurus* would cause it little trouble since the tough tree vegetation it ate was easily torn from outer limbs. *Diplodocus,* on the other hand, used its long, tapering snout to thrust deep into conifer canopies. The *Camarasaurus* nostrils were in front of the eyes, only slightly posterior. But the *Diplodocus* nostrils were far back on top of the skull between the eyes, the most posterior position of any sauropod.

Sauropods were the dominant big plant-eaters during the latter part of the Jurassic Period–from about 170 to 130 million years ago. These high-browsing dinosaurs became extinct or greatly reduced in numbers at the end of that period. In the very first stages of the succeeding Cretaceous Period, new groups of plant-eating dinosaurs began to appear. These included dome-headed pachycephalosaurs, iguanodonts and duckbilled types, parrot-beaked psittacosaurs, small and large-horned dinosaurs, and heavily armored ankylosaurs. All of these Cretaceous plant-eaters had rather short necks, and their tails were too weak for them to stand on their hind legs to feed on tree branches. They must have confined their browsing to within a meter or two of the ground. The duckbills and ankylosaurs had large, square snouts much like the white rhinos and other mammals of today that feed close to the ground. The similarity between the snout of the rhinos and that of the duckbills and ankylosaurs is another indication that these dinosaurs were low browsers.

The new feeding habits may have contributed to one of the most important revolutions in land-plant history, which began shortly after this change. Flowering plants began to appear in the early Cretaceous, after the first wave of low-browsing dinosaurs appeared, and they increased in number and diversity to become a dominant plant group by the end of the Cretaceous Period.

The earliest flowering plants probably were low shrubs, not giant forest trees. Their fossils are commonly found where environmental conditions were relatively unstable–in short-lived swamps, seasonally drained flood plains, and along levees where floodwaters dumped thick loads of silt. The new wave of Cretaceous dinosaurs, feeding on low-lying plants, gave flowering plants a natural advantage because of their short life cycle and rapid growth rate. These plants quickly re-colonized the overgrazed areas, and so their proliferation may be the result of natural selection abetted by the low-browsing dinosaurs. The flowering plants later evolved into a wide variety of herbs, grasses, and trees, many of which grow today.

Another recent conclusion that has upset the traditional view is that dinosaurs were endothermic, or warm-blooded, animals. Since the

Long-necked *Diplodocus, overleaf,* was one of the plant-eating dinosaurs that lived in what is now the northwestern United States during the Jurassic Period, more than 130 million years ago. Other plant eaters of that time included the equally long-necked *Brachiosaurus* and the backplated *Stegosaurus* (both eating from trees), and blue *Camarasaurus.* Meat-eating *Allosaurus* (right foreground) preyed on the plant eaters.

mid-1880s, they have been generally classified as reptiles, which are ectothermic, or cold-blooded, although the evidence for this assumption has never been carefully analyzed.

Cold-blooded is a misleading term. On a bright, sunny day, the body temperature of many lizards is as high or higher than that of warm-blooded mammals, including man. Lizards and other present-day ectotherms have to use external heat sources–the sun and sun-heated rocks, soil, and plants–to raise their body temperature above that of the air. And they must quickly find shady, cool places when the temperature climbs too high. Endothermic animals have a high internal heat production that can maintain a constant high body temperature in many types of weather. This allows endotherms to live and hunt for food in a much broader range of environments than can ectothermic creatures.

In cool weather, the body temperature of ectotherms drops and the animals' movements become sluggish. And they become more vulnerable to competition and predation from endotherms. Small lizards can hide from the predators in holes and burrows. But large ectotherms are at a distinct disadvantage because they cannot hide. For this reason, giant lizards and large tortoises are found today only on a few isolated islands, such as the Galapagos Islands, the Aldabra Islands, and the island of Komodo, where there are no endothermic predators of any significance.

The land vertebrates that dominated the earth before dinosaurs were a variety of primitive reptiles and amphibians, all probably ectothermic. Three lines of evidence show this. First, their skeletons are found only within about 25 degrees north and south latitude of the equator as it existed then, in environments that were warm year-round. Second, a microscopic examination of their fossilized bones reveals a low density of blood vessel channels, a characteristic of present-day ectotherms. Third, calculations of the ratio of predators to prey indicated in fossil records show that the predators of that time were ectothermic animals.

The fact that dinosaurs were not restricted to tropical latitudes is one indication that they were endothermic animals. The late Triassic Period probably had sharp climatic zones. The big, crocodilelike phytosaurs, ectotherms, were restricted to within 40 degrees of the Triassic equator; their fossils are not found in higher latitudes, except in India where warm ocean currents would have brought a mild climate. Yet fossils of the late Triassic dinosaurs are as common and diverse in the Triassic cold latitudes, where what is now South Africa is, as they are in the Triassic tropics, where Arizona and Germany now are.

Studies of dinosaur bone microstructure also indicate these animals were endothermic. Fortunately, fossilization often faithfully preserves the structure of bone, providing a window through which we can look back at the physiology of ancient animals. Such studies began decades ago, but the published results made curiously little impact on scientists

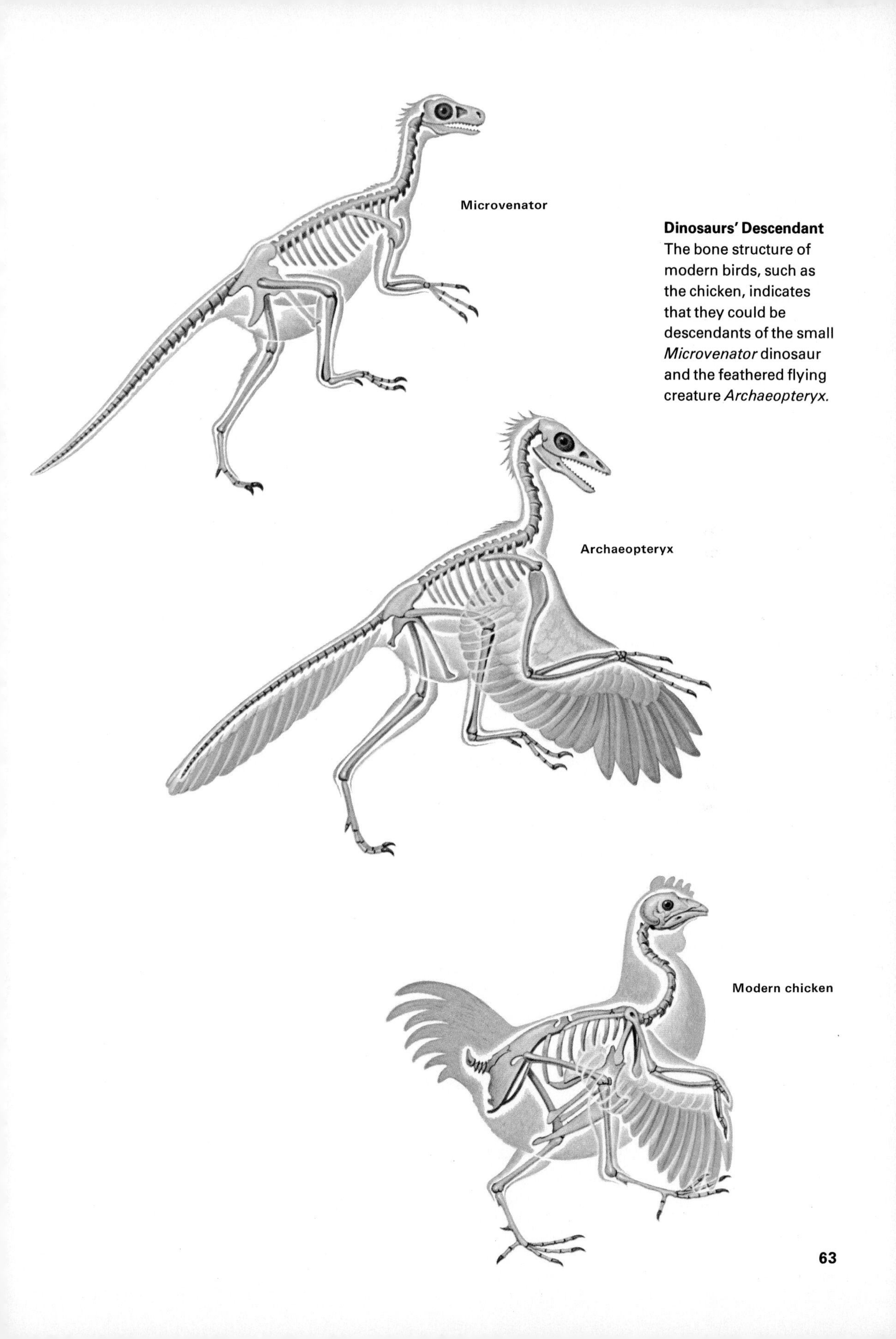

Dinosaurs' Descendant The bone structure of modern birds, such as the chicken, indicates that they could be descendants of the small *Microvenator* dinosaur and the feathered flying creature *Archaeopteryx.*

until paleontologist Armad de Ricqles began making comprehensive reviews of fossil bone tissue in Paris in the early 1970s. His insistence that the density of blood vessels in dinosaur bones indicated that the animals were endothermic is finally swaying much of the scientific community to this view. The density of blood vessels reflects higher metabolic rates in the bone tissue of dinosaurs than the metabolic rates in the bone tissue of ectotherms.

Last, but not least, the evidence of predator-prey ratios among dinosaur fossils indicates that they were endothermic. This ratio is the live weight represented by all the predator specimens divided by the live weight represented by all the potential prey, for a single local environment. Some sedimentary formations yield hundreds of individual animals. Their body weight when alive can be calculated by reconstructing their skeletons and making precise scale models. The predator-prey ratios can then be easily worked out.

It takes many more prey animals to satisfy the food needs of endothermic animals than it does to satisfy ectothermic animals. In a steady-state population, the production of new individuals equals the mortality rate. In a weasel, a present-day endotherm, the production of each new weasel, by reproduction and growth, requires 40 times its own weight in meat. In this case the predator-prey ratio would be 1 to 40. Production of a mountain boomer lizard, which is an ectotherm about the same size as a weasel, needs only 2.5 times its weight in meat, a ratio of 1 to 2.5. These lizards eat smaller lizards.

Big species are most reliable for measuring predator-prey ratios of prehistoric animals because fossilization processes tend to preserve large specimens more consistently than small ones. The predator-prey ratio for endotherms is, on the average, only one-tenth to one-twentieth of the ratio for ectotherms. The ratio can be expressed as a percentage. Thus the predator-prey ratio of endothermic weasels and mice is 2.5 per cent and that of ectothermic lizards is 40 per cent.

Of course, all of the predators do not die in exactly the same place as all of their prey. But the maximum ratios for endotherms are so much lower than the minimum ratios for ectotherms that we should be able to distinguish between the two. Predator-prey ratios in big fossil mammals that lived during the Cenozoic Era (beginning 65 million years ago) are mostly from 0.5 per cent to about 5 per cent, which is similar to that in today's mammal communities. The ratio was much higher for the early Permian Period reptiles, ranging from about 10 per cent to about 110 per cent.

In the large Morrison Formation dinosaur communities that we have measured, and in similar-age fossil groups found in Africa, predator-prey ratios range from 1 to 4 per cent, much the same as the range for fossil mammals. Most big collections of dinosaur fossils from Triassic and Cretaceous deposits found elsewhere show similar ratios.

The only locale with a high predator-prey ratio for dinosaurs is an unusual quarry in Utah that is packed with predator skeletons. This

Special Report text continues on page 65.

A Science Year Special Feature

The Linked Voyages of Land and Life

The supercontinent Pangaea contained most of the world's land 245 million years ago, then it broke up into smaller continents that slowly drifted to the positions they are in today. At the same time, the dominant vertebrate animals on the land continued their evolutionary trip into the modern world. Animal groups rose and fell in importance. They changed into more adaptable and intelligent forms, and the moving continents played a role in the changes.

On the following pages are scenes from these great related journeys, and maps which show how the continents moved. On the maps are geologic clues that tell about climate, which also affected evolution. Coal beds, made of decayed plants, suggest a wet climate; evaporite seas, a hot, dry one. And glaciation means bitter cold.

How to Use This Unit

To see how the continents have changed and moved over the past 245 million years, first turn the next three pages to the foldout acetate map of the present-day world. Unfold this map and then turn back two pages to the paper map of the Permian Period. Place the foldout map on the Permian map to compare the two. Repeat the process for the other three maps.

Late Permian Period

245 million years ago

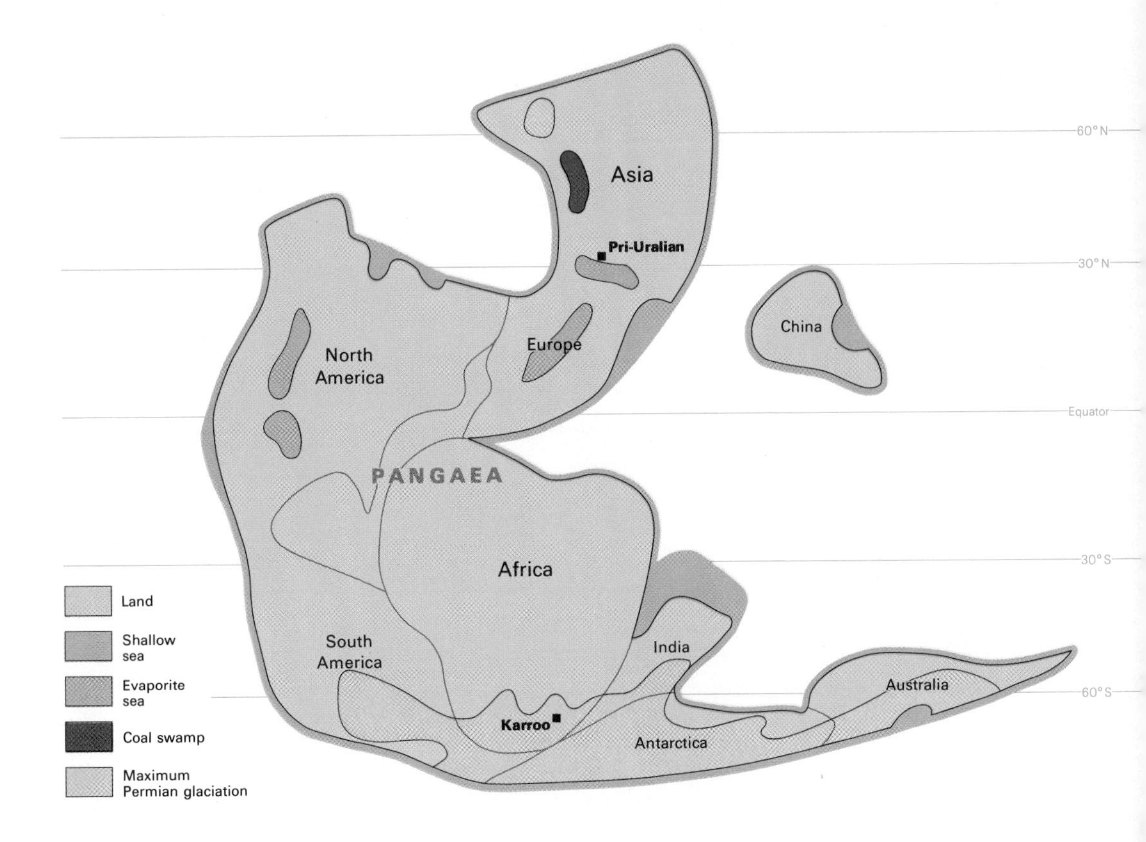

The Permian Period was a time of great climatic variation. Huge ice sheets covered the southern end of Pangaea. By the late Permian, the ice sheets had receded. But in what is now the Karroo region of South Africa, *opposite page, top,* the winters were still cold. Here, on a late winter morning, two carnivorous, or meat-eating, male anteosaurs battle for the control of a pack of females by banging bony snouts and brows. Two smaller carnivores, called gorgonopsians, flee the scene.

These animals are all therapsids, mammallike reptiles that were the ancestors of true mammals. That they could live in the cold regions of South Africa is strong evidence they were warm-blooded. They would have been hairy as well because mammals grow hair to keep warm in cold climates.

Thousands of miles to the north, nearer the equator in a quite different climate, *bottom,* a fierce, gaping, saber-toothed gorgonopsian defends its recent kill against another gorgonopsian. The dead pareiasaur is a cold-blooded reptile. This region, called Pri-Uralian, in Russia, was a harsh land of baking salt flats and evaporating salt seas. Therapsids had little need for hair in this climate.

Therapsid fossils have been found on every continent and in many different ancient climates. They were a numerous, varied, and adaptable animal group for many millions of years before they died out during the Triassic Period, which followed the Permian Period about 225 million years ago. Mammals were their descendants.

Early Cretaceous Period

110 million years ago

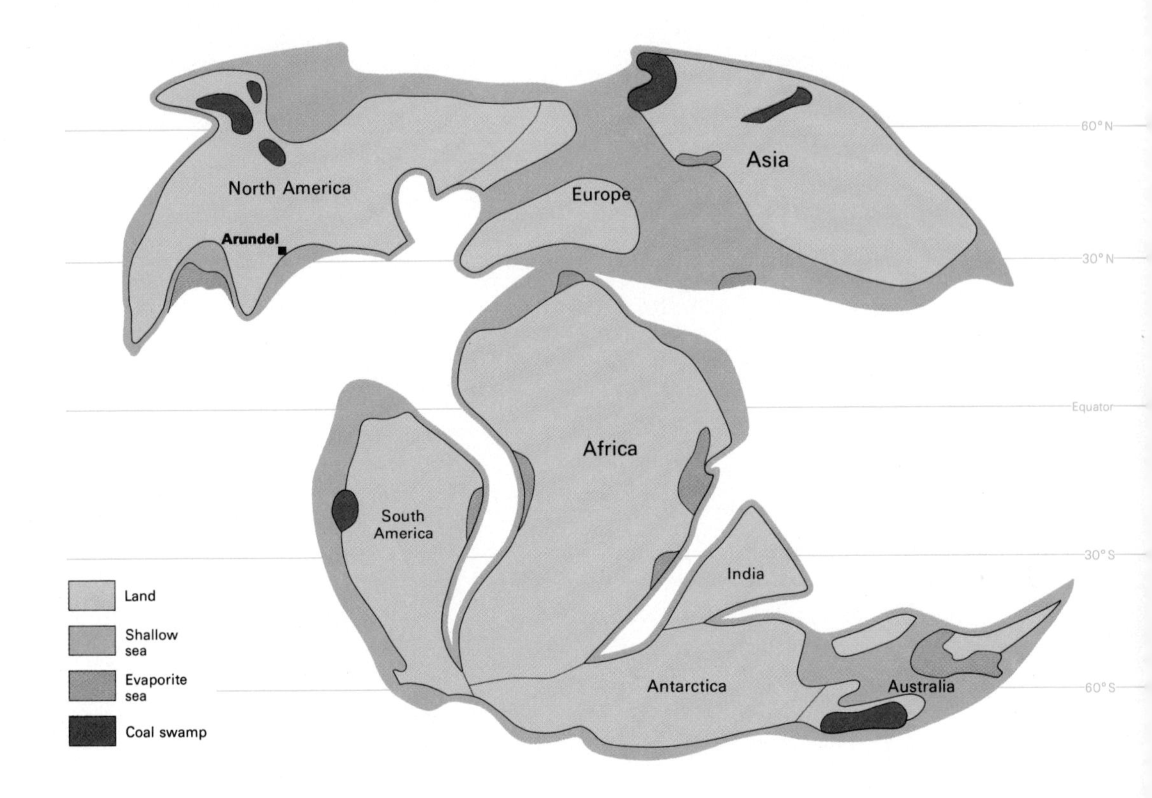

Pangaea began to break up about 200 million years ago. The northern continents—North America, Europe, and Asia—twisted away from the southern land masses with a clockwise motion and drifted off on their own. World climate was uniformly warmer and more stable during the Cretaceous Period than it had been in the Permian.

On the warm North American seacoast, at a site now called Arundel, in Maryland, herbivorous, or plant-eating, dinosaurs graze peacefully, *opposite page.* The long-necked brachiosaurs reach high to feed on pine trees while a squat ankylosaur munches ground plants, including the newly evolved flowering plants.

Both kinds of dinosaurs had close relatives on almost every other continent because the land masses were still relatively close together, and land bridges between various continents often appeared when the shallow continental seas that divided them disappeared.

At about the same time that mammals evolved from therapsids, the warm-blooded dinosaurs evolved from reptiles called thecodonts. Dinosaurs spread throughout the world and developed many different forms and life styles. Some were herbivorous; others, carnivorous. Some became the largest land animals that ever lived. For about 130 million years, dinosaurs were the undisputed masters of the land. Then, quite suddenly, at the end of the Cretaceous, they disappeared. Exactly why the great dinosaurs died out so suddenly is one of the great mysteries of paleontology. But they left descendants—the birds.

Early Oligocene Epoch

36 million years ago

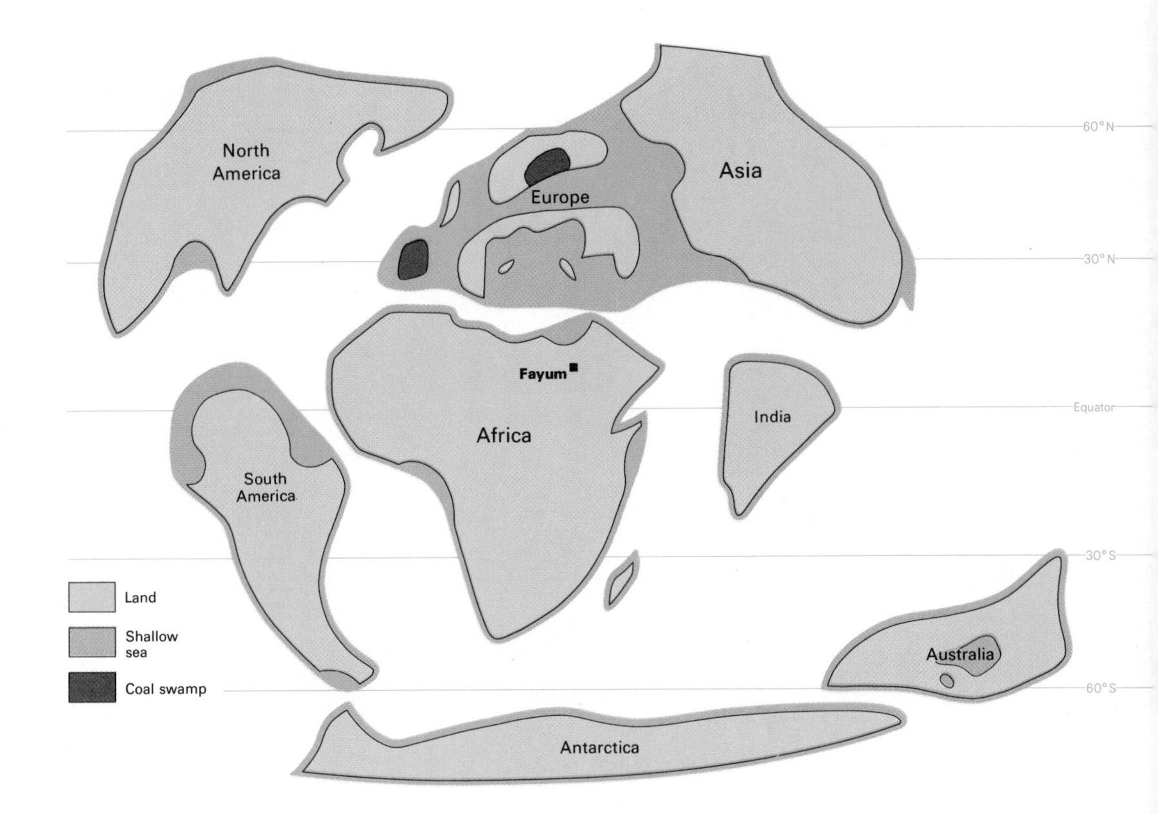

By the early Oligocene, in the middle Tertiary Period, most of the continents were well separated. A few land bridges connected Asia with North America and Africa from time to time. The Atlantic Ocean was growing as the Americas drifted west and north. South America and Australia were completely isolated, and India was nearing the collision with Asia that would form the Himalaya. The world's climate remained warm.

Two grotesquely horned, heavy arsinoitheres, about as big as rhinoceroses, send turtles and crocodiles scrambling for safety on a stream bank in the Fayum district of northeast Africa, *opposite page.* Primitive apes chatter in the tree above. Paleomastodons, secure in their number and size, drink on the jungle stream's far bank.

Partly because of the drift and isolation of the continents, mammals produced many more forms than did the dinosaurs and the therapsids. For instance, many of the Fayum animals evolved when Africa was isolated. Some of these animals came to a dead end. The arsinoitheres, for example, have no known ancestors or descendants. Others, such as the paleomastodons, have recognizable modern counterparts, the elephants. Turtles and crocodiles, on the other hand, remained relatively unchanged for several hundred million years.

One kind of mammal traveled a particularly important road. The apes of the Oligocene had fair-sized brains and quick, dexterous hands, as well, for grasping fruit and climbing trees. In later times, similar hands, guided by far larger brains, would make tools and weapons.

Pleistocene Epoch

1.7 million to 20,000 years ago

The Pleistocene Epoch of the Quaternary Period began about 1.7 million years ago. Once again, after more than 200 million years of relative warmth, ice sheets spread from the poles over much of North America, Europe, Asia, and Antarctica. The continents were roughly in their present-day positions, but so much water was frozen in ice sheets thousands of meters thick that many areas now under water were dry land.

In Africa's Great Rift Valley, about 1.6 million years ago, *opposite page, top,* a band of early men called *Homo erectus* drive hyenas away from their prey, a huge wild pig. A few australopithecines, close relatives of humans, watch warily.

Humans may have evolved in Africa, but they spread and adapted to many different places and climates. About 30,000 years ago, Cro-Magnon men, clad in animal skins, *bottom,* stalk a woolly rhinoceros and her calf on the Polish steppes of central Europe. Wild asses run across the tundra, and long-haired mammoths graze in the distance.

The time of human evolution from apelike ancestors is still obscure. Recent fossil discoveries have pushed the date back to at least 2 million years ago. Early humans seemed ill-equipped to rule the world. They had no long teeth or slashing claws for attack, like the great cats. They were not built for flight, like the antelope, or for defense, like the thick-skinned horned rhinoceros. But humans used their well-developed brains and agile hands to make their own tools for attack and defense.
In doing this, they began to alter the world. They are still doing it.

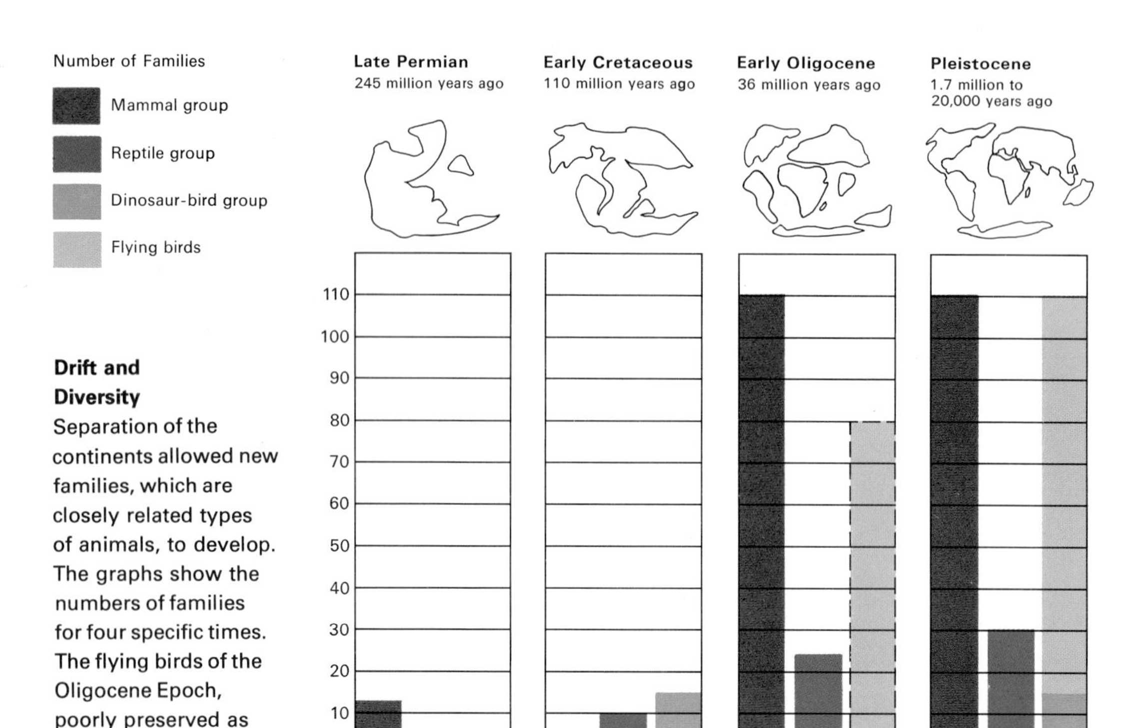

Drift and Diversity
Separation of the continents allowed new families, which are closely related types of animals, to develop. The graphs show the numbers of families for four specific times. The flying birds of the Oligocene Epoch, poorly preserved as fossils, are estimated.

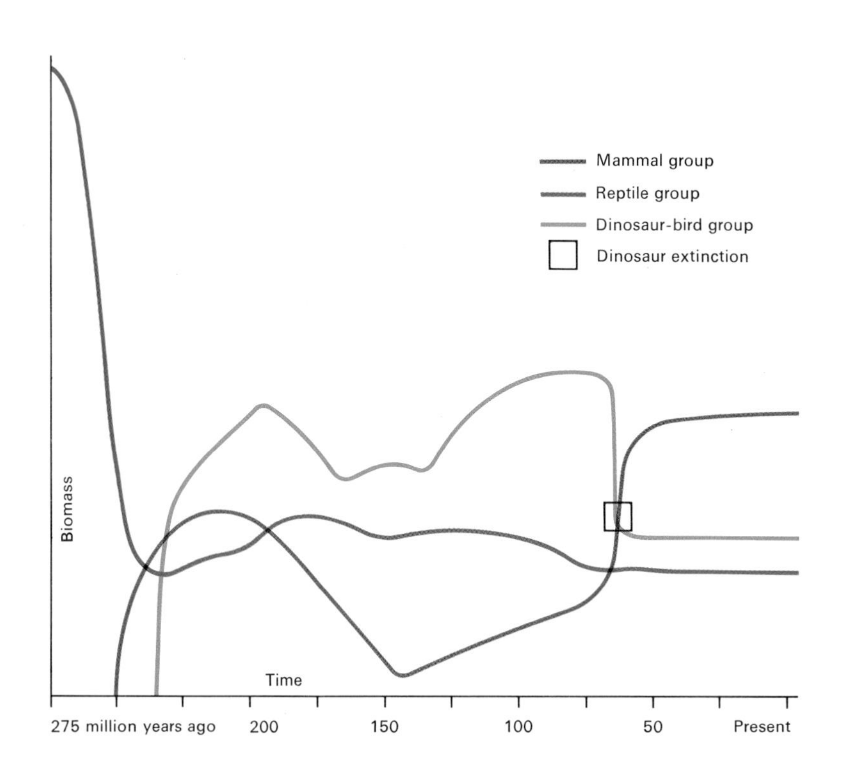

The Weight of the Evidence
Biomass, the total weight of a group of animals, is estimated from fossil remains. It is a useful measure of the relative success of animal groups at given times. The graph traces the rise and fall of the three major groups by measuring the total biomass of each. The rise of mammals to world dominance is closely linked to the sudden and mysterious dinosaur extinction 65 million years ago.

Prepared by the editors of *Science Year,* The World Book Science Annual. Artist: Jack J. Kunz.
Consultant: Robert T. Bakker, assistant professor of vertebrate paleontology and geology, Johns Hopkins University.
Printed in U.S.A. by the Trans-Vision® Division, Milprint Incorporated, and Kingsport Press, Incorporated.

quarry may represent a phenomenon similar to the La Brea pits in Los Angeles, an Ice Age deposit of treacherous tar-soaked sand. La Brea attracted, and preserved in tar, vast numbers of Ice Age predators, mostly wolves and saber-toothed tigers. The Utah site must have attracted dinosaur predators, which became stuck in the mud. The dead and dying carcasses would then lure more predators to their doom.

Warm-blooded animals living in seasonally cold climates are always insulated with hair or feathers. Fossil skin impressions show that the big dinosaurs did not have hair or feathers. But big endotherms do not need insulation in warm climates because their size alone reduces the rate of heat loss and gain. Tropical elephants, hippos, and rhinos, for example, have no hair. But even in the tropics, small endotherms, such as rats and birds, have hair or feathers to protect them from the various fluctuations that occur in the climate.

Skin impressions from two types of small vertebrates closely related to the dinosaurs tell an intriguing story. One species of small thecodont reptile had long, overlapping scales that appear to have functioned as insulation. Superbly preserved specimens of the pterosaurs, or flying reptiles, show a dense body covering of hair or hairlike feathers. So it is quite possible that some dinosaurs had hair, while others might have had scales or feathers.

The handful of specimens of the first known bird, *Archaeopteryx,* have feather impressions. They come from rocks of the late Jurassic Period, about the time the last layers of rock were deposited in the Morrison Formation. When the *Archaeopteryx* specimens were examined in the late 1880s, Marsh and others saw that they closely resembled carnivorous dinosaurs. But in the early 1900s the idea of a close link between dinosaurs and birds lost favor. In the late 1960s, Yale University paleontologist John H. Ostrom carefully reanalyzed both the old and the newly discovered *Archaeopteryx* specimens and he has now proved beyond a reasonable doubt that birds are direct descendants of dinosaurs. He has shown that even the birds of today retain a host of characteristics in their skeletal structures that are typical of the bone structure of the ancient dinosaurs.

Birds probably inherited their endothermic physiology, many parts of their bone structure, and their feathers from small carnivorous dinosaurs. Indeed, we could properly classify *Archaeopteryx* as a predatory dinosaur. So the relationship of birds to dinosaurs is like that of bats to mammals–they are aerial versions of a basically terrestrial endothermic stock. It is misleading to put birds in one class, Aves, and dinosaurs in another, Reptilia. It would be far more accurate to classify birds as a subgroup of dinosaurs. After all, they are a living, aerial expression of basic dinosaur biology.

This leads us to the logical conclusion that dinosaurs are not extinct. The evolutionary developments that made large dinosaurs spectacularly successful during the Triassic, Jurassic, and Cretaceous periods continue in the great diversity of today's birds.

Prospecting for Green Gold

By Noel D. Vietmeyer

With little-known plants, we can harness more of the sun's power in arid lands, rain forests, and even in sewage lagoons to grow food, fuel, and raw materials

We depend on plants for the food on our plates and the shirts on our backs. When a plant crop fails, the results may range from the inconvenience of slightly higher prices to the disaster of famine. When southern corn leaf blight struck the United States in 1970, corn production dropped 10 per cent. Some farmers were ruined, and prices went up. When blight struck the potato crop in Ireland in 1845 and 1846, the country was ruined because the Irish depended almost exclusively on this single crop. More than 750,000 of Ireland's 6 million people died of starvation and disease. Hundreds of thousands of others emigrated to Canada and the United States.

Although there are more than 350,000 species of plants on earth, people have used only from 3,000 to 4,000 of them–about 1 per cent–for food and raw materials. In recent centuries, civilization has concentrated on fewer and fewer plant species. Today, we depend on only about 20, ranging from corn and wheat to cotton and rubber trees. In the last 10 years, the Green Revolution has focused much of agriculture's attention even more narrowly on just the three major grains–corn, rice, and wheat.

This concentration on a few plants is growing increasingly hazardous. Today, when a drought in Kansas may cause a bread shortage in

Jojoba seeds ripen in the hot desert sun. Their oil, almost identical to sperm whale oil, has many potential industrial uses.

Moscow, we are more vulnerable than ever to the unpredictability of weather and the ravages of pests and disease. An additional problem, with the world's rapidly growing population, is that most of the good farmland is already in use. Only enormous expenditures for costly fertilizer or massive irrigation projects can make much more land available for conventional crops. And basic crops simply will not grow in many areas because of climate or soil conditions, or both.

Aware of these facts, plant scientists have begun to examine some of the thousands of unused plants to see which might be made to work for people. These are plants that could be grown, for instance, on marginal land, in difficult climates. Scientists conducting an obscure-plants survey for the U.S. National Academy of Sciences (NAS) have already identified more than 36 for further study. Some of these may help start a new Green Revolution in which unconventional plants will be transformed into important crops.

Some plants have been deliberately neglected for political or cultural reasons. For example, conquering Spaniards ruthlessly halted cultivation of the sorghumlike amaranth plant 400 years ago in Central America. A basic staple of the Aztecs, the amaranth was also important in their religion. By substituting barley products, the conquistadors struck a blow for Christianity, but at the cost of good nutrition. Recent scientific analysis shows that some plants of the amaranth family are rich in lysine, an amino acid that is crucial for good nutrition, while barley–like most plants–is deficient in it.

The author:
Noel D. Vietmeyer, a professional associate of the U.S. National Academy of Sciences, directed its study of the economic value of tropical plants.

Some root crops, such as yams and cassava, have suffered because of the European taste for potatoes, a temperate plant that grows poorly in the hot tropics. Because it was part of the cultural tradition of colonial rulers, the potato still graces affluent dinner tables in many tropical countries, though the local roots are more nutritious and yield better in these climates.

Some potential crop plants were not developed because of technical problems. Ramie (*Boehmeria nivea*), an Asian plant related to the nettle, is an example. The stems of this tall herb produce lustrous fibers as much as 1.8 meters (6 feet) long that are six times stronger than cotton. Five ramie crops can be harvested each year in favorable locations, but a brown, resinous gum clings tenaciously to the fiber. More than 100 years ago, Queen Victoria of Great Britain offered a reward of 100 pounds to anyone who could develop a process to remove the gum without harming the fiber. The prize has never been collected. Some scientists believe that ramie fiber would be more widely used than cotton if this problem could be solved. And because of ramie fiber's superior strength, fabrics made from it would last longer.

Some potentially valuable plants must be moved to new habitats before they become useful. In a new home, such a plant is freed from the diseases, parasites, and predators that have developed over thousands of years to check its growth. A few sickly Monterey pines (*Pinus radiata*) struggle against extinction by disease on California's Monterey

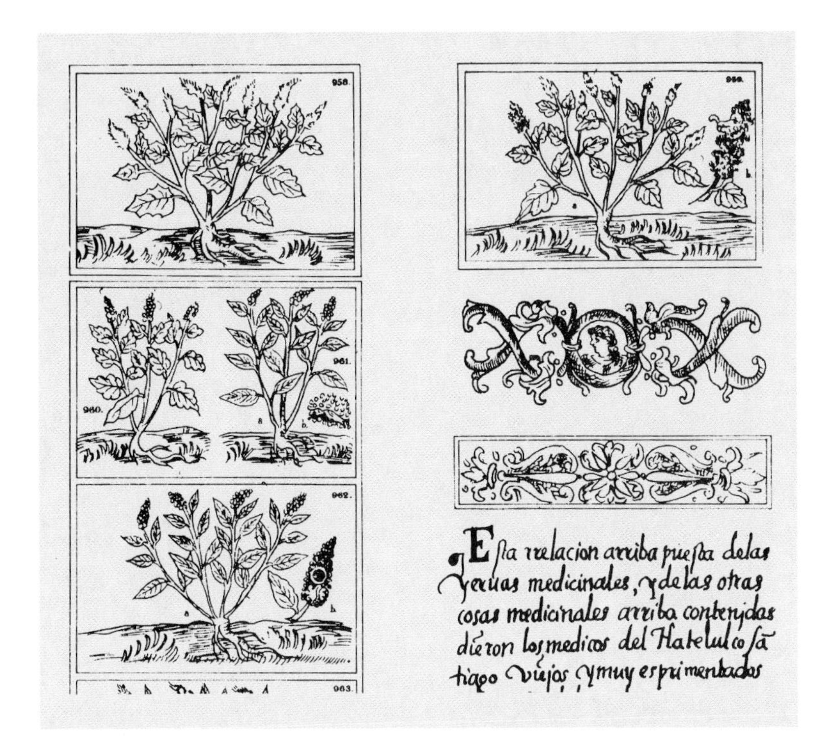

Amaranth plants are shown in the Florentine Codex, a book of Aztec knowledge compiled by a Spanish priest in the 1500s. The Spaniards stopped the cultivation of this nutritious crop for religious reasons.

Peninsula. But Monterey pine seed, introduced in New Zealand in the 1920s, has produced the largest cultivated forests in the world.

Much the same thing happened with rubber. Seeds from the wild Amazonian rubber tree (*Hevea brasiliensis*) were smuggled out of Brazil by Sir Henry A. Wickham, an English botanist, in 1876. Their majestic descendants now produce major crops in such Asian countries as Malaysia, Indonesia, Sri Lanka (formerly Ceylon), and Thailand. There, halfway around the world, rubber trees yield up to 7.4 metric tons of rubber per hectare (3.3 short tons per acre) annually. In Brazil, diseases affect the plant and it cannot now be cultivated to compete successfully with Asian plantations.

This tree has made natural rubber a billion-dollar industry for Asian countries. The United States alone imports about $500 million worth each year. Contrary to popular belief, natural rubber has not been made obsolete by synthetic substitutes. It is still used in tire casings–automobile tires contain about 20 to 40 per cent natural rubber, truck tires about 40 per cent, and airplane tires 90 per cent.

Ironically, while the United States imports all its natural rubber, another rubber-producing plant is being neglected in the deserts of Texas and Mexico. Guayule (*Parthenium argentatum*), a tangled, weather-beaten, unimpressive bush with silver leaves and yellow flowers, seldom grows more than 60 centimeters (2 feet) tall. Yet a mature guayule (an Indian word pronounced wy-*oo*-lee) bush contains up to 26 per cent rubber.

A guayule bush grows wild in the North American desert, *above.* In experiments to determine its rubber potential, it is ground up and processed in vats, *above right.* As the crude rubber floats to the surface, it is skimmed off for further refining.

Like the rubber from a rubber tree, guayule rubber is suspended in a milky fluid called latex in cells within the roots, trunk, and branches. To extract it economically, the whole shrub must be ground up in a processing mill. Guayule rubber contains sticky resins that lower its quality. But simple methods for removing these resins were developed in the mid-1940s. The purified guayule rubber had much improved properties. However, hevea rubber was cheap and plentiful right after World War II, and the development of petroleum-derived synthetic rubbers reduced the need for natural rubber. But with petroleum now in increasingly short supply and Asian rubber subject to price increases and political uncertainties, guayule takes on new importance.

Guayule was one of the 36 plants selected for their promise by the NAS panel. In a government storeroom in Washington, D.C., I discovered a large block of guayule rubber left over from one of the old research programs. I gave samples from the block to Robert M. Pier-

Crude guayule rubber, *left,* is almost identical to rubber that comes from Asian rubber trees. To make it useful to industry, it must undergo several refining steps. Adding pigment, *above,* is one such step.

son, director of research, Goodyear Tire and Rubber Company, Akron, Ohio, and Frank A. Bovey, Polymer Research Department, Bell Telephone Laboratories, Murray Hill, N.J. They analyzed the samples and found that in molecular structure guayule and hevea rubber are indistinguishable.

Spurred by this, the NAS convened a second panel to discuss guayule's commercial potential. At the meeting, Enrique Campos, director, Centro de Investigaciones en Quimica Aplicada, in Saltillo, Mexico, told us about a new guayule extraction factory that he had just built for the Mexican government. Samples extracted in this pilot-sized facility indicated that purified guayule appears to match hevea rubber in properties important in making tires and rubber goods.

So perhaps we relegated this scruffy desert shrub to obscurity too hastily. Few researchers believe that guayule production in the United States could be economically competitive immediately, but most believe that, with research to develop higher-yielding varieties and cheaper extraction processes, guayule could become a valuable crop for Arizona, California, New Mexico, Texas, and northern Mexico.

Although such desert regions have abundant sunshine, they are among the least productive on earth. Few plants can withstand the scorching temperatures, low humidity, and lack of rainfall. Those that can, such as guayule, are called xerophytes. The watermelon with its big water-storing fruit is one of the few xerophytes that have been domesticated. Scientists working on the NAS obscure-plants program have reported others that are potentially useful.

One is the tamarugo tree (*Prosopis tamarugo*), which grows in a part of Chile's desolate Atacama Desert where no rain falls in most years. Salt deposits, which are death to most plants, are so thick and widespread there that, except for the tamarugo groves, the area looks like a desolate moonscape. Chilean scientists have found that tamarugo leaves and pods are rich in protein, and large herds of sheep now graze on them. Thus, one of the most inhospitable regions on earth–the reflected glare of sunlight from the salt flats is so fierce that many of the sheep are partially blinded–has become productive. An experimental tamarugo plantation to feed goats was established in an arid area of the Canary Islands in 1973. The tamarugo may also be useful in developing livestock industries in other salt desert regions.

The buffalo gourd (*Cucurbita foetidissima*) is a wild squashlike xerophyte that grows in the deserts of Texas and Mexico, producing prodigious amounts of seeds that contain oil and a nutritious high-protein meal. It also has huge roots as big as a man that are filled with starch, like a potato. The plant could become an important source of food, animal feed, and oils in many arid parts of the world.

One of the most exciting xerophytes is the jojoba (*Simmondsia chinensis*), a wild desert shrub of Mexico, Arizona, and southern California that produces a liquid wax almost identical to the oil of the sperm whale. Sperm oil is a valuable commodity that is used in special lubricants such as those needed for automobile transmissions, heavy and precision machinery, and gears that operate under extreme pressure. It is so valuable that sperm whales have been the most heavily hunted whale species in recent years; one is killed every 39 minutes.

Sperm oil's molecular structure is difficult to synthesize commercially and until the jojoba (pronounced ho-*ho*-ba) was resurrected from obscurity, there was little hope of finding a substitute that could match all of sperm oil's applications. However, many researchers now believe that desert farmers will soon be producing "sperm oil" much more cheaply than the whalers can.

Jojoba bushes are unimpressive. Slate-green and twiggy, stunted by the desert's harsh aridity, they grow only 90 to 122 centimeters (3 to 4 feet) tall. They bear dense clusters of brown, olive-sized seeds that contain about 50 per cent oil that can be extracted by the methods used for soybeans, cottonseed, peanuts, and other oilseeds.

Although jojoba bushes grow wild in dense stands in Arizona and southern California, the plant must be cultivated to reduce harvesting costs and to improve yields. Indians on the San Carlos Apache Reservation are working with scientists at the University of Arizona's Office of Arid Land Studies to develop a plantation. A small test plantation has been started by the University of California, Riverside. Another plantation that has been operating for 10 years in Israel has some plants that yield 5 kilograms (11 pounds) of seed per year.

Estimates of annual oil production on a jojoba plantation vary from 1,120 kilograms per hectare (1,000 pounds per acre) to twice that amount. Since an average sperm whale yields 6,800 kilograms (15,000 pounds) of sperm oil, a minimum of 6 hectares (15 acres) of jojoba could save one whale per year.

While plant scientists are working to improve the jojoba plant through seed selection and breeding, industrial researchers are searching for new uses for the oil. For instance, adding hydrogen to jojoba oil produces a hard wax. This hydrogenated wax has a high melting point and compares favorably with expensive imported industrial waxes such as carnauba and candelilla wax. It may prove useful in the manufacture of polishes, paper, textiles, wax coatings to preserve food, and

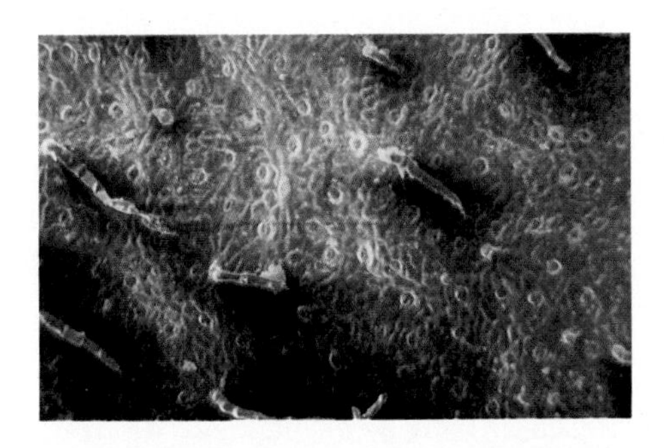

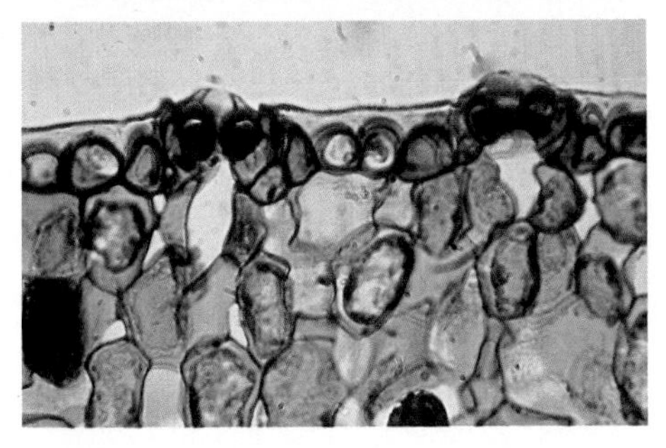

Tiny hairs on its leaf surface, *top,* slow air motion and reduce evaporation to help jojoba survive in deserts. A wax layer around the leaf pores, *above,* also reduces water loss. (Both photos are magnified 500X.) Plants growing in low spots catch rain runoff, *right.*

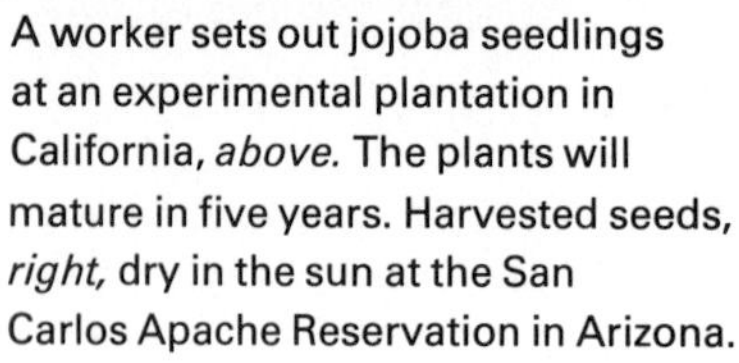

A worker sets out jojoba seedlings at an experimental plantation in California, *above.* The plants will mature in five years. Harvested seeds, *right,* dry in the sun at the San Carlos Apache Reservation in Arizona.

many other products. Partial hydrogenation produces a variety of soft waxes and creams. Jojoba oil can also be broken down into certain acids and alcohols used in disinfectants, detergents, and fibers. Jojoba meal, rich in protein, may eventually be used as animal feed.

In addition to species that grow in arid regions, the NAS obscure-plants program has considered plants that are adapted to the equally sun-rich but humid tropics. These rain forest regions of the earth cover huge areas in Latin America (including the vast Amazon Basin), central Africa, and South Asia. Intense tropical sunlight and the humid climate combine to produce prolific plant growth. Even so, the humid tropics encompass some of the most poorly fed nations on earth. In Bangladesh, for example, half of the people suffer from malnutrition.

Rain forest soils are generally poor despite the lush plant growth they support. Most of the nutrients in a rain forest are in the plants themselves. When a plant dies, its nutrient materials are recycled so quickly by other plants that the rich humus that characterizes many temperate soils can never build up. Consequently, farmers who clear

Hydrogenated wax and lubricating oil are just two of jojoba's products. The plant's brown seeds also contain a protein that may help someday to feed cattle.

jungle for agriculture find they can grow crops for only two or three seasons without fertilizer. This petroleum-derived commodity is often too expensive for widespread use in developing countries.

Nitrogen gas is the key ingredient in manufacturing fertilizer, but complicated chemical processes are required to convert it into fertilizer. One family of plants, the legumes, have found a way to use the nitrogen naturally present in the atmosphere. They form a partnership with certain bacteria whose enzymes take nitrogen from air and convert it into nitrate, the main ingredient in nitrogen fertilizer. In this way, legumes use their bacterial partners to produce fertilizer for them, which they then convert into protein. The highly nutritious peas, beans, soybeans, and peanuts are the major legumes used today. Unfortunately, they grow poorly in the humid tropics.

But the NAS study turned up an obscure legume, the winged bean (*Psophocarpus tetragonolobus*), which is native to the humid tropics. The winged bean grows in such jungle countries as Burma, Indonesia, Papua New Guinea, and Thailand, and it would seem to have a great future where it is as yet unknown–in Bangladesh, the Caribbean, central Africa, Latin America, and the Pacific Islands.

The winged bean plant resembles an ordinary runner bean–a bushy pillar of greenery, with wirelike shoots that twist upward, searching for something to grasp to hold the plant upright. Its leaves are shaped like the spades in a deck of playing cards, and its blue or purple blossoms look like sweet pea flowers. They are self-pollinating and begin forming succulent green pods within a day of pollination. The pods grow as much as 60 centimeters (2 feet) long in some varieties, and have four green flanges or "wings" running the length of the pod. The pods are crisp and snappy in texture, like Chinese pea pods, and can be eaten raw or boiled briefly. If left on the vine, they become fibrous and dry as the seeds inside mature. In two months, the dark-colored seeds are peanut-sized. They contain up to 37 per cent protein and 20 per cent oil, exactly the same as soybeans.

It is unusual for legumes to produce oil. But the two others that do–soybeans and peanuts–play an important role in commercial food products. Most margarine and many of our prepared foods are made from their oils. Like soybean oil, winged bean oil is polyunsaturated–an important dietary benefit.

The winged bean is also a root vegetable. If it is a kind of a soybean above the ground, it is a kind of potato below. Only a few varieties are known to produce roots big enough to eat, but they hold great promise, because the roots are about 20 per cent protein. People who live in the humid tropics commonly use root vegetables–sweet potato, yams, cocoyams, and cassava–as staple foods. The winged bean's roots contain 20 times as much protein as cassava, the most widely consumed root crop in the tropics, and 7 times as much as the other root crops.

The winged bean's main limitation is its weak stem. It is too feeble to support the plant upright, and if it is left to trail over the ground, the yield of pods and seeds declines. Winged bean plants must be supported on poles or trellises, which restricts them to backyard gardens or small farms. However, most of the farming in tropical nations is on just such a small scale, so this is not a serious problem.

We know about as much about the winged bean now as we knew about the soybean 60 years ago when it first drew attention in the United States. Research could make the winged bean the soybean of the tropics. It could then help to overcome the world's food problem by producing highly nutritious food in the area of greatest need–the backyard garden of even the poorest peasant. Small research programs on the bean have begun at the University of Ghana and at the International Institute of Tropical Agriculture in Nigeria. Nigerian scientists now believe that seed yields of 2,240 kilograms per hectare (2,000 pounds per acre) are possible, which compares favorably with U.S. soybean seed production. More research is needed to develop erect varieties, to study plant diseases and pests, and to increase the yield of tubers and bean pods in winged bean varieties.

The tubers, pods, beans, and even the flowers and leaves, *left,* of the winged bean can be eaten. The plants produce best when supported by poles or by trellises, *below left.* Winged bean pods are a common sight in many vegetable markets in Thailand, *below.*

Water hyacinth used to rid sewage lagoons of pollutants can be harvested, *above*, and put to other uses. In an experimental process, *above right*, the plant is converted to methane gas and fertilizer.

Not all potentially useful plants are as little known as guayule and the winged bean. The water hyacinth, for example, is well known as a pernicious weed throughout the world's wet tropics. It infests tropical lakes, rivers, and canals from India to Florida. Like a verdant froth 90 centimeters (3 feet) thick, it floats, clogging irrigation canals and shipping channels. Water hyacinth blocks oxygen from passing through the water surface, which suffocates the fish below, and it shelters the mosquitoes and snails that transmit parasitic diseases such as malaria and schistosomiasis to people.

But now we are beginning to appreciate that the water hyacinth has value. Some far-sighted scientists are learning how it can clean up our polluted environment and become a source of fertilizer.

While plant leaves are converting solar energy and carbon dioxide into usable carbon compounds for growth, their roots extract nutrients —mainly nitrogen, phosphorus, and potassium—from the soil. Water hyacinth roots, however, extract these nutrients from the water in which they dangle. Now scientists at a National Aeronautics and Space Administration (NASA) research station in Mississippi are using the water hyacinth to extract pollutants from the sewage of Bay St. Louis, Miss., a city of more than 6,000 inhabitants.

Sewage is rich in nitrogen, phosphorus, and potassium. NASA project leader Bill C. Wolverton reported at a November 1975 NAS meeting that, given a couple of weeks of growing time, water hyacinths suck virtually all of these elements out of a sewage lagoon. The Bay St. Louis effluent now pours into a hyacinth-filled, 17-hectare (42-acre) lagoon and emerges so low in pollutants that it passes federal standards for drinking water. The harvested water hyacinth, loaded with nitrogen, phosphorus, and potassium, is an ideal fertilizer.

Water hyacinth will also absorb such toxic minerals as mercury, lead, copper, and silver from industrial wastewaters, and it will remove and neutralize some organic pollutants, pesticides, and such cancer-causing compounds as chloroform.

The plant captures sunlight efficiently with its broad, flat leaves. The leaves are single storied, so that they do not shade each other the way the layered leaves of many plants do. Water hyacinth constantly produces new plants, and rafts of hundreds of thousands of plants quickly expand across the water to form a vast sunlight absorber.

Wolverton has measured water hyacinth growth of as much as 11 metric tons per hectare (5 short tons per acre) in just one day. Most of this growth is water that is absorbed by the plant, but the dry matter has sometimes increased as much as 560 kilograms per hectare (500 pounds per acre). This far surpasses sugar cane which, at a growth of 111 kilograms per hectare (100 pounds per acre) per day, is the most productive plant previously measured.

So water hyacinth may soon be used to treat sewage and industrial wastes in tropical and subtropical areas. But, as the mountains of harvested water hyacinth build up on the lagoon banks, what will we do with them? Researchers are working on that, too. Animal scientist James F. Hentges and engineer Larry O. Bagnall of the University of Florida in Gainesville have learned how to convert water hyacinth to cattle feed–even to human food. Furthermore, Wolverton and his colleagues have found that if water hyacinth is chopped up and placed in an airtight tank with bacteria it will ferment to a gas rich in methane, the natural gas used for cooking and heating.

We need to increase our kit of useful plants. Luckily, we are better equipped to do so today than ever before. We have the knowledge of plant genetics and tools such as the computer and other space age electronic instruments. We have already produced the "miracle" rice breeds that began the Green Revolution in Asia. We can now manipulate plants and plant products to meet our needs and take greater advantage of the sun's inexhaustible energy by selecting a wider range of plants for use as crops. We can select plants that will grow in unproductive areas–in deserts, in the humid tropics, in water, and in soils that are too poor or too toxic for conventional crops.

New crop plants will also make the world's food supply less vulnerable to disease and disastrous climatic change. Among the thousands of plant species, there are still undiscovered treasures.

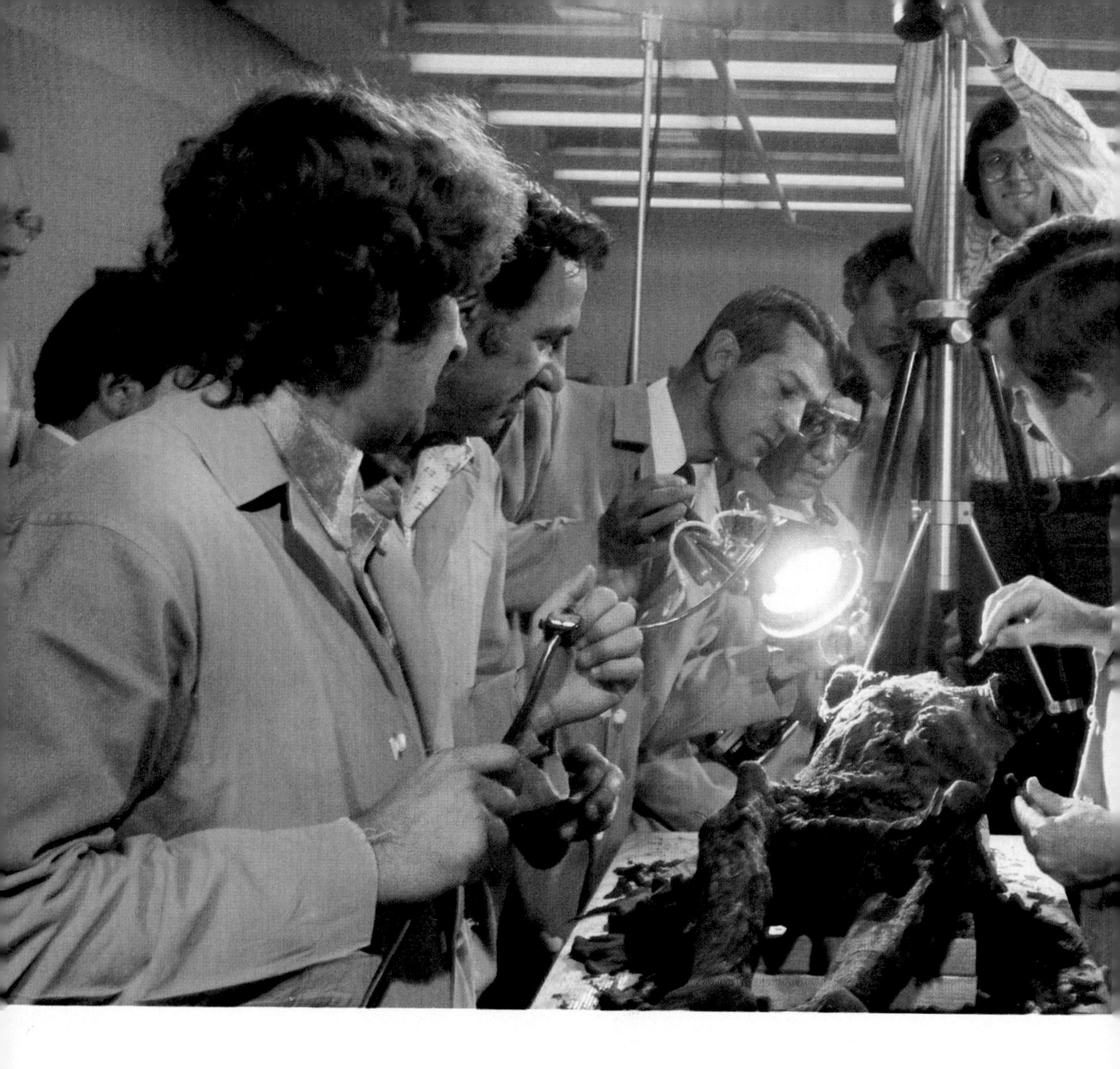

Dead Men Do Tell Tales

By Theodore A. Reyman

Mute for many centuries, mummies still have something to say about their culture, environment, and the history of human disease

Working under bright television lights, scientists conduct an autopsy of PUM-III, an Egyptian mummy more than 2,000 years old.

Whoever she was—and we will never know her name—she could never have imagined the crowded laboratory at Wayne State University School of Medicine in Detroit on Aug. 21, 1975. More than a dozen doctors, scientists, and technicians crowded around her. Pathologists, including myself, and other medical specialists took samples of her bones, skin, and vital organs. Anthropologists studied and measured her braincase and teeth. Cameras and microphones recorded the scene and the scientists' comments.

The world she knew, a little more than 2,000 years ago, was bounded by the green, fertile strips of land watered by the Nile River. When she died, probably in her twenties, embalmers performed rites that were ancient even then and mummified her. She lay in Egypt for a score of centuries. Now, far from home, she was one of the subjects in a series of unusual autopsies that hold great promise for tracing the history of human disease. The work has also led to an interdisciplinary study of the scientific, cultural, environmental, and historical information that mummies can provide.

Because she was the third mummy loaned to us by the University Museum of the University of Pennsylvania in Philadelphia, we called her PUM-III. X-ray examination before the autopsy showed a healed fracture of the second left rib and indicated that the internal organs had apparently not been removed. While unwrapping the body, William H. Peck, curator of ancient arts at the Detroit Institute of Arts, found linen with fragmentary hieroglyphics that read "in the 24th [or 28th] year of the reign of the king." The heart and lungs were in their normal positions, as was the liver. But the remainder of the abdominal cavity had been stuffed with wads of linen. The uterus was present in the pelvic cavity, but the Egyptian embalmers had removed the intestines and other organs.

PUM-III's body tissue was so poorly preserved that little could be learned from a microscopic study, but the autopsy provided other useful information. Her teeth were severely worn by a coarse and gritty diet. A perforated eardrum suggested that she had suffered middle ear disease or inflammation. Researchers at the Wayne State medical school later isolated both red and white blood cells–the first time intact white blood cells had been reported in a mummy. But we could not determine what caused her death.

The author:
Theodore A. Reyman, director of laboratories at Mount Carmel Mercy Hospital in Detroit, was one of the founders of the Paleopathology Association in 1972.

Aidan Cockburn, a Detroit physician and a research associate at the Smithsonian Institution in Washington, D.C., originated our mummy-autopsy project. Cockburn, an expert on infectious diseases, saw a way to study their evolution. He reasoned that if we knew how diseases change over long periods of time, we might learn how they develop and spread, and be better equipped to treat, cure, and perhaps even eradicate them. What better source for this type of data was there than ancient mummies?

Besides direct evidence of disease, we hoped to find protein antibodies in the mummified tissue that could be tested for evidence of such illnesses as polio, syphilis, and typhoid fever. Cockburn would have preferred to study frozen bodies such as those buried hundreds of years ago in the Siberian permafrost, because frozen tissue is usually well preserved. But world politics being what they are, this was not likely. Egyptian mummies were the next best choice and when Peck offered us DIA-I, a mummy from the Detroit Institute of Arts, in 1971, the project took shape. The original group of researchers included Peck; Cockburn and his wife, Eve, who served as the group's secretary; physiologist Robin Barraco of Wayne State University School of Medicine; and me. Barraco's special interest is the study of protein and other chemical compounds in mummified tissue.

A mummy is simply a preserved body with soft tissue remains, such as skin, muscle, and other nonbony parts. The Egyptians believed that body and soul would eventually be reunited after death, and Egyptian mummies were prepared in a complex, ritualistic manner, in an at-

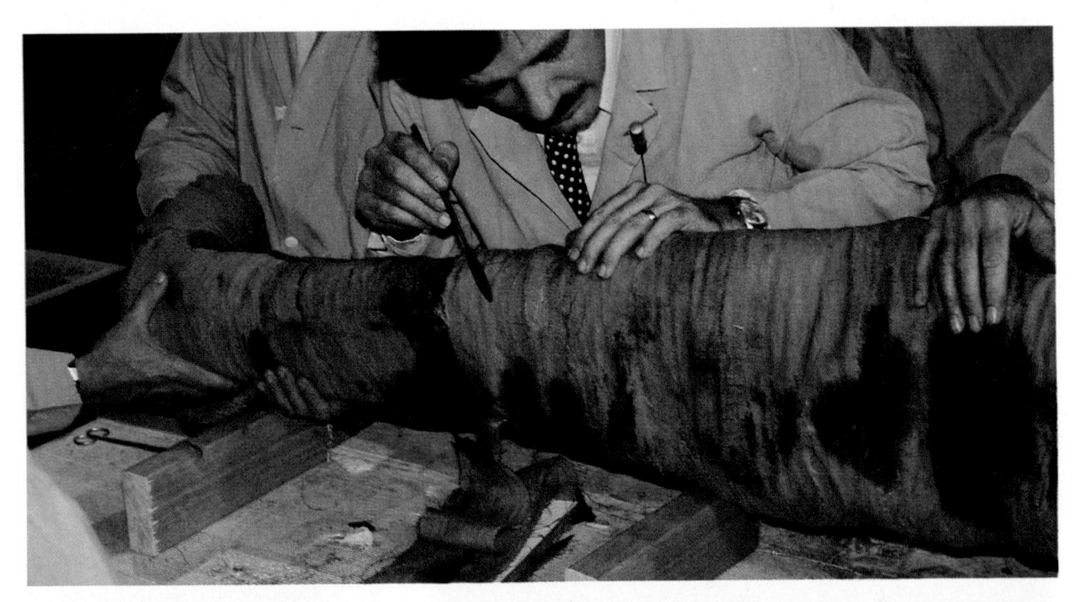

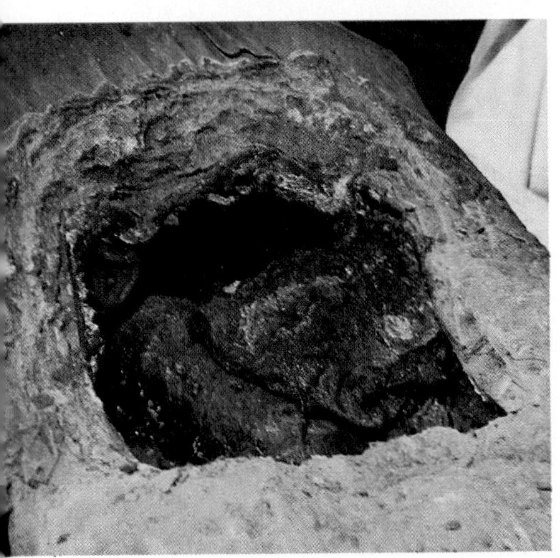

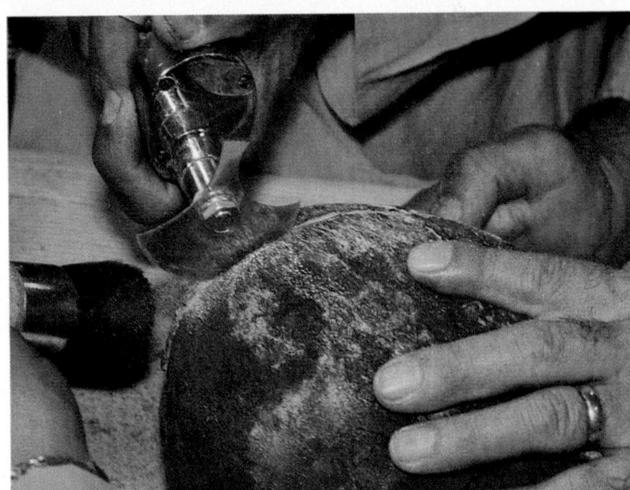

Careful handling preserves mummy wrappings for study, *top.* Vital organs in resin are often in the abdomen, *above left.* A surgical saw is used to open the skull, *above right,* so that scientists can take bone and inner ear samples and search for brain tissue. Later, in prepared laboratory samples, microscopic study reveals blood cells such as a red cell, *far right,* magnified 18,600 times, and a white blood cell, *right,* the first found in a mummy, magnified 16,000 times.

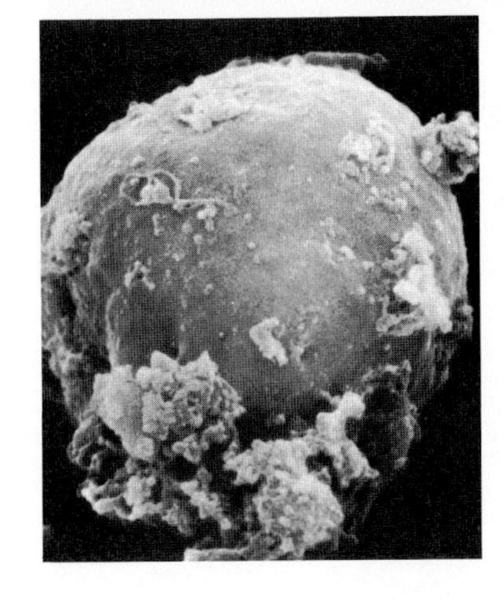

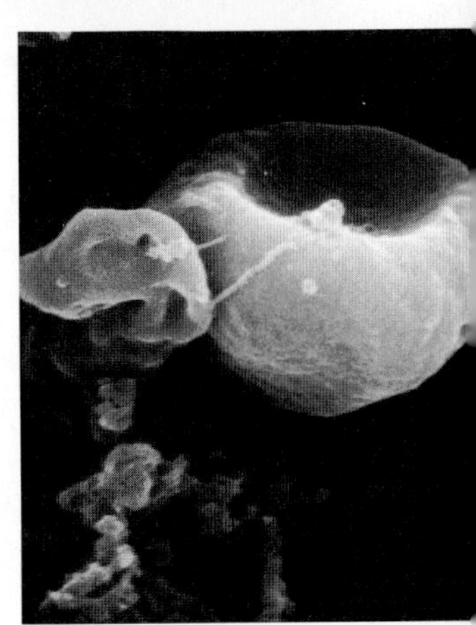

PUM-II lies on a table in the laboratory, awaiting the autopsy that yielded important information on disease in ancient Egypt.

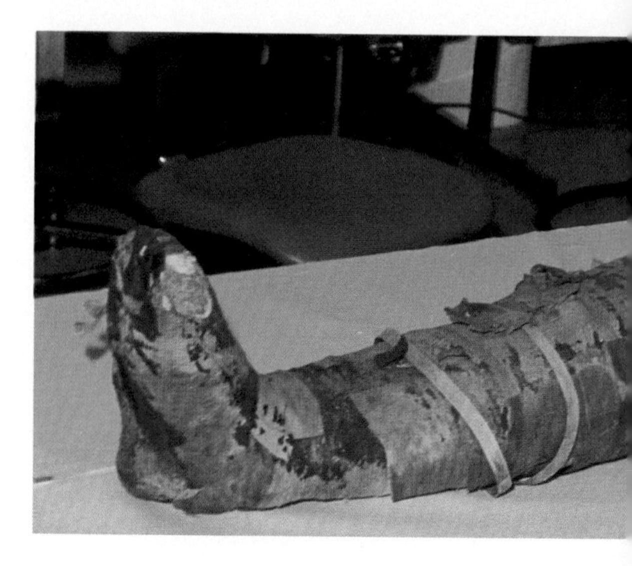

Religious symbols and writings about death on the outer coffin of an Egyptian woman who died about 1000 B.C. reflected a deep concern about the afterlife.

tempt to preserve the body. The Egyptians even placed food, clothing, and personal possessions in the tombs so that the dead would be adequately provisioned for the afterlife.

Embalming halted decay of the tissue and allowed it to dehydrate rapidly. Embalmers usually removed the internal organs, generally through an incision in the left groin or left flank. They did this because the dead body could not be dehydrated rapidly enough to prevent the organs' decay. The softened brain was drained through a hole punched into the skull through the left nostril. The internal organs were placed in a solution of natron, a naturally occurring salt similar to that found on the Bonneville Salt Flats in Utah. The natron solution stopped decay. Later, embalmers wrapped the organs in linen soaked in plant resin, generally sap from coniferous trees, and put them back into the body cavities just before the final wrapping. The resin fought bacteria and promoted dehydration.

The skin and the body cavities were cleansed with wine and oils, and the body was then packed in dry natron. The embalmers put resin into body and cranial cavities and smeared it on the skin and on each layer of the wrappings.

The Greek historian Herodotus wrote the only surviving detailed record of the mummification process, during his travels in Egypt from 460 to 455 B.C. The techniques he described were the result of a long series of modifications that began 3,000 years earlier. At first, only the pharaoh and his family were mummified. Centuries later, lesser nobility and priests were also mummified, and from about 1000 B.C. on, in the declining years of the empire, anyone who could afford it could be mummified. By then, embalmers had abandoned much of the original mummification process. Instead of applying resin, an expensive import from Lebanon and Syria, they began using dirt or mud. Many bodies were simply wrapped in large sheets of linen and buried with-

out any preparation. The practice of mummification ceased when Christianity spread to Egypt in the first three centuries after Christ.

Although I have performed thousands of modern autopsies, my first experience with mummy DIA-I in 1971 was fascinating—even though the scientific results were disappointing because of the mummy's poor condition. The skin was black and fissured, looking very much like cracked plastic, and almost as hard. In some places, the skin was missing, and I could see the underlying dried muscle. Considering the tattered wrappings and the amount of exposed skin, I was amazed that the body had lasted all these years. I gingerly snipped small pieces of skin and muscle from inconspicuous areas, because the mummy was to be displayed at the Detroit Institute of Arts later. I could hardly contain my desire to run back to the histology laboratory and immediately start preparing the meager samples for examination.

While we were still studying the tissue from this mummy, Cockburn arranged with anthropologist David O'Connor of the University Museum in Philadelphia to conduct an autopsy on one of the mummies there. Anthropologist Solomon Katz and pathologist Michael Zimmerman of the University of Pennsylvania would collaborate with us. The Cockburns and I flew to Philadelphia in May 1972, and found a great deal of excitement and public interest. There was extensive news coverage of the autopsy. We knew from the X rays taken of PUM-I before the autopsy that the mummy, a male about 30 years old at death, was not well preserved. Almost every joint of the body was distorted or had fallen apart. Large sheets of linen had been wound carelessly around the body, and there were holes in the wrappings over the tips of the toes and the right shoulder. The skin was decomposed and the tissue in the body cavities was so poorly preserved that we could not tell if the internal organs had been removed. Only a part of the brain remained intact.

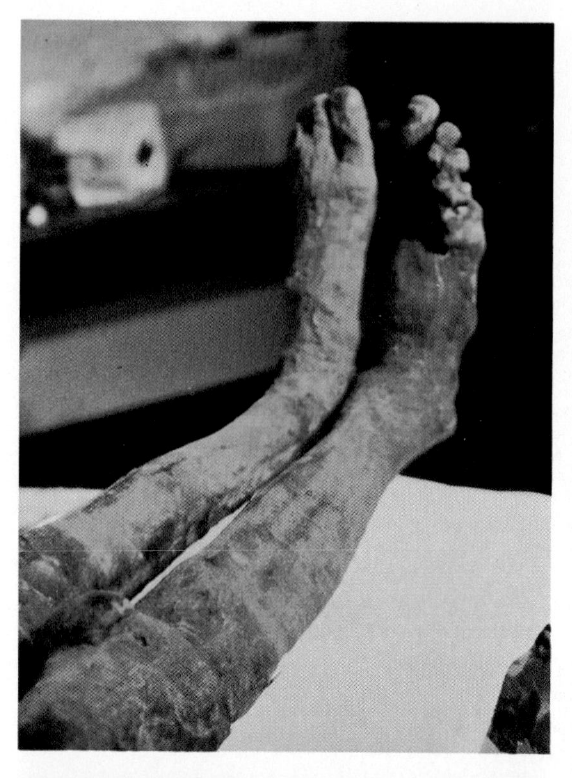

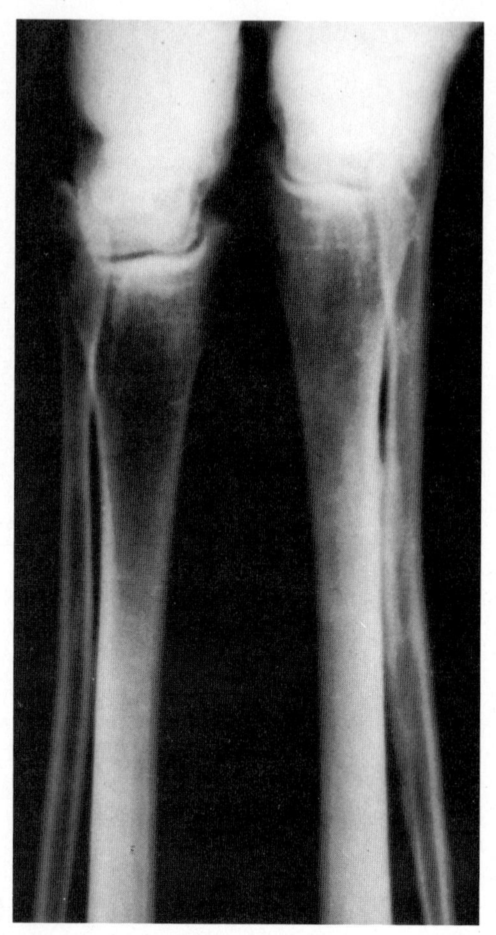

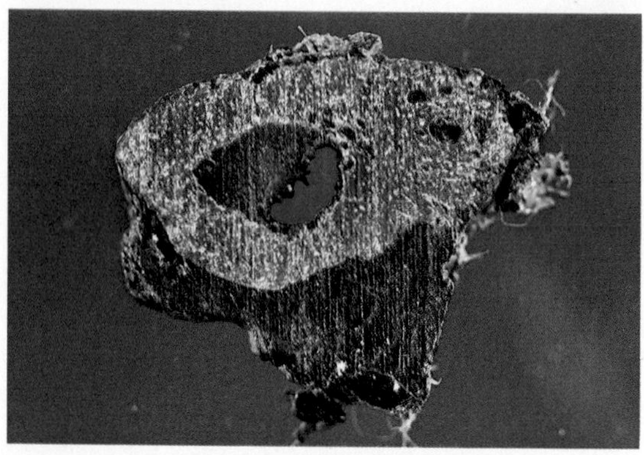

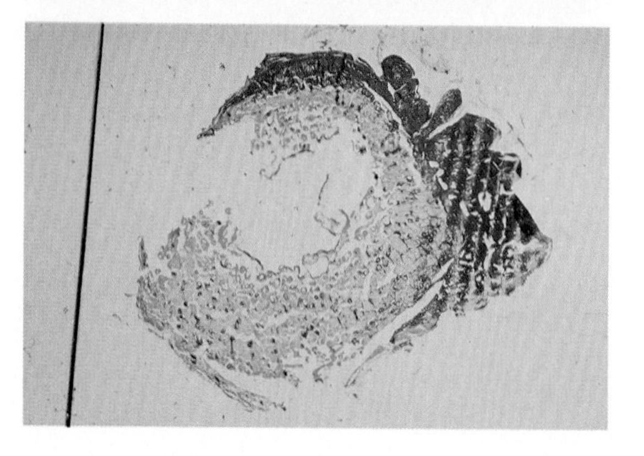

The right leg of PUM-II was swollen, *top left.* X rays, *above,* showed that his right fibula, smaller of the two lower leg bones, was abnormal. A cross section of the fibula, *middle left,* clearly showed swelling on one side of the bone. In a very thin slice of bone tissue, *bottom left,* the swollen area turned an abnormal red when stained, suggesting periostitis, an inflammation.

Excitement ran high when we found a small piece of yellowed paper in the chest cavity, but it proved to be only a scrap from a 1904 Cairo newspaper giving the time of the tides in French. A worker must have stuffed the scrap of paper into a hole in the shoulder wrappings just before the mummy was shipped to the United States. We took samples of tissue from the chest and abdominal cavities where we knew the internal organs should be. We also took other bone, skin, and muscle samples for laboratory examination.

As word of our autopsies spread, we received queries from museums and laboratories and information about mummies from other investigators in the United States and many other countries. Cockburn suddenly realized that there was already a vast amount of data on ancient diseases in the hands of scientists in many disciplines. As a result, the Cockburns, Peck, Barraco, and I formed a group called the Paleopathology Association in 1972. Its primary function is to gather data on ancient diseases, organize it, and redistribute it in a quarterly newsletter edited by Eve Cockburn. By March 1976, membership in the association had grown to more than 250 in the United States and a dozen other countries.

Our first two mummies yielded little medical information, but we learned a great deal about preparing and processing the tissue so that we could study it under the microscope. We also learned that we must tailor each dissection to the method of mummification. Fortunately, our third autopsy provided us with better medical results. The autopsy on PUM-II, donated by the University Museum and the Philadelphia Museum of Art, was performed as part of a two-day symposium entitled "Death and Disease in Ancient Egypt," held in Detroit in February 1973. Extensive X-ray and external examination indicated that the mummy was well preserved. What we did not realize was that the wrappings on the mummy were literally glued together with plant resin. The wrappings and resin formed a hard shell that had to be chiseled off, a task that took eight people almost seven hours.

When the body was finally uncovered, we found the embalmer's incision in the left flank. But we could not use standard scalpels and forceps on PUM-II; the mummified tissue was so hard that we had to use electric saws and large, stiff-bladed knives. Once into the body cavities, we found four large packets containing the internal organs. Part of the heart was still in the chest, but we could find no trace of the brain in PUM-II's skull.

The X rays showed no fractures, but they did show a congenital abnormality in the lower spine and a thickening of the bones in the lower right leg. Including the lower spine abnormality, we identified six disease processes in this mummy, a male about 35 years old when he died 2,200 years ago. Microscopic examination of tissue revealed atherosclerosis, hardening of the arteries due to fat accumulation, in the aorta; periostitis, an inflammatory thickening of the bone surface, in the lower right leg; a perforated eardrum, indicating an ear infection; an intestinal parasite; and deposits of carbon and silica in the lungs. The carbon probably came from soot given off by fires and oil lamps, and the silica from exposure to sandstorms. Yet, despite all we had found, we could not determine the cause of PUM-II's death.

We performed our next autopsy in Toronto, Canada, in August 1974, in conjunction with physicians Gerald Hart and Peter Lewin, and several University of Toronto scientists. From the inscription on his coffin, we knew the mummy was a teen-aged boy named Nakht who had been a weaver in the royal temple at Thebes about 1200 B.C. Egyptologist Nicholas Millet of the Royal Ontario Museum in Toronto, who arranged for the autopsy, presided over the unwrapping.

There had been very little preparation of this body; it had simply been wrapped in the boy's clothing. Yet the body was well preserved by natural dehydration. We found all the organs in their normal position, and the brain and heart were so well preserved that they have been put on display at the museum, along with the body. Microscopic examination of the remaining tissue revealed that it too was well preserved, and we identified several maladies. In the intestine, we found two different types of parasite eggs, one from *Taenia*, a tapeworm, and the other from *Schistosoma haematobium*, a parasite that migrates from the bloodstream and lodges in the intestine and urinary bladder. The organism also lodged in Nakht's liver, producing scarring, or cirrhosis. The spleen was enlarged and may have ruptured. The schistosoma infestation probably caused the boy's death. This particular finding is of great historical interest, because schistosomiasis is still a widespread medical problem in Egypt, particularly since the completion of the Aswan High Dam on the Nile River in 1968.

From the beginning, our studies were planned as a multidisciplinary investigation. As Cockburn pointed out, we must approach each mummy as if it were the last one on earth. Therefore, we must investigate and analyze it carefully from every conceivable area of interest. As a result, nonmedical specialists have helped us gather many important facts during the autopsies.

Before anything is done to a mummy, archaeologists try to determine the type of wrapping. Then the body is thoroughly examined by X ray, not only for evidence of disease but also for signs of objects within the wrappings or in the body cavities. The variations in the mummification process and the materials used are studied by archaeologists and cataloged for their historical value. Samples of resins or plant material, such as small fragments of wood and fir tree needles that we find within the wrappings, are carefully collected and sent to botanists for identification.

Lewis Nunnelley, a physicist at the University of Colorado in Boulder, and I performed chemical studies of heavy metals such as mercury and lead found in mummy tissue. Environmentalists studying industrial pollution to determine what constitutes normal heavy metal content of tissue have expressed interest in these tests. Since all of our

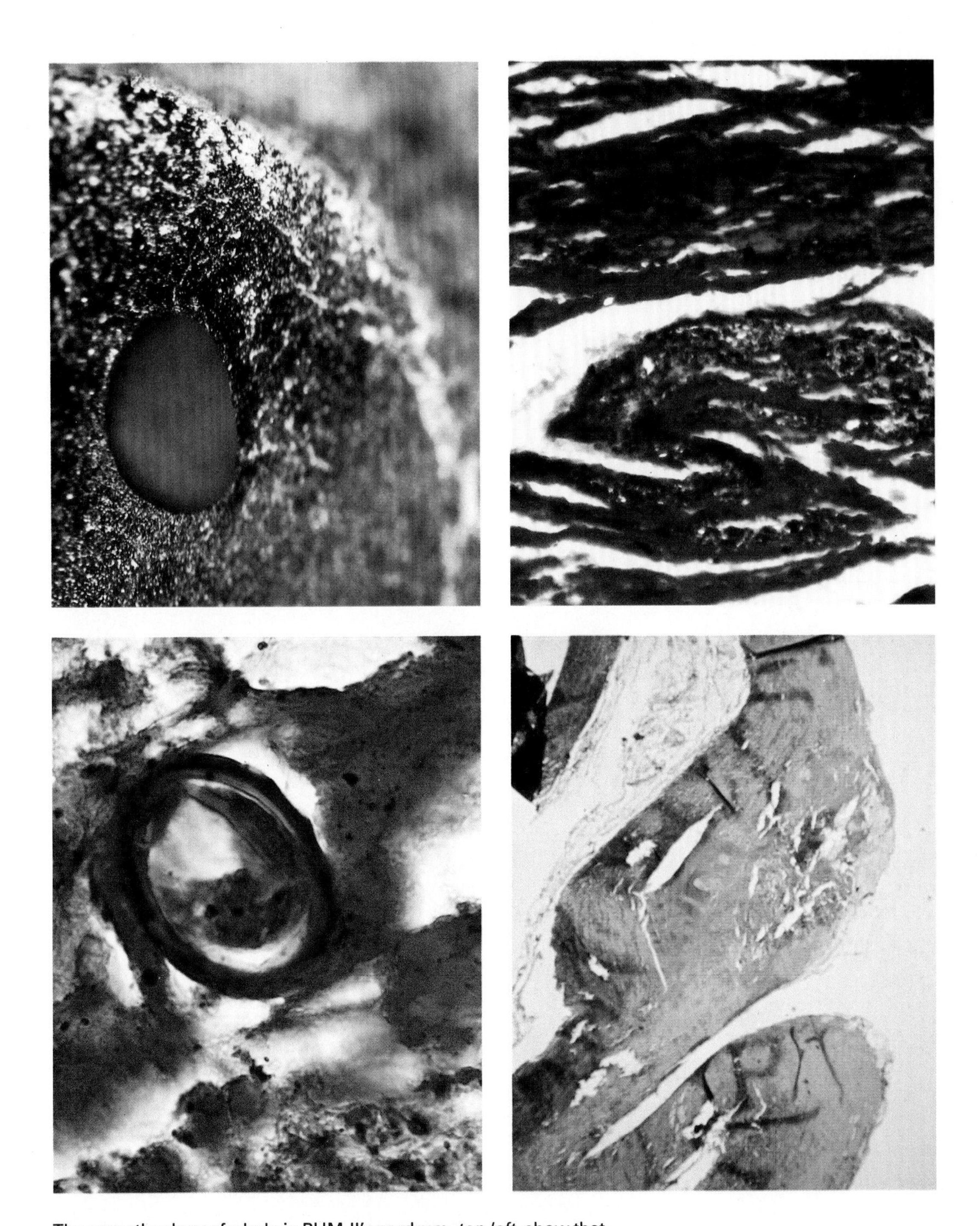

The smooth edges of a hole in PUM-II's eardrum, *top left,* show that it was caused by infection, not an accidental puncture. The shiny grains of silica in his lung tissue, *top right,* suggest that he had silicosis, a disease like coal miners' black lung. Doctors also found the egg of *Ascaris,* a roundworm parasite, *above left.* PUM-II's aorta, *above right,* was dark and twisted. The lighter bulge at middle right contains the hardened fat deposits that cause atherosclerosis.

mummies come from a preindustrial population, they should be a good source of comparative data. An unprocessed cotton boll within the linen wrapping of a mummy who died about 200 B.C. was an unexpected find of great interest to archaeologists and historians. The boll proves the existence of cotton in Egypt at least 400 years earlier than was previously known.

Entomologists at the U.S. Department of Agriculture have now identified at least six different beetle and fly larvae mummified with the bodies. Some of these ancient insects differ slightly from present-day species, suggesting that we are seeing some evolutionary changes.

In addition to Egyptian mummies, scientists have studied mummies from many other parts of the world. Often, they have been mummified naturally by the dry heat of the desert or the cold of permafrost areas such as Alaska and Siberia. Zimmerman and his associates autopsied two Alaskan mummies in 1971 and 1975. The first was a middle-aged male who died about 300 years ago and was placed in a volcanic cave. The condition of his lungs indicated that he died of pneumonia. The second, an elderly female, died about 1,600 years ago. Eskimos found her frozen body on St. Lawrence Island in the Bering Strait. She had extensive atherosclerosis, but particles of moss and other plant material found in her lungs suggest that she was buried in a snowslide and died of asphyxiation. We have been examining her frozen tissue for intact blood proteins, including the elusive antibodies that would identify various diseases, but our preliminary data have been disappointing.

The Indians of Peru mummified their dead for thousands of years, until about A.D. 1500, when Spain conquered the empire of the Incas and brought Christianity to the country. Medical researchers Marvin Allison and Enrique Gerzsten of the Medical College of Virginia in Richmond have been performing autopsies on Peruvian mummies for several years. They have examined more than 100 mummies of adults and children and found evidence of many diseases, including atherosclerosis, parasitic infestations, and tuberculosis. They also saw many instances of serious injury, some apparently fatal.

A beetle larva found preserved in a mummy may provide useful information about ancient insect life.

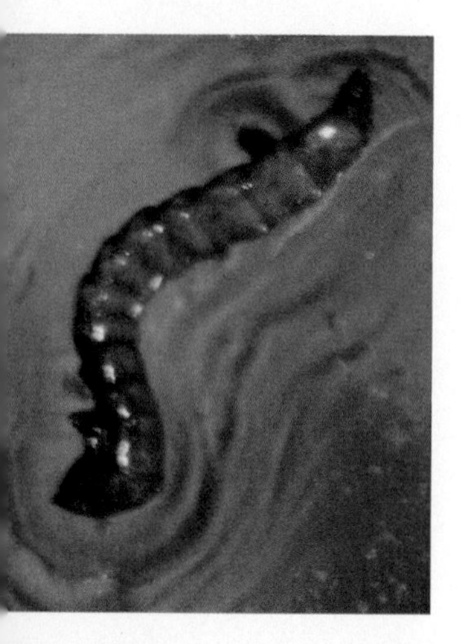

An astonishing example of mummification was found in China. Physician Ou-wei of Peking reported the autopsy of the 50-year-old Lady Ch'eng, who died 2,100 years ago. Her elaborate tomb was discovered in 1972. Unlike the Egyptian mummies, her skin and most of her internal organs were soft and pliable. The body had been bathed in a solution containing mercury salts, and the underground tomb had been covered with charcoal, which absorbs oxygen in the air, thus preventing decay. Her body was so well preserved that the investigators were able to identify more than 100 melon seeds that were in her stomach. Examination revealed that she had probably died of a heart attack caused by severe coronary artery disease. The Chinese scientists also found evidence of gallstones, tuberculosis, and two intestinal parasitic infestations.

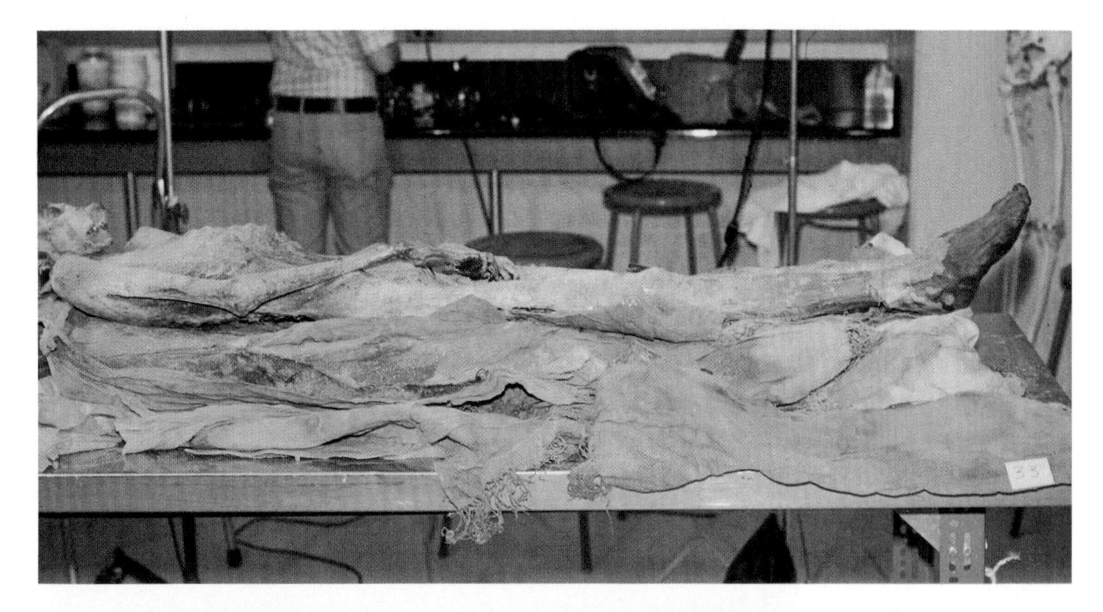

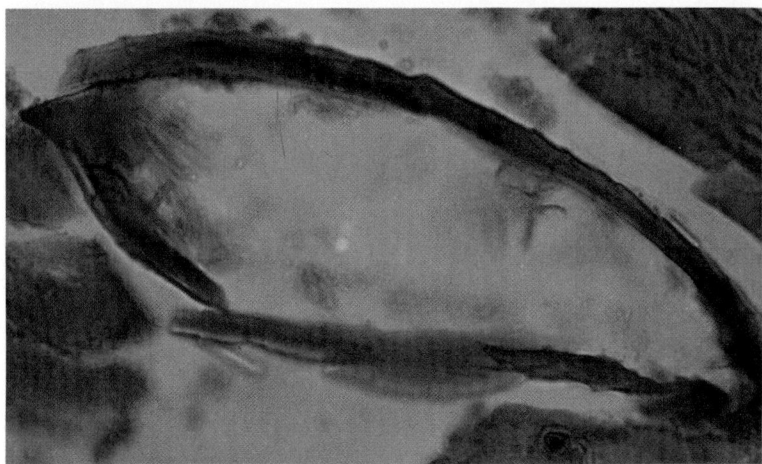

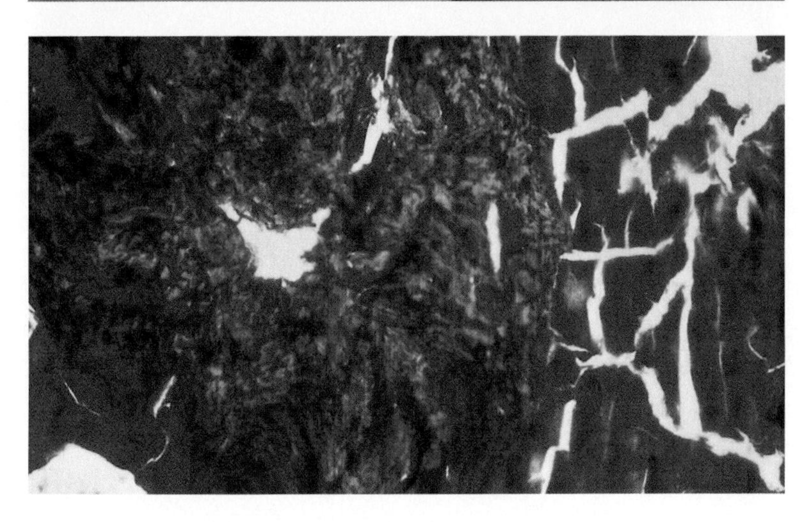

Nakht, a teen-aged weaver from Thebes, *top,* was examined in a University of Toronto laboratory. An egg in his intestine, *middle,* came from the parasite causing schistosomiasis. The blue connective tissue in his liver, *bottom,* is an indication of cirrhosis brought on by the parasite. This probably killed Nakht.

Other modern afflictions also have ancient counterparts. For example, in 1975, pathologist E. Tapp and his associates in Manchester, England, found evidence of silica, or sand, in the lungs of an Egyptian mummy, a condition that we also found in PUM-II. Tapp's data complement ours and verify an ancient counterpart to what medical researchers call "Negev Desert Lung" in today's desert nomads. The silica probably is sand inhaled during desert storms. Similar changes can be seen in coal miners' lungs; the coal dust they inhale contains varying amounts of silica.

Researchers George Lynn and Jaime Benitez of the Wayne State medical school are conducting a highly specialized investigation, studying the ears and hearing apparatus of Egyptian and Peruvian mummies for evidence of disease. They have identified several cases of chronic infection in the mastoid bone and perforation of the eardrum caused by acute ear infections. All the changes are similar to those that are commonly seen in cases today.

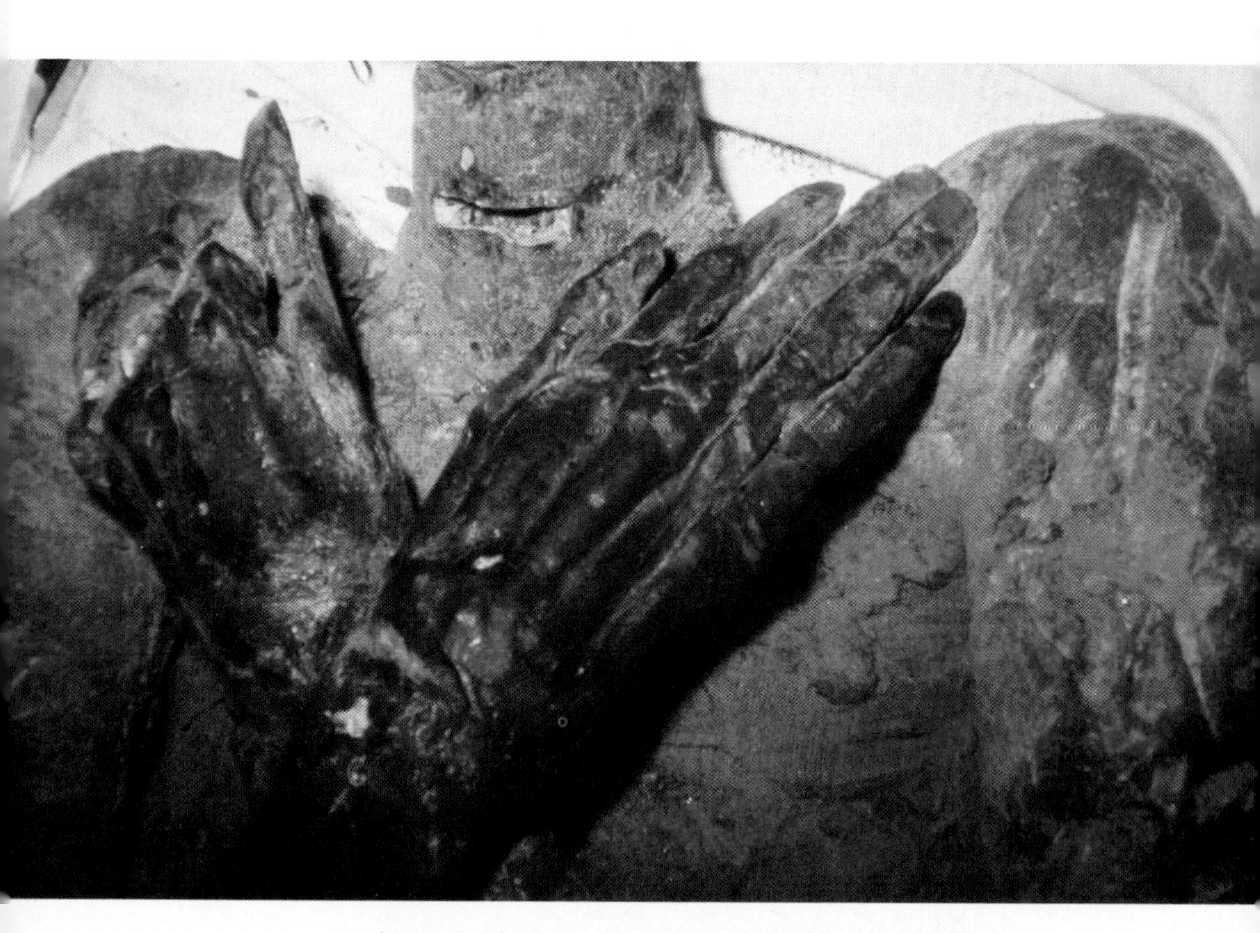

Arms crossed ceremonially on his chest, PUM-II, silent for 2,000 years, has something to tell us about life, sickness, and death in the ancient world of pyramids and pharaohs—and perhaps in our own.

We have found that infection caused much of the illness that afflicted the people we have studied. But injury was also a prominent factor in their lives. We found no cancer, nor has anyone else. Cancer tends to be a disease of older people, and life expectancy in ancient Egypt was only about 30 years. Yet atherosclerosis, long thought also to be a disease of old age, was common in the mummies that we examined. Scientists will now have to reconsider its cause because it is quite apparent that old age and the great stresses of modern technological society are not the only culprits.

Study of the fats found in the mummy tissue may have direct bearing on this problem. Barraco has identified cholesterol and a number of other fats in the tissue of several mummies. Analysis of the fats may shed light on the diet of these ancient people as well as its relationship, if any, to the development of atherosclerosis and other diseases.

We have even found that air pollution produced some disease in Egyptians and other ancient populations. What began as a simple attempt to find disease patterns and to isolate intact protein antibodies from the tissues of these ancient remains has been expanded into a dozen fields of research. The wisdom of Cockburn's original idea that the autopsy be a multidisciplinary exercise has become apparent.

We have examined and cataloged methods of mummification and the materials used by the ancient Egyptians. If we can determine the source of the plant resins and the cotton we found in the mummy wrappings, we may have firsthand evidence of the trade routes used by pre-Christian Egypt and other countries, the goods they traded, and the economic life of the ancient world.

Barraco is also investigating a new method of establishing the date of death of the mummified bodies. The new method, called amino-acid dating, not only may prove more accurate than radiocarbon dating, but also requires much smaller samples of tissue and wrappings. Amino acids are small protein units found in equal amounts as d- and l- forms in the living body. After death, however, the d- form slowly but constantly changes to the l- form. The amount of change is proportionate to the time the person has been dead. Barraco is presently testing the method with mummies of known dates to see how accurate it is.

Our work has produced a large amount of educational material. Each autopsy is extensively photographed, including several hours of color videotape. These visual aids allow us to share with others the detailed steps in each autopsy.

While these autopsies have provided a wide range of medical, cultural, and historical data, my primary objective as a pathologist is to determine disease profiles in a given human population. This information should simplify the task of studying the effects of different and changing environments on human diseases. Comparing the illnesses of ancient and modern man may give us the answers we seek about the evolution of disease.

The Fate of The Cosmos

By P. James E. Peebles

Either through eternal expansion or catastrophic collapse, most of the universe is doomed to a violent end

Ever since the first person gazed up into the night sky, people have wondered about the nature of the universe. How did it begin? What is it like today? Is it changing? In what way? And, perhaps the toughest question, how will it all end?

To try to unravel this last puzzle, we are drawing upon the answers we have found to the other questions. Most astronomers believe that the universe started with a big bang, a gigantic explosion of matter some 10 or 20 billion years ago, and that it has been expanding ever since. Supporting this idea was the discovery in 1965 of a type of radio waves called microwaves coming from every direction in the sky, the remnant of that primordial fireball. We

Galaxies formed from the radiation and gas of the big bang continue to move apart in an expanding universe.

also think we know what the universe is like on the large scale, ignoring individual stars and planets. Our observations fit into a theory that tells us how the universe is changing. If we knew more details of its present state, we could use the theory to predict how the universe will end. Put simply, the alternatives are two: contraction and total collapse to infinitely high density; or expansion forever.

Maps of the sky show that galaxies–swarms of billions of stars, such as our own Milky Way–tend to clump together in clusters and groups of clusters. Our nearest large galactic neighbor, the Andromeda Nebula, is about 2 million light-years away, unusually close as these things go. The Milky Way, the Andromeda Nebula, and several smaller objects belong to the Local Group of galaxies, which in turn is part of many knots and groups of clusters centered on the Virgo cluster of galaxies. Similar clusters and groups of clusters seem to be quite uniformly spread throughout space. This vast spread of galaxies, clusters, and groups of clusters holds the clues that may enable cosmologists to determine the fate of the universe.

In the 1910s and 1920s, two American astronomers, Vesto M. Slipher in Arizona and Edwin P. Hubble in California, began a systematic study of galaxy motions. Galaxies are so far apart that we could never hope to see actual changes of position in our lifetime, or in many lifetimes. But we can determine how they move in an indirect way by studying an effect similar to the one that gives the distinctive rising and falling pitch to the sound of a passing car. An approaching car seems to have a high pitch because it is moving toward the sound waves it is making. To a stationary observer, the wavelength appears to be shortened, causing a higher frequency, or pitch. As the car moves away, the sound waves seem to be stretched out, decreasing the frequency and giving the sound a lower pitch.

The author:
P. James E. Peebles, a professor of physics at Princeton University, studies clusters of galaxies for clues to how the universe will end.

Light waves work the same way. If a galaxy is moving away from us, its light appears stretched into longer wavelengths, toward the red side of the spectrum, making it appear redder than if the galaxy were stationary relative to us. If a galaxy is moving toward us, its light waves are shifted toward the blue. By studying this light shift, Slipher discovered that most galaxies are moving away from us.

Hubble discovered the pattern behind this motion–it is as if the universe is expanding. A convenient way to visualize Hubble's discovery is to consider the surface of a balloon. Suppose that dots painted on the balloon represent the galaxies. As you blow up the balloon, each dot moves away from every other one. A creature on one of these dots would see the others moving away in exactly the same pattern that Hubble found for real galaxies.

This balloon model should not be taken too literally. However, it is convenient in attempting to answer such questions as: "Where is the center of the expanding universe?" Note that the pattern of motion a creature sees on our two-dimensional balloon is the same no matter what dot he happens to be sitting on. The model universe is not expanding from a point in space–rather, space itself is expanding. The center is everywhere and nowhere.

The expansion of a balloon is easy enough to visualize, but can we imagine that our three-dimensional universe is expanding? And if the universe contains all space, into what can it expand? There is no direct answer to the second question, and the best mathematical theory declares the question meaningless. This is not as unreasonable as it seems. Intuition or experience is a guide to understanding physical processes that operate on the human scale. But it is impossible to experience directly the properties of phenomena on vastly different scales, such as the fantastically small atom or the immeasurable reaches of the universe. In these cases, modern physical scientists must draw upon mathematical theory as a necessary aid to their own common sense and intuition.

The mathematical theory that describes the expanding universe is the general theory of relativity, published by Albert Einstein in 1915. Although Einstein then attempted to modify his theory to fit an unchanging universe, the Russian mathematician Alexander Friedmann showed in 1922 that the original theory was quite reasonable if we accept that the universe is either expanding or contracting. Hubble discovered the expansion effect in 1929, and by 1930 scientists recognized the connection between the behavior of the universe implied by Einstein's general theory of relativity and Hubble's effect. The combination seemed irresistible to most astronomers, and they concluded that the universe is indeed expanding.

If the universe is expanding, Einstein's theory points to our two possible endings–either the universe keeps on expanding forever, becoming ever more dilute, or else it stops expanding someday and begins to contract, eventually collapsing to infinite density. The difference is simply that the galaxies have enough velocity to escape from one another's gravitational pull in the first case, but not in the second.

Escape velocity from a planet is determined by the planet's mass and radius–the greater the mass and the smaller the radius, the greater the velocity needed for an object, such as a rocket ship, to escape from the planet's gravitational pull. If the rocket ship does not reach escape velocity, it will eventually fall back to the planet's surface. The same principle applies to the expanding universe. We know the speed at which galaxies are moving apart by observing their motions. Whether they are moving fast enough to escape one another eventually depends on the mass of the universe or, more exactly, the mean mass per unit volume.

Scientists have followed two main routes in trying to determine how the universe will end–measuring the rate at which the expansion is slowing down and measuring the mass of the universe. The first route makes use of the fact that it takes light a specific amount of time to travel a certain distance–in one second, it travels about 300,000 kilometers (186,000 miles). This means that we see a distant galaxy not

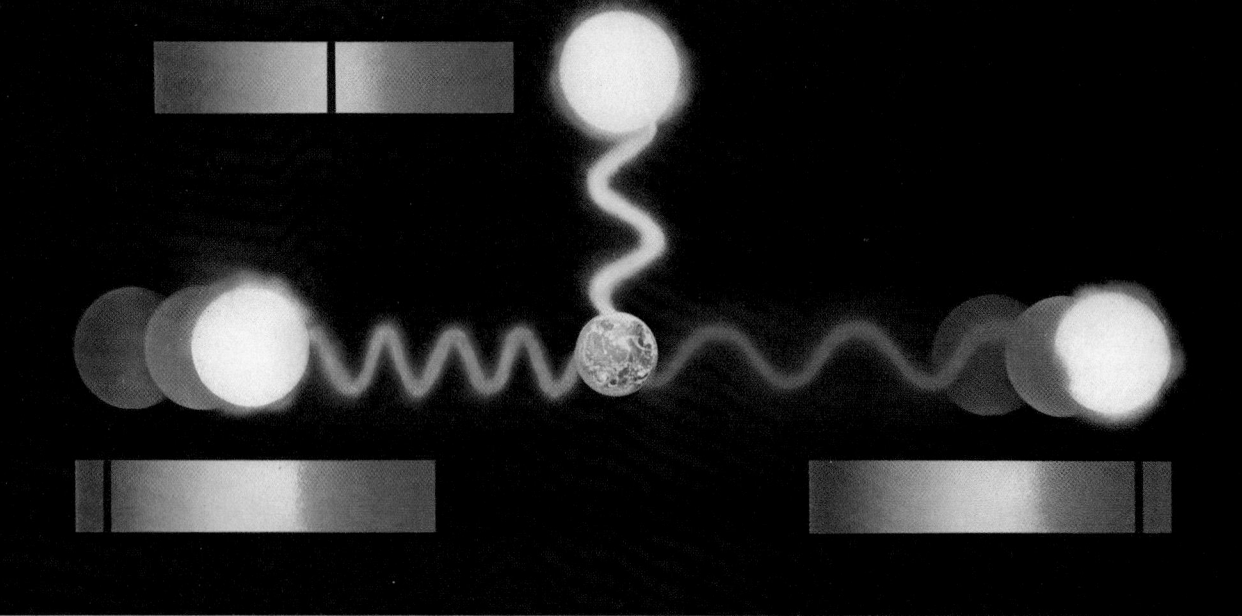

Clocking the Stars By Light Shift

If a galaxy is moving toward us, its light appears shifted toward the shorter wavelengths on the blue side of the spectrum. If it is moving away, the light seems stretched toward the red side. The size of the shift depends on the speed of the object. Light from a stationary source does not change.

as it is now, but as it was long ago when the light we are now receiving left the galaxy. By comparing the expansion of the universe then and now, we hope to determine the precise rate at which the expansion has been slowing. We can then decide whether the decrease is rapid enough to cause the universe eventually to collapse.

This approach has been thoroughly studied for more than 15 years by astronomer Allan R. Sandage of the Hale Observatories near Pasadena, Calif. To conduct his study, Sandage first located a number of galaxies that have about the same intrinsic luminosity–that is, if they were all lined up at the same distance from us, they would be equally bright. He then measured the apparent brightness of each galaxy to determine its relative distance from Earth. By a law of physics, if Galaxy A is one-fourth as bright as Galaxy B, it is twice as far away. Sandage also determined the red shift of each galaxy, which told him how fast the galaxy is moving away from us. According to Hubble, the velocity of recession (deduced from the red shift) for fairly close galaxies is proportional to the distance (deduced from the apparent brightness). Much more distant galaxies deviate slightly from this relationship because they are seen as they were at a much earlier time, when they were moving faster. By plotting the deviation for galaxies at different distances, Sandage hopes to determine the rate at which expansion is slowing down.

Unfortunately, Sandage is looking at distant galaxies when they were much younger than those nearby, and it is reasonable to suspect that young galaxies would look different from old ones. They presumably are brighter simply because young stars tend to be brighter than older ones. If Sandage cannot figure a way to correct for this difference, he might seriously underestimate the distances of faraway galaxies because a young, bright galaxy would seem to be closer than it actually is. This also would produce a deviation from the Hubble effect that would indicate that the expansion of the universe is slowing down faster than it really is. This problem has delayed the completion of Sandage's project.

Although Sandage has not yet been able to determine the ultimate fate of the universe, he has contributed a great deal to our knowledge of galaxies. He discovered that the brightest galaxies in clusters with the same red shift, therefore at the same relative distance from us, all have nearly the same apparent brightness. This suggests that the brightest galaxies in all clusters, no matter what the distance, are all about equally bright. The fact that the brightest galaxies in each cluster have such uniform luminosities must be an important clue to how galaxies formed. Perhaps there is some fundamental limit to the amount of material that can be assembled in a space as compact as that occupied by a galaxy.

Sandage and others are now attacking the theoretical problem of predicting how rapidly the brightness of a galaxy changes as its stars evolve. When they solve this, cosmologists might be able to determine definitely the rate at which the universe is slowing down.

The second route to predicting how the universe will end is based on its mean mass density. If we knew the mass, we would know whether the present expansion exceeds escape velocity.

To measure the mass of the universe, cosmologists must first determine the masses of individual galaxies. This is done by the same method astronomers use to weigh the Sun. Although gravity is constantly pulling the Earth toward the Sun, the Earth does not fall into the Sun because it is moving sideways around it just fast enough to stay in a roughly circular orbit. Once we know how far the Earth is from the Sun, and how long it takes to travel around its orbit, we can compute how strongly it is being pulled toward the Sun. This pull is determined by the mass of the Sun. Thus, we can weigh the Sun by studying the motions of the Earth or any of the other planets. Astronomers can weigh a galaxy in a similar way. They observe the motions of bright stars in a galaxy and ask how massive the galaxy must be to have a gravitational force strong enough to hold the moving stars in their orbits around the center. The combined mass obtained in this way for all the galaxies does not appear to be enough to hold the universe together–the pieces of the universe would fly apart forever unless there is much more mass present than we can see.

Such unseen mass could exist. Imagine a hollow planet, with all of its mass concentrated in its outer shell. The gravitational attraction of this mass would cause an apple outside the planet to fall to the surface just as an apple falls on a solid planet like the Earth. But an apple inside the hollow planet would not react to any gravitational force. Each bit of mass in the outer shell would exert a gravitational pull on the apple inside, but the strength of the pulls in different directions would balance one another, leaving a net effect of zero. This curious

Motion Measures Mass
The Sun's pull balances Earth's sideways travel so the orbital motion measures the mass of the Sun (top). Similarly, a star's motion weighs the galaxy within its orbit (middle). The motion of two galaxies around each other detects mass hidden in dim halos (bottom).

property of the gravitational force was discovered by the English scientist Sir Isaac Newton, who first published his law of gravity in 1687.

What if a massive halo, a kind of shell, of dim matter surrounded the bright part of a galaxy? According to Newton's theory, this halo mass would not affect the motions of the stars that we see inside the halo; only the matter lying inside their orbits would affect their movements. By using only the motions of these stars to weigh the galaxy, we would miss all the weight in the dim halo mass.

This is not an idle speculation. For example, a standard astronomical photograph of one large galaxy–M87–shows a bright blob in the center, with a rather fuzzy edge. But in 1971, Halton Arp of Hale Observatories and Francesco Bertola of Padua Observatory in Italy used an isodensity tracer at the Jet Propulsion Laboratory in Pasadena to map very slight variations in light in a photograph of M87. This sensitive machine scanned the plate in hairline bands, about 0.2 millimeter wide. The technique produced a sort of contour map outlining the changes in light level around the galaxy, with bands closer to the center representing greater light levels. They found that the image of the galaxy sprawled across the entire photographic plate.

The light from the enormous halo presumably comes from stars. We do not know what else this halo might contain. Perhaps there are burned-out star remnants such as white dwarfs, neutron stars, or black holes. Or maybe there are low-mass stars that glow too weakly to be readily detected. Perhaps there are even unborn stars similar to the planet Jupiter, with masses too low to ignite the thermonuclear reactions that cause stars to shine.

We can get some idea of the amount of mass that may be hidden in such halos by playing the weighing game on a larger scale. The motion of one galaxy around another can be used to weigh the halo, or a good part of it, because each galaxy can be outside the other's halo. Just as an apple outside a

When a conventional photograph of galaxy M87, *above,* is scanned by a sensitive light tracer, light differences appear across the entire plate, *below.* This indicates the galaxy may be surrounded by a halo of stars.

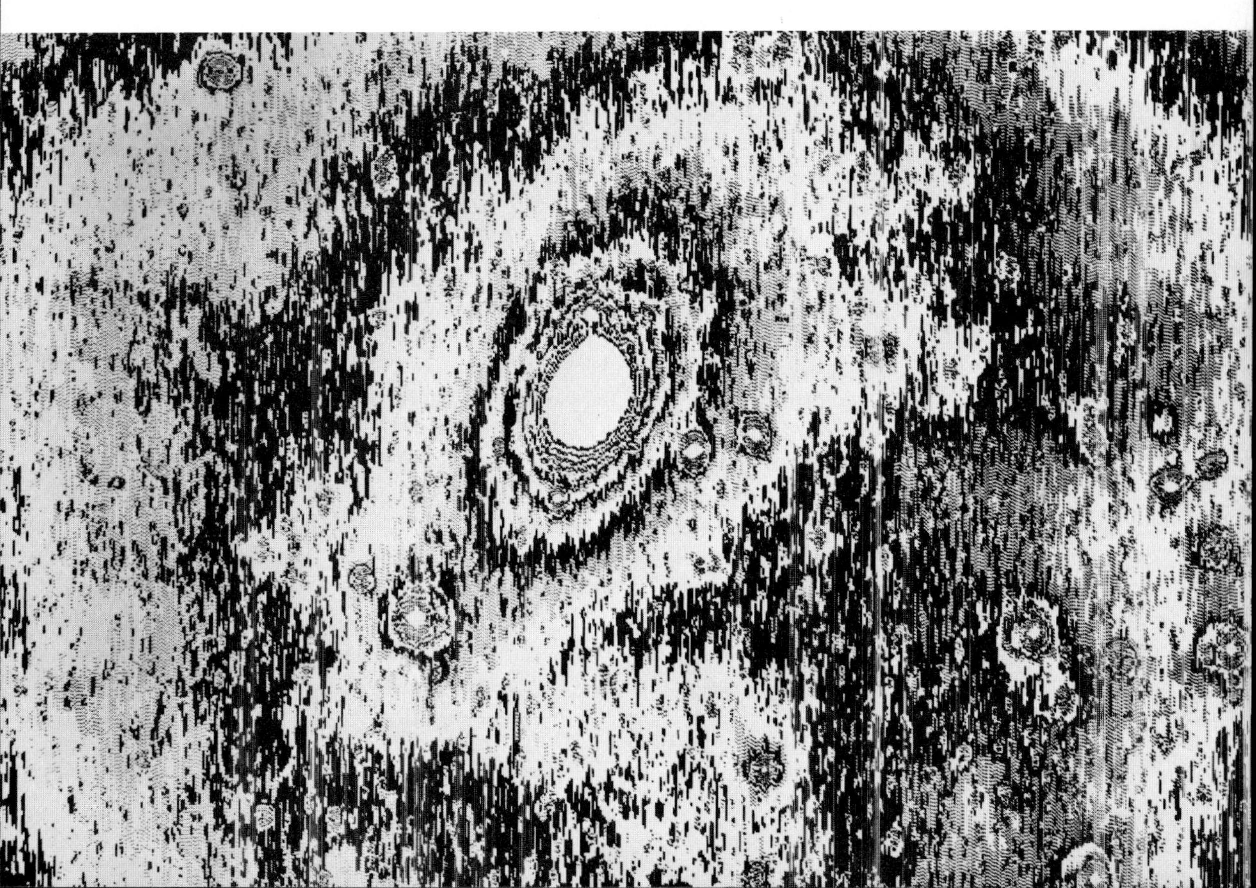

hollow planet reacts to the gravitational attraction of the planet while an apple inside does not, a distant galaxy reacts to the halo mass missed by an inner star.

An interesting example is the motion of the Andromeda Nebula, our nearest neighbor galaxy. The light from this galaxy is shifted toward the blue, suggesting that, unlike most other galaxies, it is moving toward us. Probably this is because both the Andromeda Nebula and the Milky Way are quite massive and were fairly close together when they formed. Thus, the gravitational force between them would have reversed the initial tendency to move apart. We know how far apart the pair is today, the speed at which they are coming together, and approximately how long ago expansion started at the big bang, so we can tell how strongly gravity is pulling the pair together. The combined weight found in this way is about 10 times greater than the presently accepted figure, which suggests that a large amount of mass exists in undetected halos around these two galaxies. Still, there is a possible flaw in this assumption. Perhaps the Andromeda Nebula is just accidentally passing by, and not gravitationally bound to us at all.

We can rule out the idea of such an accidental encounter by examining other clumps of galaxies to see if there is a general indication of undetected mass. In 1933, astrophysicist Fritz Zwicky at the California Institute of Technology (Caltech) in Pasadena made the first attempt to understand the dynamic motions of a cluster of galaxies. After studying the motions of the galaxies in the Coma cluster, a very tight knot of galaxies, Zwicky concluded that if the galaxies in this cluster had only as much mass as the accepted estimate, their gravity could not hold the cluster together. His conclusion still holds–apparently there is much hidden mass in that cluster.

Nevertheless, astronomers have been cautious about assuming that all galaxies have a lot of hidden mass. The galaxies in extremely dense clusters such as Coma are probably not typical. It may be dangerous to assume that the properties of galaxies found in these rather special environments are characteristic of galaxies in general.

Another well-studied group is the Virgo cluster. Here, too, mass as estimated is not sufficient to hold the cluster together. This cluster is much less dense than Coma and its galaxies look fairly typical, reducing concern that Virgo may be a special case. But there is another problem. The Virgo cluster is so spread out that it could have been freely expanding for a long time. Perhaps a sort of explosion occurred earlier and the cluster is still freely falling apart. If so, astronomers argue, we should not expect its mass to hold it together.

Arguments like these have been raging since the 1930s, and we still do not have a really convincing answer to the problem. We need to study many more clusters, concentrating on tight knots of galaxies that would quickly fly apart if they were not held together by gravity. Only through such a broad study can we determine if there is a general, reasonable, and consistent case for missing mass.

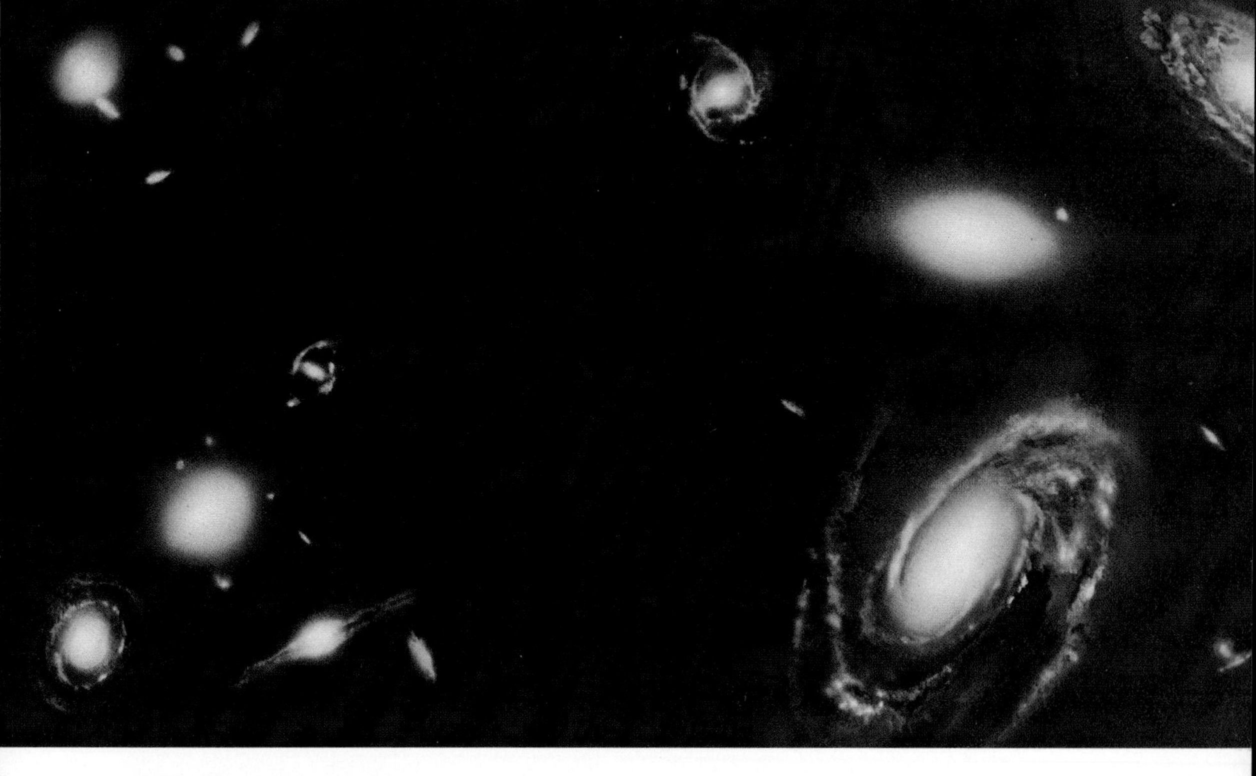

Rushing Toward Doomsday

The universe today contains swarms of galaxies that gather in clusters, *above.* The clusters move away from each other in an ever-expanding space. Many are relatively young, with blue stars and spiral arms. As the universe ages, *below,* the stars will redden and the spiral arms will become less evident. Neighboring clusters will continue to move apart, but the galaxies within individual clusters will pull together slowly. Gravitational attraction will draw off streams of matter as they get closer, and smaller galaxies will start to merge with larger ones.

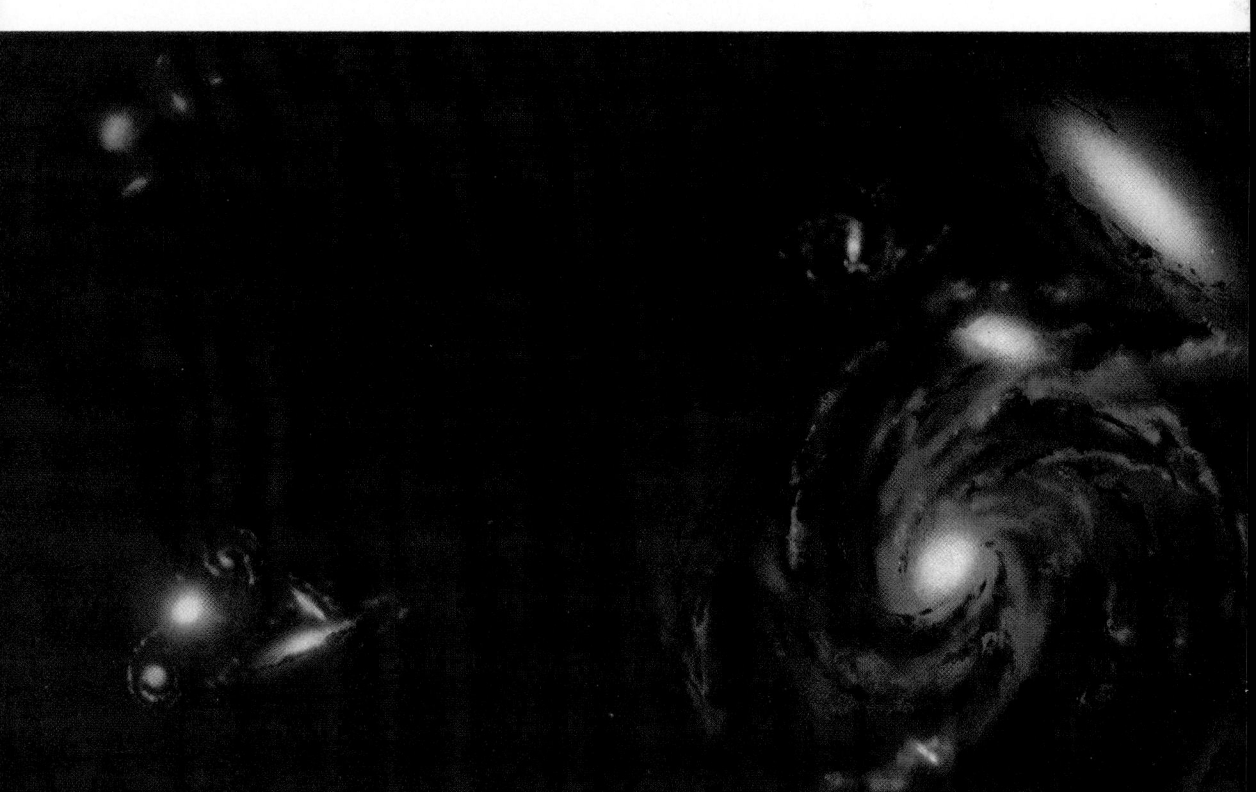

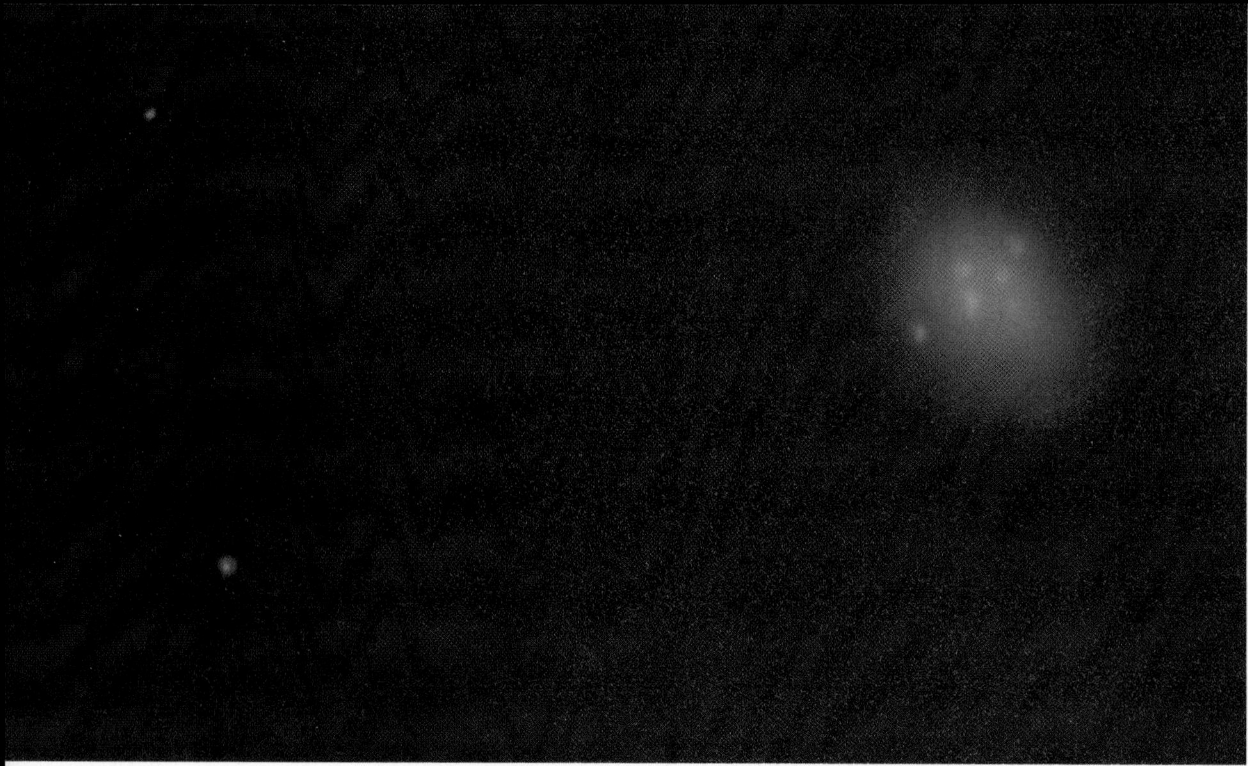

Eternal Expansion
In an expanding universe, clusters of galaxies will continue to retreat from each other, *above.* But those within each cluster will draw together, eventually merging in a supergalaxy. Individual stars may be flung out to form a dilute halo. In the final stage, *below,* the halo will become more and more dilute as almost all matter collects in the dense central nucleus, which by then may be a black hole. Other clusters will be so far away they will have disappeared, and the galactic object will drift forever alone in increasingly empty space.

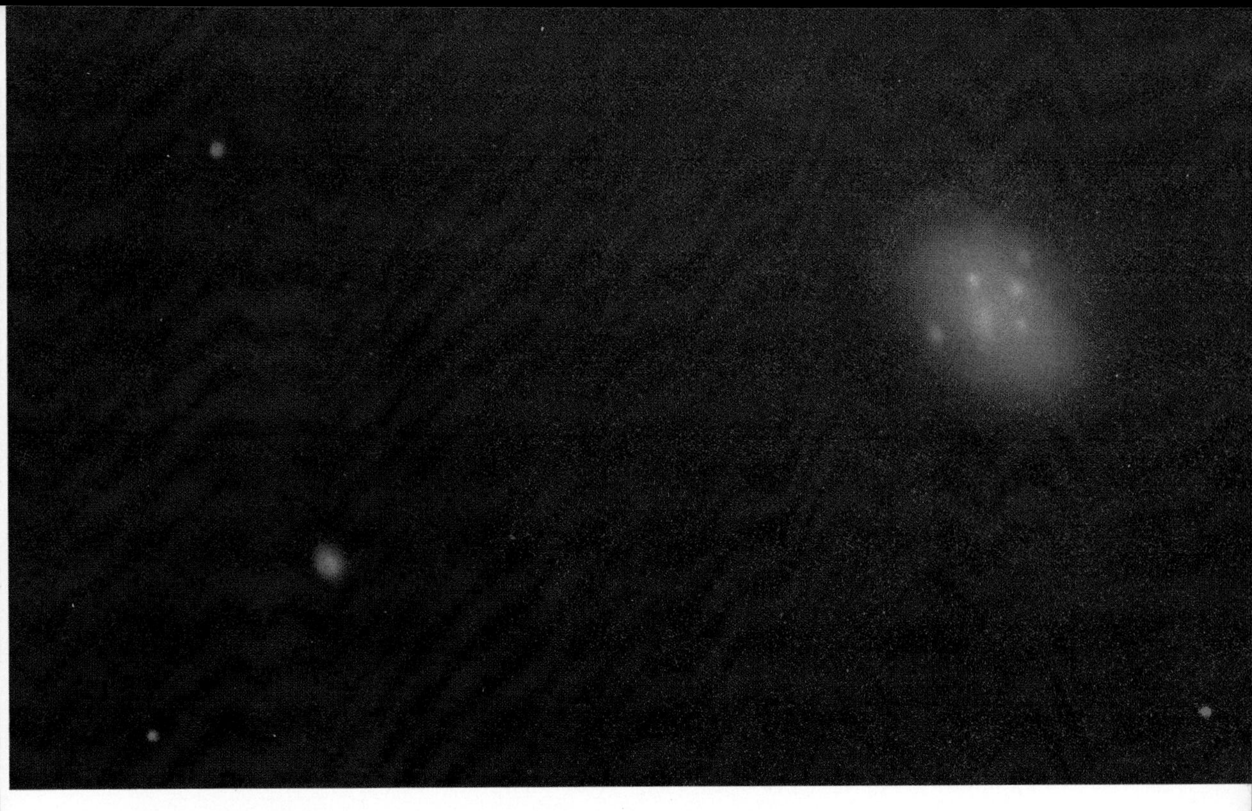

Catastrophic Collapse
In a closed universe, gravitational attraction will eventually slow the separating clusters of galaxies enough to reverse the expansion, *above.* Neighboring clusters, then condensed into supergalaxies with multiple nuclei, will come rushing together as all parts of the universe draw close again. In the final chaos, *below,* galaxy clusters will tear through each other, trailing wispy filaments of gas. Stars will crash and explode. Ever more violent collisions will continue until all matter will be packed together in the infinite density of a single black hole.

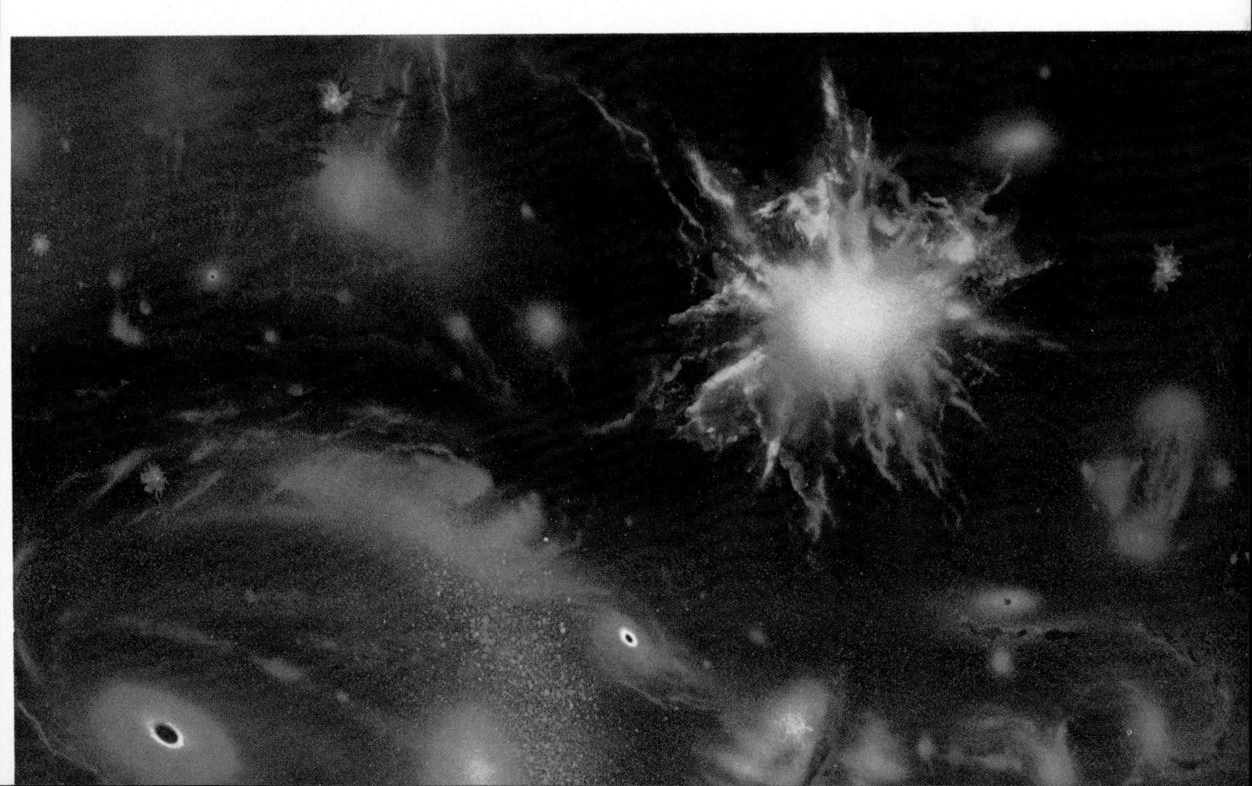

One project of particular interest is the study of isolated pairs of galaxies reported in 1975 by Edwin Turner of Caltech. These are cases where two galaxies are close together and no other bright galaxy is nearby. Pairs of this sort are so common that most of them could not be accidental encounters. Also, they are so close together that it is hard to believe they are flying apart. If they were, we would have to suppose that in the not-so-distant past a good many galaxies abruptly flew apart, which seems unreasonable. Consequently, there are enough safeguards on Turner's project to eliminate many questions raised by earlier attempts to weigh galaxies. Turner's data also indicate that galaxy masses are about 10 times the accepted value.

The ultimate study of mass would require an examination of the positions and motions of all the galaxies within, say, 500 million light-years of us. This would provide an enormous amount of information about how galaxies are distributed and how they move. A few years ago, such a project would have been impossible because the measurements would have taken so long. However, new techniques of light detection and data handling now make such a project much less forbidding. I am convinced that it will happen in the next five years and that it will cause a mini-revolution in the concept of galaxy clustering.

But even if we accept the missing mass idea as fairly reasonable, we are still a long way from determining how the universe will end. Cosmologists once thought that the mass in galaxies was 30 to 50 times too small to be able to stop the expansion of the universe. Current studies, such as Turner's, increase galaxy masses only about 10 times. Although it is hard to judge whether any additional mass will show up, there is certainly still lots of room for more undetected mass, for example, outside the halos that presently interest us.

Even though the answer is still out of reach, it is interesting to look at the two possibilities for the end of the universe. If the universe does not have escape velocity, the expansion will eventually stop and reverse, and the universe will begin to collapse. Galaxies will draw together, overlap, and break up, until the density becomes so great that stars and planets will begin to collide. These violent collisions will continue, sucking the bits and pieces of the universe into ever more massive lumps. After a long period of time, perhaps 50 billion years, the density will reach infinity.

Some physicists have considered a more modest version of this kind of violent collapse. When a star several times more massive than the Sun exhausts its nuclear fuel, Einstein's general relativity theory predicts that the star must either shed most of its mass or else collapse so completely that its density becomes infinite. Left behind is a black hole, with gravity so high that not even light can escape. The collapse of the universe would be similar, except that no black hole would be left behind because there would be no outside–all of space would be involved. What happens then is pure speculation. It might be the end of the universe or it might be only the end of a cycle, after which the universe would expand again in a new big bang, heralding the start of a fresh round of expansion and collapse.

In the other possibility, the universe will continue expanding forever. This does not mean that each bit of matter will end up remotely separated from every other piece. By and large, matter is gravitationally bound up in giant clumps, such as galaxies and clusters of galaxies. As the universe expands, these clumps remain intact while the space between grows ever greater. In our own neighborhood, the Milky Way and the Andromeda Nebula will continue to orbit around each other as the universe expands, but as they slowly lose energy, the orbit will gradually contract until the two galaxies merge. With still further loss of energy, this lump eventually will become so compact that much of it will collapse into a black hole. If our motion away from the Virgo cluster is less than escape velocity, as seems likely, the combined Milky Way and Andromeda Nebula–or their black hole remnant–eventually would drift into a much more massive black hole forming at the center of the Virgo cluster.

The violence of such collapses might fling a few stars out of galaxies to end up floating free in the ever-increasing empty spaces. Such a drifting star might never totally collapse, though it would eventually exhaust its nuclear fuel and become a burned-out remnant.

Either way–by contraction or expansion–matter will suffer a violent end. Inevitably, people have strong personal and philosophical opinions about which ending seems most reasonable. To those who argue that the universe will expand forever, an attractive feature is that the universe is not doomed to suffer a total collapse from which there is no return, if Einstein's general relativity theory is to be believed. Others fail to find this argument compelling, because much of the matter does come to a violent end.

The idea of a cyclically expanding and contracting universe seems particularly attractive to some astronomers because it points a way out of the awful problem of what came before the big bang. Perhaps the universe really is eternal, passing through endless cycles of expansion and contraction. There are, however, some details that do not fit. According to the theory, matter in the universe tends to clump together more and more as time goes on. This means that just after the big bang the distribution of matter must have been extremely uniform, with no more than a hint of irregularity. Otherwise, mathematics tells us, things would have clumped together much more by now than they actually have. So the cyclical model needs a way to even out an irregular universe in order to send it out uniform again. None of this can be predicted by Einstein's general relativity theory; perhaps some new, more fundamental theory will point to an answer.

I am suspicious of the arguments on both sides. The physical universe has provided many surprises and we must expect many more. Until we understand the universe and its parts much better than we do now, all pronouncements about its end are premature.

Disabling the White Death

By Robert H. March

Snow researchers have developed an early-warning system to lessen the destruction of avalanches

It begins on a seemingly tranquil, snow-covered mountain slope. Then suddenly, a low-frequency, barely audible "whump" signals a change. A crack appears near the top of the slope and a huge slab of snow begins to slide, gaining speed as it descends. By the time it has dropped 1,000 meters (3,000 feet), it has become a churning, cascading wall of snow moving at up to 300 kilometers (185 miles) per hour on a cushion of air and snow crystals.

From a distance, a slab avalanche, as this type of snowslide is known, may look no more threatening than a fluffy summer cloud. But it can pack the destructive energy of a small atomic bomb. Huge trees snap like matchsticks. Nothing man-made can stand before it. Houses and bridges are crushed by the blast of wind that precedes it down the slope, then buried under thousands of tons of snow. Even well clear of its path, its wind can hurl large trucks off a highway. In a minute or so the avalanche is over, its fury spent on the valley floor.

Centuries ago, Swiss villagers christened these terrifying snowslides the "white death," and fearful Alpine travelers blamed them on vengeful mountain spirits they called trolls. Until very recently, scientific understanding of what causes avalanches was hardly more satisfactory than this ancient superstition. But research on avalanches in the mountains of the Western United States has uncovered an important fact–a troll never strikes without warning. This clue has enabled the researchers, using modern scientific tools, to learn ways to predict and thereby better control the white death.

Their efforts come none too soon. In the early 1900s, the mountain ranges of the Western United States were sparsely populated. The handful of miners and trappers who wintered in the high country learned to shun the avalanche tracks–the steep slopes where avalanches repeatedly occur. But the explosive growth of skiing and other winter sports has brought to the hills an army of flatlanders who revel in the beauty of a virgin, snow-covered slope, but know little about the menace that lurks beneath its surface.

Even a tiny avalanche no more than a few centimeters thick can be lethal. An unwary skier, tripped up in the swirling snow, becomes an obstacle over which a drift forms. Snow is much stronger than it looks, and the buried skier is immobilized as surely as was Gulliver by the Lilliputians. If rescuers do not find him within an hour, he will almost certainly suffocate.

Protecting the unwary tourist has become a major burden of ski-resort operators and the federal and state agencies that manage outdoor recreation. As more and more areas are opened to winter sports, finding more reliable methods of avalanche prediction and control has become an urgent necessity. This responsibility rests with a small band of scientists, working closely with ski patrols and highway crews.

The author:
Robert H. March is a professor of physics at the University of Wisconsin, Madison. He wrote "A Subatomic Surprise" for the 1976 edition of *Science Year.*

Snow researcher Richard A. Sommerfeld counts himself lucky to be a member of this band. He has a unique opportunity to develop his scientific interests in the behavior of solid materials, of which snow is one, in some of the most spectacular surroundings in the world. A man who would never be comfortable in a routine laboratory setting, he is far more at home somewhere above the timber line, his instruments and field notebook in a backpack.

Sommerfeld works for the United States Forest Service out of the Rocky Mountain Forest and Range Experiment Station in Fort Collins, Colo. Educated as a geochemist, he is part of the Alpine Snow and Avalanche Project, an interdisciplinary team of five scientists and three technicians. The other project members are trained in meteorology and hydrology. They all work on many phases of snow research in addition to avalanches. It is a measure of the modest scale of snow science that the project is the largest of its kind in the United States. In fact, this field counts less than 200 specialists throughout the world. In this small, friendly discipline, communication is usually informal, and many of Sommerfeld's colleagues have discussed their latest work over gourmet dinners he will whip up if he is given half an excuse.

The scientific challenge faced by snow researchers is an old one in the applied physics of solid materials. Physicists try to explain nature through simple laws. Unfortunately, nature is rarely simple. It takes more variables to fully describe the snow that falls in a single storm than anyone can hope to measure and catalog. The applied physicist needs a fine sense for the judicious art of compromise. He must elimi-

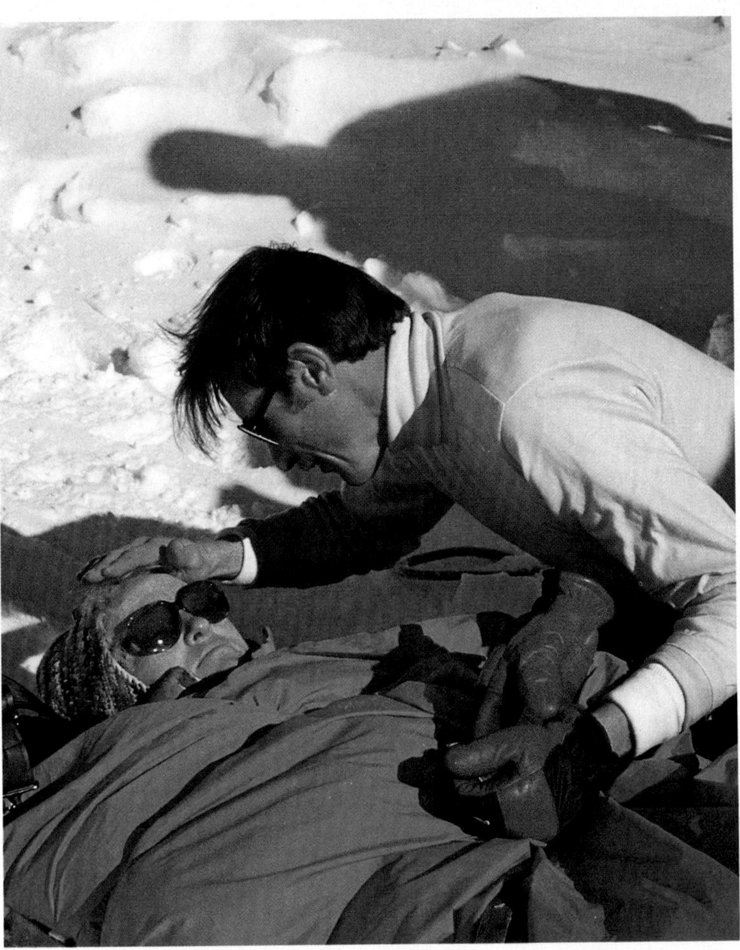

Ski-patrol members form two lines, *above,* and methodically probe an avalanche track during a training session. A "victim" receives aid, *left.*

nate or ignore enough variables to make the problem manageable, yet include enough to make accurate predictions.

The basic physics of an avalanche is simple enough. It is a contest between gravity, which tries to pull the snow down the slope, and the internal strength of the snow, which enables it to cling together and stay on the mountainside. When something happens to shift the balance in favor of gravity, the snow comes down. The worst avalanches take place on slopes of between 30 degrees and 45 degrees. Steeper slopes cannot build up a sufficient snow cover, and the snow on shallower ones rarely slides. Unfortunately, the most dangerous slopes are also those preferred by expert skiers, which is why it is often the best skiers who fall victim to avalanches.

The complex part of avalanche physics is understanding the snow. As anyone who has wielded a snow shovel can appreciate, snow comes in many forms–dry or wet, light or dense, fine-grained or in cottony

Fresh snow, *right,* forms a strong, lacy web of overlapping and interlocking filaments. As the snow ages, *below right,* the interlocking filaments become strong, stubby bridges between increasingly rounded crystals.

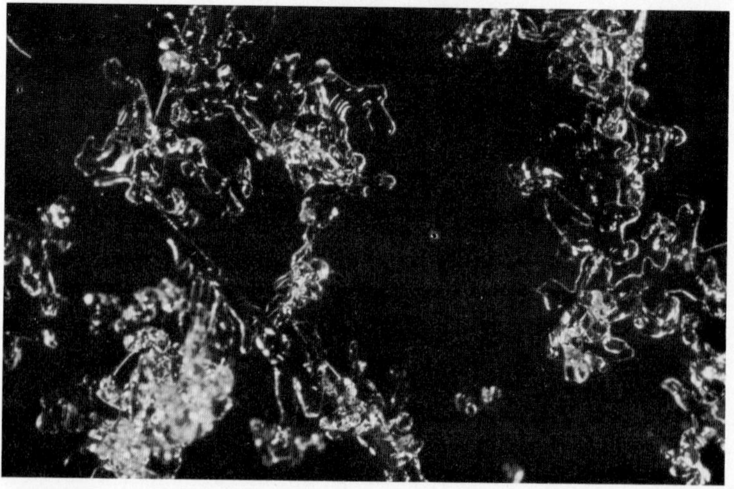

puffs. The storm that deposits it can range in intensity from a flurry to a blizzard. All these factors affect the strength and weight of the snow. And once it is on the ground, the story has just begun. Sun, wind, and the daily cycle of temperature changes transform and reshape it. Even deep within the snow, changes are taking place in the layers that turn out to be crucial in the formation of avalanches.

To the complexity of the snow, one must add the difficulty of studying it in place. Most avalanche tracks begin high on a mountain face, in terrain that is all but inaccessible in winter. A team like Sommerfeld's can watch only a handful of these tracks at a time, and each track breaks loose just a few times each winter. Finally, the very presence of observers moving about on a snowfield and probing it, can alter the delicate balance of forces that holds the avalanche back.

Avalanche research is so new that most of its pioneering scientists are still alive and active. It began in Switzerland in the middle 1930s with the work of Robert Haefeli, a soils expert who hoped to apply what he knew about landslides to avalanches. But snow is so different from soil that it soon became evident a new approach was needed. Civil engineers, geologists, meteorologists, and physicists joined in the task, launching the field's present interdisciplinary character.

A snow researcher uses a shear frame to measure the snow's shear resistance, or ability to withstand sliding apart in layers, *above left.* A spin tensile tester—a type of centrifuge—*left,* measures tensile resistance, the snow's ability to withstand being pulled apart.

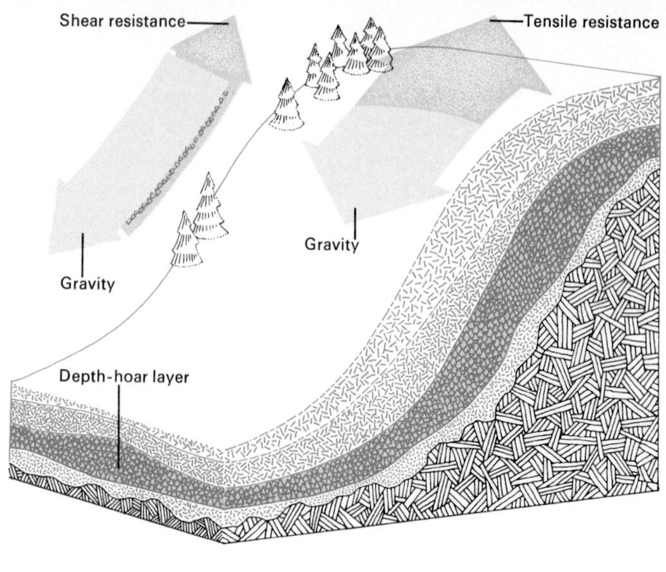

Gravity Wins a Tug of War
Snow's tensile resistance enables it to hang from the top of a slope (top) while its shear resistance enables layers to cling to each other and the mountainside. Gravity wins the tug of war (middle) when a weak, buried layer of highly transformed snow, called depth hoar, crumbles as the overlaying snow layers, or slab, pull apart along a tensile tear. The slab begins to slide atop the loose depth-hoar granules almost at once and crushes everything in its path until it finally comes to rest on the valley floor (bottom).

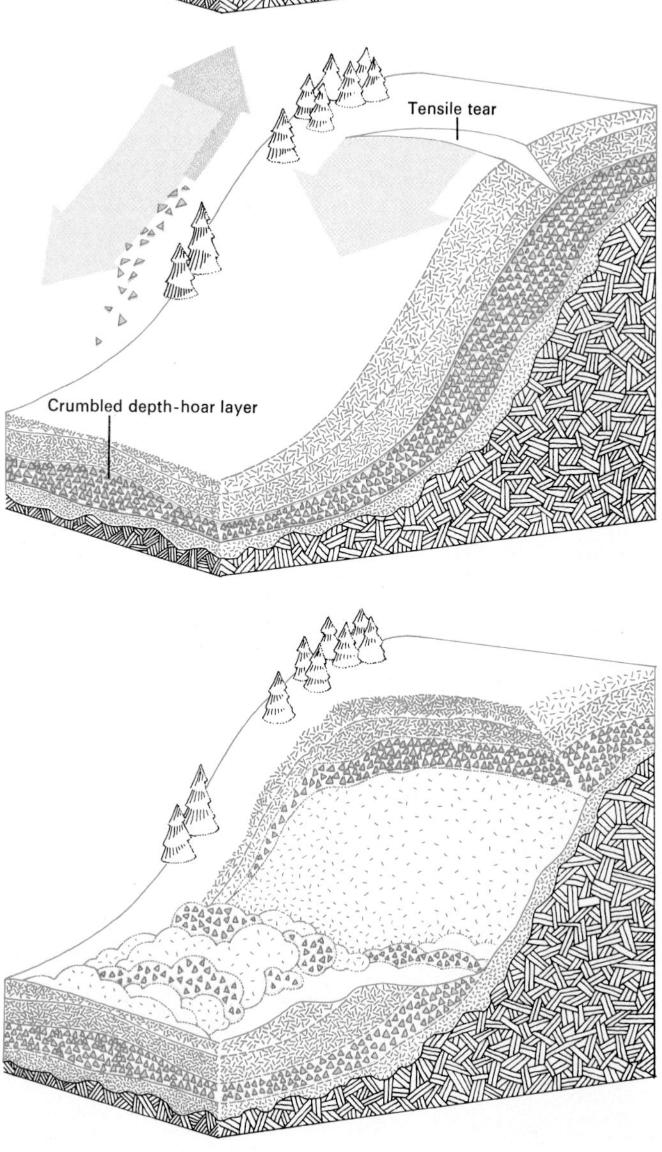

Shortly after World War II, the Forest Service opened a ski area at Alta in Utah's Wasatch Mountains and realized that the steep slopes and deep powder that attracted skiers frequently caused avalanches that closed the roads leading into the area and threatened skiers' lives. The service hired Montgomery M. Atwater, an expert deep-powder skier, as a snow ranger and hoped he could "do something about" the snowslides that threatened Alta. Atwater soon discovered how little was known about the phenomenon he was supposed to control. Although untrained in science–he originally wanted to be a writer–Atwater got some solid research underway. And he used his flair for communication to alert the public and the Forest Service to the need for a more systematic attack on the problem.

Research was launched at Fort Collins in 1962 by Mario Martinelli, who now directs the project on which Sommerfeld serves. Martinelli's responsibilities on the project now extend far beyond the basic snow research. He also offers land-use advice to keep property owners from building in avalanche tracks. And he supervises the operation of a warning network whose avalanche alerts are broadcast to people in Colorado's ski country.

Sommerfeld joined the Alpine Snow and Avalanche Project in 1967. "Nobody here was really trained for snow research," he says with a twinkle in his eye. "I guess we just drifted in."

Nonetheless, he had ample credentials for the job. While completing his doctoral thesis, he worked from 1961 to 1964 on the University of Washington's ice-island project, spending months at a stretch adrift on an ice floe in the Arctic Ocean. His next stop was a three-year stint at the University of California, Los Angeles, doing laboratory studies of the chemical and physical processes that form and transform minerals. But his Arctic experience had whetted his appetite for adventure, so when Martinelli's group advertised for a researcher, Sommerfeld leapt at the chance. "If you look at snow as a mineral," he explains, "then I didn't completely change fields. Only now I study changes that take weeks, rather than millions of years."

The changes Sommerfeld studies occur in the crystal texture of snow, which starts as the open, lacy pattern of filaments that we recognize as a snowflake. In new-fallen snow, these filaments overlap and interlock, binding the snow into a light but strong web. As the snow ages, unconnected sharp filaments gradually disappear and the snowflakes become more rounded and more compact. Interlocking filaments grow into strong bridges between crystals. Rounding occurs as water molecules pass off as vapor from sharp points of the crystal and condense as solids on rounded parts.

To understand avalanches, Sommerfeld and his colleagues also study how snow resists being torn apart. They distinguish between two forces. One force is tension, a simple, direct pull. The other force is shear, which makes one layer of a material slide over another. When a material pulls apart, it is said to fail in tension. If it slides apart in layers, it fails in shear. Both tension and shear resistance go hand in hand in snow, which generally has a layered structure resulting from successive snowfalls. In fighting gravity, snow's ability to resist tension enables it to literally hang from the top of a slope while its shear resistance enables layers to cling to one another and to the slope.

The key to a slab avalanche is the presence of a weak, buried layer that is ready to fail in shear. But the overlying snow, or slab, cannot break loose unless it fails in tension. Failure usually occurs at the top of the slope where the tension is greatest because the snow there must hold against the pull of all the snow downhill. The shear failure extends downhill from the tensile tear in a fraction of a second. So the entire slab is set in motion almost at once.

Sommerfeld believes that depth hoar, a form of granular snow, is the weak, buried layer responsible for most large slab avalanches. Depth hoar forms because the rocks of a mountain slope, buried under snowy blankets since early autumn, are usually warmer than the outside air. The heat stored in the rocks vaporizes part of the snow from the deep, densely packed layers. As the water vapor percolates upward through the snow, it condenses, releasing heat and causing other snow crystals to vaporize. The process repeats itself many times before the heat reaches the surface and escapes to the outside air.

When snow researcher Richard A. Sommerfeld squeezes a seemingly solid chunk of depth hoar, it collapses into unattached granules.

Depth hoar forms because vaporization and condensation do not happen at the same spots on snow crystals. Water vaporizes from the top of a crystal and condenses on the relatively colder bases of those above it. Unlike the process that leads to round granules and strong bridges between them, this percolation process creates stepped, pyramid-shaped granules that do not cling well to one another. Because the temperature difference between layers is greatest near the ground, percolation usually produces hoar granules in the deepest, therefore the oldest, snow layers.

A chunk of depth hoar seems solid enough. Yet at the slightest pressure, it collapses into a loose mass of unattached, compact crystals. Once a depth-hoar layer fails in shear, its crystals act like rough ball bearings for the slab of snow above it. Initially, the faster the slab moves, the less resistance the hoar crystals offer to its motion. This makes hoar crystals an ideal lubricant for a slab avalanche.

To find the weak underlying hoar layer in an avalanche track, researchers must rely on volunteer observers on the spot, most of them members of ski patrols. However willing they might be, these volunteers have neither the time nor the training to measure the strength of snow layers daily in their check for potential avalanches. So Sommerfeld and glaciologist Edward R. La Chappelle, now at the University of Washington in Seattle, devised a simple classification scheme that enables a ski patroller to estimate the strength of snow simply by looking at it through a magnifying glass and weighing a small sample. Their system can be learned quickly and gives estimates of snow strength that are within about 20 per cent of the actual strength.

Brewing a Weak, Buried Layer

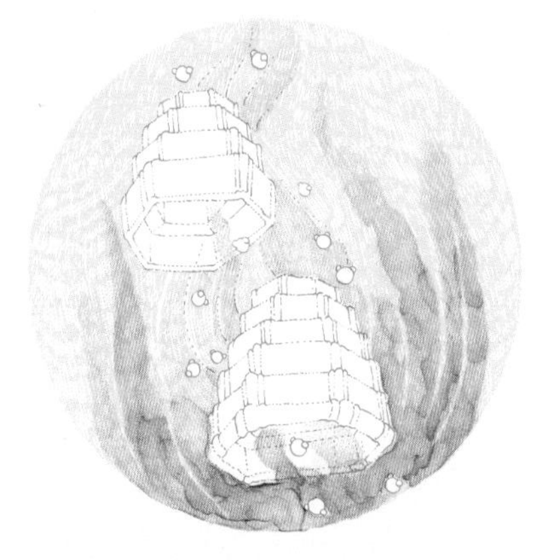

Warmed by rocks, water vaporizes from deep layers of snow, condenses, and causes more vaporization, *right.* At each step, water vapor from the top of a granule condenses on the relatively colder bases of granules above it, *above.* This changes snow into stepped, pyramidal granules of depth hoar, *below.*

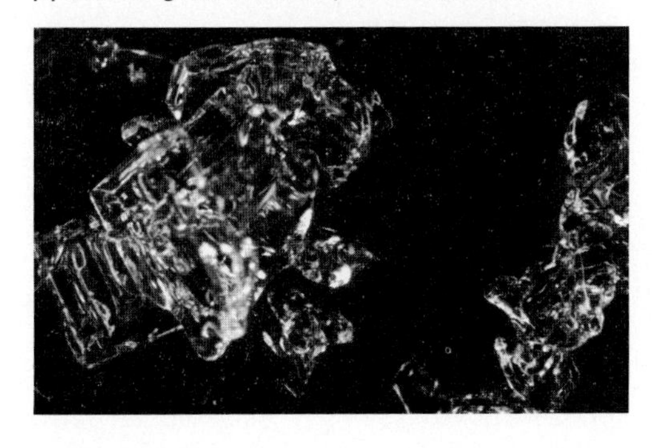

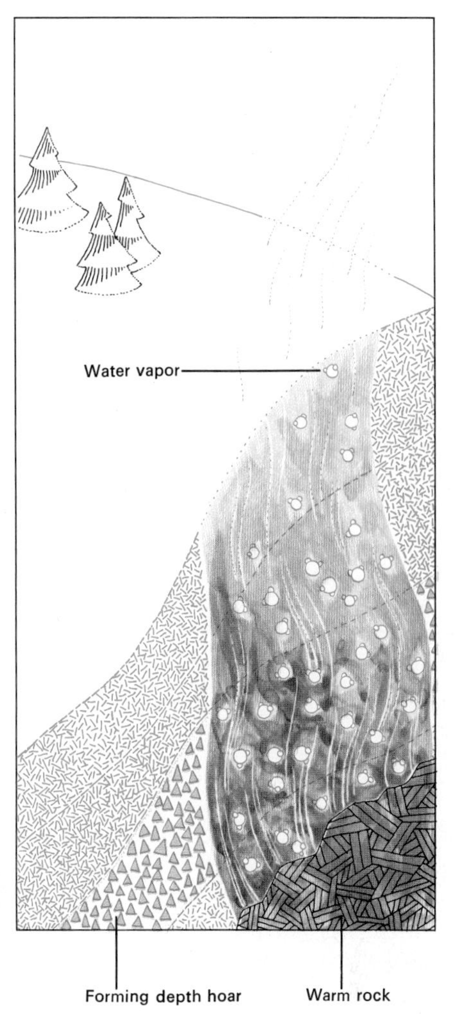

Measurements of actual strength in different classes of snow take much more time and effort. Researchers must dig down through as much as 3 meters (10 feet) of overlying snow, then collect the samples in standardized metal cylinders. They rush the snow-filled cylinders to a cold laboratory–an unheated shack close to the snowfield–where they quickly weigh the samples and spin them on a centrifuge before the snow changes from its original condition. The speed of rotation at which centrifugal force causes the snow to fly apart gives a measure of its tensile strength.

Sommerfeld relies to some extent on the talents of yet another group of volunteers in the strength-measuring phase of the project. A number of college students have found avalanche research an exciting way to gain practical experience while earning college credits. In early 1976, 20-year-old Deborah Levine, a geology major from Colgate University in Hamilton, N.Y., was busy collecting snow samples and

A student volunteer digs snow samples, *right,* outside a research laboratory which is almost lost (center right) in the white vastness of Berthoud Pass in Colorado, *above.*

testing them in the cold lab. An expert skier who is at home in the outdoors, Levine finds the rigors of high-altitude snow research a bonus rather than a hardship. She hopes eventually to go into oil exploration, and her work in the Rocky Mountains has proved that she has a natural inclination toward field research.

Thanks to his own efforts and those of volunteers like Levine, Sommerfeld has assembled a detailed picture of the properties of depth hoar and how it changes. Sommerfeld guessed that a crumbling of the layer into unattached crystals might well produce some characteristic sound that could be picked up by a buried microphone. It also seemed unlikely to him that the entire depth-hoar layer could collapse at once. He reasoned that wiring an avalanche track for sound might provide advance notice of its intention to cut loose.

To test this idea, Sommerfeld buried about a dozen microphones in avalanche tracks above Berthoud Pass, about 90 kilometers (55 miles)

In the cold laboratory, Sommerfeld monitors his troll counter, an early warning system linked to microphones that are buried in snow on an avalanche track at Berthoud Pass.

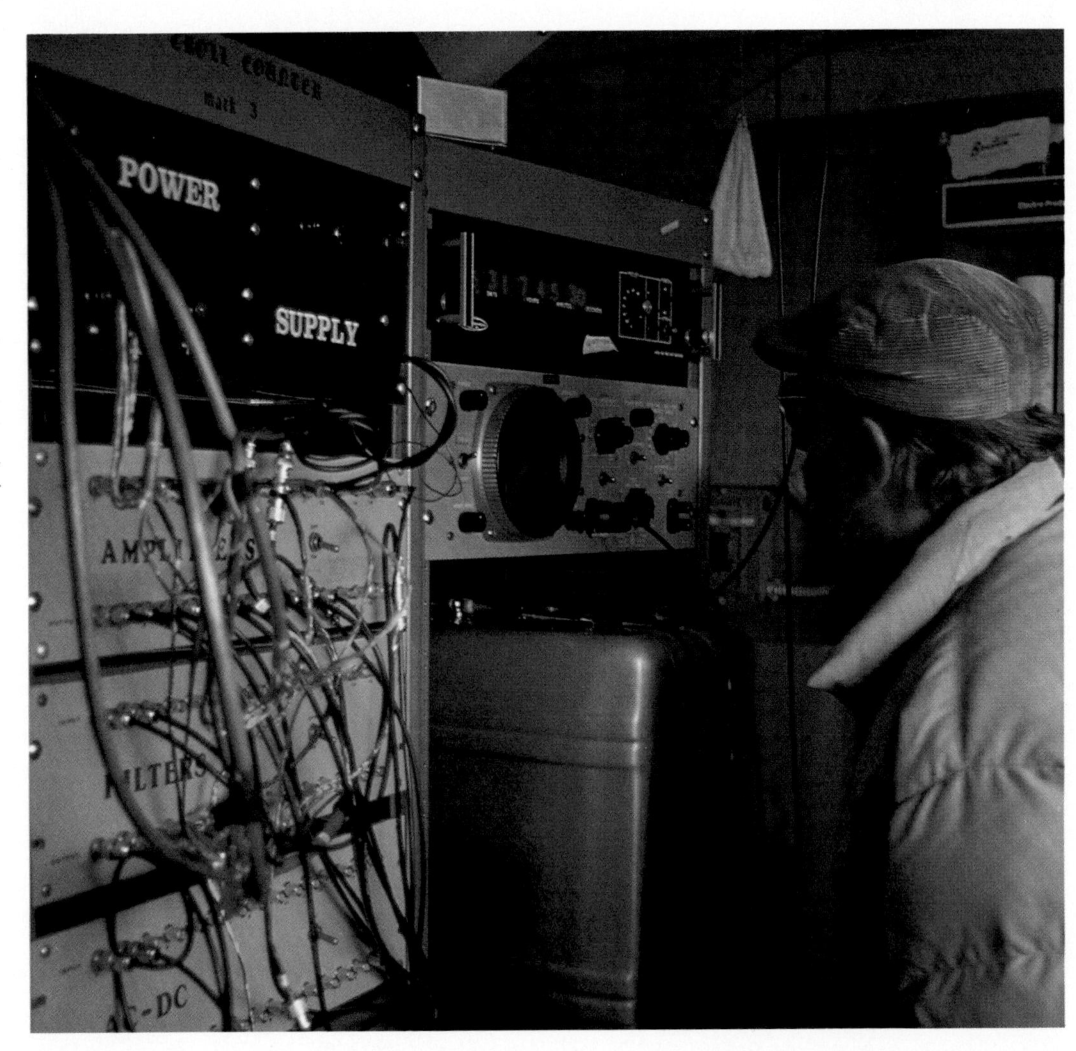

A ski patroller lights an explosive charge he will throw downhill to trigger an avalanche. This is an effective, but increasingly costly, control method.

west of Denver. The microphones selected work best at low frequencies near the bottom of the range of hearing, around 20 hertz. He then hooked these microphones to a multitrack tape recorder. Remembering the old Alpine superstition, Sommerfeld dubbed his listening device a "troll counter."

Two winters of testing have shown that Sommerfeld's intuition was good. The warning rumble of an avalanche is a series of short bursts of sound that is too faint for the human ear, but can be easily distinguished from normal background noise. These bursts continue for 16 to 30 hours and then subside. Examining the recorded evidence on the wired tracks, Sommerfeld reports that no avalanche has occurred without giving such a warning, and none of the warnings proved to be false alarms.

The troll counter cannot predict the exact moment when an avalanche will slide, but it does not have to. Once a crumbling depth-hoar layer signals an unstable track, a small explosive charge set off near the top of the slope will bring it down at a time when no one is around to be caught in it.

Explosives have long been the accepted method of avalanche control. The most experienced members of the ski patrol, specially trained in avalanche safety and explosives handling, set and detonate the charges. Without sure knowledge of which slopes are ripe to slide, however, it is a tricky, expensive, and dangerous business. The ski patrol must blast every track that might threaten ski trails, and blast it often. If no avalanche comes down, the track is presumed stable–but that is no guarantee that it will not come down of its own accord before the next blast. Moreover, this sort of concentrated attention can only be lavished on the busiest downhill ski runs. Cross-country skiers, whose trails range far and wide through the high country, must learn to spot and avoid potential avalanche tracks on their own.

The troll counter may change this picture. But first this bulky, one-of-a-kind laboratory instrument must be developed into a reliable, cheap, and compact alarm. The current microcomputer revolution, which has created inexpensive computers on a single tiny chip of silicon, may hold the answer (see INVASION OF THE MICROCOMPUTERS). A microphone, microcomputer, transmitter, and battery could be built into a small troll counter that could stand guard at even remote avalanche tracks, silently listening for the sound of a crumbling hoar layer, then quickly relaying an alert to a control center.

Perfecting a self-contained troll counter is a challenge for engineers and technicians who like to work indoors. Dick Sommerfeld will doubtless move on to new problems, probably somehow contriving always to remain in his beloved outdoors. And he will still be holding barbecue feasts that attract mountain people and old friends from throughout the continent. And perhaps Debby Levine, as she skims through deep-powder snow on some future winter holiday, will remember that she once played a small part in making her trail safer.

Our Chemical Ancestors

By Michael J. Dowler

Clues from space, from living cells, and from earth's early rocks suggest how simple chemicals may have sired the first forms of life

This region of Baja California in Mexico looks as desolate as the entire earth was before life appeared more than 2 billion years ago.

It was the poisoned planet's last gasp. In desperation, its inhabitants were about to launch a spaceship to another part of the galaxy. Ironically, the technology that caused their demise made possible this attempt to preserve the one thing they felt certain was sacred–life itself. Unable to escape themselves, they had chosen the essence of life, an adaptable and hardy bacterial spore that could grow and evolve to generate more complex forms of life. Their astronomers had found a distant site, a sort of womb in which to plant these bacteria. Forty light-years away, Star 670911 had a system of planets that included one with an environment where the bacteria should flourish.

A future tale of our planet earth? Not if a proposal by British biologist Francis H. Crick and biochemist Leslie E. Orgel is correct. This would have been a prehistoric tale in which the earth was the spaceship's target. Their concept–the Directed Panspermia theory–suggests that the first life on earth may have been sent here deliberately by intelligent beings somewhere else in the universe.

Crick, who is with the Medical Research Council at Cambridge University in England, shared the 1962 Nobel prize for physiology or medicine for his role in determining the double-stranded helical structure of deoxyribonucleic acid (DNA), the molecule that transmits genetic information from one generation of living cells to the next. Orgel is a researcher at the Salk Institute for Biological Studies in San Diego. Crick and Orgel note that the Directed Panspermia theory has little experimental support. But they published the idea in 1973 because they believed it should not be rejected without serious consideration. Thus, it is one possible, though not necessarily probable, answer offered by scientists to the question: How did life originate on earth?

The best-known answer in the Western world is the Judeo-Christian story of creation as outlined in the first chapter of Genesis in the Bible. Other religions also teach that a divine being created life. As symbolic tales, such beliefs are not in direct conflict with scientific theories. Historically, both theological and scientific explanations have coexisted. However, because they attempt to account for all the known facts, scientific explanations compete with one another and some fail as scientific knowledge grows.

While working in Orgel's laboratory in the early 1970s, I came to realize that a more fundamental question than how life started had to be addressed first. It is: What is life? We think we can tell if an object is alive by looking at it. But when we try to list simple criteria to precisely define life, we run into problems. For example, most biologists would list at least growth, reproduction, response to stimuli, and metabolism (the conversion of food to energy and biological building blocks). However, there are living things that do not meet all of these criteria. For example, mules cannot reproduce. Such things as fire appear to meet all the criteria, yet we know that fire is not alive.

The author:
Michael J. Dowler is an associate professor at San Diego State University where he teaches courses on the origins of life.

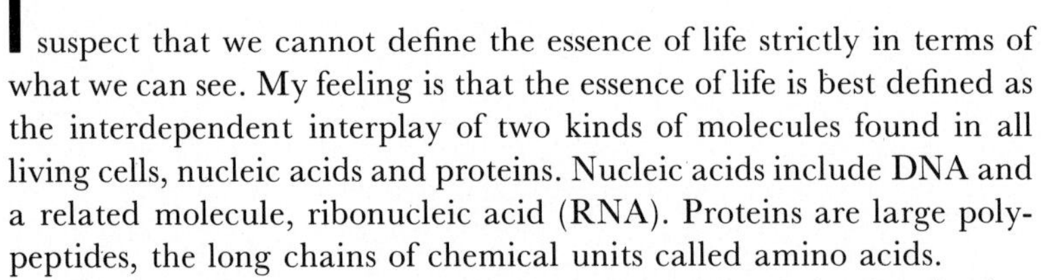

I suspect that we cannot define the essence of life strictly in terms of what we can see. My feeling is that the essence of life is best defined as the interdependent interplay of two kinds of molecules found in all living cells, nucleic acids and proteins. Nucleic acids include DNA and a related molecule, ribonucleic acid (RNA). Proteins are large polypeptides, the long chains of chemical units called amino acids.

In living cells, the nucleic acid-protein interaction is fundamental to growth and reproduction. DNA and some forms of RNA store information in their spiral-staircase structures in the form of codes based on the alternating sequence of only five chemical units, the organic bases. Other RNA molecules transfer these codes to places in the cell where amino acids are linked together in the proper sequence to form all the different proteins that become part of the cell's structure and play key roles in many of its processes. This basic understanding of how life goes on helps us to understand what life is, but it says very little about how life may have started.

Crick and Orgel are not the first to suggest that life could have reached the earth as a sort of "infection." Swedish chemist Svante A.

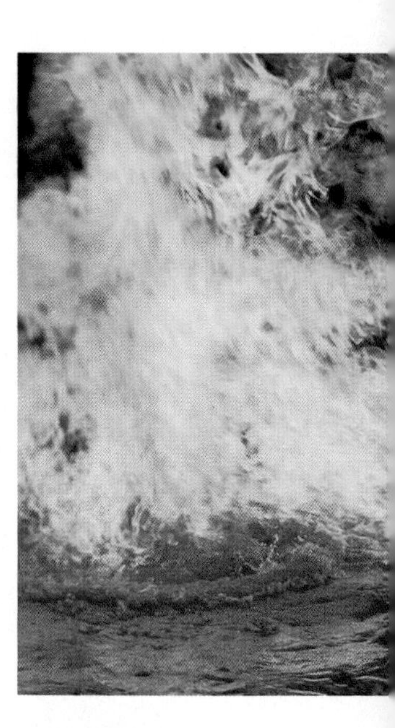

Most plants and animals, including people, *left,* meet all the criteria on a biologist's check list for life—growth, reproduction, response to stimuli, and conversion of food to energy and usable molecules. However, although a mule, *center,* does not reproduce, it is alive, while fire, *right,* seeming to meet all the criteria, is not.

Arrhenius proposed the Panspermia theory in 1908. He held that life on earth arose from spores accidentally pushed here by the pressure of light from a star in another solar system. Most scientists, however, believe that no bacteria could survive the journey without planned direction and protection.

Arrhenius and others were trying to find a better answer than the discredited theory of spontaneous generation, which held that life arose from decaying organic matter. For example, maggots were believed to come from decaying meat, and mice were thought to issue spontaneously from piles of dirty rags in a cellar. The belief was so strong that Belgian chemist Jan Baptista van Helmont in the early 1600s published a "recipe" for making mice from wheat and sweaty underwear. The theory was first questioned when Italian biologist Francesco Redi showed in 1668 that maggots did not arise spontaneously from decaying meat, but from previously deposited fly eggs. But when Dutch biologist Anton van Leeuwenhoek in the late 1600s saw microorganisms in nearly everything he examined with his crude microscope, spontaneous generation again found favor as the best explanation for how they formed. French bacteriologist Louis Pasteur ended the controversy in 1860 with his brilliant experiments showing that microorganisms grew in an organic broth only after contamination by the air. In other words, living things came from other living things like them, and not from dead matter.

In the early 1920s, Russian biochemist Alexander I. Oparin formulated a new idea. At about the same time, British biologist J. B. S. Haldane independently developed the same theory, now known as chemical evolution. They held that living things developed from the natural interaction and evolution of chemicals in the earth's early

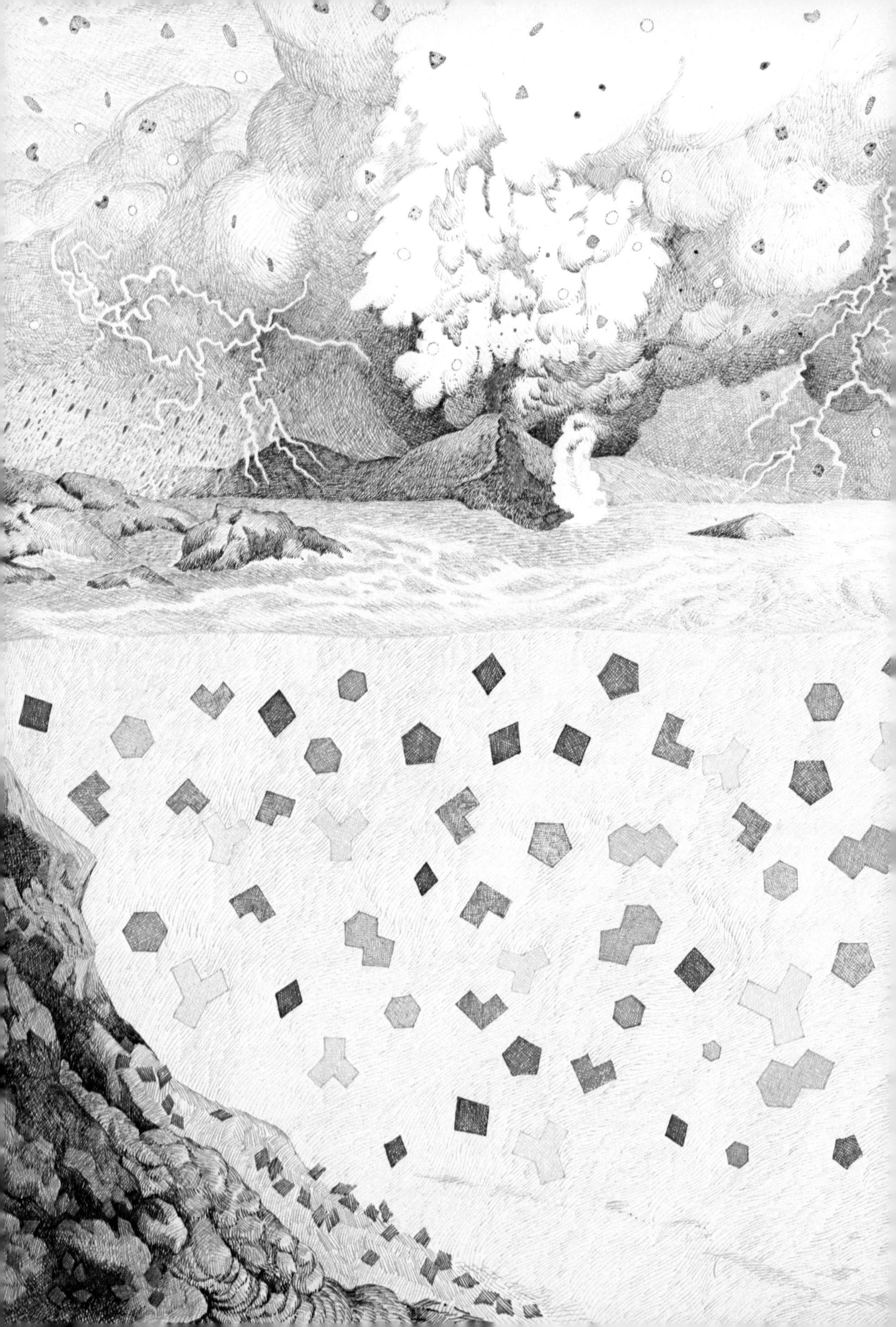

Our Chemical Ancestors

A primitive earth atmosphere of ammonia, methane, water vapor, and other gases (excluding oxygen) was likely the fertile fog in which life began. Lightning, sunlight, and other forms of energy sparked the formation of such active intermediate molecules as formaldehyde. These collected in oceans and reacted further to form the basic building blocks—amino acids, organic bases, and the sugars ribose and deoxyribose—of important molecules in living cells.

Elements

● Hydrogen
(C) Carbon
(O) Oxygen
(N) Nitrogen
(S) Sulfur
(P) Phosphorus

Primitive Atmospheric Molecules

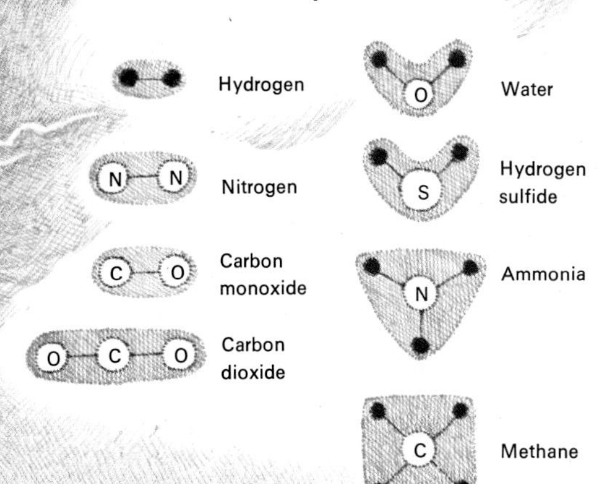

Active Intermediate Molecules

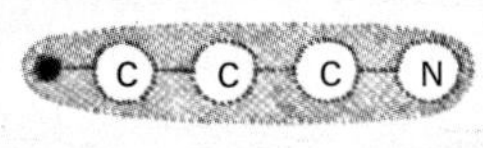

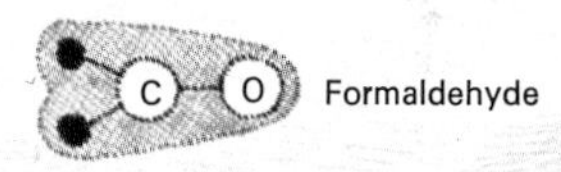

Molecular Building Blocks of Life

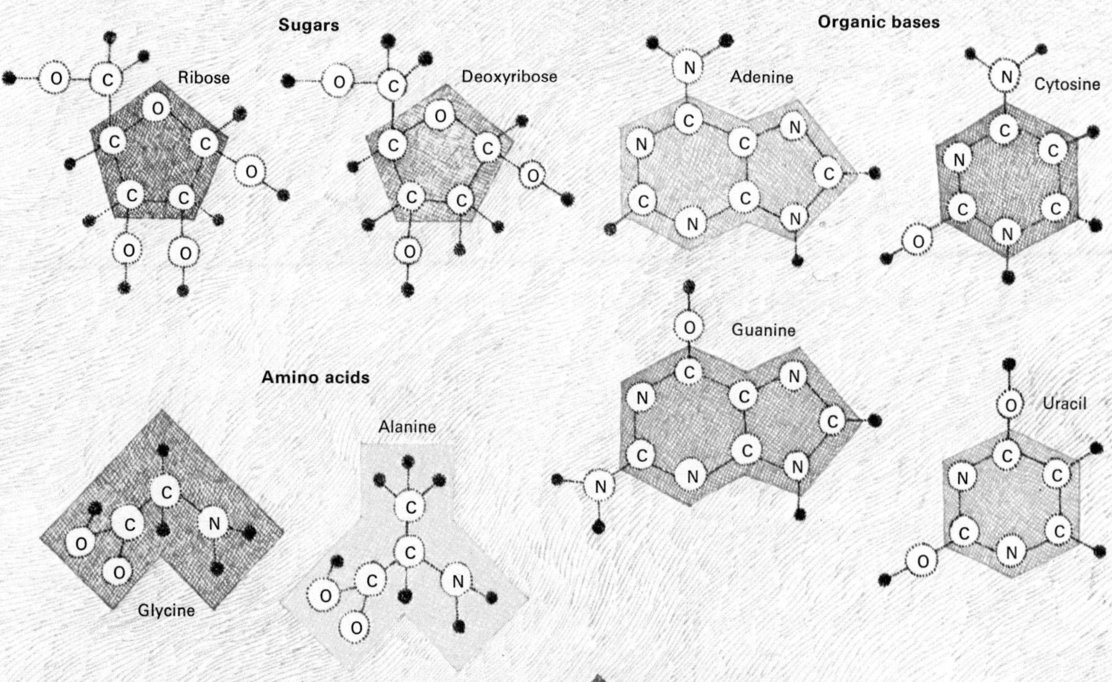

atmosphere and oceans. Energy in the environment, such as lightning, sunlight, and volcanoes, provided the driving force for the chemicals to react and form more complex molecules. These molecules then interacted and organized into still more complex molecules including nucleic acids and proteins. Their interaction is, by our definition, life.

The Oparin-Haldane theory of the origin of life from inorganic matter is, in fact, a theory of spontaneous generation. Pasteur had disproved only the spontaneous generation of whole organisms during a short time span. Chemical evolution involves much more time. From radioactive decay dating measurements and other evidence, geologists believe that the prebiotic, or before-life, earth lasted between hundreds of millions and billions of years, an enormous length of time.

If we could travel back more than 3 billion years in a time machine to the prebiotic earth, we would probably find a landscape as desolate as the Southwestern United States is today. Of course, there would be no life–not even the simplest microscopic animals or plants. Volcanoes would probably be erupting, filling the sky with dust. Because sunlight would scatter off the particles floating in the smog-filled atmosphere, sunrises and sunsets would be spectacularly colorful. Oceans and lakes would appear brown, colored by tarry chemicals washed from the sky by the rains. Quite possibly, oil slicks would cover large portions of the ocean surfaces.

Like astronauts exploring in space, we would have to carry a supply of oxygen in our time machine because the prebiotic atmosphere would not have any. We have evidence that the earth's early atmosphere probably consisted mostly of some combination of hydrogen (H_2), water (H_2O), nitrogen (N_2), ammonia (NH_3), methane (CH_4), carbon monoxide (CO), and carbon dioxide (CO_2). We have several reasons for believing there was little or no free atmospheric oxygen (O_2). Hydrogen is the dominant element in our universe, and therefore probably dominated the primitive atmosphere, forming water with whatever oxygen was present. Also, compounds rich in oxygen are rarely found in very old rocks, and other planets thought to have been formed in the same way as earth lack free atmospheric oxygen. Most important, oxygen is not one of the gases spewed out during volcanic eruptions, which is how we believe the earth's prebiotic atmosphere most likely originated.

Sunlight probably directly produced the oxygen that is so abundant now by splitting water molecules into oxygen and hydrogen in the upper atmosphere and, indirectly, by photosynthesis–the food-making process of plants that produces oxygen as a by-product. Calculations indicate that photosynthesis was the major source of oxygen, hence this gas became an abundant part of our atmosphere only after the origin of plant life.

Unlike some theories, chemical evolution can be partly tested in the laboratory. Chemist Stanley L. Miller, while a young graduate student at the University of Chicago, performed the first such experiment

Brown solids and the active intermediates hydrogen cyanide and cyanoacetylene formed in this glass vessel when electric discharges passed through a mixture of ammonia and methane gas. The basic building blocks of life's most important molecules are made in such experiments that simulate the primitive earth's atmosphere and its oceans.

in 1953. He passed a lightninglike spark through a mixture of water, ammonia, methane, and hydrogen continuously for a week, then examined the compounds that formed. The results were astounding. Although relatively simple, many of the compounds were important constituents of living things. In particular, Miller's experiment produced a large number of the amino acids that are important to life. By testing the mixture of gases, Miller showed later that hydrogen cyanide (HCN) and formaldehyde (H_2CO) form first. Then they react to form the amino acids.

Miller's work launched a new experimental field, prebiotic chemistry. Today, in a handful of chemical laboratories throughout the world, investigators are amassing evidence to check the theory of chemical evolution. So far, most of the 20 amino acids known to be essential to human life have been formed from many different oxygenless mixtures of gases in reactions driven by a variety of energy sources, including electric discharges, ultraviolet light, and heat. Researchers have also synthesized under possible prebiotic conditions the building blocks of nucleic acids–the sugars ribose and deoxyribose and the organic bases adenine, cytosine, guanine, and uracil. The sugars form when formaldehyde and limestone, which was probably abundant along with phosphates in rocks, are warmed in water. Researchers have synthesized the organic bases, which together with sugars and phosphates make up nucleic acids, from hydrogen cyanide and other prebiotic molecules. And they have made vitamins and other small biological molecules.

Unlikely Underwater Weddings
The building blocks cytosine, phosphoric acid, and ribose must give off a molecule of water as each bond of a ribonucleic acid unit is formed. This would have been an almost impossible event in the prebiotic oceans. The amino acids alanine and glycine, *opposite page,* faced the same problem in linking up to become polypeptides.

From this work and from geologic evidence we have pieced together a reasonable story of how life may have originated. As the early earth cooled, volcanoes spewed out gases that formed its early, oxygen-free atmosphere. In time—perhaps over millions of years—energy from such sources as lightning, hot lava, thermal springs, or ultraviolet sunlight transformed the simple gases into the water-soluble hydrogen cyanide, formaldehyde, and other prebiotic chemicals. Dissolved in rain, these were washed into the oceans and lakes, where they mixed and reacted to form the biological building blocks—amino acids, sugars, and organic bases.

Thus far, the story of chemical evolution is not only possible but also inevitable, according to the many experiments. This view was bolstered from an unexpected source in the late 1960s when radio-astronomers discovered a series of organic molecules in interstellar space. Besides hydrogen, water, and carbon dioxide, these scientists found many molecules, including formaldehyde, hydrogen cyanide, and cyanoacetylene, another biochemical molecule that forms from methane and nitrogen. So, the first step on the road to life is neither very difficult nor very rare. In fact, the universe appears to be full of the necessary complex organic molecules.

Unfortunately, the second step is much more difficult. Scientists are still puzzling over how the simple building blocks may have linked together in the prebiotic oceans and lakes to form the long chainlike molecules of life. Not only must energy be available to power such reactions, but also, as each link is forged, a water molecule must be

driven off. For example, when amino acids join to make polypeptides, a water molecule forms. Likewise, when a deoxyribose or ribose sugar links to an organic base, a molecule of water forms. And another forms when this unit bonds to a phosphate molecule to form a nucleotide, the basic unit of nucleic acids. These reactions are called dehydration reactions. They would be as likely to happen spontaneously in a prebiotic ocean as an apple would be to dry out in a bucket of water.

It would seem that as soon as the abundant building blocks began to link together in prebiotic lakes and oceans, water molecules would react with the short chains and break them apart. Thus, the water that is so essential for mixing the units also acts as a powerful barrier to their forming the more complex molecules of life. This water problem represents a major stumbling block for prebiotic chemistry researchers. Some have chosen to step over it by proposing unique environments that favored dehydration reactions on the prebiotic earth. Others have suggested chemical paths by which proteins and nucleic acids might have formed in water without dehydration reactions.

In October 1975, for example, chemist Clifford N. Matthews and his co-workers at the University of Illinois at Chicago Circle suggested a scheme to form polypeptides from an organic compound called poly-alpha-cyanoglycine. Their approach by-passes the dehydration problem, but it runs into others. Poly-alpha-cyanoglycine can be made from hydrogen cyanide, but no one has yet shown in experiments how it might have been formed in prebiotic times. In addition, these polypeptides will break up unless they quickly find a water-free haven.

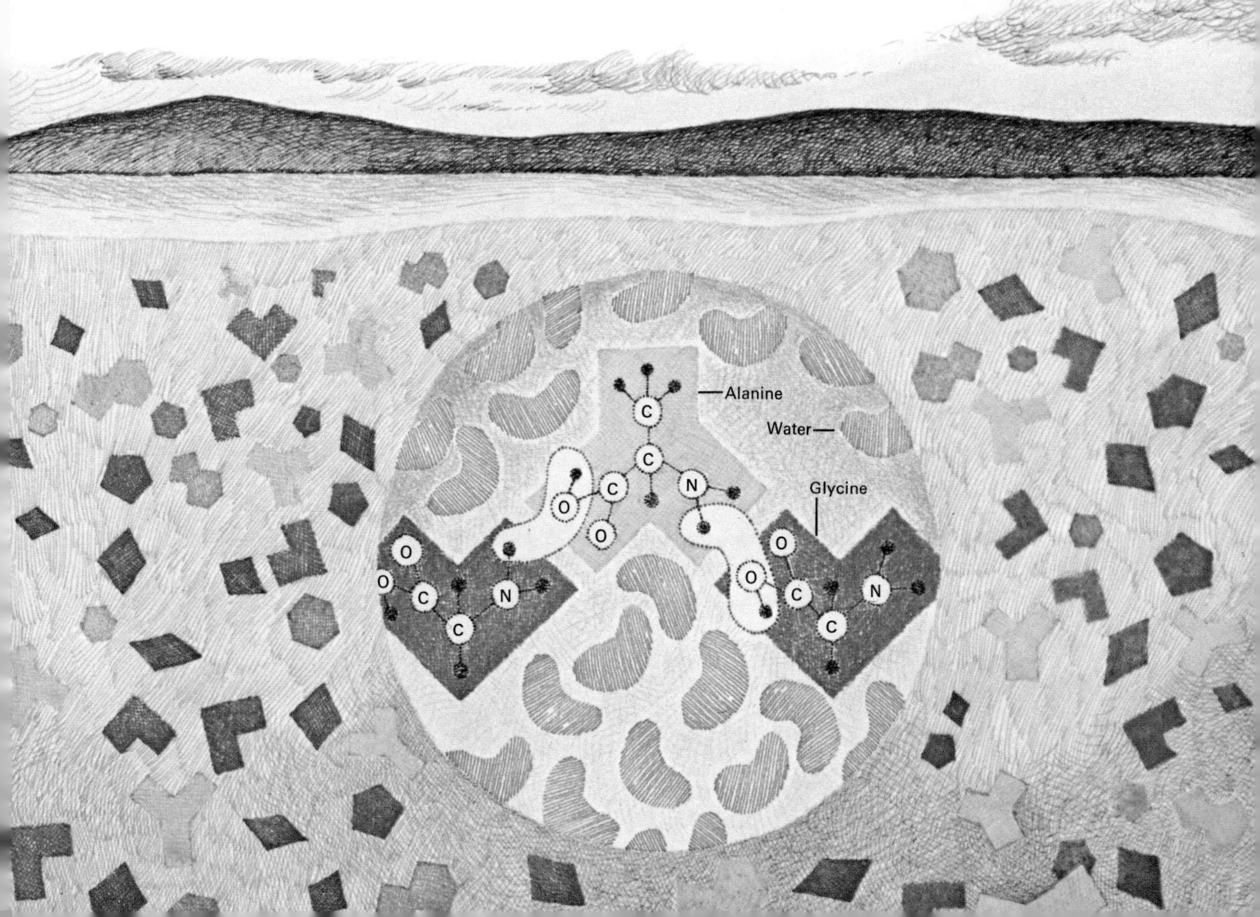

Several environments might have provided the proper conditions for the basic molecules to give up water and join together to form the more complex molecules of life. Certain clays near undersea rift valleys, *above,* might have offered suitable surfaces for the reactions. Deserts, *below,* and ocean tidepools, *right,* would have alternately been hot and dry and cold and wet. High temperatures near volcanoes, *far right,* could have stimulated the hookup of hot, dry amino acids.

Polypeptides also break up into amino acids in our body cells. But they are built at a faster rate, powered by the energy-providing molecule adenosine triphosphate (ATP). Prior to polypeptide assembly, ATP reacts with the amino acids to form active compounds called amino acid adenylates. These do not lose energy because they are bound in the crevices of large enzyme molecules that protect them until they transfer to the growing polypeptide chain and quickly become part of it. But the prebiotic ocean did not contain such complex protective molecules, themselves proteins, so most chemists believed that a process like body chemistry was impossible.

But in 1970, Israeli chemist Aharon K. Katchalsky and his co-workers at the Weizmann Institute of Science in Rehovot made a polypeptide with essentially the same chemistry that takes place in the human body. Instead of enzymes, two types of common clay, a zeolite and a montmorillonite, provided the favorable environment. These clays occur in abundance on the ocean floor. Katchalsky showed that the zeolites provide a water-free haven for amino acids to react with ATP to form amino acid adenylates. These active amino acids then form polypeptides on the surface of montmorillonite clay particles. Katchalsky's brilliant work led to the hope that nucleic acid synthesis might have occurred in a similar way. Unfortunately, before he could determine the exact mechanism involved in these reactions, he was tragically murdered in 1972.

Biochemist Sidney W. Fox of the University of Miami in Coral Gables, Fla., suggested in 1960 that the dehydration problem could be overcome by heating a mixture of dry amino acids at a temperature of 170°C (340°F.). Such conditions may have existed near volcanic lava flows, where high temperatures would have vaporized all the water from a prebiotic pool and forced the synthesis of polypeptides from amino acids. Using these high temperatures in the laboratory, Fox and his co-workers have achieved some stunning results.

The heating produces polypeptides that Fox calls proteinoids, and which are chemically similar to present biological proteins. In addition, the proteinoids, when they are boiled in water and then cooled, appear to aggregate into cell-like spherical structures that Fox calls microspheres. Microspheres look like simple bacteria and seem to have many amazing cell-like attributes. For example, photomicrographs show that they have a double-layered shell, and they appear to grow branches and even reproduce by dividing in half as do present-day microorganisms. In 1975, Fox and his colleague Ferencz Denes announced that certain microspheres can make ATP from related low-energy chemicals, phosphate, and sunlight in a primitive kind of photosynthetic process.

Although Fox is admired for his innovative and pioneering research, many scientists claim that the extremely high temperatures and dry conditions needed to create microspheres would rarely be found on the early earth. They believe that life arose on earth through the action of widespread conditions rather than from a few isolated events. Fox's work is also controversial because it is not clear how cells evolved to make nucleic acids and proteins in water if they originated as microspheres on hot, dry land near lava flows.

Orgel and his Salk Institute co-workers have been suggesting since 1970 that a form of wet-dry cycling may have played an important role in forming simple compounds that resemble both protein and nucleic acids. The scientists evaporated the water from a solution of the chemicals that they believed prebiotic oceans and lakes contained, then gently heated the solid remains at about 65°C (150°F.). Their method yielded both polypeptides and chains of nucleotides. "An alternation of hot-dry and cold-wet conditions occurs in annual summer-winter cycles and daily day-night cycles in many deserts," Orgel wrote in 1975. "In coastal areas, tide pools that dry out between tides could also have provided the required alternation of conditions." His group measured temperatures of sand in deserts. The temperatures ranged from 65° to 90°C (150° to 195°F.). Orgel's theory appeals to many researchers because the necessary cyclic hot-dry and cold-wet conditions for forming both proteins and nucleic acids would have been widespread on the early earth.

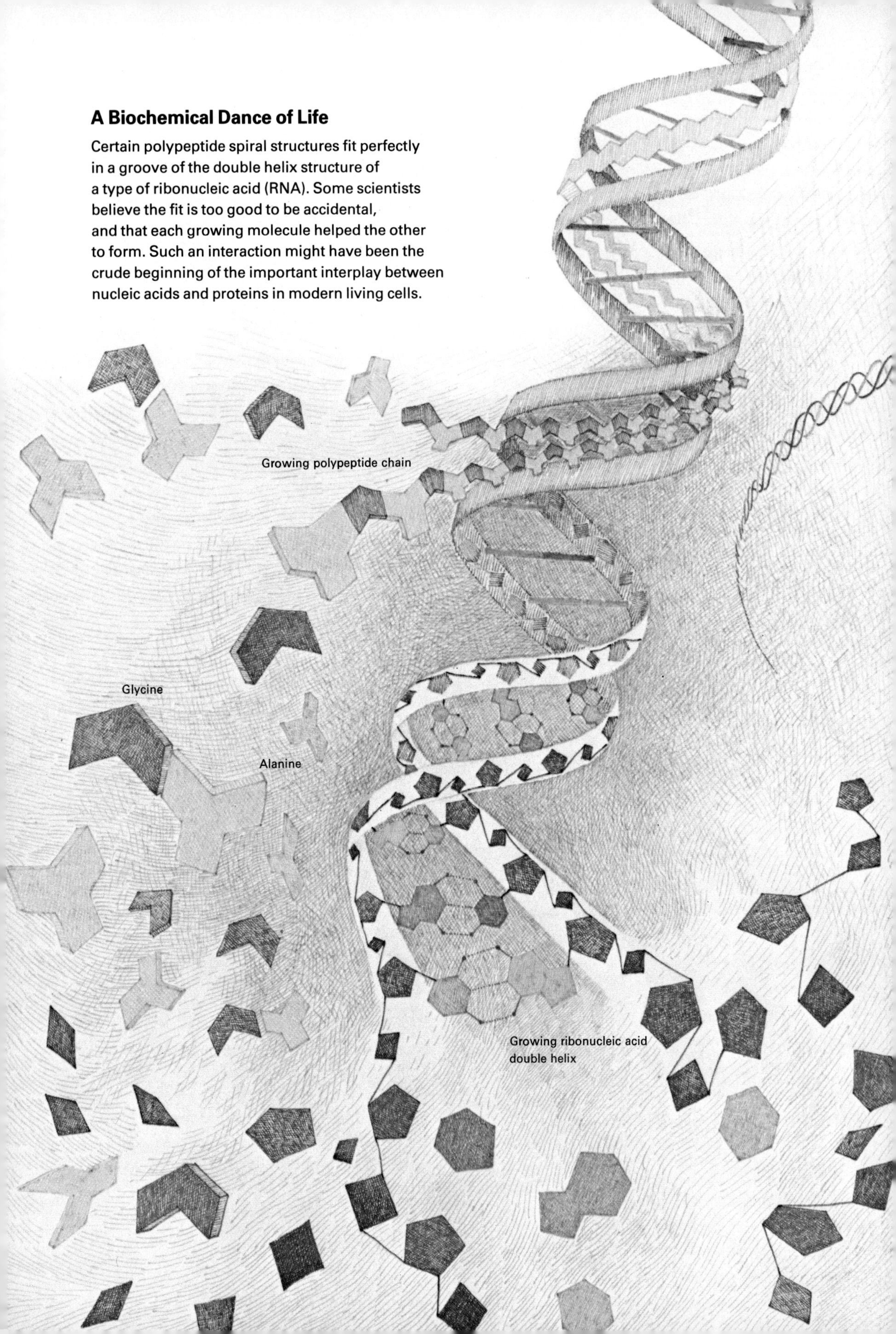

A Biochemical Dance of Life

Certain polypeptide spiral structures fit perfectly in a groove of the double helix structure of a type of ribonucleic acid (RNA). Some scientists believe the fit is too good to be accidental, and that each growing molecule helped the other to form. Such an interaction might have been the crude beginning of the important interplay between nucleic acids and proteins in modern living cells.

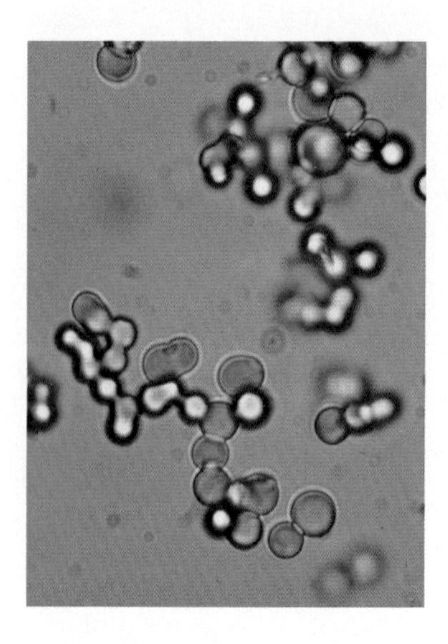

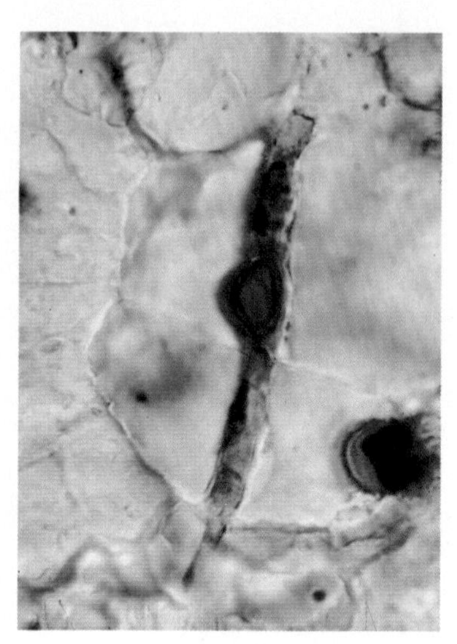

Proteinoid microspheres, *right,* can be made in a laboratory from mixtures of amino acids that probably abounded on the prebiotic earth. They look like cells, and some even divide in half like cells. Although they may have evolved to cells, there are many unexplained differences between them and such ancient microfossils as a blue-green alga, *far right,* believed to be 2.3 billion years old.

One of my colleagues, geologist Dale E. Ingmanson, and I are now investigating another unique environment that may have played a role in chemical evolution–the brine, or saltwater, pools found in the Red Sea rift valley. Such underwater valleys mark the places where the earth's continents are inching apart. We believe that these environments may be common sites for most of the conditions suggested by the various theories of chemical evolution. Located along fault lines where hot magma and gases from the earth's interior escape onto the ocean floor, the extremely salty pools lie in large holes about 2 kilometers (1.2 miles) in diameter and 200 meters (650 feet) deep. The pools are warm, 56°C (133°F.), contain montmorillonites, and lack oxygen.

We believe processes similar to those that led to life billions of years ago may still be at work today in the brine pools. Because microorganisms are so abundant, scientists have long believed that any crude molecule that might originate on the road to life anywhere on earth today would be quickly eaten up by these relatively complex life forms. But the brine pools appear to be free of life, even microorganisms. They also contain methane and may hold the other ingredients needed for life to evolve anew. We hope to examine this possibility before the exploration of these geologically intriguing regions leads to inadvertent contamination of the pools with life.

Even though researchers have suggested several ways in which proteins and nucleic acids may have overcome the water problem on the prebiotic earth, they are still struggling with the most difficult question that they must answer–how did these molecules first organize to form cell-like creatures capable of reproducing and passing on a pattern for new growth to their offspring?

Fox's microspheres, though they look like cells, do not contain nucleic acids. He believes that the proteinoids came first and provided a unique environment for the subsequent development of nucleic acids.

In 1975, Orgel and biophysicist André Brack of the Centre de Biophysique Moléculaire in Orléans, France, suggested that the structure of early polypeptides helped in bridging the gap between chemical and biological evolution. They cited evidence that the most common amino acids made in prebiotic experiments tend to form stable polypeptide structures, called beta pleated sheets, that resemble the bellows of an accordion. These sheets consist of rows of zigzag polypeptide chains that form readily when the amino acids in them alternate between two types: hydrophobic, or water hating, and hydrophilic, or water loving. Interestingly, the two amino acids that form in the largest amounts in spark-discharge experiments are glycine, which is hydrophilic, and alanine, which is relatively hydrophobic. Orgel and Brack believe that if prebiotic polypeptides formed stable structures at all, they formed beta pleated sheets. Pairs of such sheets could have stacked together to form a kind of sandwich, the hydrophilic surfaces facing outward and the hydrophobic surfaces facing one another to form a water-free interior.

Brack and Orgel also believe that the alternating structure of these polypeptides may have been crucial for the first step in the origin of a genetic code. They argue that the most common nucleotide chains that would have formed under prebiotic conditions would be those with alternating organic bases. Those nucleotides containing the single-ring bases cytosine and uracil would alternate with nucleotides containing the double-ring bases adenine and guanine. They suggest that the first form of genetic coding occurred when short chains of such alternating nucleotides built short polypeptides in which hydrophilic and hydrophobic amino acids alternated.

Other researchers have investigated the role that the structures of primitive nucleic acids and polypeptides may have played in the earliest events of biological evolution. In 1974, Charles W. Carter, Jr., and Joseph Kraut worked on the nucleic acid-protein interaction problem at the University of California, San Diego. Using computer modeling techniques, they observed that two zigzag polypeptide chains could form a double helix that would fit perfectly in a groove of an RNA double helix. Moreover, the atoms within this combination would be perfectly positioned to hold such an organization together. These two molecules fit so perfectly that Carter and Kraut believe that each must help the other form.

It would seem simply incredible to me if all the evidence in support of chemical evolution from computer models, astronomy, geology, and chemical laboratories was only a giant web of coincidence. Man had his first chance to check the theory of chemical evolution experimentally on another planet in July 1976, when a Viking I instrument package landed on Mars and tested its soil. If the analyses of these tests show life, or the chemicals that are suspected to lead to life, on Mars, this will be one more piece of evidence that life is not rare but, rather, the inevitable result of the chemistry of our universe.

Invasion of the Microcomputers

By Gene Bylinsky

Once concealed in research laboratories and insurance offices, versatile computers are now appearing in shopping centers, classrooms, and even in many homes

A housewife walks up to a computerized deposit box in the entryway to the State National Bank in Evanston, Ill. She inserts a plastic card in a slot, types some numbers on a keyboard, and waits for instructions to appear on a video screen. When the screen flashes "go ahead," she touches a key marked "withdrawal," types some more numbers, and waits. A moment later, five $10 bills and her plastic card come out of the machine.

The housewife then drives to the Dominick's supermarket in nearby Morton Grove to do her weekly shopping. At the computerized checkout counter, the clerk passes each of her purchases over a glass plate that contains the scanner for a computer. The computer totals her bill and provides an itemized checkout slip.

Computers are appearing in classrooms in such devices as this learning module that asks arithmetic questions to which the student can respond.

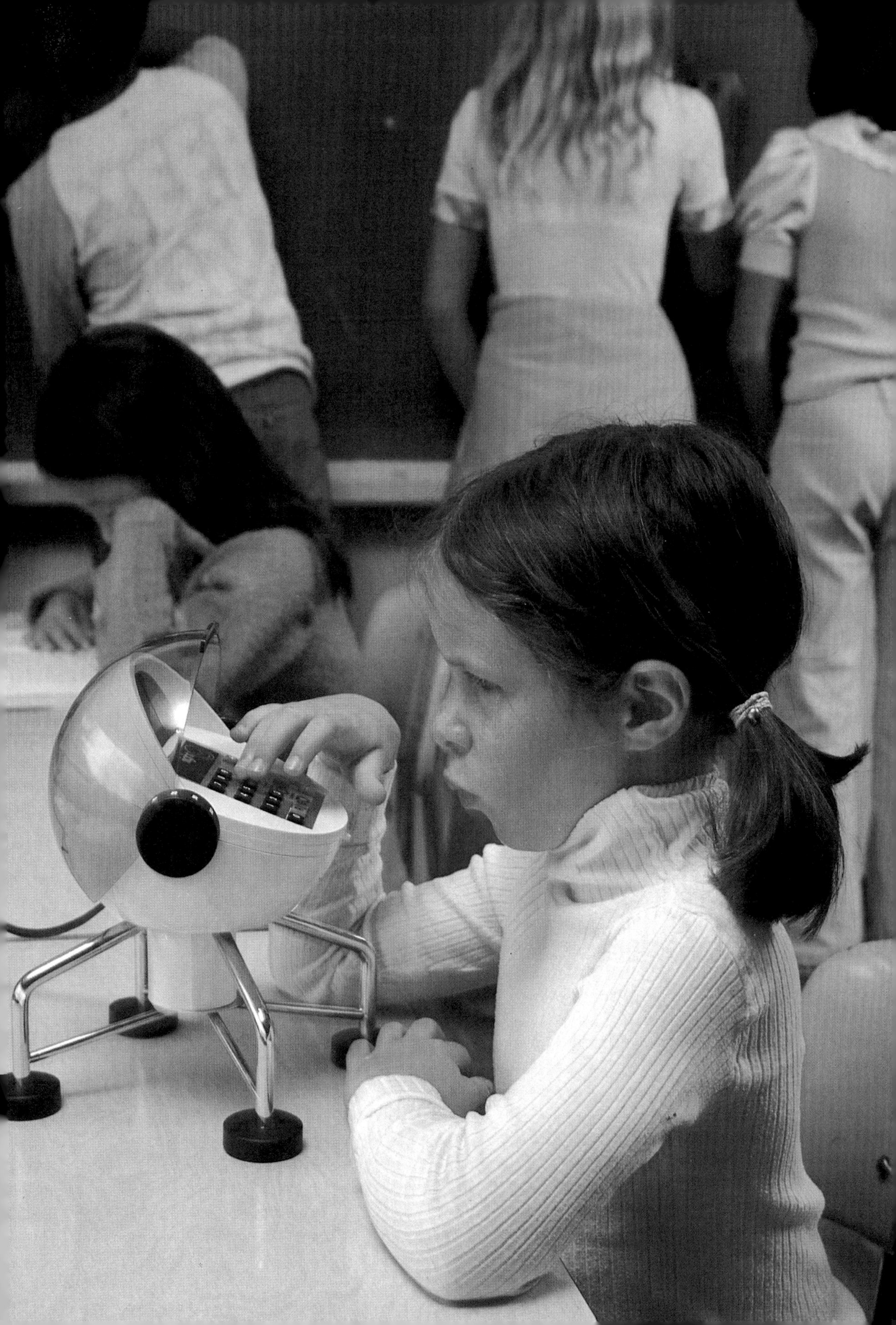

Coming face to face with such computers in the course of her day is becoming common for the housewife, and for all of us as well. Computers are appearing everywhere. Miniaturized models are in daily use as digital wrist watches, in portable calculators, and as the control in microwave ovens. They serve as the "brains" in electronic games and the regulators in traffic lights. Computers are now also found in all sorts of scientific and medical instruments. They measure the flow of fuel through gas-station pumps, control telephone switching devices, and prepare bank statements. And they will soon be used in electric typewriters, washing machines, and refrigerators. As a bemused Tom Bennett, electronics engineer for Motorola Communications & Electronics Incorporated, recently put it, "Suddenly, everything sitting out there is getting smart."

The new computers bear almost no resemblance to the first ones built about 30 years ago. Those early metal monsters bristled with thousands of glowing vacuum tubes. They were used to integrate complex elements in biological and physical models, compute the orbits of comets, and set up actuarial tables for insurance companies. They had little impact on our daily lives. But they started an accelerating drive to make the faster, less expensive, and–most important–smaller computers that touch each of us personally today.

The electronics industry has fueled the computer revolution with three generations of components–each more efficient and smaller in size–as vacuum tubes gave way to transistors and transistors to microprocessor chips. The first electronic computer, the Electronic Numerical Integrator and Computer (ENIAC), was built in 1946 at the University of Pennsylvania. It ran on 18,000 vacuum tubes, weighed 28 metric tons, and was as large as a six-room house. It took highly trained engineers to operate ENIAC.

The invention of the transistor in 1947 really started the proliferation of computers. The transistor is more dependable and uses less electricity than the vacuum tube it replaced. And it is much smaller. By the late 1950s, engineers had jammed hundreds of transistors onto a square of silicon as small as a baby's fingernail. They called such transistor assemblies integrated circuits (ICs). In effect, ICs are small, complex switching systems. They were used to build the first minicomputers, compact machines that are as small as the top of an office desk. The new technology dramatically reduced the cost of computers, and their use was greatly expanded in science, mathematics, business, industry, and education.

The development in the late 1960s of advanced integrated circuits, known as Large Scale Integration (LSI) chips, started a new kind of computer revolution. LSI chips were more than switching systems. They were, in themselves, elementary computers that served as basic subsystems–such as guidance controllers or memory storehouses–for more sophisticated computers. They soon appeared in a variety of roles in big computers as well as tiny calculators.

The author:
Gene Bylinsky, science writer for *Fortune* magazine, wrote the article on Carl Djerassi for *Science Year,* 1975.

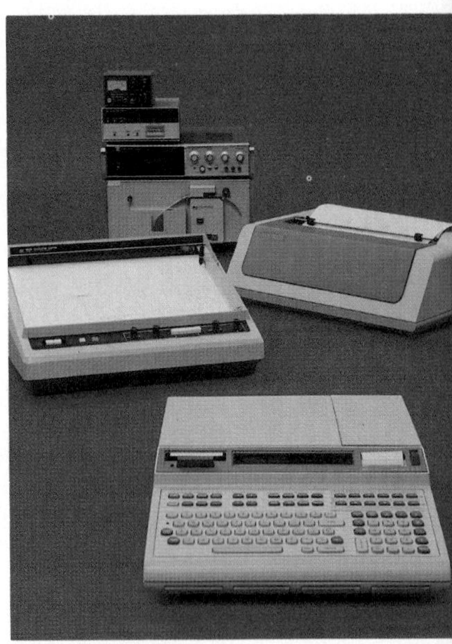

The rise of miniaturized components, *left,* started the trend to increasingly compact computers. The first room-sized computer, ENIAC, *top left,* was run by many electron tubes. Transistors enabled development of the desk-top computer, *top right,* and silicon chips led to pocket calculators, *above.*

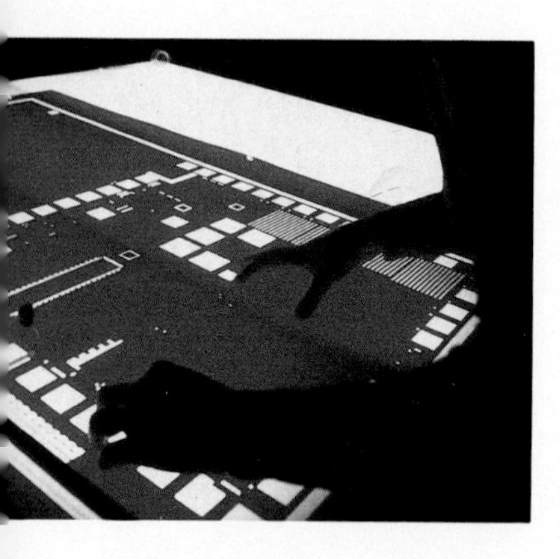

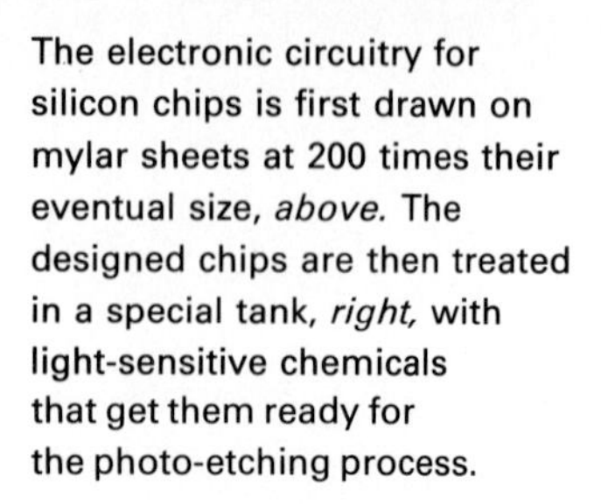

The electronic circuitry for silicon chips is first drawn on mylar sheets at 200 times their eventual size, *above.* The designed chips are then treated in a special tank, *right,* with light-sensitive chemicals that get them ready for the photo-etching process.

LSI chips were at first designed for the specific needs of the particular computer in which they were used. But then in 1969, Marcian E. (Ted) Hoff, Jr., a young engineer at Intel Corporation, in Santa Clara, Calif., came up with a new concept for the design of LSI chips. Hoff visualized a computer smaller than a minicomputer–a microcomputer. It would be slower than its big brothers, but just as versatile because it also could be designed so that its tasks could be changed.

Unlike most other computer engineers, Hoff thought in terms of systems rather than in terms of component parts. He based his concept on the fact that, regardless of its size, any computer consists of three basic systems. First, there is the arithmetic system that performs the mathematical, or logic, operations. Second, there is the memory system, where information is stored. Finally, there is the control system, which guides the computer in working on the stored data. In a pocket calculator, for example, the control system directs the electrical signals through the arithmetic system, which does the calculations, and into the memory system, which stores the results until they are released on the display panel as a series of numbers.

Hoff condensed the computer's arithmetic system so that it could be put on one tiny silicon chip. This was called a microprocessor. He attached two other chips to the microprocessor, one holding data used in the arithmetic system and the other–the control system–to drive the computer. This combination of chips, his first rudimentary microcomputer, could perform 10,000 calculations a second–as many as ENIAC could do with its 18,000 vacuum tubes.

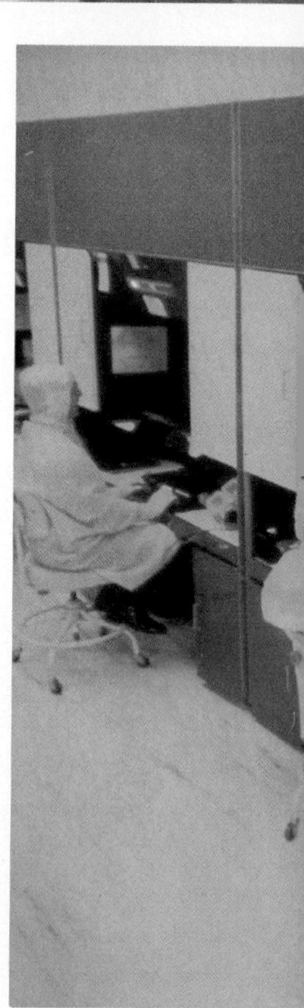

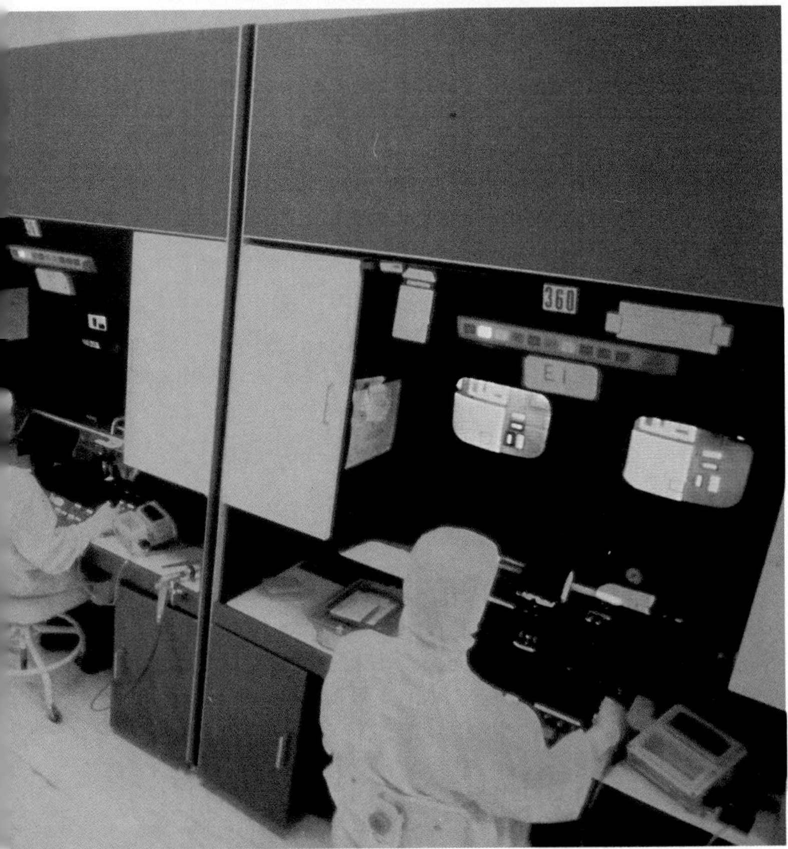

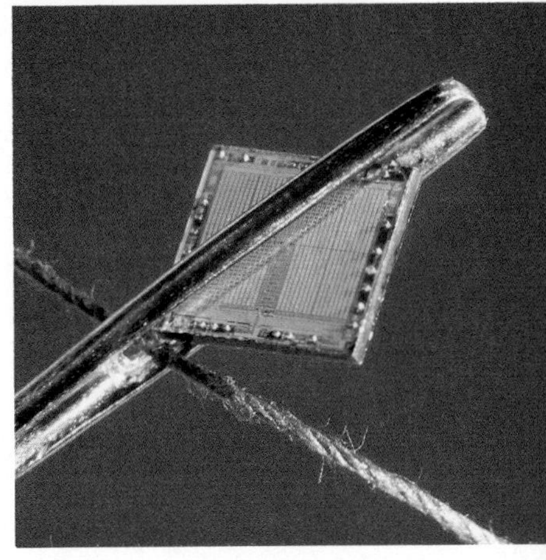

Silicon chips are removed from a furnace, *top,* where they have been heat-treated after receiving the light-sensitive chemicals. Operators then use television, *left,* to align chips with the printed circuitry before they are exposed to light. A finished chip is so tiny it goes through the eye of a needle, *above.*

The first microcomputer is held by its inventor, Ted Hoff, *right.* The complex circuitry on a microcomputer chip, *far right,* has been enlarged about 40 times.

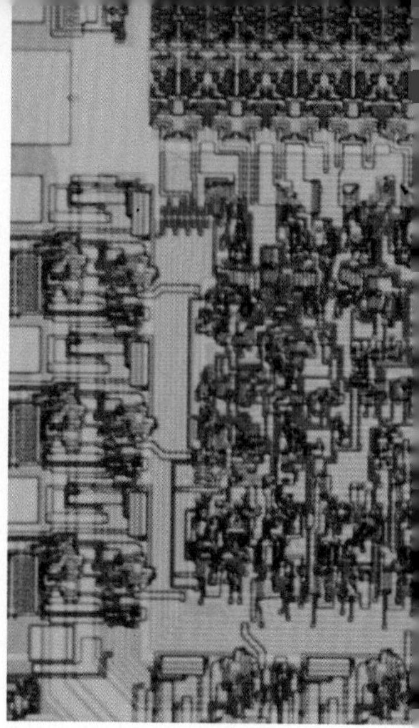

Hoff's microcomputer was so compact that it could easily fit in a matchbox–or be tucked into the side of an electric typewriter, a traffic light, a vending machine, a butcher's scale, or any other device that needed precise controls. In addition, because its single chips were complete systems or major portions of systems that could now be changed at will, the microcomputer offered a previously unknown flexibility for thousands of machines, instruments, and other devices.

This flexibility is apparent in a demonstration apparatus built in 1974 by Pro-Log Corporation of Monterey, Calif. In its basic form, it is a digital clock. But when the memory chip is replaced, the "clock"–equipped with a loudspeaker–suddenly starts playing a tinny version of the theme from the motion picture *The Sting*. With still another memory chip, it becomes a rudimentary piano that the operator can play by pressing a few keys. Pro-Log's device hints at the potential of the microcomputer to produce versatile analytical instruments, desk calculators that can perform many different jobs, and countless other products that designers haven't even thought of yet.

The memory chips contain the instructions, or program, that tell the machine exactly what to do. In the first big vacuum-tube computers, these programs had to be painstakingly written out by hand as endless strings of "bits,"–the 1's and 0's of the basic binary computer code. The 1's and 0's function as a collection of electronic switches, or "traffic lights," that guide the electrical signals to and from their destinations.

All kinds of data can be reduced to 1's and 0's. Depending on the programmed instructions, these bits can be reconstituted within the computer so that they come out as music, a series of printed numbers, a plot on a graph, or pictures on a video screen. Engineers can feed specifications for a proposed structure into such a computer and have the program displayed as a picture on a screen. The computer then

becomes a sort of electronic wind tunnel. An engineer can design a computer model of a bridge, for example, then subject it to simulated stresses and loads to see if his creation stands up or collapses. This procedure saves the cost of constructing real models and purchasing expensive test equipment.

By the mid-1960s, computer programmers no longer needed to write their instructions by hand in 1's and 0's. They manipulated the computer's operations with computer "languages" such as FORTRAN (FORmula TRANslation), COBOL (COmmon Business Oriented Language), and PL-1 (Programming Language 1). They could write the programs in these kinds of computerized pidgin English and have special programs called compilers translate them into the machine code of 1's and 0's.

Still, writing programs and inserting them in the computer is time-consuming. With the new microcomputer chips, it is possible to make immediate changes in total computer programs. To change a program, the computer operator simply inserts a different chip that is inscribed with the new program.

"In the past," says Gordon E. Moore, president of Intel, "a machine was built to do one particular task. If you wanted it to do something else, you started over again and designed a new machine. Now you can give a different set of instructions to the same machine and it will do something different."

What is more, microcomputer chips can be mass-produced as inexpensively as are other electronic parts, and on the same assembly lines used to make other types of integrated circuits. This is possible because the electronics industry in the 1960s developed a way to inscribe very complex electrical circuits on tiny surfaces. First, engineers draw the intricate circuits for a microcomputer chip on sheets of paper, usually on a scale about 500 times larger than the actual chip. The drawings are then reduced photographically to miniature size and photoengraved on a chip of silicon. Technical improvements in this process and mass production have dropped the price of microcomputer chips to under $5.

Nowhere is the flexibility of microcomputer chips being applied more dramatically than in scientific and medical instrumentation. For example, a blood-chemistry analyzer built by Chemetrics Corporation of San Mateo, Calif., performs 19 preprogrammed basic clinical tests that are selected by keyboard control. The keyboard operator can also punch in new instructions to change any of the computer programs. Such operating advantages save time and increase the accuracy of the chemical analyses. Large numbers of tests can now be done inexpensively, improving the preventive and diagnostic aspects of medicine.

Microcomputers also improve the quality of scientific and industrial instruments. New instruments, driven by microcomputers, can perform their tasks quickly and accurately. For instance, an infrared spectrophotometer to analyze light rays, built by the Perkin-Elmer Corporation of Norwalk, Conn., analyzes a spectrum in 14 minutes compared to the 50 minutes that are needed by a conventional machine. The instrument also constantly calibrates itself, and it does not require a highly skilled operator. As William Bennett of Perkin-Elmer explained, "The skill now resides in the microcomputer. This frees scientists for more creative tasks."

Some new desk calculators use easily replaced cartridges that contain different memory chips to do statistical, installment loan, invoicing, or a host of other specialized computations. Banks formerly used large computers for such calculations or did them by hand. A new postal scale can be updated with a fresh memory chip to take postal-rate changes into account. Erasable memory chips even make changes within a program possible. These are chips with inscribed circuits that can be altered by giving them new instructions. In this way, the chip's program can be updated without changing the inscribed circuit itself.

Improved gasoline mileage, lower repair costs, cleaner exhaust emissions, and simple maintenance are big inducements for using microcomputers in automobiles. Simpler versions of computer chips–not complete microcomputers–are now being used in some automobiles to constantly supervise performance factors such as ignition and voltages. Some 1976 Chrysler models have been equipped with a lean-burn computer system to produce cleaner exhausts and a 5 per cent reduction in fuel consumption by constantly adjusting the carburetor for the right mix of air and fuel. For this system, Chrysler also built a mechanic's diagnostic device, about as big as a portable TV set, that contains a microcomputer. Complete microcomputers are expected to start appearing in all cars by about 1980. Ford Motor Company, for example, has found that microcomputer controls can cut fuel consumption by as much as 20 per cent, and the company plans to install such controls in its 1979 models.

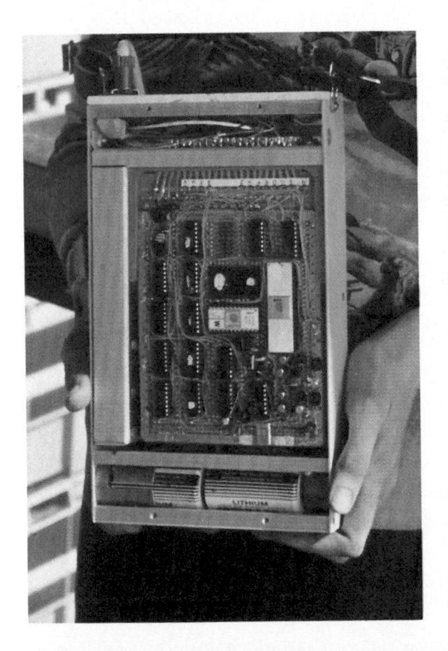

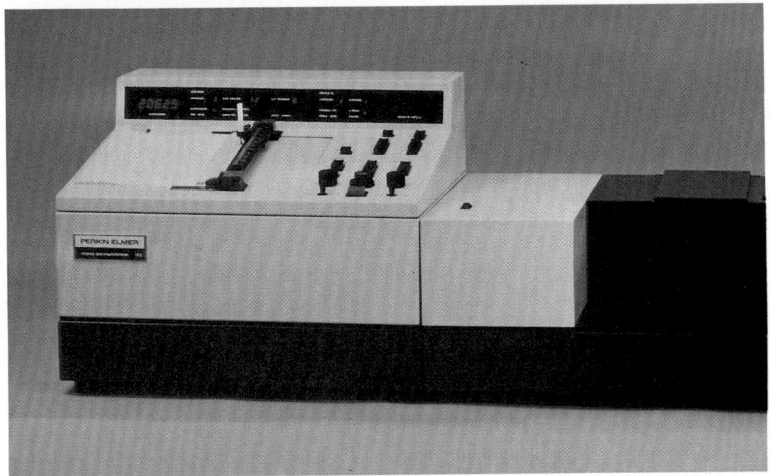

Scientific instruments like a portable heartbeat analyzer, *above left,* a laser-beam particle analyzer, *above,* and a spectrophotometer, *left,* make use of one or more microcomputer chips.

One reason microcomputers haven't been put into cars and appliances sooner is lack of inexpensive temperature, radiation, and motion sensors that function as their eyes and ears. At Frigidaire, for example, engineers had difficulty developing heat sensors for the microelectronics in their microwave oven. The first electronic oven controls they tested had 150 individual arithmetic systems. That kind of appliance would have been impossible to market because of its complexity and cost. But microprocessor chips have since cut costs to acceptable levels. Now the arithmetic and memory systems are embedded in three or four chips, and some microwave oven manufacturers use a single-chip microcomputer which has both arithmetic and memory systems on the same square of silicon.

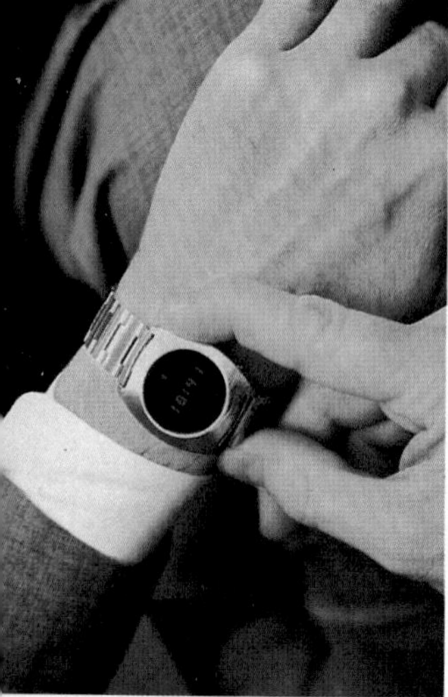

A new electronic game called Pong, a fast-paced version of table tennis, is made possible by a tiny microcomputer attached to a TV set, *above.* Microcomputers also control the digital wrist watch, *left,* and the grocery checkout system, *below,* where a sensing device registers the price of each item marked by a series of short vertical bars. The checkout system totals the bill and provides an itemized checkout slip.

From filling station to auto repair shop, to kitchen, to office, microcomputers pervade—to run gasoline pumps, *top left,* check engines, *left,* control microwave ovens, *top,* and check credit by phone, *above.*

Electronic cash registers, checkout terminals that are hooked up to a store computer, appeared in 1975 in supermarkets. Some of them are equipped with lasers that can "read" the Universal Product Code (UPC)–the series of short vertical bars printed on cans and packages. As a checkout clerk passes the product over the terminal's scanner, the laser "reads" the code, identifies the item, and relays the coded information to the store's computer. The system retrieves the programmed price from the terminal's memory, displays the product name and price on a lighted panel, and prints this information on the customer's receipt. The system also improves inventory control.

But consumers' groups and retail clerks' unions have opposed use of the terminals. Because the price is not needed on the product, consumers' groups fear that prices might be manipulated at the expense of the customer. The unions fear that jobs will be eliminated because prices can be changed simply by a keystroke that registers in the memory chip instead of by stamping each can or package by hand.

Because of the opposition, several states have passed mandatory price-marking legislation. Nevertheless, by the end of 1976, about 60 supermarket chains were scheduled to try out the scanners in about 150 stores in the United States.

Some supermarket terminals will even allow customers to automatically make bank deposits, transfer cash from their bank accounts to pay for groceries, and conduct other transactions without going to their banks. The First National Bank of Chicago, for example, made such an arrangement with 12 Jewel Companies food stores in the Chicago area in November 1975. This "electronic fund transfer" service will attract more customers to supermarkets, while it offers banks a myriad of new "branches" at minimal expense. Legal questions remain, however, as to whether such service is a form of branch banking, which is illegal in many states.

In the initial stages of introducing any new computerized system, mistakes are made and frustration follows. One such problem has arisen in department stores using an electronic checkout system similar to that found in supermarkets. A sales clerk may fail to remove the electronic garment tags after they are read by the new cash register. Then, as the customer leaves the store, a detection system designed to protect against shoplifters spots the forgotten tag and sounds a shrill alarm. Irate customers, falsely arrested as shoplifters, have sued merchants and recovered damages.

At the other extreme are the computer criminals, skilled professionals who have learned how to cut through computer codes and other protective devices to embezzle funds. Some students at the Massachusetts Institute of Technology in Cambridge managed to improve their computer-stored grades in 1970. But where most people have become aware of the potential for computer abuses is in the handling of their personal credit accounts. Many customers have been mistakenly billed for merchandise and been unable to get the errors corrected easily.

The rapid advance in electronics widens the gap between those who know how to use the new technologies and those who do not. Even though the computer industry has created thousands of new jobs, most of them require specialized skills that less-educated workers do not have. Yet, these are often the workers displaced by the computers.

The widening use of calculators in schools raises perhaps the most disagreement over the new computer technology. Although the National Council of Teachers of Mathematics has endorsed their use–as valuable aids, not substitutes–for teaching math, acceptance of calculators has been far from universal. Many schools allow their use; others prohibit them, fearful that the new machines will create a generation of mathematical illiterates. Some educators are concerned that children who cannot afford to purchase calculators will be at a disadvantage. See ELECTRONICS, Close-Up.

To determine the calculator's effectiveness as a teaching tool, the National Science Foundation has given grants to the University of Denver and the San Francisco school system for research at different grade levels. Early results favor the continued use of the machines. Other studies indicate that at least one group of students is definitely benefiting. Tests in California and Massachusetts showed that children with learning disabilities such as *dyslexia* (a brain disorder that results in the inability to read properly) learned basic math better with calculators.

More educators might use calculators as teaching aids if manufacturers used more imagination. One small company, Centurion Industries Incorporated of Redwood City, Calif., builds the Digitor, a computerized learning aid shaped like the lunar landing module for first- and second-grade students. Unlike a calculator, the device requires the student to give an answer before it responds. Students must answer each problem correctly before going to the next one. Each correct answer is rewarded with a "happy face" on the readout display, and a "sad face" appears for a wrong answer.

It is, of course, too early to evaluate fully all the problems and possibilities the computer revolution is bringing to us. But compare it to the development of electricity, which also began with unwieldy devices–huge arc lamps that could only light city streets. Then Thomas A. Edison's incandescent bulb began to light up homes, factories, and offices. The electric light changed man's life style immeasurably, and lengthened the time he could devote to intellectual pursuits. Now, electronic "intelligence" in the form of computers promises to extend the reach of man's mind in new directions by freeing him from tedious, time-consuming chores and making his life easier.

Whether the spread of computer electronics produces cultural and social disorder or helps boost man's intellect to new heights depends largely on how intelligently we use this new power. One thing is certain–there is no turning back from the new electronics, just as there was no turning back from the electric light.

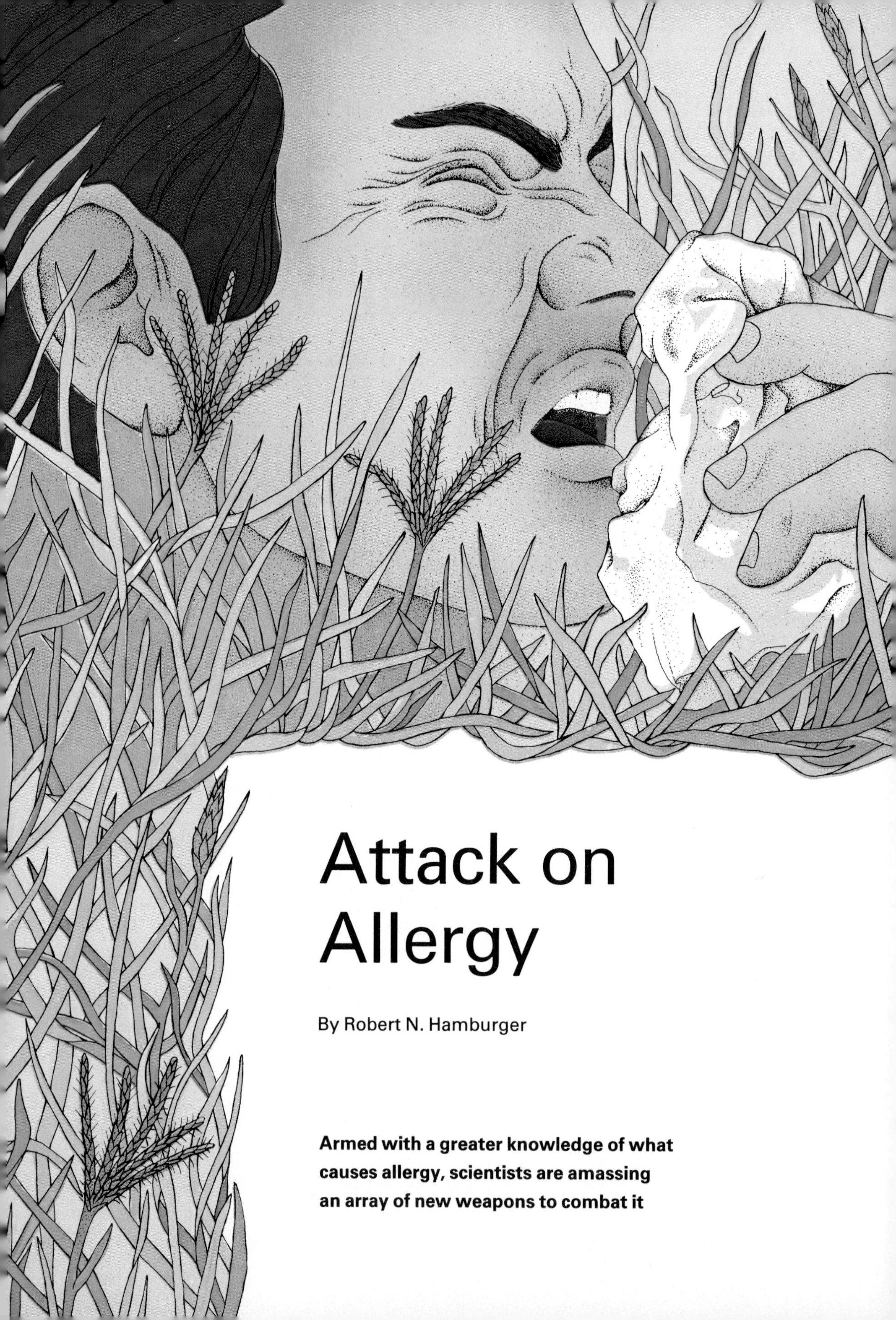

Attack on Allergy

By Robert N. Hamburger

Armed with a greater knowledge of what causes allergy, scientists are amassing an array of new weapons to combat it

I knew there was some risk involved as I prepared to inject myself with one of several substances my research indicated might block allergy in human beings. The chemical structures of the substances were reassuring, not characteristic of any poisons that I knew. Besides, I had injected them into test mice, and they had shown no ill effects.

Yet, I was the first human test subject. Twenty minutes after the injection, I became nauseated, developed a headache and stomach pains, and left my laboratory at the University of California School of Medicine in San Diego to rest at home.

The next day, I began to suspect what had really happened. Several of my friends and associates had recently had some minor flulike malady with symptoms that were the same as mine. By coincidence, my illness began just after I injected myself. Once I recovered, I tested and retested the substances on myself without any problems.

After establishing their safety, I began testing the substances against allergy, and on Aug. 1, 1975, reported the first success with one of them. My work is just one example of the growing amount of research on allergy underway in many laboratories. This activity stems mainly from a new understanding of the biological mechanisms that trigger allergies, an understanding that is based on details that we have uncovered only in the past 5 or 10 years. Until then, we had only sketchy knowledge of what causes reactions such as hay fever, asthma, rashes, and anaphylaxis, or allergic shock.

The ancient Egyptians may have been first to record an allergy. Hieroglyphics in the tomb of King Menes indicate that he died in 2641 B.C. from the sting of a wasp or hornet, a probable case of allergic shock. Hay fever and asthma were described several times by physicians over the next 3,500 years. But man did not begin to understand allergy until he started to study it scientifically.

One of the first studies loosely tied allergy to the immune system. In 1902, French physiologists Paul Portier and Charles R. Richet injected a small quantity of poison from a sea anemone into dogs. The scientists hoped to provide prophylaxis, or protection, against future exposure to the poison. When they injected the poison a second time, however, the dogs died suddenly from allergic shock.

By 1921, all forms of allergy had been observed in a variety of animals, including man. That year, German physicians Carl Prausnitz and Heinz Küstner performed a classic experiment that marked the beginning of our knowledge of how allergies work.

The author:
Robert N. Hamburger is a professor and head of the Pediatric Immunology and Allergy Division, University of California, San Diego.

Küstner was allergic to fish, and demonstrated this with a simple test. He scratched his arm, daubed fish broth on the scratch, and an inflamed swelling appeared. Prausnitz was not allergic to fish and showed no such reaction to the same test. With this knowledge, the scientists proceeded to inject some of Küstner's blood serum into Prausnitz' arm. They waited 24 hours, then scratched and daubed the injection site. Inflammation and swelling on Prausnitz' arm showed that a substance in the bloodstream of an allergic person like Küstner produces allergy. Further experiments indicated that this substance, called reagin, is produced in allergic people in response to such allergens as molecules from pollens and foods. It is similar to the way immunoglobulins, or antibodies, are produced. However, most immunoglobulins are produced by the body in response to harmful bacteria or other "foreign" agents such as poisons and help to destroy these agents. In allergy, it seems, the body mistakes the allergen as a harmful agent, and produces reagin. Although the reaction causes discomfort rather than protection, the parallel was strong enough to indicate to early researchers that reagin might be an immunoglobulin.

Scientists knew the structure and behavior of several types of immunoglobulins, and a number of investigators tried to show that reagin was one of these types. Then, in 1967, immunologists Kimishige and Teruko Ishizaka, a husband-and-wife team at the National Jewish Hospital and Research Center in Denver, proved that reagin was a new type of immunoglobulin, now called immunoglobulin E, or IgE.

However, to formulate even the sketchiest theory of the mechanisms behind allergy, researchers still had to show how IgE fits in with what was already known. For example, several substances, of which histamine is most significant, produce the inflammation and other characteristic responses at the site of the allergic reaction. Most of these substances are produced and stored in a kind of package within two types of cells–the mast cells, which lie within the body tissues, and the basophils, blood cells that circulate throughout the body.

IgE turned out to be the perfect link between allergens and the mast cells and basophils. In the same year that the Ishizakas identified reagin as IgE, immunologists S. G. O. Johansson and Hans Bennich of the University of Uppsala in Sweden discovered a patient with multiple myeloma whose diseased cells produced large quantities of IgE, enough for scientists to determine its structure. They found that an IgE molecule is composed of four chains of chemical units called amino acids. The chains are bound together in a Y shape. The ends of the two arms of the Y vary chemically according to the specific allergen with which the molecule is associated, and these ends bind to that allergen. The rest of the Y is chemically identical for all IgE molecules. A portion on the tail just below the juncture of the arms binds to receptor sites on the outer membrane of mast cells or basophils.

With these details in mind, we can trace the sequence of events that leads to an allergy attack. A person's first exposure to an allergen triggers the formation of specialized blood cells called plasma cells.

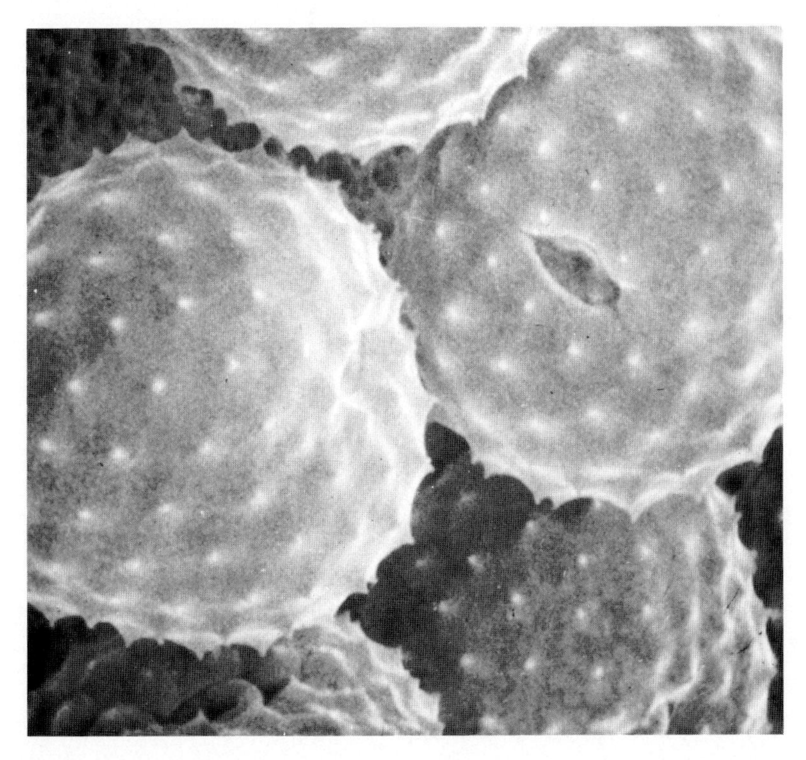

Ragweed pollen grains, magnified about 3,000 times, are the major culprits for many hay fever victims. The opening in the grain at top right is a germ pore, through which the molecules that trigger the allergic response may exit.

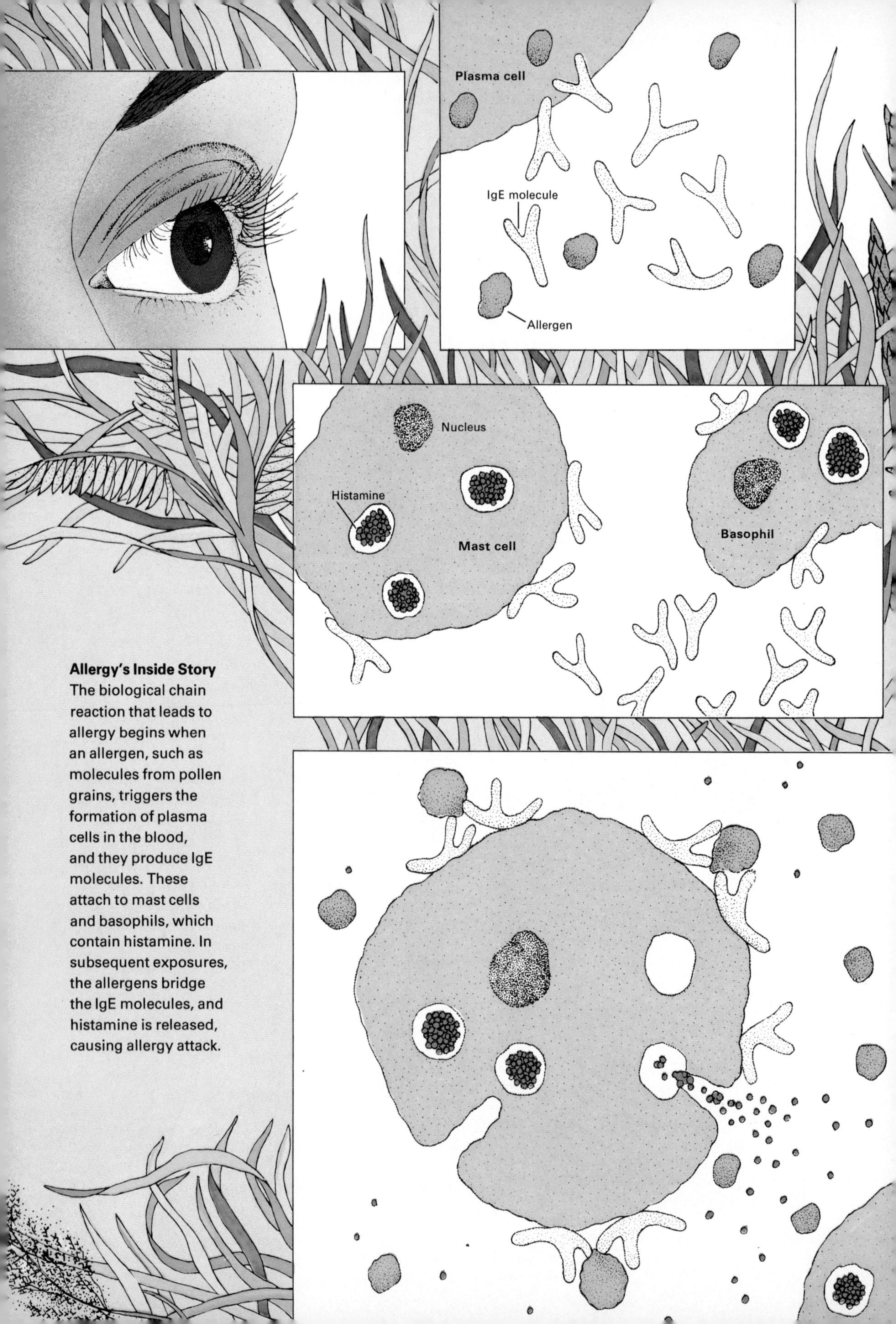

Allergy's Inside Story
The biological chain reaction that leads to allergy begins when an allergen, such as molecules from pollen grains, triggers the formation of plasma cells in the blood, and they produce IgE molecules. These attach to mast cells and basophils, which contain histamine. In subsequent exposures, the allergens bridge the IgE molecules, and histamine is released, causing allergy attack.

These cells produce millions of the specific IgE molecules that will recognize and react with that allergen. The IgE molecules spread through the bloodstream and attach to the special receptor sites on the membranes of mast cells and basophils throughout the body.

Scientists have tried to determine how many such receptor sites there are on a single cell. In 1974, for example, immunologists Henry Metzger and Anthony Kulczycki, Jr., of the National Institutes of Health in Bethesda, Md., reported experiments in which they added great numbers of rat IgE molecules that they had made radioactive to laboratory cultures of rat basophils. Then, by washing away free IgE molecules and measuring the radioactivity emitted by the basophils, they calculated that between 250,000 and 1.5 million IgE molecules had attached to each cell.

With such large numbers of IgE molecules bound to basophil and mast cell surfaces within the body, the stage is set for an allergic reaction. The second time the allergen molecules enter the body, each one binds to an end of an IgE molecule arm. Then each allergen binds to an arm of a second IgE molecule, forming a bridge between it and the first IgE molecule. Somehow, this bridging generates a signal through the cell membrane, and the cell releases histamine, which causes the allergic reaction.

These, then, are the basic steps in the biological chain reaction that leads to an allergy attack. However, there are other factors that determine the extent and severity of the attack. For example, researchers have found evidence that the basic energy level – the metabolism rate – of the cells at the allergy site, such as the nose and lungs, is very important. The more energy (higher metabolism) there, the less these organs are affected. Emotional factors are also important. A person under stress, for instance, is much more likely to have an allergy attack, or to have a more severe attack, than one who is not under stress. All these details clarify how traditional allergy therapy works and suggest several new forms of treatment.

Of the traditional therapies, the easiest to understand is elimination of allergens. Unfortunately, this is not practical in many cases, such as those in which the patient would have to part with a loved pet or move to a new locality. Worse, such steps often provide only temporary relief, for the patient can become more sensitive to other allergens.

Doctors now treat most allergy patients with one of the almost 100 oral antihistamines. These drugs do not prevent the release of histamine from the mast cells and basophils, but they keep it from acting on the organs normally affected by allergy. When avoidance and antihistamines do not provide enough relief from the allergic reaction, many doctors turn to immunotherapy, one of the oldest, but still least-understood treatments. This therapy, also called hyposensitization or desensitization, is better known as "allergy shots." Doctors use increasingly strong solutions of the patient's allergen – molecules from pollen, mold,

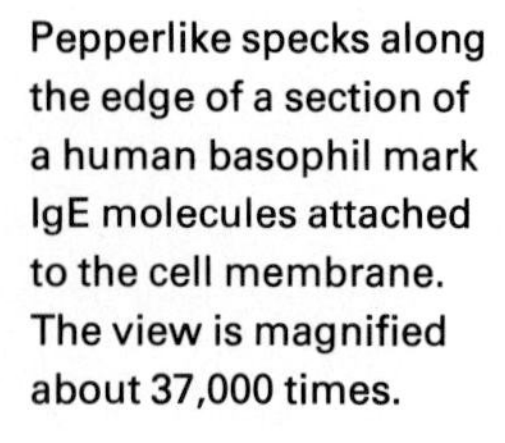
Pepperlike specks along the edge of a section of a human basophil mark IgE molecules attached to the cell membrane. The view is magnified about 37,000 times.

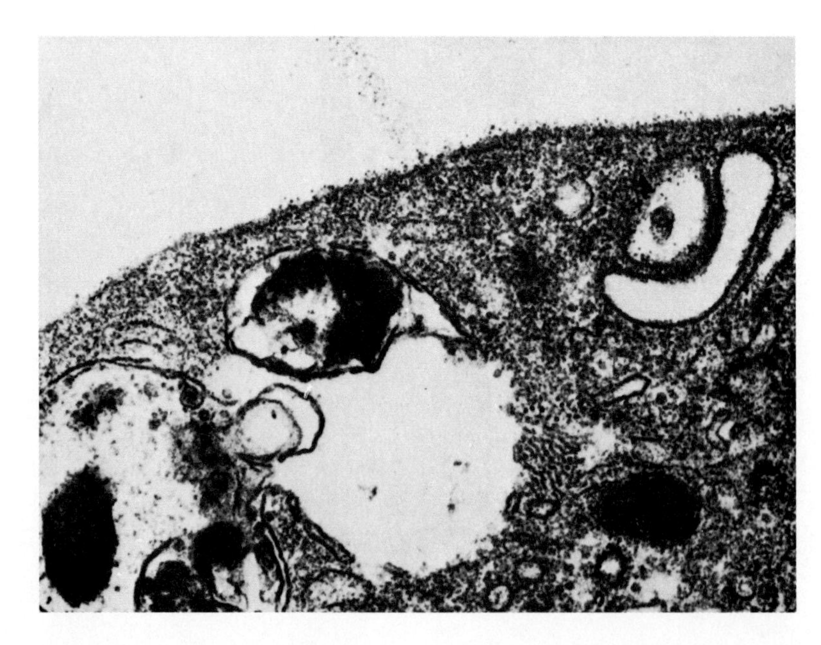

dust, animal dander (skin and hair particles), and other substances. These solutions are injected into the patient at regular intervals until the maximum dose tolerated without an allergic reaction is reached. This maintenance dosage is then injected from once a week to once a month. This treatment may provide relief from allergy attacks after anywhere between a few months and two years.

Several changes related to the mechanisms of allergy occur during the increasing-dosage period. For example, the number of IgE molecules specific to the allergen drops. But researchers cannot yet credit any of the changes with the improvement in the allergic person's condition. In fact, it was not until 1965 and 1968 that two separate research groups finally provided convincing data on the effectiveness of immunotherapy. Until then, many doctors believed that the treatment was no better than systematic injections with any known inactive substance. They based their belief on the knowledge that simply paying more attention to the patients in most cases of chronic diseases often provides measurable relief of symptoms. This, of course, is logical considering the known effects of psychological factors on allergy.

Doctors now agree that immunotherapy works fairly well for perhaps 85 per cent of their allergy patients. But until we learn how and why it works, we are not likely to improve the method significantly.

Another kind of traditional treatment uses the most potent weapons for allergy therapy–epinephrine, or adrenalin, and corticosteroids, or steroids. When injected, adrenalin raises the energy level of cells. Through this action, it remarkably reverses anaphylactic shock and is an effective bronchodilator, almost instantly opening constricted bronchioles in the lungs to relieve asthma. The steroids very effectively reduce inflammation and other allergy symptoms. However, their long-term side effects–particularly in children, where growth may be stunted and maturation delayed–are so harmful that we use steroids only for short-term therapy or to treat overwhelming allergy. We have recently been evaluating beclamethasone, a new form of steroid that can be inhaled directly into the lungs. It provides dramatic relief of asthma symptoms with no undesirable side effects. Unfortunately, however, it is not useful against any other form of allergy.

Adrenalin is also effective as an inhalant against asthma, as are isoproterenol and theophylline, the oldest bronchodilator. Since 1969, researchers have developed an entire family of oral, injectable, and inhaled bronchodilators that act very specifically on the bronchioles, with virtually no side effects. Called beta agents, these drugs are often used along with theophylline for extremely effective relief of asthma. Like adrenalin, isoproterenol, and theophylline, the beta agents raise the energy level of cells in the bronchioles.

Another useful inhalant for asthma is cromolyn, but we are not sure exactly how it works. The drug is peculiar in that it has no effect once an asthma attack is underway. However, it will prevent an attack if used before exposure to an allergen. Several pharmaceutical firms are

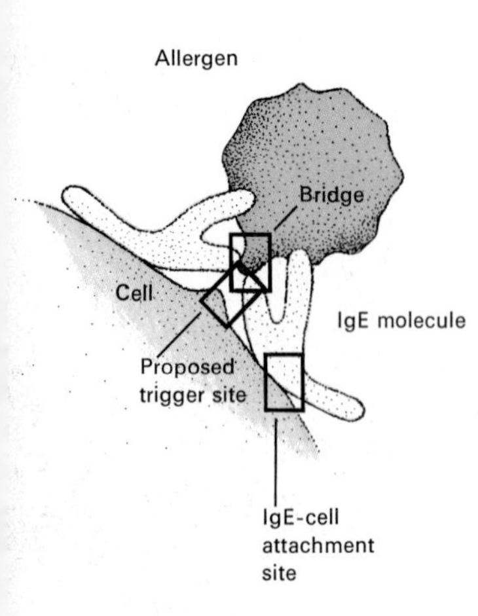

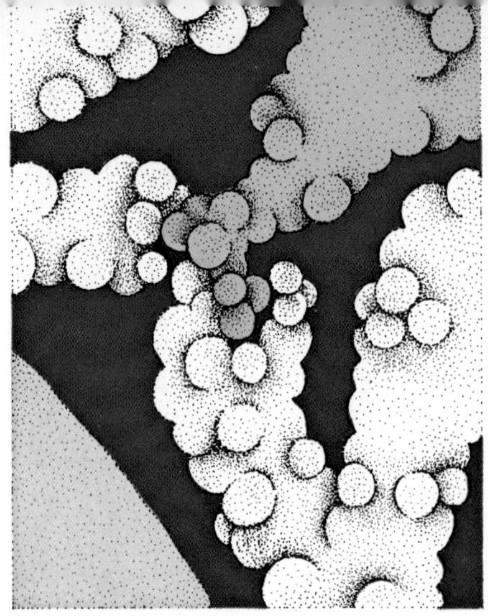

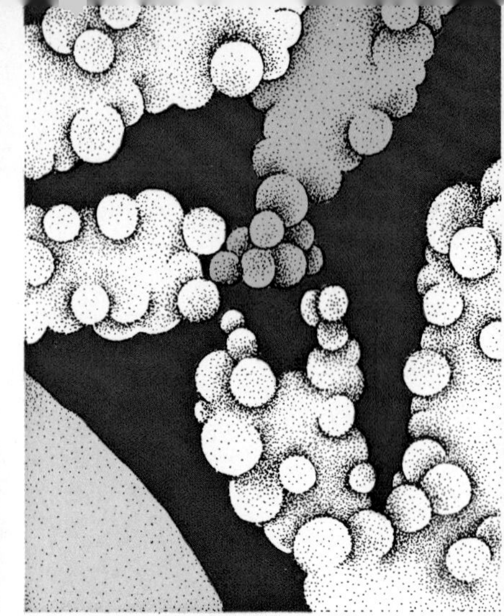

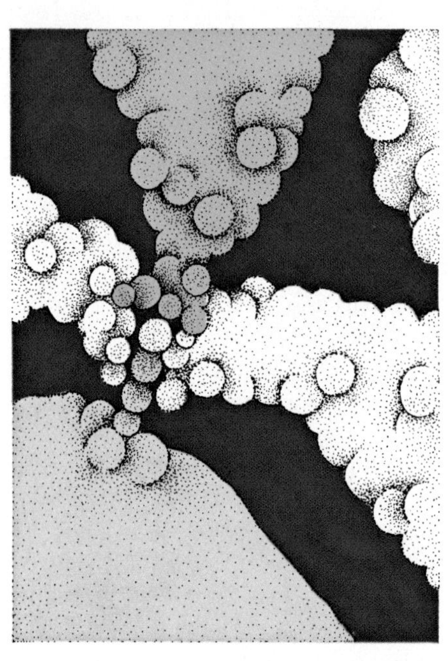

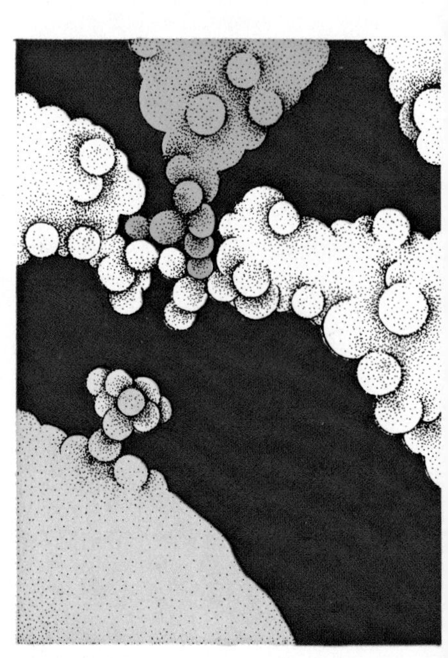

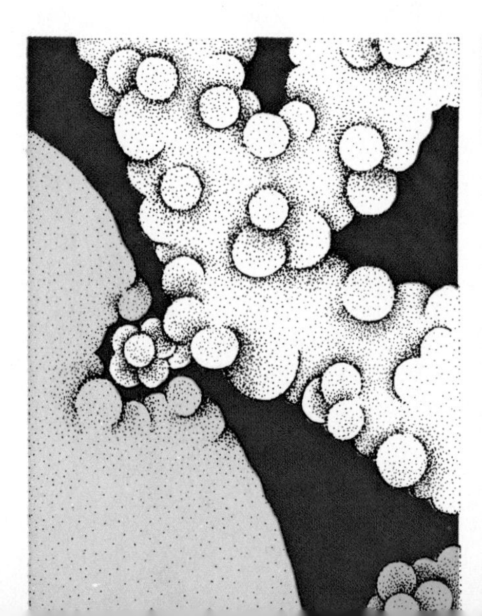

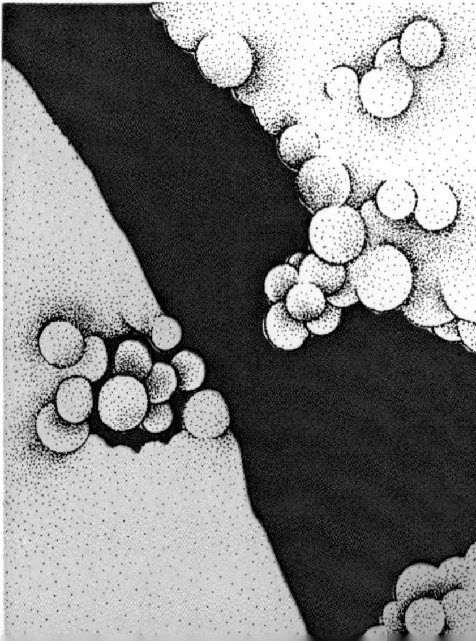

New Battle Plans
The latest knowledge of allergy's mechanisms suggests three ways to fight it. In the first, *top pair,* an allergen that normally forms a bridge between two IgE molecules is altered chemically so that it can attach to only one. This prevents histamine release and attachment of unaltered allergen. In the second, *middle pair,* a proposed trigger normally activated by the bridged molecules is chemically capped. In the third, *bottom pair,* a synthetic portion of the IgE molecule blocks the attachment of a whole, natural IgE molecule.

developing and testing substances that are chemically related to cromolyn, and may be close to some exciting breakthroughs. One breakthrough might be the development of a substance that is effective against allergies other than asthma. Another could be a cromolynlike substance that can be taken in pill form.

Perhaps the most promising therapies will spring from the more detailed knowledge that we have gained of the biological chain reaction that ends in an allergy attack. Some of the most interesting research of this particular kind is aimed at developing treatments that focus on and around the surface of the mast cells and the basophils. Such research includes hapten inhibition, trigger-site inhibition, and IgE-binding-site inhibition.

The work on hapten inhibition is based on our knowledge that each allergen molecule has several sites that fit, and therefore can bind to, various IgE molecules. A hapten is a substance with only one such site. There are two kinds of haptens, a tiny fragment of an allergen molecule and an entire molecule with all but one site chemically blocked. A hapten can, therefore, bind to only one IgE molecule at a time. This means it not only cannot cause bridging, but it also blocks binding sites that would otherwise be available to natural allergen molecules.

In 1972, immunologist Alain L. de Weck and his colleagues at the University of Bern, Switzerland, described research that proved the effectiveness of hapten inhibition. They worked with penicilloyl, the molecule within penicillin that causes most penicillin allergy. The scientists attached two chemical complexes to the penicilloyl molecule, a relatively small and simple allergen, one at one end and the other on the side. This procedure, they felt, should have blocked all but one binding site, producing a penicilloyl hapten. When injected into patients experiencing severe allergic reactions to penicillin, the material promptly alleviated the symptoms and also eliminated their previously positive tests for penicillin allergy.

But there were some problems. For one thing, about 10 per cent of the patients had allergic reactions to the hapten. For another, the substance helped only 80 per cent of the patients. Also, many allergists doubt that haptens of other allergens can be produced, because most of them are so much more complex than penicilloyl. To see if these doubts are valid, de Weck and his colleagues and physician Arthur Malley of the Oregon Regional Primate Research Center, Beaverton, are now using chemical techniques to break allergens into tiny pieces, then injecting them into patients to see if the pieces inhibit allergy.

Signal-receptor inhibition, still highly theoretical, is based on a hypothesis proposed by immunologist Denis R. Stanworth of the University of Birmingham in England in 1973. In experiments, he observed that some relatively small molecules composed of amino acids caused mast cells and basophils to release histamine when neither allergens nor bridged IgE molecules were present. Aware that IgE molecules are made up of chains of amino acids, Stanworth theorized that some segment of them acts as his small molecules did. He proposed that his small molecules caused histamine to be released by contacting a trigger site on the cell membrane and that bridging brought the proper site on an IgE molecule into contact with the same trigger. He is now searching for a small chain of amino acids that will bind to the trigger without tripping it. This would block the trigger, preventing its firing through the normal allergic mechanism.

My own research is on IgE-binding-site competition. It is based on several things we know about IgE molecules. First, the portion of the IgE molecule by which it binds to a mast cell or basophil lies in a specific part of one of two of its chains, called heavy chains because they are longer than the other two and thus have heavier molecular weights. Second, large numbers of IgE molecules formed in response to one allergen can prevent allergic response to a second allergen. Too, very large pieces of the heavy chain of the IgE molecules can also prevent allergic response. The most reasonable explanation is that the molecules or large pieces attach themselves to IgE receptors of the mast cells and basophils, thus preventing the IgE formed in response to the second allergen from attaching. Because small pieces of the heavy chain have failed to prevent the allergic response, researchers have assumed that binding requires almost the entire chain.

I reasoned differently. I noted that immunoglobulin molecules other than IgE contain a region in their heavy chains where they perform specific functions not performed by IgE molecules. Perhaps the IgE molecule's binding site, which gives it its specific ability to attach to mast cells and basophils, lay in its corresponding region. So I compared the amino acid sequences in the IgE molecule's heavy chain with the sequences in the heavy chains of the other immunoglobulins. I found several unique sequences of from 5 to 10 amino acids in the IgE molecule's chain. Some of these sequences are in areas that correspond roughly with the specialized region in the other immunoglobulins. Explaining what I had done, I convinced biochemist Russell F. Doolittle of the University of California, San Diego, to synthesize several of the sequences. One of these sequences was the substance I injected into myself before becoming ill.

In my experiments testing the substances against allergy, I used a recognized test based upon the technique that Prausnitz and Küstner used to show that allergy depends on some material in the blood. I first injected myself and other test subjects with serum from a patient allergic to guinea pig dander. Then, as expected, the guinea pig dander raised a full allergic response–inflammation and swelling–at the injection sites. However, when I first injected one of my substances, then the serum, the allergic reaction was about 70 per cent less severe. The substance is a pentapeptide, a chain of five amino acids.

The fact that my substance inhibits allergic reaction does not prove that its amino acid sequence is the same as that of the binding site of IgE, and attaches to and blocks the receptor sites on mast cells and basophils. The Ishizakas and others think that its success comes from some druglike activity. These scientists could certainly be right. In our experiments with the substance on living subjects, it was impossible to keep track of what happened at the molecular level.

However, a series of experiments in 1974 and 1975 seems to indicate how our molecule works. We reasoned that even if the substance has the same amino acid sequence as the binding site of an IgE molecule, it would have difficulty binding to mast cells or basophils. This is because it would not keep the shape it has when it is part of the chain of hundreds of amino acids that are normally its neighbors. However, it should be flexible enough to assume its normal shape and fit the cell receptor site when in close proximity to it. Therefore, it should take many more molecules of our substance than of IgE to get an equal quantity of molecules attached to the cell. If we could indirectly prove this, we would have evidence that molecules of our substance, although in small numbers, were binding to the cell surface.

We used human basophils in test tubes for our experiments, adding varying quantities of whole IgE molecules to some of the test tubes and varying quantities of our molecule to others. Then we added radioactive IgE molecules to both groups of tubes. The radioactive IgE molecules could attach to the cell membranes only at receptor sites that the original IgE molecules or our molecules had not occupied. By measuring the radioactivity left on the cells after they were rinsed, we found, as we suspected, that many more of our molecules were required to get a single one attached. Indeed, the number of molecules of our substance that had to be added to the cells to get one molecule attached was 2,000 times greater than the number of IgE molecules required to attach one. This means that the binding capacity of our molecule is 2,000 times less than that of the IgE molecule.

The relatively low binding capacity of our substance might appear to make it a poor prospect as a potential drug against allergy. In fact, however, there are innumerable ways of altering a molecule's structure slightly to change its biological activity, and we are planning such work on our molecule, confident that we can increase its ability to bind by from 100 to 1,000 times. The biggest problem is determining the consequences of blocking the binding sites on all mast cells and basophils in the body. We can be relatively sure that this would eliminate all allergic response. But the biochemistry of the human body is so complex and our knowledge of it so limited that many experiments will be needed before we can be sure that such treatment will not produce harmful side effects.

Scientists in our laboratory and others will continue the search for new drugs and other methods of combating allergy. The stakes are high. There are more than 35 million allergy sufferers in the United States alone, and many millions more throughout the world. But we are headed for the day when the sneezing and wheezing, and all the discomfort and danger related to allergies will be no more.

Plants: The Renewable Resource

By John F. Henahan

As supplies of fossil fuels wane, scientists are perfecting ways to tap the sun's energy through plants and their discarded products

The Green Mountain Power Company of Burlington, Vt., considered a deliberate return to the past in 1976. Forced by environmental concern to burn low-sulfur coal costing about $50 per ton in its power plants, the company debated whether it may be cheaper and cleaner to go back to burning wood, the main source of energy in the United States until about a century ago. Their new look at an old fuel was spurred by an ample supply of surplus timber in Vermont's cool forests and a new harvesting technique that turns a tree into a pile of uniform wood chips in seconds, at a cost of about $5 per ton. The company estimated that enough wood is available to fuel up to one-fourth of Vermont's electric generators.

The Russell Corporation in Alexander City, Ala., ordered by the United States Environmental Protection Agency (EPA) to reduce pollution from its coal-burning boiler or close down operations, bought a $1.7-million power plant to burn wood wastes from nearby sawmills. Eventually, this apparel manufacturer plans to burn wood chips.

These and an increasing number of other companies located in forested regions are turning to wood because it is relatively cheap, widely available, and leaves a residue after burning that can be used as an agricultural fertilizer. It is also renewable. Trees and other

A Chiparvestor converts a tree to wood chips that can be burned to generate electricity.

plants turn water and carbon dioxide into burnable carbohydrates in the continuing sunlight-powered process of photosynthesis. In contrast, the mainstays of our energy supply–coal, oil, and natural gas–are the products of photosynthesis that was completed eons ago. As such, they are doomed to extinction. Known reserves of oil and natural gas in the United States may not last beyond 1990.

Although wood is a good fuel, many scientists believe that some substances found in wood and agricultural products are too valuable to burn. They are particularly intrigued by the carbohydrate cellulose, the woody part of trees and other plants found mostly in the walls of their cells. Cellulose is already widely used to make paper, fabrics, and plastics. Now scientists are studying new biological and chemical ways to turn cellulose into protein supplements to feed the hungry people of the world, as well as into the chemicals needed for literally thousands of products now made from coal, oil, and natural gas. The more optimistic forecasters envision a future world economy with cellulose as the key resource.

They base their forecasts on some staggering numbers. Plant life throughout the world makes an estimated 100 billion tons of cellulose each year by capturing and storing solar energy. Through photosynthesis, some of this energy is stored in the bonds of sugar molecules, especially glucose, which the plants forge into long, chainlike cellulose molecules. Cotton fibers are about 90 per cent cellulose, and about 45 per cent of a typical tree's cell walls are cellulose. The long cellulose chains form three-dimensional structures with shorter hemicellulose chains that contain sugar molecules other than glucose. The structure is held together by a complex substance called lignin.

The author:
John F. Henahan, a free-lance science writer, wrote "James Gunn" for the 1976 edition of *Science Year.*

Many scientists are concerned because much of the world's solar energy supply, captured and stored in trees and other plants, is literally going to waste. They have proposed several schemes to grow, harvest, and release that stored energy on a massive scale. In 1970, George C. Szego of InterTechnology Corporation in Warrenton, Va., suggested a concept that he called the energy plantation. Land that marginally produces food crops, forage, or trees for lumber would be used to grow "energy bushes," fast-growing trees or other plants raised expressly to fuel nearby electric power plants. InterTechnology scientists estimate that about 40 million hectares (100 million acres) of such land is available in the United States.

Converting solar energy to electric energy in this way is cheaper than using the most efficient man-made solar collectors. But to compete with oil and coal, energy bushes that capture and store more solar energy are needed. To this end, InterTechnology scientists and foresters at Pennsylvania State University are growing hybrid poplar trees on test plots in central Pennsylvania. One of the most efficient hybrid poplars the foresters have developed is Clone 388. It converts about 0.6 per cent of the sunlight to chemical energy and can be planted as densely as 1 tree every 4 square feet (0.4 square meter). The trees grow rapidly, and after they are chopped down, new trees spring from the stumps. This cycle can be repeated several times.

InterTechnology biologist Charles W. Vail reports that Bermuda grass, certain relatives of the bamboo plant, and sugar cane might be

A tall stand of mature, regularly spaced, 7-year-old sycamores, *far left,* is ready for harvesting. New growth quickly sprouts from the stumps, *left,* and the regenerated trees grow taller than a man in less than a year, *below.*

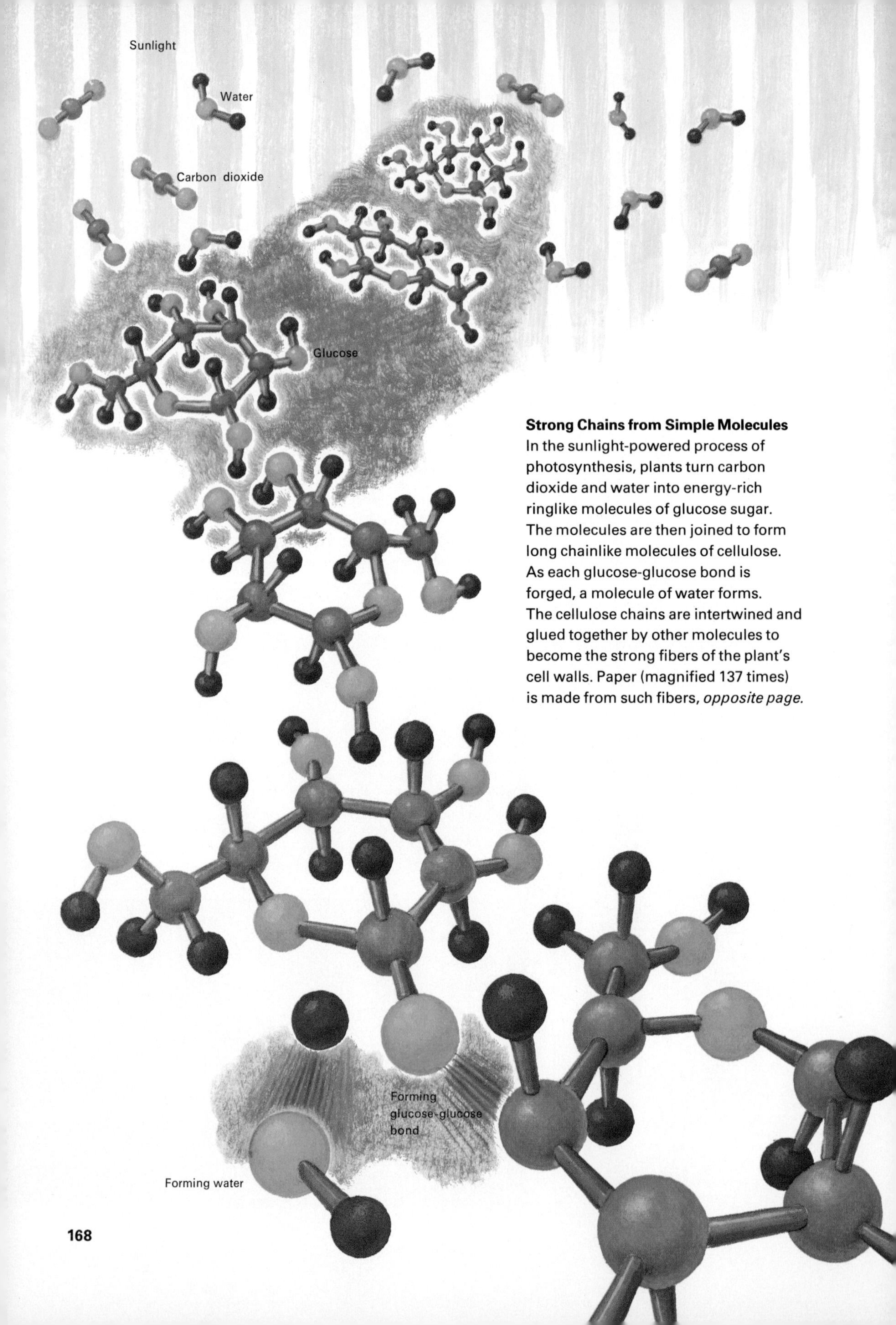

Strong Chains from Simple Molecules
In the sunlight-powered process of photosynthesis, plants turn carbon dioxide and water into energy-rich ringlike molecules of glucose sugar. The molecules are then joined to form long chainlike molecules of cellulose. As each glucose-glucose bond is forged, a molecule of water forms. The cellulose chains are intertwined and glued together by other molecules to become the strong fibers of the plant's cell walls. Paper (magnified 137 times) is made from such fibers, *opposite page.*

good energy bushes in warm climates. Sugar refiners in Louisiana now fuel some power plants with bagasse, the cellulosic pulp left after the juice is squeezed out of sugar cane. InterTechnology scientists estimate that an energy plantation could produce electricity as economically as can coal. But their estimates have yet to be tested. Vail believes the first energy plantations should be located in areas that already have large amounts of cellulosic wastes from farms and forests, which could be used to supplement the energy bushes.

Many scientists think we should first tap the vast quantity of waste wood and plant material available in the United States before planting large tracts with energy bushes. Biochemist Arthur E. Humphrey of the University of Pennsylvania estimates that such agricultural and food wastes as bagasse, corncobs, cornhusks, potato peelings, and cereal straw and husks add up to 400 million tons of dry material each year. Livestock manure, which contains a high proportion of undigested cellulose, accounts for another 200 million tons, and it is easy to collect because much of it is concentrated in feed lots. When the other solid wastes are added in, the total figure reaches 1 billion tons annually, half of which is cellulose.

Garbage, old newspapers, and other municipal wastes–150 million tons a year are collected in U.S. towns and cities–are accessible sources of cellulose. Such solid wastes have become extremely difficult to dispose of in environmentally tolerable ways. Big-city incinerators add to air pollution; garbage dumps pose health and fire hazards; and landfill sites are dwindling near many urban areas.

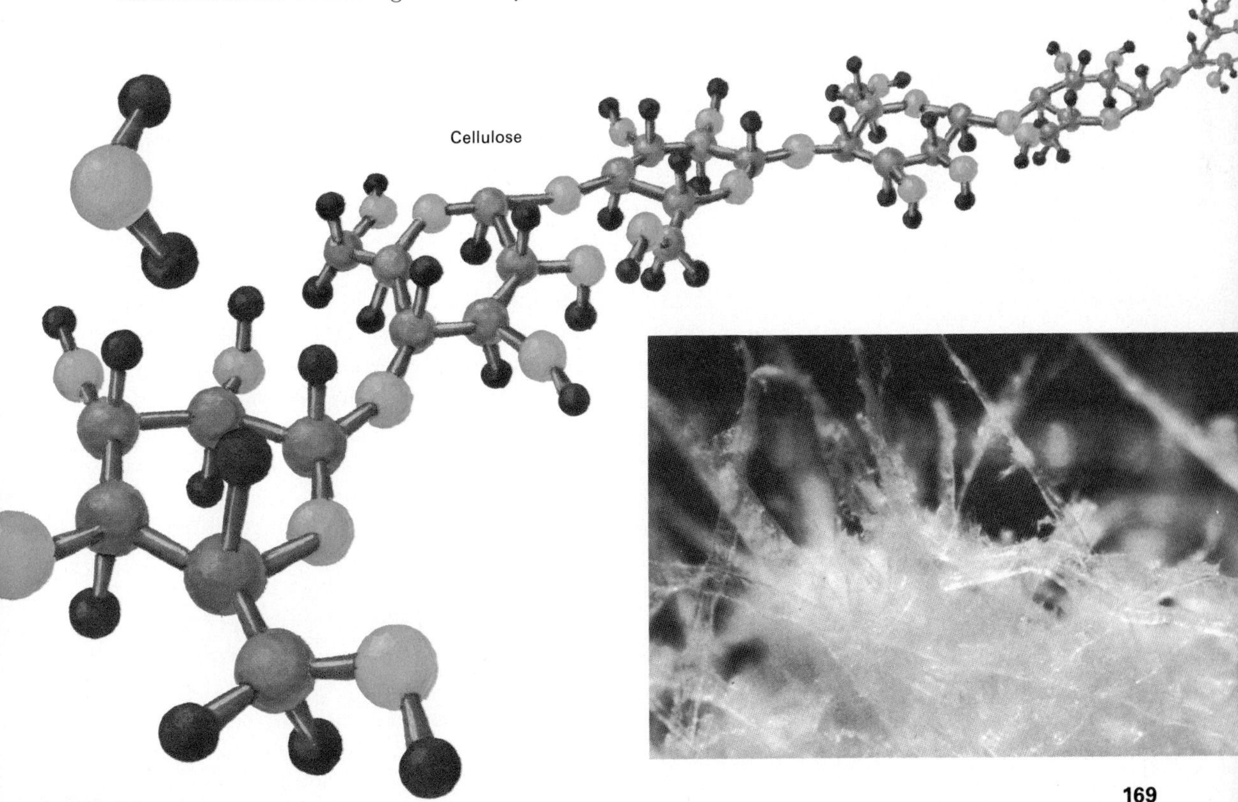

A loader pushes urban wastes onto a conveyor, *left.* Separated from reusable glass and metals, shredded cellulose waste is burned to generate electricity, *above,* or pyrolyzed to make synthetic fuel oil, *below left.*

Scientists and engineers have developed almost 50 processes to convert municipal wastes to burnable fuels. For example, the Union Electric Company in St. Louis, Mo., now separates metal and other heavy materials from the city's municipal wastes, then shreds the remaining cellulose and organic waste and burns it in a specially modified steam boiler to generate electricity. The company plans to spend $70 million to expand its 300-ton-per-day trash-processing plant to supply all St. Louis electric power plants. It will reduce dependence on coal by as much as 20 per cent. Similar schemes are operating or being planned in other cities including Ames, Iowa; Chicago; Bridgeport, Conn.; Philadelphia; and Menlo Park, Calif.

San Diego County, California, is completing a 200-ton-per-day waste-processing plant to make fuel oil that the local power company will burn in its oil-fired boilers. David Anderson of the San Diego County Department of Community Service says that it makes more sense to convert the wastes to oil, because oil is easier to store, transport, and burn. The cellulosic waste is converted to fuel oil by pyrolysis, a process in which the waste is partially decomposed in an oxygenless atmosphere at a temperature of 700°C (1300°F.). The heating value of a barrel of the resulting oil is about 75 per cent that of a barrel of commercial fuel oil.

The San Diego pyrolysis demonstration plant is a joint venture with the EPA and the Occidental Research Corporation of La Verne, Calif. It converts about 36 per cent of the energy input–wastes, electricity, and a small amount of oil needed during pyrolysis–into pumpable energy in the form of fuel oil. Overall costs were estimated originally at about $13 per ton of wastes, but have risen to over $20 per ton, more than the projected cost of transporting the wastes to distant landfills. The designers of the pyrolysis test program, which ends in

1977, claim that larger pyrolysis plants would be more economical. To help offset the costs, the waste is separated into its valuable components—200 tons of waste yields about 24 tons of iron, 10 tons of glass, 2 tons of aluminum, and about 100 tons of organic wastes, half of which is cellulose. The metal and glass can be sold to scrap dealers and bottle and can makers. When converted by pyrolysis, the cellulosic waste yields 34 tons of liquid fuel.

Many scientists, convinced that wood and cellulosic wastes are too valuable to burn, argue that chemical compounds locked in these materials can provide the building blocks for many manufactured items we take for granted. Or these compounds could be turned into food for the world's hungry. Glucose is especially important because fermentation by microorganisms can turn it into ethyl alcohol, and chemists know how to turn ethyl alcohol into thousands of the chemicals needed to make plastics, synthetic fibers, rubber, and drugs.

Unfortunately, much of the glucose a plant produces is chemically locked into the cellulose of the cell walls and is difficult to extract. As each glucose molecule joins several thousand similar units in forming the giant, chainlike cellulose molecule, one molecule of water is released. To break the chains apart into glucose units, water molecules must be put back in by hydrolysis, a process that requires energy.

Lignin, the polymer that glues parallel chains of cellulose and hemicellulose into tough, chemically resistant fibers, adds to the problem. Some trees contain as much as 30 per cent lignin. Its tough protection is a good quality if the wood is to be used to build a house or its furnishings. But it is a nuisance to scientists who want to unravel the cellulose molecules and break them apart into glucose units. The scientists must use strong chemicals and high temperatures and pressures. Fortunately, nature has provided a biological way to break the cellulose chains into glucose molecules. Certain microorganisms produce enzymes that hydrolyze cellulose. Many animals, such as cows and other ruminants, benefit directly from these microorganisms, which thrive in the animals' digestive tracts.

Researchers at the United States Army's Natick Laboratories in Massachusetts found a way to harvest enzymes from a fungus and then use the enzymes to turn wastepaper and other cellulosic materials into sugar syrup. Their work traces back to the South Pacific during World War II, when soldiers discovered that cartridge belts and other cotton fabrics were mysteriously being eaten away. Microbiologists identified the culprit as *Trichoderma viride,* a fungus that makes the enzyme cellulase. Cellulase breaks cellulose into glucose, on which the fungus feasts.

The Army solved its fungus problem by developing sprays and switching to synthetic fabrics, and *T. viride* was all but forgotten. Then, in 1970, researchers in the Pollution Abatement Division of Natick Laboratories suggested that the fungus might help solve the mounting waste-disposal problem. They designed a system to convert trash to sugar and developed more efficient strains of *T. viride.*

The Army scientists created the new strains by exposing groups of *T. viride* to cobalt radiation. The radiation caused many genetic mutations, some of which changed the fungus' ability to produce cellulase. Some of the mutants produced no cellulase and were discarded, but others produced more. A mutant called QM9414 made four times more cellulase than did the original *T. viride*. Later, the Natick scientists found that at least two cellulase enzymes, working together, had been responsible for converting cellulose to glucose efficiently. They now know that one loosens the bonds between the rigid, interlocking cellulose fibers, allowing the second to break apart individual chains.

In the first step of the Natick process, *T. viride* QM9414, feeding on pure cellulose and several growth-promoting nutrients and hormones, makes large amounts of cellulase in a 30-liter (8-gallon) bioreactor. When enough cellulase has been produced, the fungus is filtered off. The remaining yellow liquid, which contains the enzymes, is fed into a second vessel, a 250-liter (65-gallon) bioreactor, and begins attacking cellulose-rich wastes there.

To produce the most glucose, the researchers first soften up the cellulose for the attacking enzymes. They shred newspapers or other cellulosic wastes, then pulverize them in a ball mill to partially break down the interlocked fibers. Added to the large bioreactor, the treated wastes are converted to a sugar syrup containing as much as 15 per cent glucose. This is done at atmospheric pressure and temperatures of only about 50°C (122°F.). The lignin in the cellulosic wastes is not attacked by the enzymes. Along with other by-products, it is filtered off, dried, and thrown away. However, chemical engineer Leo A. Spano believes that the lignin could be burned as fuel or converted to useful chemicals economically if the Army system expands as planned to produce 9,100 kilograms (20,000 pounds) per month.

Spano is enthusiastic about the Army's enzymatic hydrolysis process. In theory, it can produce about 10 per cent more glucose by weight than the waste cellulose originally contained because the cellulase enzymes put back the water molecules that the plants took out to form cellulose. Moreover, because cellulase acts only on cellulose, the process can be used for all kinds of wastes. Harsh chemical methods of putting back the water chew up the lignin and many other compounds as well as the cellulose. Enzymatic hydrolysis, however, produces a relatively pure glucose syrup.

Economic realities strike the only sour note in the story. Spano estimates that it would cost about 24 cents per kilogram (11 cents per pound) to make sugar by the Natick process, compared to about 17 cents per kilogram (8 cents per pound) for commercial sugar in 1976. However, he is confident that costs can be reduced substantially by varying such reaction conditions as temperature and acidity, by lowering the pretreatment costs, and by using cheaper, more available forms of waste cellulose. He is also convinced that researchers may yet discover *T. viride* mutants that will increase enzyme production.

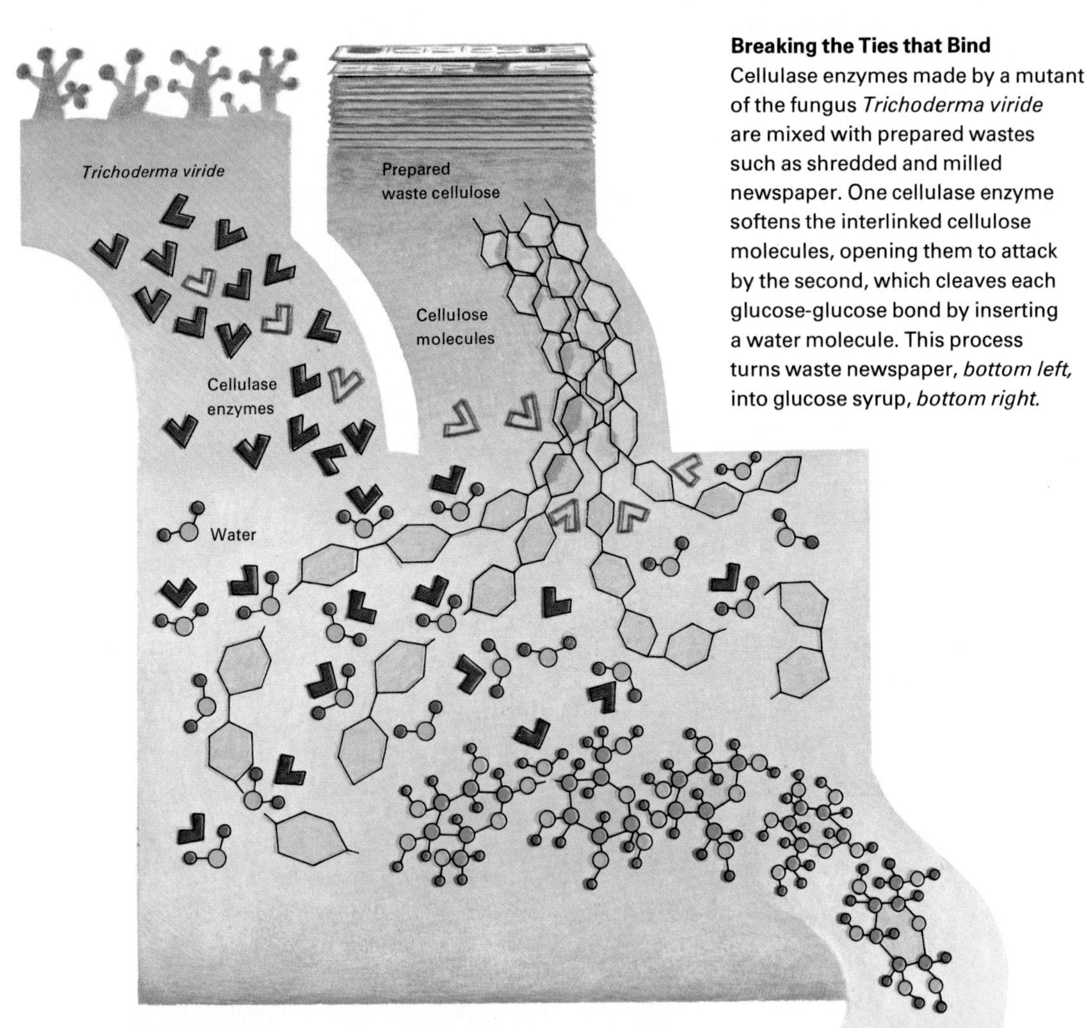

Breaking the Ties that Bind
Cellulase enzymes made by a mutant of the fungus *Trichoderma viride* are mixed with prepared wastes such as shredded and milled newspaper. One cellulase enzyme softens the interlinked cellulose molecules, opening them to attack by the second, which cleaves each glucose-glucose bond by inserting a water molecule. This process turns waste newspaper, *bottom left,* into glucose syrup, *bottom right.*

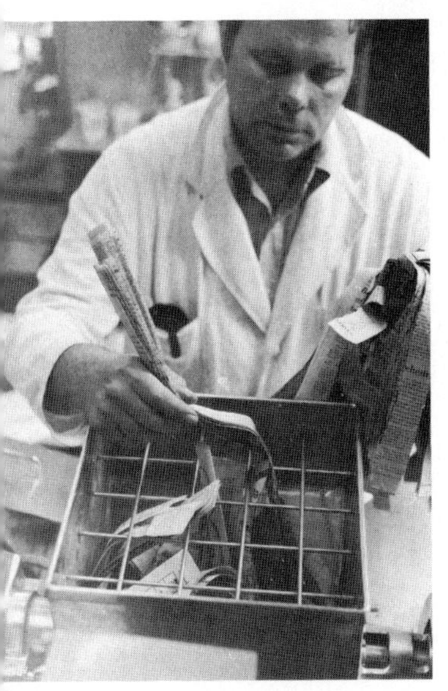

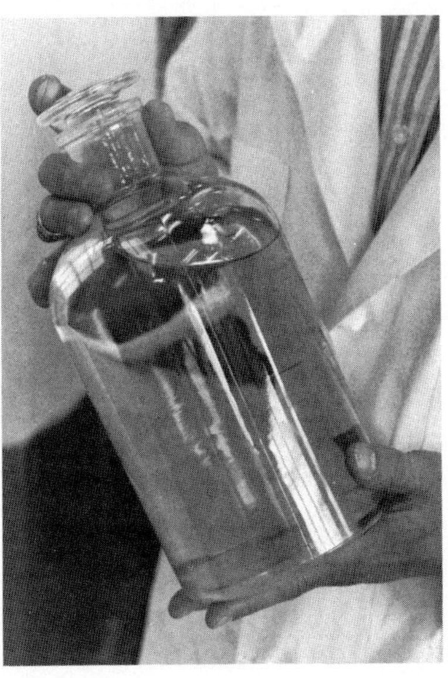

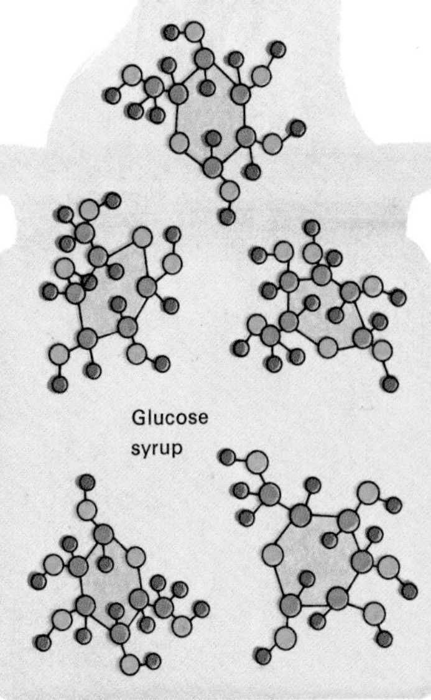

Citrus waste, *above,* is fed to edible fungi in a fermenter, *above right,* in Belize, in Central America. The fungi, which thrive on cellulose, are harvested and fed to animals as a protein supplement.

If costs can be lowered, biologists, chemists, and food technologists are prepared to convert the glucose into still other products. Certain yeasts and other microorganisms that feed on sugar produce ethyl alcohol. This important yeast fermentation process daily converts billions of kilograms of starches and sugars in cereal grains to ethyl alcohol in the world's breweries. Chemical engineer Charles R. Wilke and his associates at the University of California, Berkeley, have tacked a yeast fermentation process onto a *T. viride* enzymatic hydrolysis process and produced a high grade of ethyl alcohol.

About 1.1 billion liters (300 million gallons) of ethyl alcohol are produced each year for industrial uses from ethylene, an important chemical that comes from oil or natural gas. If sugar can be produced cheaply enough from cellulosic wastes, fermentation might replace chemical synthesis of ethyl alcohol and even open up new uses. For example, some scientists propose a gasoline blend containing 10 per cent ethyl alcohol to stretch our dwindling petroleum supplies. Ethyl alcohol by itself might make a clean-burning, low-knock fuel for modified automobile or bus engines.

If cellulose-derived ethyl alcohol becomes cheap enough, it could be used to make the chemicals that are now used to make it. For example, it could be converted to ethylene, which is used to make polyethylene, a plastic used in products ranging from bottles to space-vehicle components. Butadiene, the starting chemical for synthetic rubbers, and other chemicals can also be made from ethyl alcohol.

Food technologists have plans for the yeast and other microorganisms that multiply as they feed on the glucose during the fermentation process. These microorganisms are a concentrated source of protein, and they can be harvested and fed as high-protein food supplements to animals–perhaps, ultimately, to humans.

Chemical engineer Clayton D. Callihan and his co-workers at Louisiana State University (LSU) in Baton Rouge take a direct approach to this cellular husbandry. They produce cellulomonas, a high-protein bacterium that feeds on the cellulose in bagasse. Cellulomonas has a powerful appetite for straw, grass, corncobs, and many other agricultural wastes. LSU microbiologist V. R. Srinavasan first isolated the strain in the soil of a Louisiana sugar-cane field about eight years ago.

To make the bagasse and other agricultural wastes more digestible for the cellulomonas, the LSU researchers boil the wastes in an alkaline solution. The pretreated cellulose, other nutrients, and cellulomonas are combined in a fermenter where the bacteria grow rapidly. The solids left after fermentation, mostly bacteria, are filtered and dried to a granular material that is more than half protein. It is a high-quality protein and appears to be a safe food supplement. According to Callihan, rats fed a diet containing as much as 40 per cent of it gained weight steadily with no ill effects.

Spano also claims that cellulose-derived protein is inexpensive; he estimates 31 cents per kilogram (14 cents per pound). This cost compares favorably with the cost of other microbial proteins, including those produced by the fermentation of petroleum. Soybean protein, considered one of the most nutritious plant sources of protein, costs about 80 cents per kilogram (35 cents per pound).

Although agricultural wastes may be converted profitably to high-protein animal feed on a large scale, the question remains whether this process can benefit the world's hungry people. "What is really needed in the underdeveloped nations," says Frazer K. E. Imrie of Tate & Lyle, Limited, a British sugar producer, "is a village technology that would convert readily available agricultural wastes into animal feed using only inexpensive fermentation equipment." Many African and Central and South American nations must import such traditional animal feed as soybeans and fish meal from more developed nations. For example, the British protectorate Belize in Central America imports about 2,200 tons of animal feed each year at a cost of nearly $300 per ton. At the same time, Belize generates each year about 2,300 tons of citrus wastes, which Imrie believes could easily be converted into animal feed for about $200 per ton.

To test the village-technology concept, Tate & Lyle built a waste-to-protein plant in Belize that works on the same principle used in the LSU process. Instead of cellulomonas, a fungus variety called *Aspergillus niger* ferments the cellulosic citrus wastes, producing about a ton of dried protein supplement per week. If the test proves successful, similar conversion plants may be built to help reduce feed imports, create jobs, and improve the diet in many underdeveloped nations.

Although the sun's energy–locked in the chemical bonds of plant cellulose molecules by photosynthesis–can be released in power plants or converted to food and industrial chemicals, many of the conversion techniques are too costly now. But scientists are motivated to continue their cellulose-utilization research by a look at the energy hourglass through which natural gas, oil, and even coal and nuclear fuels are relentlessly passing. They know that the time will come when plant life may be the only earthly energy resource we have left.

Unveiling Venus

By Michael J. S. Belton

Probing through the clouds that shroud Earth's nearest neighbor, astronomers find clues to a stark new picture of that inhospitable planet

Venus is like no other planet in the solar system. Highly acidic mists form the thick clouds that cover it. Violent winds whip through the upper reaches of its deep, oppressive atmosphere. Beneath the clouds, a torrid landscape lies cracked and parched, openly scarred by activity that begins deep in its interior.

Our image of Venus has changed vastly in the last few years as the result of a spectacular expansion in our ability to explore it. The United States has sent three Mariner spacecraft flying by Venus, and Russia has sent seven Venera probes into the atmosphere, including four that successfully landed and sent back information from the surface. On Earth, new equipment and techniques make telescopic observations of Venus far more efficient and sensitive than before.

Venus was once considered the Earth's twin. In size and density, it is similar to the Earth. Venus is 12,104 kilometers (7,521 miles) in diameter with a density of 5.2 grams per cubic centimeter; the Earth has a diameter of 12,756 kilometers (7,926 miles) and a density of 5.5 grams per cubic centimeter. But the two planets have little else in common. Venus moves in an orbit about 108 million kilometers (67-

million miles) from the Sun. Because it is 40 million kilometers (25-million miles) closer than the Earth, Venus receives twice as much solar energy. Venus rotates in the opposite direction to the Earth, and makes a complete rotation only once every 243 days, while it takes 225 days to go around the Sun. Because of the slow, backward rotation, a night on Venus lasts for 58 Earth-days, and there are less than two Venus days in one Venus year. Also, by some strange coincidence, Venus' rotation period seems to be related to that of the Earth. Whenever Venus is closest to the Earth, it presents the same side to us.

Many of the known facts about Venus have been derived from Earth-based telescopic observations. Two techniques in particular seem likely to continue to add to our knowledge. Physicists Albert Betz, Robert McLaren, Michael Johnson, and Edmund Sutton of the University of California, Berkeley, have developed a highly sensitive instrument called the infrared heterodyne spectrometer. They used it in 1976 to measure emissions from carbon dioxide molecules in the high atmosphere of Venus. Because the wavelengths of the emissions seem to shift depending on the velocity of the source (a phenomenon known as the Doppler effect), Betz and his associates were able to measure wind speeds on Venus by comparing the wavelengths of carbon dioxide at different points on the planet. Their measurements are so precise that they can determine wind speeds with an accuracy never before possible from Earth.

The author:
Michael J. S. Belton, astronomer at Kitt Peak National Observatory, specializes in studies of the planets. He is particularly interested in atmospheric dynamics.

In 1968, the French astronomer Pierre Connes first used another instrument–the Fourier Transform Spectrometer (FTS)–in the study of the planets. Connes estimated that the FTS would be 10 billion times more precise at infrared wavelengths than a conventional planetary spectrometer, thus permitting observations that would otherwise be impossible. The immediate result of using this instrument was the discovery of three important molecules on Venus–hydrochloric acid, hydrofluoric acid, and carbon monoxide.

The FTS is only now coming into widespread use. It works on the principle of interference, using high-performance mirrors. Incoming light from a planet is divided into two equal beams that travel over separate, carefully arranged paths before being recombined and focused on a detector. By introducing a precisely known delay in the path of one beam, scientists can make the two beams interfere with each other when they come together again, changing the intensity of the recombined beam to produce a graph called an interferogram. From the interferogram, scientists can derive a spectrum used to analyze the characteristics of the planet's atmosphere. The FTS is more than an extremely efficient detection device. When it is coupled with the increasingly sophisticated detector arrays that are now becoming available, it also will provide a very precise spectral picture of the entire planet. Previously, an astronomer could obtain the spectrum of all places on the disk of a planet only by laboriously observing the spectrum point by point.

Because it had been difficult to study Venus from Earth until recently, the Mariner and Venera space missions, which started in the early 1960s, have provided much of our information about the planet. The spacecraft carried spectrometers–which provide data on chemical composition, pressure, temperature, and motion in the atmosphere–and other specialized scientific equipment to study Venus. The Mariner fly-by missions sent back data on the planet's size and mass, and information about conditions in the upper atmosphere. The Venera craft, which entered the atmosphere, sent back even more detailed information on temperature, composition, and wind speed.

The atmosphere at the surface of Venus is 50 times as dense as that of the Earth and the atmospheric pressure is almost 100 times as great. The average surface temperature is 475°C (890°F.), compared with 14°C (57°F.) on Earth. The atmosphere is almost entirely carbon dioxide; there is little free oxygen and virtually no water.

Why should Venus' atmosphere be so different from the Earth's? The planets are next to each other in the solar system, and their masses and densities are almost the same, suggesting that they are made up of much the same kind of material. Thus, we would expect their atmospheres to be similar also.

Venus has very little free oxygen, probably because the planet has no plants. In the solar system, abundant oxygen is found only on the Earth where it is thought to be the waste product of plant life processes. This makes it a relative latecomer in the atmosphere.

The large amount of carbon dioxide in Venus' atmosphere might well result from the planet's lack of water. When carbon dioxide dissolves in water, it combines much more readily with other substances. On Earth, the extensive oceans provided the water necessary for this process. Once dissolved, the carbon dioxide rapidly became chemically trapped as limestone, chalk, and other carbonate deposits. This could not happen on Venus, and the carbon dioxide remained in the atmosphere. Less than 1 per cent of the Earth's present atmosphere is carbon dioxide; more than 4,000 times that amount is trapped in the rocks. If all the trapped carbon dioxide were released, Earth's atmosphere might become quite like that on Venus.

But why is there no water on Venus? On Earth, water originally formed from the hydrogen and oxygen released from minerals in the hot interior. Perhaps Venus simply did not have enough hydrogen-bearing minerals to form large amounts of water vapor that could then escape through the surface to the atmosphere. Many planetary astronomers consider this unlikely because primitive Venus and Earth were so close together and apparently similar in composition.

Another possibility is that there were large amounts of water in the early Venus atmosphere. But, because Venus receives twice as much solar energy as the Earth, the two atmospheres evolved along entirely different lines. Astronomers expect atmospheres normally to develop until they reach an equilibrium state, and then remain stable and

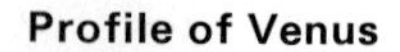

Profile of Venus

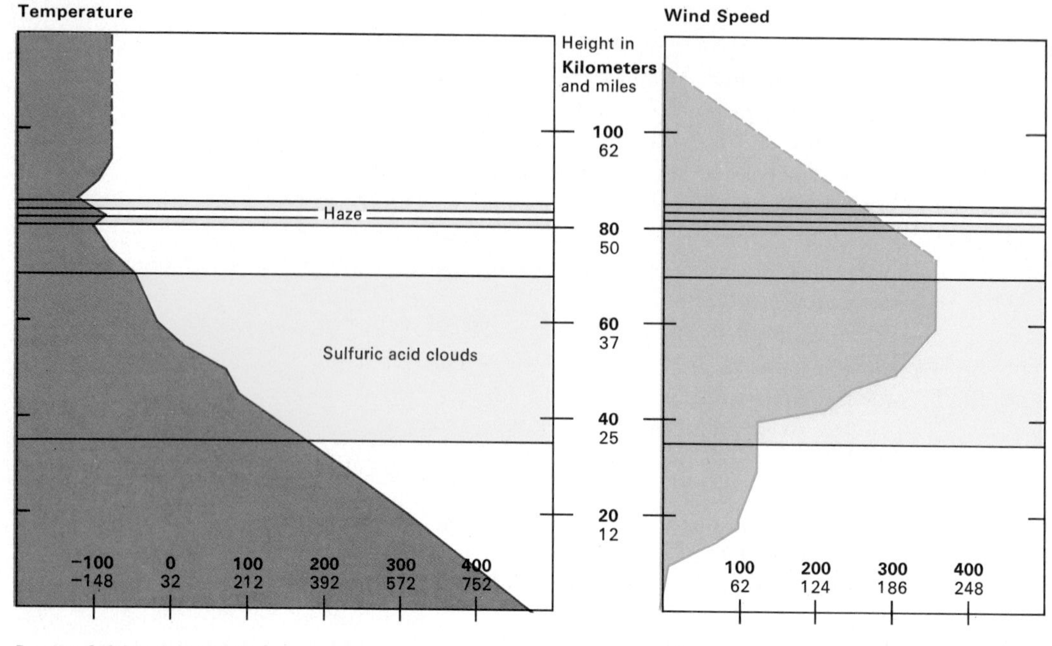

Temperatures in Venus' atmosphere get steadily hotter with decreasing altitude, reaching close to 480°C (900°F.) at the planet's surface. Wind speeds peak near the cloud tops and drop to almost zero at the surface. Thin layers of haze float above the sulfuric acid cloud deck.

predictable over long periods of time, with the energy received from the Sun and that lost to space more or less in balance. However, this stable state may have been impossible on early Venus.

Here is what might have happened. In the first billion years after the Earth and Venus formed, their interiors would have settled into states much like they are now, with heavy materials sinking to the center and light ones floating to the outside crust. In the process, gases would have escaped from the interior, perhaps through volcanoes, to form the atmosphere. At this early stage, both atmospheres would contain large amounts of carbon dioxide and water. Stimulated by the Sun's energy, they would develop until they reached a state where input and loss of energy balanced. As Earth settled into this equilibrium state, most of the water condensed out of the atmosphere and ended up in the oceans. But Venus had a much more difficult time reaching equilibrium. The temperature was just too hot for water to condense, so it remained a gas. The water vapor acted as a very efficient insulator and prevented heat from escaping into space, thereby forcing the temperature to keep rising. In the outer atmosphere, sunlight changed water into hydrogen and oxygen; the hydrogen escaped into space, and the oxygen settled back to the surface to slowly combine with the surface rocks. Not until Venus almost entirely lost the water in its atmosphere did it finally settle into equilibrium.

The atmosphere of Venus contains many other substances, including those that have condensed to form the clouds. Astronomers so far know little about the extent of the cloud layers or whether the layers differ in composition. Mariner 10 found several layers of haze about 80 kilometers (50 miles) above the surface, but these are quite thin and insubstantial. The main cloud layer has its top between 60 and 70 kilometers (37 to 44 miles) above the surface. How far down it goes is still a matter of speculation.

The chemical nature of the clouds was a mystery for many years. Only lately have scientists reached any strong theoretical agreement about what they are–planetwide mists of concentrated sulfuric acid droplets. Similar mists occasionally occur on Earth as the result of urban pollution, smelter fumes, and fumes from catalytic converters on automobiles. In short, the visible clouds on Venus can be considered an extreme example of global smog.

On Earth, rain washes most of the sulfuric acid and associated sulfates out of the atmosphere. This cannot happen on Venus, so the sulfuric acid remains suspended in planetwide mists.

But why sulfuric acid on Venus? Ronald G. Prinn and John S. Lewis of the Massachusetts Institute of Technology in Cambridge theorize that the high temperatures on Venus release sulfur com-

Astronomers at Kitt Peak National Observatory adjust the Fourier Transform Spectrometer. It sends light from a planet along carefully plotted paths to produce the spectrum needed to analyze the planet's chemical characteristics.

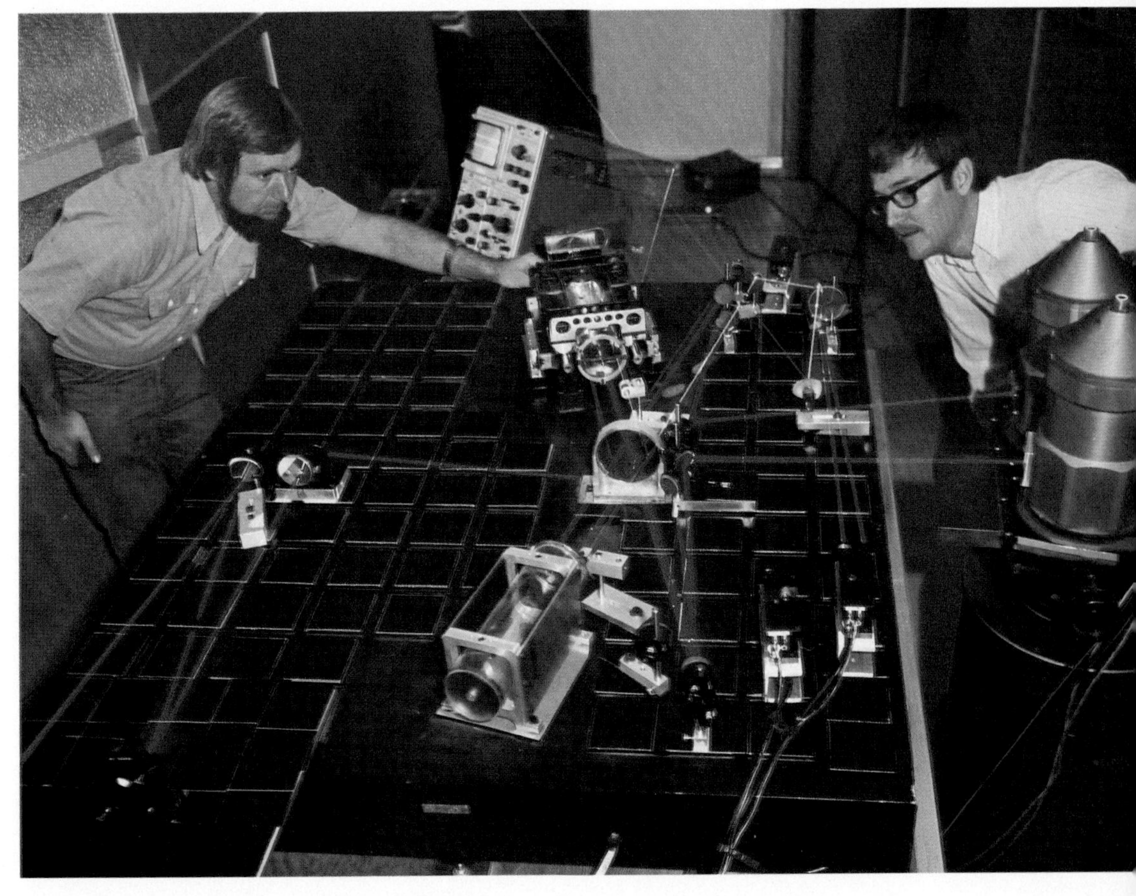

Winding Up the Atmosphere

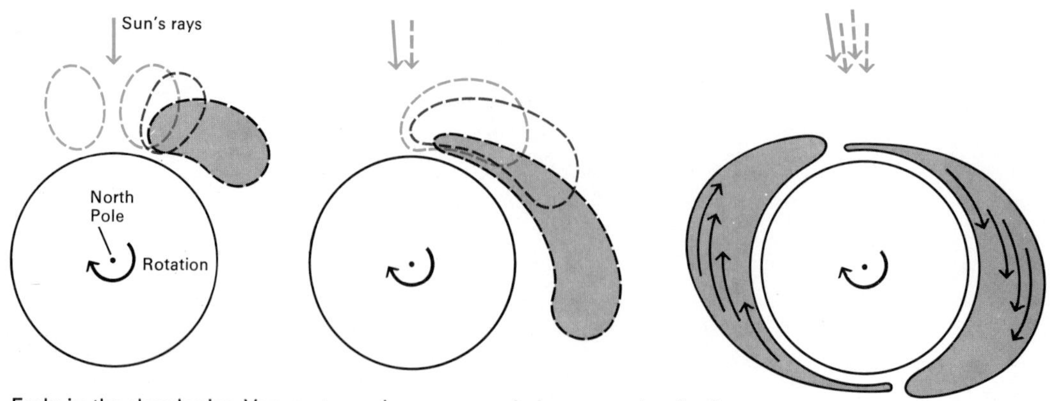

Early in the developing Venus atmosphere, warmed air rose under the Sun in two large convection cells. As the Sun's rays slowly shifted, the cells tilted and stretched (action of only one shown), creating an atmospheric flow that today zips around Venus much faster than the planet rotates.

Modeling the Clouds

Cloud patterns that swirl around Venus, *above,* may be caused as atmospheric waves interact with the winds, *right.* The winds move west, circling Venus in 4.8 days at the equator and in 3 days at higher latitudes (**A**). At the equator, a wave moving west seems to add to the wind velocity; at higher latitudes, a different type of wave appears to counteract part of the wind velocity (**B**). The apparent net velocity is equal in both cases (**C**), giving the illusion of a permanent horizontal Y (**D**).

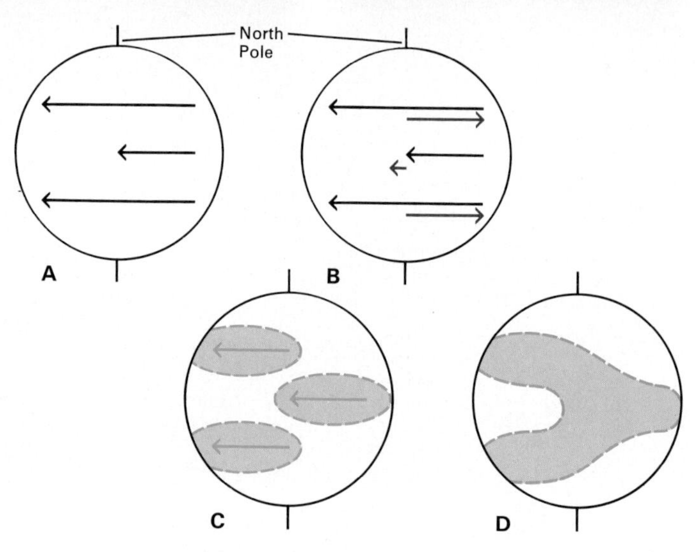

pounds from the surface rocks. These compounds combine with the carbon dioxide in the atmosphere to form carbonyl sulfide. This gas does not combine easily with other substances and is carried by atmospheric currents into the high atmosphere. Exposed there to ultraviolet radiation from the Sun, the carbonyl sulfide rapidly breaks apart into sulfur atoms and carbon monoxide.

Prinn theorizes that the sulfur then may join other sulfur molecules to form pure sulfur crystals, or it may combine with traces of water and the tiny amount of free oxygen formed by the breakdown of carbon dioxide to form sulfuric acid molecules. High in the Venus atmosphere, where temperatures are quite low, sulfuric acid molecules would collide and combine into the liquid droplets of the mists.

Pictures from Mariner 10 and Earth-based photos taken over many years seem to show that the Venus atmosphere is rapidly swirling around the planet. Recent observations show that the actual situation is more complex. Parts of the atmosphere are indeed swept around by swift winds, but the actual motion appears to be combined with other influences that give the illusion of even greater motion.

Data from the Venera probes and Earth-based observations show that the wind speed rises steadily from about 3.5 kilometers (2 miles) per hour at the surface to about 360 kph (225 mph) at about 65 kilometers (40 miles) above the surface. The magnitude of the high winds is quite startling–the entire upper atmosphere seems to flow around the planet 50 times faster than the planet itself rotates.

Prodigious forces must be constantly at work to produce and maintain such motion in the atmosphere. On Earth, such high wind velocities occur only in limited regions and are usually generated by steep temperature variations. The jet streams, for example, are usually associated with weather fronts, the narrow regions that separate cold and warm air, but such fronts have not been observed on Venus.

The high-speed winds on Venus might be caused by the same physical processes that give rise to motions in a fluid when it is heated by a moving flame in a laboratory. Gerald Schubert of the University of California, Los Angeles (UCLA), suggested this in 1969. He noted that when liquid mercury is placed in a circular channel and heated by a slowly moving Bunsen burner, the mercury begins to flow around the channel four times faster than the flame moves, and in the opposite direction to the flame.

Schubert proposed that a similar phenomenon might occur on Venus, with the Sun as the flame and the atmosphere as the fluid. Imagine that it is dawn on Venus and that the atmosphere, completely still, begins to heat up as the Sun slowly rises. An outside observer would see a very large thermal wave, or tide, moving through the atmosphere following the position of the Sun. To offset this effect, the atmosphere sets up giant convection cells that carry hot air to cooler regions and cool air to warm regions. If the Sun remained fixed in the sky, this would be the final result. But, as the planet moves in its orbit, the Sun

A parched, sweltering landscape fades into the haze of the dense Venus atmosphere, *overleaf.* Sharp rocks dot the land and the Sun glows feebly through sulfurous clouds. An erupting volcano may add an eerie light.

Sharp-edged boulders, apparently young, litter the surface of Venus in a photograph sent back to Earth by Russia's Venera 9.

appears to move across the sky, and the convection cells are stretched out in the direction of the Sun's apparent motion. Depending on the nature of the atmosphere and the rate at which the planet rotates, the atmosphere could move much faster than the Sun's track moves, and in the opposite direction.

A second possibility, proposed by Rory Thompson of the Woods Hole Oceanographic Institution in 1970, is closely related to the moving-flame proposal, except that it does not require the Sun's motion. In Thompson's theory, there is a shear force in the atmosphere between two layers that are moving at different speeds. This also creates an unstable distortion in the convection cells, which in turn increases the shear. Richard Young and James B. Pollack of the Ames Research Center, Mountain View, Calif., reported in April 1976 that they had studied this effect in three-dimensional computer simulations of the Venus atmosphere. They found that, as Thompson predicted, the original shear was amplified, leading to the build-up of rapid rotation in the atmosphere. Also, the atmosphere continued to move rapidly around the planet, even when Young and Pollack reduced the Sun's motion to zero in their calculations.

In addition to actual motions in the atmosphere, astronomers are also studying another influence on Venus that could give the illusion of atmospheric motion. This centers around combinations of certain types of internal gravity waves and Rossby waves. An internal gravity wave is a fluctuation in the density of the atmosphere that is influenced by gravity. It spreads horizontally and vertically at substantial speeds. As the wave moves, the progressive changes may give the illusion that the atmosphere is moving, but these changes are temporary. A Rossby wave is similar except that it is caused by the rotation of the atmosphere, rather than gravity.

Both types of waves are prominent in the Earth's tropical atmosphere and have been studied extensively since 1970. The waves on Earth are thought to be self-propagating–once started, they keep going on their own. Because these waves carry large amounts of energy and momentum, they are important phenomena in maintaining the Earth's atmospheric circulation.

On Venus, these two types of waves might explain a dark Y-shaped feature that shows up in ultraviolet pictures of the clouds. This enigmatic feature wraps itself in a broad band completely around the center of the planet, and moves around the planet about every four days. Schubert, Anthony Del Genio of UCLA, and I found in 1976 that specific types of internal gravity and Rossby waves, when present together, could produce the appearance of a long-lived, horizontal Y as the waves interact with background winds. Even though the winds move west at the same speed all over Venus, the angular velocity–the rate at which they travel completely around the planet–is much greater at higher latitudes than at the equator, because the circumference is greater at the equator. The winds circle the planet in 4.8 days at the equator and in about 3 days at higher latitudes. An internal gravity wave at the equator also appears to travel west, thus increasing the apparent angular velocity of the wind. At latitudes of about 40° north and south, Rossby waves are generated eastward, thus appearing to diminish somewhat the angular velocity of the wind. The apparent net angular velocity moving west is equal in both cases, giving an outside observer the illusion that the waves form a permanent Y.

These waves may set up the pattern of ultraviolet cloud markings seen in photographs of Venus. The darker parts of the pattern absorb

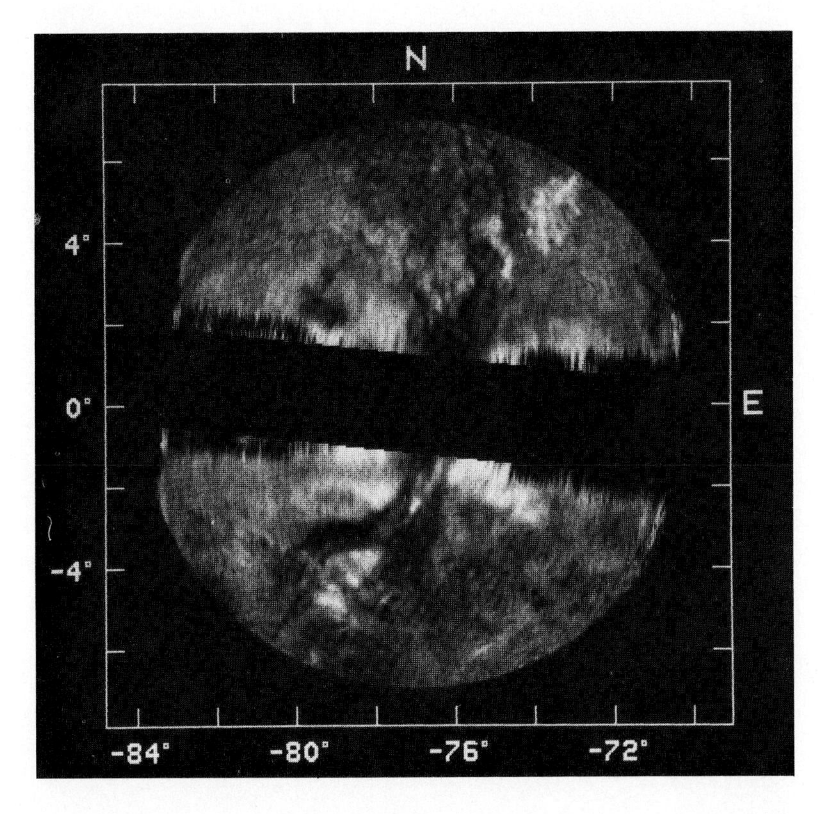

A radar image of a tiny portion of the Venus equatorial region shows a massive rift canyon that hints at tremendous underground activity.

more sunlight than the lighter regions, thus maintaining the waves. The waves continually carry energy and motion from the lower atmosphere to the stratosphere, which may well be the source of the real motions that we observe.

Earth-based radar and the Russian Veneras have supplied fragmentary information about the surface of Venus. The most detailed information comes from two panoramic pictures of the surface, sent back to Earth in October 1975 by Venera 9 and Venera 10. The probes, which landed 2,200 kilometers (1,365 miles) apart, sent back pictures of quite different types of terrain. Venera 9 landed in a field of angular, sharp-edged boulders lying on what appears to be a firm surface. The rocks show no obvious signs of erosion or of being partly buried. The Venera 10 picture shows a totally different scene–large, flat, worn rocks that resemble an eroded outcropping of broken granite lie partially buried in a fine, dark material.

Similar scenes can be seen on Earth. Clusters of boulders are often formed by weathering in mountainous areas. But the type of weathering that produces such rock fragments on Earth requires moisture and wide variations in temperature, conditions that almost certainly cannot occur on Venus. Venus has no water and the temperature varies no more than a fraction of a degree.

The sharpness of the boulders in the Venera 9 picture and their exposed position on the surface strongly suggest that the rocks must be relatively young. Because Venus' atmosphere is so thick and heavy, the surface wind speed of roughly 3.5 kph reported by the Venera landers is high enough to blow dust around. Small rocks on the parched surface would be partially buried or eroded in a relatively short period of time–as seen in the Venera 10 picture, which appears to show a much older area.

If the rocks in the Venera 9 picture are as young as they appear to be–perhaps no more than a few hundred years old–then how were they formed? Clues from radar pictures strongly suggest that they were created by crustal movement.

Radar astronomers using the Deep Space Network at Goldstone, Calif., and the newly resurfaced 305-meter (1,000-foot) antenna at Arecibo, Puerto Rico, are now making detailed, high-resolution pictures and altitude maps of the Venus terrain that show glimpses of a highly active landscape. The first detailed picture of a region called Maxwell, made with the Arecibo antenna, shows that the area has a highly irregular outline and a complex surface structure quite unlike anything familiar to us. Scientists at Arecibo speculate that Maxwell may be a huge field of highly porous lava.

Meanwhile, another surge in the results from Earth-based radar came in 1975 when Richard Goldstein and his associates at the Jet Propulsion Laboratory (JPL) in Pasadena, Calif., released several new pictures and altitude maps made with the Goldstone antennas. One of the pictures shows a massive channel resembling a rift valley, perched at the top of a ridge. The channel is 1.5 kilometers (1 mile) deep, 120 kilometers (75 miles) wide, and at least 1,300 kilometers (800 miles) long. Another picture gives a tantalizing glimpse of a curved formation that, if completed, could form a circular feature 1,500 kilometers (930 miles) in diameter. Other pictures show extremely rough, possibly mountainous, regions, and one reveals a small elevated area of calderalike features, the kind of craters found on top of volcanoes.

According to Michael Malin of JPL, the parts of the surface seen in the pictures indicate a tremendous amount of geologic activity on Venus. These exciting pictures inspire barely restrained speculation. They hint at long faults slicing across the surface, volcanoes that may still be active, and great shiftings of the ground.

These pictures, while a tremendous achievement, are only the first samples of an enormous amount of information that should soon be available as the Arecibo radar data is processed. The outlook for the future is frankly astounding. Astronomers at Arecibo predict that they will map most of the Venus surface in the next few years, picking out features as small as 1 kilometer (0.6 mile) across and 100 meters (330 feet) high. This would surpass the best photographic maps of the Moon ever made from Earth.

By the time the Arecibo radar completes the first maps of Venus, we will begin replacing our current speculations with firm knowledge. In the summer of 1978, two Pioneer spacecraft will blast off for Venus. Reaching the planet in December, one will inject four probes into the atmosphere and then enter the atmosphere itself. The second will move into Venus' orbit to study the atmosphere for up to 300 days. Meanwhile, the new telescopic techniques born of today's electronics revolution will continue to yield data of unsurpassed quality.

These marvelous techniques are exciting, but the picture of Venus that they are producing is increasingly inhospitable. So why study Earth's "twin planet"? Why should Venus, or any other planet, be of particular interest to us?

Beyond the satisfaction of the simple search for knowledge, the exploration of Venus and other planets has a practical importance. We live on a planet that is constantly changing. The Earth's crust moves and, with it, the continents on which we live. The air we breathe is constantly evolving; it is not the same as it was 2.5 billion years ago when oxygen was a minor constituent, or even 100 years ago when it contained much less carbon dioxide.

In years to come, Venus–inhospitable as it clearly is, with its sulfurous clouds, choking atmosphere, and parched and distorted landscape–will become as familiar to us as Earth. With this familiarity will come understanding. By careful study of Venus and the other planets, we may yet learn to conquer the terror of the storm, the threat of a changing global environment, and the catastrophe of earthquakes and volcanoes. By learning more about the other planets, we will come to know more about the forces that shape our own.

Airships Make a Comeback

By Lee Edson

Lighter-than-air craft may be one solution to a number of modern transportation problems

The huge, gas-filled airship glides silently over the city, casting a shadow that moves slowly across the buildings and streets below. Inside, passengers relax in the lounges, nap in their staterooms, linger over lunch in the dining room, or enjoy the panoramic view from observation-deck windows.

A vision of the future? Yes, but it is also a scene from the past. Many airships made such flights, and perhaps the most memorable involved the German airship *Hindenburg*. On May 6, 1937, it passed over New York City for the last time. As the 247-meter (812-foot) airship glided by the Empire State Building, sightseers waved excitedly and the *Hindenburg* passengers waved back. Then the cigar-shaped craft headed toward its mooring at Lakehurst, N.J.

derek grinnell

Moments later, as the huge airship nosed up to the mooring mast, the highly explosive hydrogen gas inside its hull burst into flame with a blinding flash. Thirty-six of the 97 persons on board were killed in the blaze that destroyed the *Hindenburg* and also signaled the end of an era of great airships.

For the next 40 years, most people believed that airships, like clipper ships and horse-drawn coaches, were symbols of times that would never return. The huge, slow-moving, lighter-than-air craft were made obsolete by the smaller and much faster airplane. But the current energy crisis, air pollution, and the problems of transporting goods efficiently to remote areas may trigger a comeback for airships.

The United States Navy and Coast Guard and the National Aeronautics and Space Administration (NASA) funded special studies in the early 1970s that indicate advantages in using airships for many purposes. A number of other countries, including Russia, have designed airships and actually built some models.

Because airships do not need runways to land–in fact, they do not have to land at all–enthusiasts envision dozens of uses for them. Lighter-than-air ships could transport lumber and bulky pieces of equipment to and from remote areas. As luxury passenger liners, they could carry tourists on photographic safaris to the deep jungles of the Amazon or to ice floes in the Arctic. Some supporters see airships as giant floating hospitals capable of delivering lifesaving services to remote outposts or to areas made inaccessible by natural disasters such as floods or earthquakes. Others believe airships could serve as ocean research platforms, hovering above rough seas while lowering equipment to the ocean bed.

Not everyone shares this striking view of the future, however. Some people believe the slow-moving airships would never fit in with today's fast-paced life style. "In a world of supersonic Concordes, who wants to ride in a balloon?" asks an airship critic. "As long as people are keyed to speed and to schedules, the airship is not likely to make it, except in a very limited way."

Nonetheless, dreams have motivated airship builders and designers since French inventors lofted the first hot-air and hydrogen-filled balloons in 1783. Also in the 1780s, Jean Baptiste Meusnier, an officer in the French Army Corps of Engineers, designed a balloon called a dirigible, from the Latin word meaning to steer. Even though Meusnier never built a model from his blueprint, it incorporated the principles that airship designers applied 100 years later.

His blueprint called for a cigar-shaped outer bag with an inner bag to contain hydrogen gas. The space between the two bags was to contain air inflated to a pressure higher than that of the atmosphere to maintain the shape of the craft.

Power, which is necessary to control the airship, was a more difficult problem. Meusnier suggested his airship could be driven by three propellers turned by hand by 80 men. Other early inventors pondered

The author:
Lee Edson is a science writer. He wrote "X Rays from the Sky" for the 1974 edition of *Science Year.*

Lighter-than-air craft evolved from the first hot-air balloon, *left,* flown in France in 1783. The airship era began in 1852 with Henri Giffard's gas bag, *below.* The first dirigible, *La France, bottom,* flew in 1884.

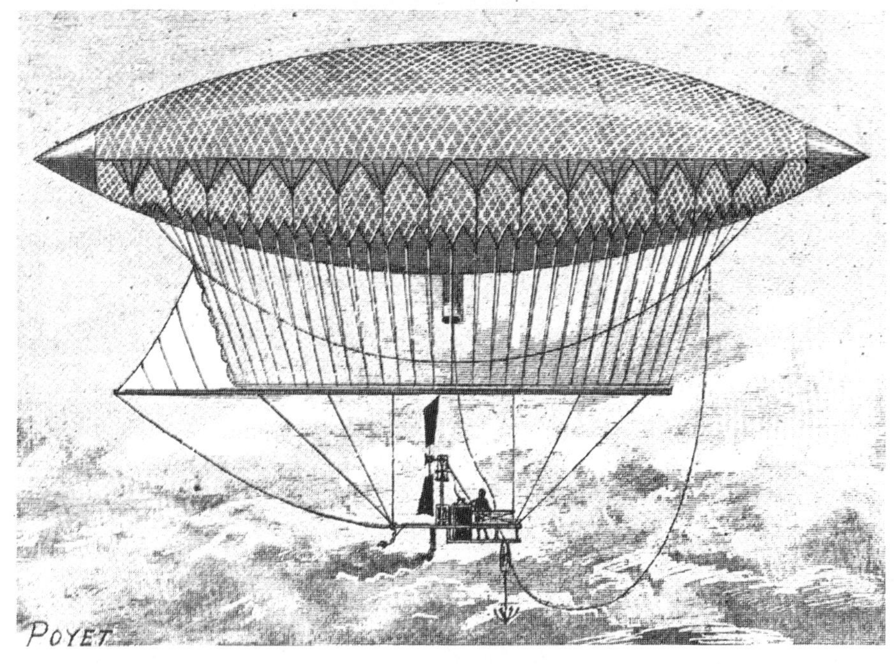

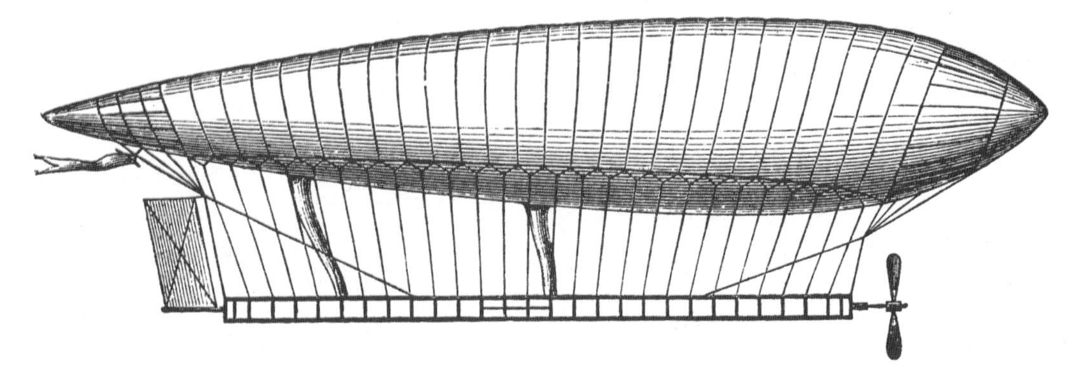

The *Graf Zeppelin,* one of the great airships of the 1920s and 1930s, glides silently over a town in Germany.

using sails or paddles. Some even came up with the fanciful idea of harnessing flocks of birds.

The airship era actually began in 1852, when French engineer Henri Giffard flew over Paris in a cigar-shaped gas bag 44 meters (144 feet) long that was powered by a 2.2-kilowatt (3.2-horsepower) engine. However, Giffard could do little more than turn slightly from a straight course with the bag's saillike rudder. In 1884, inventors A. C. Krebs and Charles Renard solved the problem of control with a 51.8-meter (170-foot) cigar-shaped balloon, *La France*, which was steered by a rigid rudder and powered by a battery-driven motor. It was the first airship to fly to a given destination, turn around, and return to the starting point on its own power.

Thereafter, airships grew bigger and better. In the early 1900s, airship builders in France, Germany, Great Britain, Italy, and the United States concentrated on three types: nonrigid bags, commonly called blimps, that were held in shape by the pressure of the contained gas; rigid airships, which had a covering material stretched over a metal or wooden framework with the gas contained in several compartments inside; and semirigid craft, which had a keel, or backbone, holding the gas bag stiff. The crews controlled the upward and downward movements by moving weights, transferring gas from one end of the ship to the other, or by opening valves and letting some of the gas out. Some airships even used swiveling propellers for control.

Count Ferdinand von Zeppelin, a German, was indisputably the biggest name in early airship building. In 1900, the count unveiled the first of his rigid, cigar-shaped, aluminum-frame airships. He believed that the airship's lifting power depended upon how much gas the craft could contain. The first zeppelin, as these craft came to be called, was 128 meters (420 feet) long and held about 9,910 cubic meters (350,000 cubic feet) of hydrogen.

During World War I, the British, French, and Germans drafted both blimps and rigid airships into military service to scout for submarines and carry out air-sea rescues. The Germans even used zeppelins for bombing raids on Great Britain. After the war, Hugo Eckener, an associate of Count Zeppelin, built several luxury passenger airships, complete with electric kitchens. In 1928, he completed the first of the truly great airships, the *Graf Zeppelin*. It was 240 meters (800 feet) long and had five engines that produced a total of 1,976 kw (2,650 hp). It could fly at more than 100 kilometers per hour (70 miles per hour) and carried 50 passengers. The dining room, lounges, observation deck, and staterooms were inside the hull. After its maiden voyage across the Atlantic in 1928, the *Graf Zeppelin* was used for regular commercial passenger service and traveled more than a million miles without an accident. But Eckener also built the ill-fated *Hindenburg*.

Transatlantic passengers relax in part of the luxurious lounge that was inside the hull of the *Graf Zeppelin.*

Storms and turbulent weather destroyed many of the big, rigid airships by snapping their frameworks. Others, like the *Hindenburg*, were consumed in flames fed by the hydrogen gas they carried. One by one, governments and private companies abandoned airships and concentrated on developing the airplane. Although a few blimps remained for advertising and military surveillance, the giant rigid airships had disappeared from the skies by the start of World War II.

Today's airship visionaries do not intend to repeat the mistakes of the past. For one thing, the huge craft would all be filled with helium, which provides almost as much lift as hydrogen but will not burn. In addition to the gaseous helium, some designers propose liquefying some of the gas and carrying it aboard airships as a backup supply. They estimate that 454 kilograms (1,000 pounds) of liquefied helium could provide 2,800 cubic meters (100,000 cubic feet) of gas for additional lifting power. However, they must solve the economic problem of carrying the heavy equipment needed to liquefy helium.

Today's airship designers also have a wide variety of new and better building materials available to them. Many early airships had cumbersome frameworks covered with varnished or rubber-coated cotton or linen that often came loose and caused unnecessary drag. Advances in space age technology have made it possible for future airship designers to use computer-designed, prestressed, metallic skins with a lightweight honeycomb structure. Such coverings could withstand strains a hundred times greater than early designers ever envisioned. Other possible hull coverings include composite sheets of plastic and beryllium alloy or stainless steel.

Modern instrumentation is also important to the revival of the airship. Airship pilots in the 1930s had few instruments on board to monitor weather conditions or the ship's position. Human beings had

The Shape of Things to Come?

A cigar-shaped luxury passenger airship of the future has cells in its huge hull that contain noncombustible helium gas. A nuclear reactor furnishes power. Several levels of decks contain staterooms, dining areas, and lounges. Passengers can even take an elevator, *below right,* to an observation deck, *below left,* atop the airship's rigid hull.

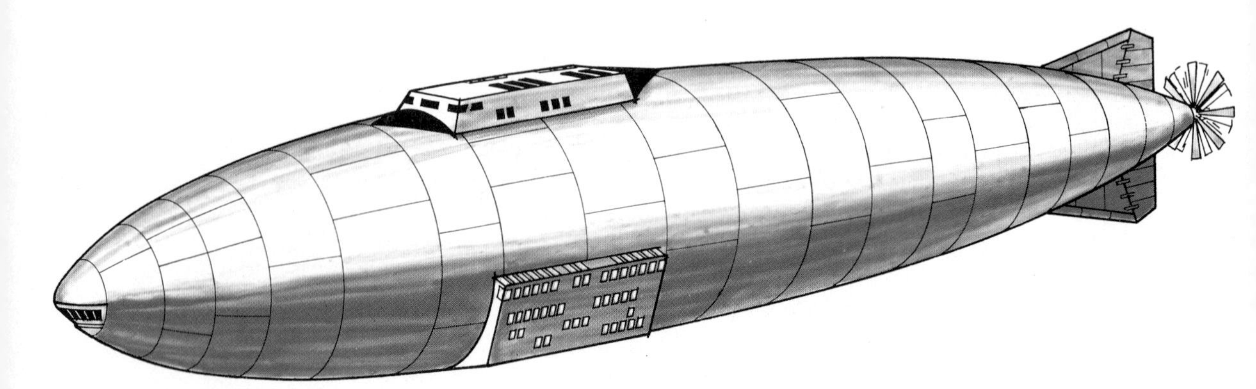

to check conditions inside and outside the ship. Now, computers can do this quickly and more accurately. Modern navigational equipment could automatically and continuously check the airship's position; sensors could provide emergency warnings of storms and also potentially dangerous conditions inside the ship; radar could ensure safe approaches to moorings.

One rigid-airship design based on modern technology has been drawn up by Francis Morse, professor of aerospace engineering at Boston University. His cigar-shaped craft would be about as long as the Empire State Building is tall. Its hull would hold 354,000 cubic meters (12.5 million cubic feet) of helium in 17 gas compartments. Its most efficient operating speed would be 160 kph (100 mph), and it would have unlimited range because it would be nuclear powered.

In the passenger version of his craft, Morse envisions three decks of staterooms with private bath along the sides of the airship, a cocktail lounge, and a 15-story elevator to a grand ballroom at the top of the hull with a huge window facing the stars. Passengers could be ferried to the airship by airplane. The plane would slow down to match the speed of the airship and then a skyhook under the airship would catch the plane and haul it inside. In the cargo version of the ship, the three passenger decks would be replaced by five holds, each with 2,300 cubic meters (80,000 cubic feet) of space.

The nuclear reactor in the Morse design would be housed deep in the hull of the airship. The reactor would heat helium, and as the gas expanded it would pass through a turbine, powering a 3,000-kw (4,000-hp) turboprop engine, which would turn two propellers on a single shaft in the stern.

Other modern airship designs are hybrids of the airplane and the airship. One is a deltoid, or triangular-shaped craft. When in motion, the airship's shape would act as a giant wing helping to keep it in the air. It was designed by William Miller, who heads Aereon Corporation, a Princeton, N.J., firm that has experimented with deltoid shapes since 1959. Miller tested a helium-filled 8.4-meter (27.5-foot) prototype of his deltoid in late 1970. He claims the triangular ship is more stable and maneuverable than other lighter-than-air craft and can reach greater altitudes than conventional airships. Since these are held aloft by gas that is lighter than air, the heights they can attain are limited because the atmosphere grows thinner at high altitudes. Dirigibles of the 1930s flew at between 610 and 1,829 meters (2,000 and 6,000 feet). Engineers at Boeing Vertol Company of Philadelphia have suggested a hybrid in an ellipsoid shape, which is like a slightly elongated football with rounded ends.

Helicopter pioneer Frank N. Piasecki, president of Piasecki Aircraft Corporation of Philadelphia, has developed a design that is perhaps closest to becoming a reality. Piasecki's airship would be powered by four helicopters–two mounted on each side of a rigid hull. For a nonrigid ship, the helicopters would be mounted on an H-shaped

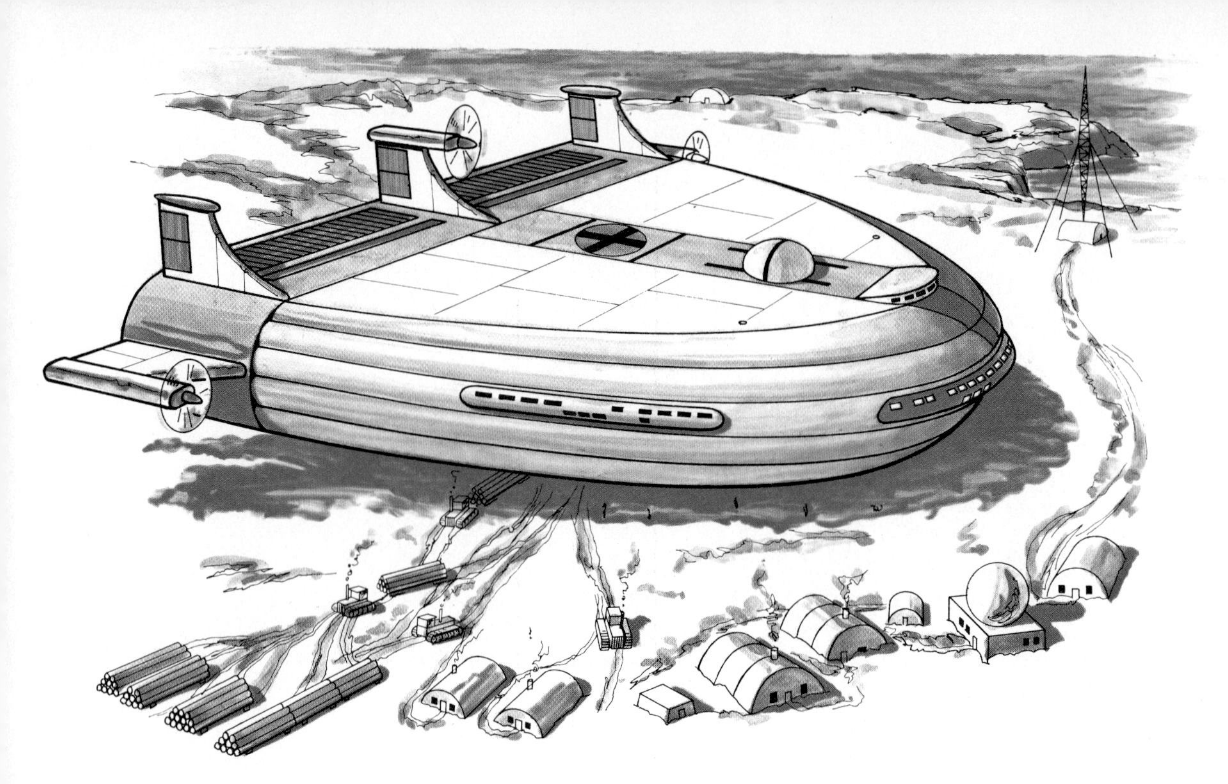

Another airship of the future with a deltoid design hovers above the ice to deliver equipment to an Arctic outpost.

frame attached to the hull. Gas inside the hull would provide lifting power, and the helicopters would control the position and movement of the airship. One pilot would control all the helicopters.

But the flurry of interest in airships is not as strong in the United States as in other countries. John West of Skyship, Limited, in London, has designed a "flying saucer" that is 213 meters (700 feet) in diameter and has a huge tailfin. Ten turboprop engines with tilting propellers are mounted around the edge of the disk. The craft would be built around a hexagonal aluminum framework. Cargo space in the center of the structure could hold up to 360 metric tons (400 short tons). A 9-meter (30-foot) model of this flying saucer was flown successfully in 1975 in a hangar at Cardington, England. The designers plan a 60-meter (200-foot) model before building a full-sized craft.

British designer John Wood of Aerospace Developments, Incorporated, in London, has been working on the problem of efficiently transporting natural gas. Wood envisions using the natural gas to lift the airship and to fuel it along the way. At the craft's destination, the natural gas–except that needed to fuel the return trip–would be pumped out and helium pumped in. His airship's hull would be a stiff shell made of some lightweight plastic or metal. The hull itself, rather than a framework, would bear the major stresses. It would carry 2.8-million cubic meters (100 million cubic feet) of gas.

To minimize the hazards of carrying the explosive natural gas, the hull material would be fireproof and virtually punctureproof. Electronic devices would constantly watch for abnormal conditions. Also, the natural gas would be stored in several compartments surrounded by an outer lining. Helium in the space between would serve as a barrier to stop a fire from spreading. Another British firm, Airfloat Transport, Limited, is working on a gas-transporting design in which the main airship would tow blimps loaded with fuel.

Since the 1930s, Russia has experimented with lighter-than-air craft. In the mid-1960s, the Russians designed one that used hot air for lift. Exhaust gases from turbojet engines in the nose were piped through the hull to heat the air and also provided propulsion and maneuverability through a swiveling nozzle in the tail. The Russians reportedly are now working on designs for nuclear-powered airships.

Despite all this worldwide interest, the airship still has a lot of head wind to overcome. One major obstacle is economics. No one seems willing to chance large-scale development because little is known about the costs of airship transport. "You can't really make a hard-and-fast economic survey of airships, because there are no airships to compare with other modern transportation forms," says William L. Kitterman of the U.S. Energy Research and Development Administration, who has attempted one of the few economic studies of the problem. However, his figures show that if you assume reasonable capital costs for airship building–for example, the nearly $25 million needed to build a Boeing 747–then airships appear to be more economical than jet planes for hauling large loads over long distances. Kitterman estimates that an airship can carry as many as 2,000 passengers and as much as 900 metric tons (1,000 short tons) of cargo across the Atlantic Ocean for a round-trip fare of $100 per person and a cost of 3 cents per ton mile. This is considerably less than the current 23 cents per ton mile by airplane.

Building hangars for the large dirigibles was a major expense in the early 1900s. But present-day airship boosters argue that this is no longer a problem. "One does not really need a hangar today," says J. Gordon Vaeth, head of systems engineering at the National Environmental Satellite Service of the National Oceanic and Atmospheric Administration. "The modern airship can be kept flying almost endlessly; it can be repaired and refueled in flight and can be kept on the mooring mast without any danger. For building and major overhauls you can use an outdoor turntable so that the ship can always be turned into the wind for stability. You might even dig a large hole in the ground for the airship and put a plastic cover on it for protection."

A great deal of inertia and skepticism must be overcome if these giant craft are to return to the sky. But this has not dampened the enthusiasm of airship advocates. "The time is ripe for the return of the dirigible," says Vaeth, an airship officer during World War II and one of America's most ardent airship promoters. "The airship can fly with less noise than a jet, stay aloft longer, and operate where no airports or landing strips exist," he says. "It can transport cargo cheaply over long distances using considerably less fuel than jets and with virtually no exhaust to pollute the air. Here, in one alternative transportation system, is a partial answer to the current energy and pollution crisis."

The Secrets of Winter Sleep

By Albert R. Dawe

In the search for what triggers hibernation, scientists have learned much about the changes that accompany this form of winter survival

Curled up in its burrow, the tiny thirteen-lined ground squirrel was deep in hibernation. It rested on its haunches, its head tucked into its stomach. Although its fur stood on end, fluffed to conserve body heat, its body temperature matched that of the chilly air outside. Its heart beat only once or twice a minute, and it took a slow breath perhaps every five minutes. Yet its paws and nose were bright pink, showing that oxygen was circulating easily in its blood.

This might have been a typical midwinter scene, under the snow-covered ground of the frozen forest. But it was July, and this ground squirrel was hibernating in our laboratory at Loyola University's Stritch School of Medicine in Maywood, Ill. We had succeeded in triggering hibernation artificially, at a time when it would not normally occur in the ground squirrel.

Some animals and birds survive the winters of the sub-Arctic and middle latitudes protected by heavy fur or thick feathers. Others escape by migrating to warmer climates. A few cope with the stress of winter by hibernating.

In the broadest sense, a hibernator is any animal that goes into a sleeplike state for the winter. Worms, insects, amphibians, reptiles, and some birds and mammals go into a sort of dormant state as cold weather approaches. In true hibernation, however, body temperature drops to within a few degrees of the outside temperature. Bears, for example, are not true hibernators because their body temperature remains near normal even when they are in a deep winter sleep. One other characteristic distinguishes true hibernators–they are able to arouse at intervals during the winter and warm up against the outside cold by their own means. Only certain birds and mammals can do this. Reptiles, fish, and the other winter sleepers can warm up only when the outside temperature rises.

The number of true hibernating species is relatively small. Among the mammals, it includes dormice, chipmunks, hedgehogs, marmots (including woodchucks), some types of bats, most of the ground squirrels, and perhaps some deer mice.

A hibernator goes through three states throughout the year–activity, hibernation, and arousal. During the summer, the animal is fully active. As fall approaches, it begins slipping into periods of hibernation, interrupted by brief periods of arousal that resemble summer activity. As the early weeks of the hibernation season pass, the periods of hibernation grow longer, reaching a peak in the middle of winter. Then the periods gradually become shorter until the animal finally arouses into a fully active state in the spring. During the midwinter peak period, ground squirrels usually hibernate for about 14 days at a time, alternating with brief periods of arousal lasting two or three days. Woodchucks hibernate for about 30 days at a time, also alternating with arousal periods of two or three days. There are no visible differences between the active summer state and the winter arousal.

The author:
Albert R. Dawe is deputy director and chief scientist at the Office of Naval Research in Chicago. He has been studying hibernation for more than 20 years.

During hibernation, an animal undergoes a number of physiological changes. Its body temperature drops and its heartbeat slows to as little as 1/200 of its normal rate. The heart beats irregularly, with as few as two beats per minute. However, the animal's blood pressure remains at its normal level.

Some blood vessels narrow during hibernation, and the blood undergoes a number of changes to maintain circulation. Although the number of both red and white blood cells declines, the red cells live up to three times longer than normal. Blood coagulates more slowly and the red cells do not clump together as they normally would when exposed to cold temperatures.

An active animal may breathe several hundred times per minute. In hibernation, this rate drops to as little as one breath every five minutes. The hibernating animal requires only about 1 per cent of the oxygen it needs when active, and its blood remains highly saturated with oxygen. This makes the blood bright red, giving a rosy glow to exposed skin on the animal's paws and nose.

The animal's muscles maintain their tone during hibernation, even though its body movements are restricted to slight shiftings every few hours. The digestive and excretory systems continue to operate, but their functions change.

A layer of white fat is found under the skin over the entire body, providing insulation to conserve heat. Brown fat, found mainly between the shoulder blades and around the breastbone, provides the

A Dramatic Arousal

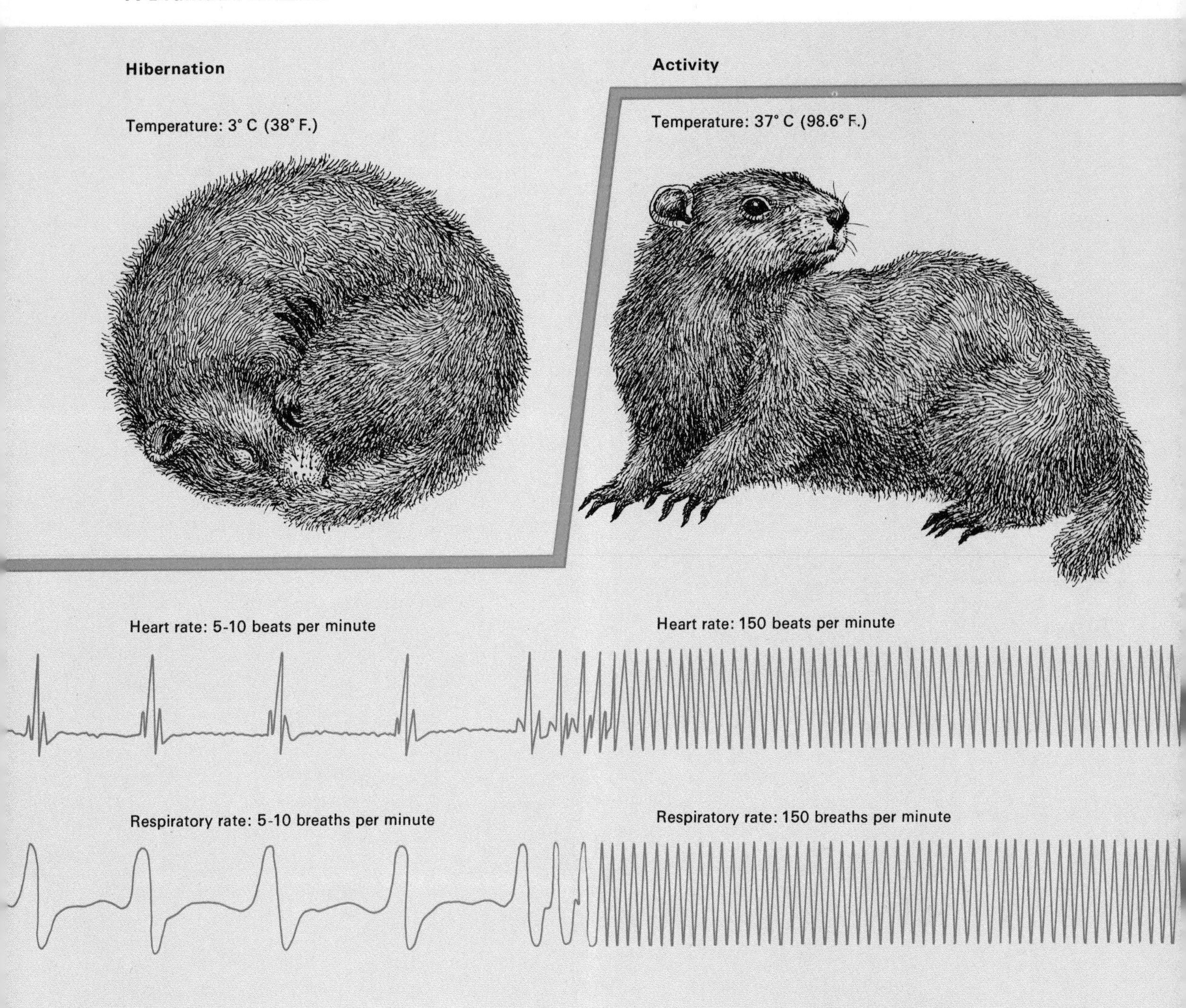

A hibernating animal, such as the woodchuck, undergoes tremendous physiological changes. Its body temperature, heartbeat, and breathing rate drop dramatically. They return to normal as the animal arouses from its deathlike sleep.

fuel to maintain life during hibernation and provide quick energy when it is time for the animal to arouse.

A hibernating animal is easily disturbed when touched and heat always arouses it. However, the animal appears oblivious to changes in light and sound that may occur around it, although the heartbeat may change in response to noise.

Arousing is a dramatic process. The animal's entire body shivers and shakes violently. Great gasps indicate an overwhelming call for oxygen from the body cells. The animal's blood rapidly darkens as breaths become much deeper and more rapid. Its body temperature climbs quickly, despite the outside cold. In a short time, the arousing animal burns up immense amounts of energy, apparently contributed

A ground squirrel uncurls and rolls onto its back in the first stages of arousing. Once righted, it struggles to get its feet under control. After two hours, eyes open and tail erect, the animal is fully alert.

by the brown fat. Not only does the brown fat provide energy to reactivate the cells, but it also rapidly releases a large amount of heat as it breaks down. After about two hours, the animal is fully aroused. Its digestive system begins to function in a normal way again and it may nose around for a snack from previously stored food before again sinking back into hibernation.

Because many investigators over the past century have studied what happens during hibernation, we now have a fairly good picture of the physiological changes that take place. But we still do not know exactly what triggers hibernation at a certain time of year.

We took a first important step in this search in the spring and summer of 1968. In March, my associate, Wilma Spurrier, and I drew 3 milliliters (about 0.1 ounce) of blood from a hibernating ground squirrel and immediately injected 1 milliliter (0.033 ounce) into a leg vein in each of two ground squirrels that had come out of hibernation for the season. When they were left in a cold room that was kept at a temperature of 7°C (45°F.), the two squirrels went back into hibernation within a few days.

Following the same procedure, we drew blood from the two hibernating squirrels in June and injected it into three other active ground squirrels, who then hibernated. Again, in July, we drew blood from the three hibernating animals and used it to trigger hibernation in another five active ground squirrels. These experiments provided the first evidence that the blood of a hibernating ground squirrel might carry a trigger substance for natural hibernation. Furthermore, the substance can be transferred to a warm, active animal, causing it to hibernate when exposed to the cold.

We extended our experiments in 1969, using the same procedure to trigger hibernation. We found that blood taken from an animal that had been hibernating continuously for two or three weeks was more effective in triggering hibernation in another animal than blood from an animal that had just entered a hibernation period. In other words, the trigger substance seemed to build up in an animal's blood during each hibernation period. However, we also found that the rare ground squirrel who was adapted to the cold–that is, one that had never hibernated, even during winter–could not be made to hibernate even when we injected a hibernator's blood.

At the same time, we also conducted experiments to determine the nature of the substance that triggered hibernation. We already knew from earlier observations that blood from a hibernating ground squirrel differed in several important ways from blood taken from the same animal when it was not hibernating. We had found three kinds of particles in the blood of the hibernating animal that we could not identify. Also, the red blood cells were tougher and did not break down as easily as those taken from an active animal. And we found that many red cells from the hibernation blood were folded over. When warmed, they unfolded to their normal shape.

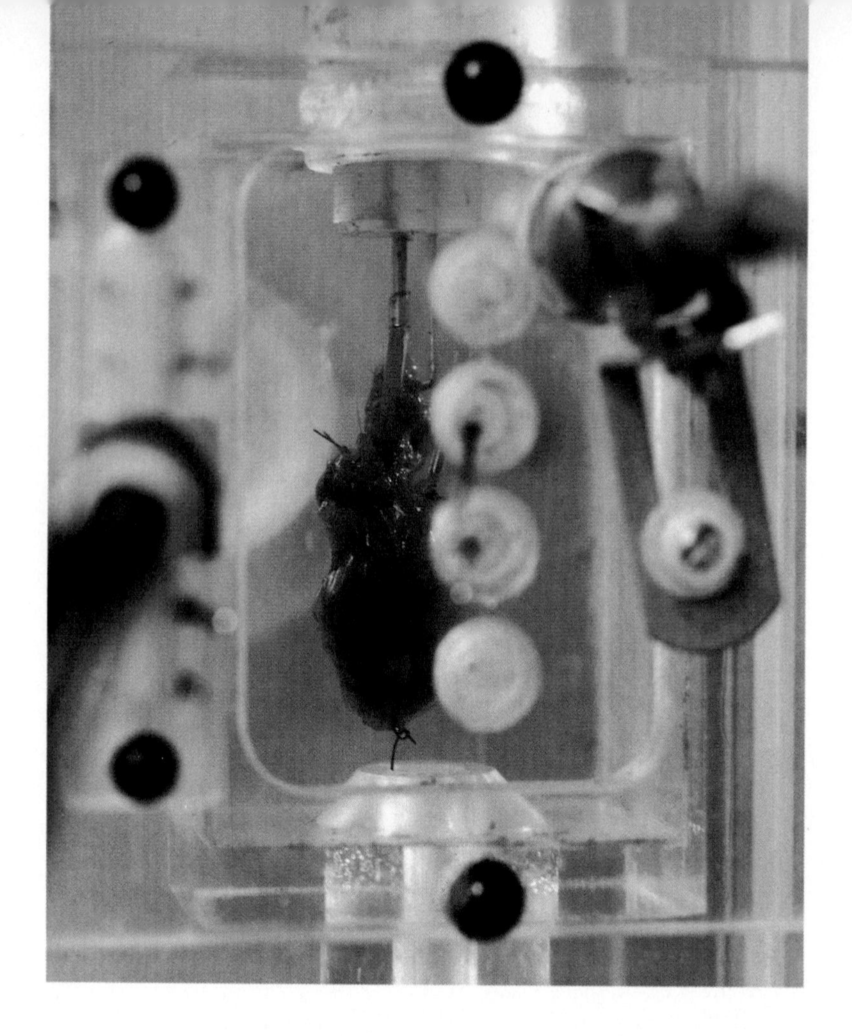

The heart of a ground squirrel is attached to electrodes of a machine that measures electrical activity in a study of how the animal's heart adapts to hibernation.

The next step was to determine whether the trigger substance was in the blood cells or the serum–the clear, liquid part of the blood. We separated these two blood components in a centrifuge and injected the cells into one group of animals and the serum into another. Both groups hibernated. We then put the serum through a molecular filter and gave one group of animals the material that went through the filter and another group the residue, or part that was left behind. We found that the filtered material triggered hibernation, while the residue did not. So the trigger material appeared to be a substance in the serum that was small enough to go through the filter. But if it was in the serum, why did the red blood cells also trigger hibernation? Apparently the trigger substance was able to cling to the red cells even when centrifuged. We also learned that we could preserve hibernation blood serum for at least six months at −50°C (−58°F.).

Since we kept woodchucks as well as ground squirrels in our laboratory, we next set out to discover whether blood from a hibernating woodchuck could trigger hibernation in ground squirrels. We injected whole blood from a hibernating woodchuck into two ground squirrels and blood serum into a third ground squirrel. All three ground squirrels hibernated. Other experiments indicated that blood from a hiber-

Pulling the Trigger: A Theory
During summer, antitrigger molecules interact with hibernation trigger in the blood to keep the animal active, *a.* But trigger production continues until, in the fall, it spills over into the tissues and starts hibernation, *b.* Trigger buildup continues, *c,* until spring, when antitrigger production resumes, *d.* When enough antitrigger is produced to override the trigger, *e,* the animal again becomes active.

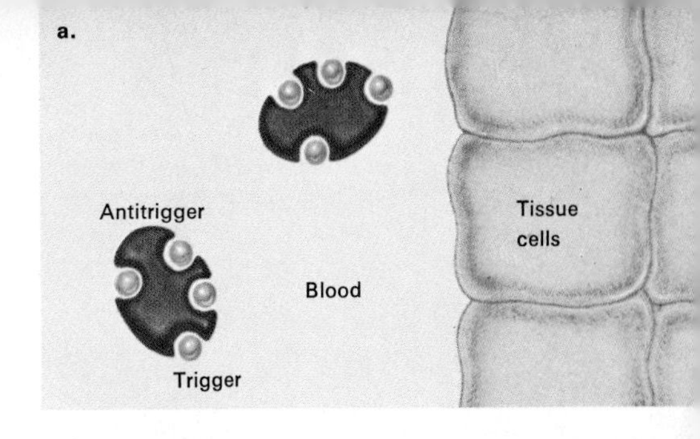

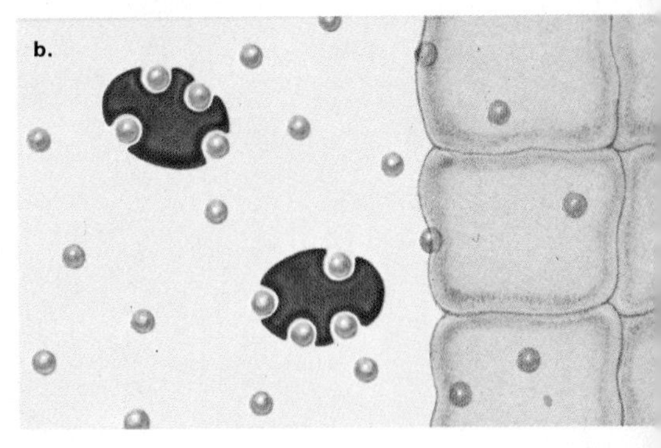

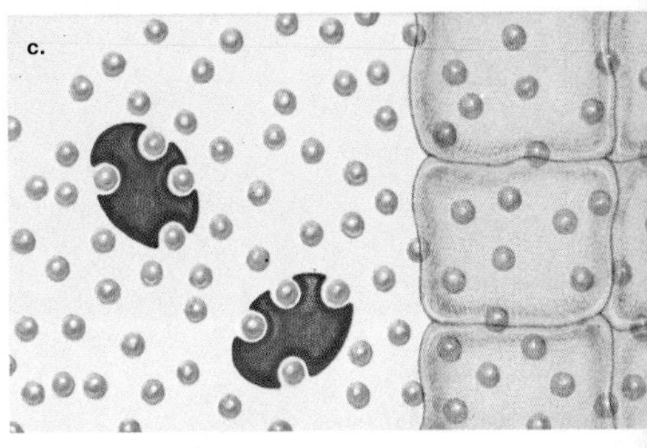

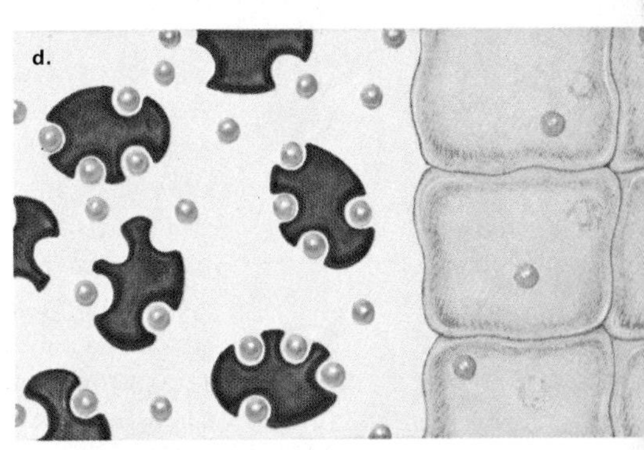

nating woodchuck was even more effective than blood from a hibernating ground squirrel in triggering hibernation in ground squirrels.

Strangely enough, none of the ground squirrels reacted adversely to the woodchuck blood, even though animals generally might be expected to reject such foreign material. We do not know the reason for this, but it might be because the two species are fairly closely related. Also, we used blood serum for most of our experiments, rather than whole blood, thereby screening out many of the proteins that cause adverse reactions.

In another experiment during the winter of 1969 and 1970, we tried to trigger hibernation in animals that were kept in a warm room during the winter. Normally, a hibernator will not hibernate if kept in a warm room, even during the winter hibernation season. We first implanted a small thermometer called a thermistor in each of 21 experimental animals. We then gave woodchuck hibernation serum to seven of the ground squirrels, ground squirrel hibernation serum to seven, and ground squirrel active serum to seven. By measuring the body temperatures each day, we found that the temperature dropped in all the animals that had received hibernation serum, and several of the animals actually hibernated. Later, when we placed all three groups in a cold room, the 14 ground squirrels that had received hibernation serum promptly hibernated, as would be expected in winter, while the seven that had received active serum did not. These results indicated that the trigger substance can act both in summer and in winter, and that it acts in winter whether the environment is cold or warm.

In the course of our experiments, we discovered that ground squirrels induced to hibernate during the summer continued to hibernate through the following winter. Moreover, some failed to become

Woodchucks in the author's laboratory awake with the arrival of spring and enjoy the chance to frolic when let out of their cages. One of them, Rascal, finds the shoulder of researcher Wilma Spurrier a convenient place to snack.

fully active the following spring. Several animals have gone through alternating periods of hibernation and arousal continuously for several years; apparently they will hibernate that way until they die.

During our research, we had encountered evidence that some blood serum samples seemed to work against triggering hibernation. We decided to investigate this further. In the summer of 1973, we mixed different combinations of filtered serum and residues from various animals in hibernating and active states. We found that mixtures containing residue from an active animal postponed the start of hibernation far beyond the normal time, even when the mixture also contained the trigger substance. This suggested that the residue contained an antitrigger substance that could cancel the effect of the hibernation trigger. We also knew that the antitrigger substance was a larger molecule than the trigger, because it could not go through the filter.

We now believe that the changing proportions of the trigger and antitrigger substances determine the time of hibernation, not the presence of trigger alone. We think the trigger is manufactured throughout the year. The antitrigger, however, appears to be manufactured only during part of the year, beginning in midwinter and gradually building up until spring when it begins to drop off. During the summer, the antitrigger seems to interact with the trigger, preventing the trigger from inducing hibernation. But the trigger substance continues to form and its concentration in the blood eventually reaches a level where it can override the antitrigger and spill over into the body tissues to start the first fall hibernation. Once hibernation occurs, the tissues no longer need the trigger and it begins to pile up in the blood. This is why blood taken from a hibernating animal, especially late in a hibernation period, can trigger hibernation in an active animal. As spring approaches, antitrigger production starts up again and eventually the antitrigger concentration becomes great enough to override the trigger. The animal then returns to an active state.

Other experimenters are trying to determine the exact nature–size, electrical charge, and chemical composition–of the trigger substance. A principal researcher in this area is Peter Oeltgen of the Stritch School of Medicine, who is trying to analyze blood components on the basis of pH values, or relative acidity, using a process called isoelectric focusing. Determining the precise nature of the trigger substance is difficult because blood serum is a complex mixture of thousands of substances. Isolating a single substance without knowing much more than its general size range is extraordinarily demanding.

According to our theory, hibernation is not simply a behavioral response to changing environmental conditions; it is also a physiological phenomenon in which the tissues and organs themselves hibernate. Current research in my laboratory is concentrated on the ground squirrel heart, an organ whose activity is related to metabolism, oxygen consumption, and general activity levels. Earlier experiments showed that the heart of a hibernating ground squirrel can survive outside the body and register activity on an electrocardiogram many times longer than the heart of an active ground squirrel. These differences may be important in explaining why the heart of a hibernator can continue activity at temperatures so cold they would stop the heart of a nonhibernating animal. We are now studying how the isolated heart of the ground squirrel responds to the trigger and antitrigger substances in each of its three states–active, hibernating, and arousal. This may eventually have important implications in preserving organs for transplant.

Meanwhile, other researchers throughout the world are studying various other aspects of hibernation. Some are studying how the normal functions of such organs as the liver, kidneys, and pancreas change during hibernation. How do these organs adapt and continue to function under the adverse conditions imposed by hibernation? Other researchers are investigating hormonal activity and the role of the adrenal gland in hibernation. Still others are interested in glucose levels and insulin production. Many aspects of hibernation are still poorly understood, but we hope that by learning more about the heart of hibernating animals, we will eventually understand more fully the metabolic and physiological functions of all animals, including man.

dr sitron
earthquake 040909 040902 15nor 15nor
89ea 89ea guatemala m7575 i9090
fedorov

Fingers on the Planet's Pulse

By Barbara L. Sleeper

A group of earth-watchers chart the short, sudden changes on this restless, lively globe, and spread the news quickly to those who need to know

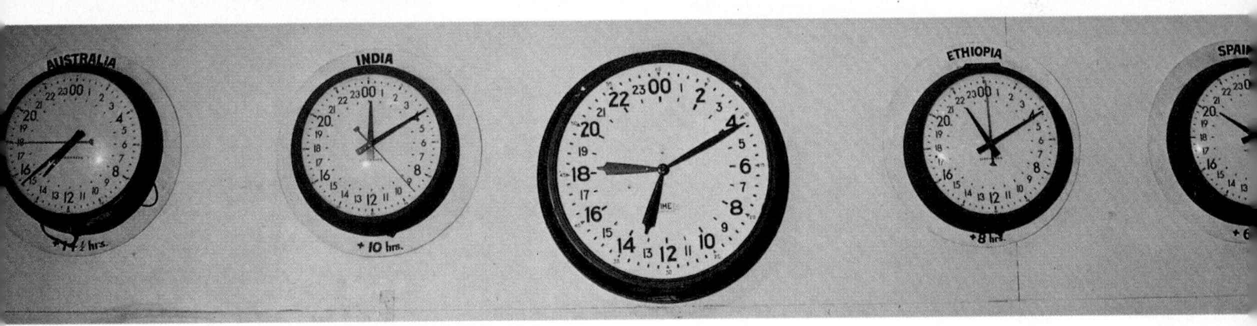

A machine coding computer-generated data for satellite observations punches a paper tape into a pile on the floor of the busy Smithsonian Astrophysical Observatory Communications Center in Cambridge, Mass. Coils of tape waiting to be decoded and analyzed hang on pegs along the walls. Clocks, covering another wall, silently report the exact time in major cities around the world. Telephones ring, and a 1,000-watt radio transmitter, ready for an international call, crackles with static. A steaming teakettle shrilly whistles on a stove squeezed into an unlikely domestic corner of the room. Lost in the din of humming, buzzing, and clicking, a gray teletypewriter (telex) machine, one of six telexes that line one wall of the communications center, types out an incoming cable.

It is 18:09 Greenwich Mean Time (GMT) – 1:09 P.M. in Cambridge – on Wednesday, Feb. 4, 1976, as the telex completes the message sent from Moscow by satellite. In cablese, it says:

dr sitron
earthquake 040902 040902 15nor 15nor
89we 89we guatemala m7575 i9090
fedorov

The message, from Evgeny K. Fedorov of the Soviet Seismic System in Moscow, reports to Robert Citron, director of the Center for Short-Lived Phenomena in Cambridge, that Russian seismographs recorded an earthquake at 9:02 GMT. The quake struck Guatemala at 15° north latitude, 89° west longitude, with a magnitude of 7.5 on the Richter scale and an intensity of 9.0 on the Modified Mercalli Intensity scale. Fedorov does not have to say it was a major earthquake, the figures speak for themselves. A quake of such magnitude and intensity will destroy poorly constructed buildings and cause many casualties. Within minutes, the message is carried upstairs from the communications center to the Center for Short-Lived Phenomena, which begins to deal with this massive disaster in a unique way.

The Center gathers data on the natural and man-induced events of brief duration that constantly occur on planet earth. It distributes the information to government agencies, universities, and other research institutions and scientists in many countries, and members of its own international student network.

As improved communications make the world seem smaller, and increasing population makes it more crowded, man has been forced to revise his view of this planet. The earth can no longer be looked upon as an endlessly exploitable resource. Dynamic, powerful, and unpredictable, it is also fragile and limited; like any home, it must be tended and cared for. But much is still unknown about its complex geologic and biological processes and the effects of civilization on the global ecosystem that supports all life.

The author:
Barbara L. Sleeper,
a zoologist, is the
education director
of the Center for
Short-Lived Phenomena.

Many important natural and man-made events are brief and, thus, hard to study, though they may have far-reaching, long-term effects. Sidney R. Galler, then assistant secretary for science of the Smithsonian Institution in Washington, D.C., was struck by this fact in 1963 as he witnessed the birth of Surtsey, a new volcanic island off the south coast of Iceland. Such an event, occurring in a remote area and lasting only a few weeks, might have gone unnoticed in another age. But Icelandic scientists alerted their counterparts around the world and many researchers went to Iceland during the eruption and collected invaluable data. Studies of Surtsey continue today.

Excited by this on-the-spot team effort, Galler began planning a permanent center that could collect and rapidly disseminate timely information to the world's scientists whenever such phenomena occurred. Using Galler's plan, the Smithsonian Institution established the Center for Short-Lived Phenomena in January 1968, in a small windowless office at the astrophysical observatory, to take advantage of the international communications network set up for the observatory's worldwide optical satellite tracking stations. In the beginning, the Center had a desk, one telephone, and an old typewriter. Citron, who

Communications worker Al Miner checks on incoming messages at the busy Center for Short-Lived Phenomena.

had managed several overseas satellite-tracking stations, was named director, and given the job of gaining the support and participation of the international scientific community.

The Center's task is to rapidly obtain and accurately catalog and disseminate scientific information on earthquakes, fireballs, meteorites, volcanic eruptions, sudden ecological changes, and other scientifically important events. Such occurrences are often newsworthy and are reported by the media, but they may not be reported until long after the phenomena occur. Many other important events are reported inaccurately or not at all. For example, the early news reports of the Guatemala earthquake had conflicting estimates of the death toll and damage it had inflicted.

Event-research coordinator Jan Connery was the first person to read the message reporting the Guatemala earthquake. Putting down a report she was preparing on a volcanic eruption in Ecuador, she conferred with Richard Golob, the Center's operations manager. Connery and Golob had been trying to complete reports on a finback whale beaching near Groton, Conn.; a volcanic eruption in Alaska; and the Ecuador eruption to make an afternoon printing deadline.

Connery assumed responsibility for the earthquake report, and placed a hurried call to find out if the Associated Press wire service in Boston had any additional information about the quake. But it was a fruitless call; the Guatemala City communications tower had collapsed, and the few telephone circuits still operating were jammed. Realizing that she was dealing with a major disaster that would produce a barrage of news reports and perhaps exaggerated statistics,

The Guatemala earthquake leveled most of the town of Patzicía, west of Guatemala City. The Center's final report on the quake said 23,000 died.

Research coordinator Jan Connery calls an international agency to verify details about the Guatemala earthquake.

Connery decided to wait for more substantial information from the Center's correspondents in Guatemala City.

Connery found 29 telegrams with details on the Guatemala earthquake waiting when she arrived at the Center the next morning. The telex machines had received quake information throughout the night via the Disaster Communications Network, which provides translations of disaster bulletins from foreign radio broadcasts. Center correspondents Markus Båth, director of the Seismological Institute in Uppsala, Sweden, and John Derr of the United States National Earthquake Information Service in Golden, Colo., had also sent telegrams.

As information poured in about the Guatemala disaster from Center correspondents, Citron and Connery planned how to gather additional facts. Citron was to try to get through to a correspondent at the Instituto Geographico Nationale in Guatemala City while Connery called embassies, wire services, U.S. geologists, and other sources.

Connery started her day by putting the telegrams in chronological order. She assembled specific bits of information–such as the time of the earthquake, location of its epicenter, quake magnitude, amount of damage, and any associated geophysical phenomena–and determined what additional facts were needed and which statistics should be verified before the Center's report could be prepared.

Connery spent the rest of the day telephoning the Honduran and Guatemalan embassies; the International Tsunami Information Service in Honolulu, Hawaii; and the U.S. Agency for International Development (AID) in Washington, D.C. Each conversation was taped on a small cassette recorder to save time and ensure accuracy. The tapes were to be transcribed later to provide a permanent document.

Between phone calls, Connery lunched on an egg-salad sandwich at her desk. Most of the people that she called offered all the information

The Alberta Fireball flashed across the skies of the Western United States and Canada on August 10, 1972. The quick reporting of this event by the Center helped scientists gather fresh meteorite samples for laboratory study.

they had; in some cases, Connery had to question them carefully to get the details she needed, such as the duration of the tremor, which, she found, lasted 30 seconds.

Although conflicting at times, details about the earthquake gradually began to accumulate. Samuel Bonis, the Center's correspondent at the Instituto Geographico Nationale, reported in a relayed telephone message that the quake caused landslides and extensive faulting and fissuring. He was sending out a team to map the geological changes and agreed to forward a report.

In the late afternoon, Connery called the League of Red Cross Societies; AID's 24-hour office in Washington, D.C.; and the New York City Liaison Office of the United Nations Disaster Relief Coordinator. This call proved the most helpful of the day, finally providing the detailed information on casualties and damage she was seeking.

About 5:30 P.M., most of the staff began to leave the office. Connery continued to make phone calls and Marie Enander, Swedish-born business manager, volunteered to help her transcribe the tapes recorded during the day. Only then did Connery remember the theater tickets she had for that evening. Although she, Golob, and Enander had agreed earlier to work all night at the Center to meet the report deadlines, she took a few hours off for another egg-salad sandwich and a trip to the theater.

Connery returned to the Center at 10:45 P.M. to find Enander still transcribing the earthquake tapes. Golob was calling sources in Alaska –where it was not yet dinnertime–for information on the volcanic eruption there. About midnight, Connery again called the AID office for the latest estimates of deaths and damage in Guatemala, which had been rising all day. Initial reports indicated only about 2,000 dead. Now it looked as though 5,000 Guatemalans had been killed, 10,000 injured, and 100,000 left homeless. Enander continued the ear-straining job of transcribing tapes. All three were exhausted, but determined to complete the reports. Drinking cup after cup of coffee, they worked through the night. Connery finally completed the last draft of the Guatemala earthquake report at 8:00 A.M. the next day. She gave it to Golob, who checked and edited it. When she looked out the window for the first time in nine hours, she was startled to see that it had snowed in the night.

After breakfast, Connery and Enander returned to find Golob typing the final copy of the Guatemala report to the beat of a hard-rock tune on the radio, his energy rallied by the rhythm. The report was sent to the print shop just as other staff members arrived to start the new day. Within a few hours, the Guatemala earthquake report was in the mail to the Center's subscribers and members of its international correspondent network.

Earthquakes and volcanic eruptions can have long-lasting effects on the environment. A severe earthquake can cause landslides and dangerous faulting and fissuring at the earth's surface. River-drainage systems may be changed, and tide levels may be raised or lowered. Volcanic dust, spewed into the air during a brief eruption, can linger in the earth's upper atmosphere for decades, affecting weather and

Richard Golob, the Center's operations manager, scans the first version of an event report that will be edited for distribution to a worldwide list of subscribing scientists.

Hundreds of mice search for food in a storeroom in Victoria, Australia, during a 1969 mouse plague. The Center reports such events to zoologists who study population explosions.

agriculture. Other events may have less disastrous effects, but they all help us understand the earth.

The Center for Short-Lived Phenomena has collected information on more than 1,200 transient geologic, astrophysical, biological, anthropological, and pollution events, and has issued some 2,500 reports. The events have included the discovery of the Wama Stone Age tribe in Surinam; the Shantung tidal wave in China; a spontaneous soil burn in the Philippines; a white-tailed rat invasion in Spain; and moonquakes recorded by equipment that had been left at Tranquility Base by Apollo astronauts.

The Center usually issues notification reports by telephone, cable, telegram, and teletype within hours after information is received. More detailed reports are airmailed as soon as possible. Usually, 5 to 10 event reports are issued weekly, and the Center's staff may be working on from 10 to 15 reports at the same time.

One of the Center's main goals is to provide information quickly so that research teams can study ongoing short-lived events in the field. For example, it contributed significantly to scientific research during a squirrel migration that occurred in the Southern United States in 1969. From correspondents' reports, the Center estimated that 20-million animals were leaving their usual range in search of food. The information enabled several teams of field biologists to study this puzzling phenomenon as it was taking place.

On Feb. 8, 1969, a blinding blue-white fireball turned night into day over a 1,600-kilometer (1,000-mile) path from southern Mexico to El Paso, Tex., as it broke up in the earth's atmosphere. It strewed more than 4 metric tons (4.4 short tons) of meteoritic material over the area. It was reported by hundreds of eyewitnesses, and the Center's prompt alert enabled investigation teams to collect meteorites and study one of the largest-known meteorite showers. Air Force meteorologists in a high-flying B-57 plane calculated the fireball's path through the earth's atmosphere and collected meteoritic dust left in the atmosphere. Within 100 hours, 20 meteorite samples were in the laboratories of the Smithsonian Astrophysical Observatory and other government groups for radioisotope and chemical analysis.

In 1971, the Center reported the discovery by Filipino scientists of the Tasaday tribe in the Philippines. As a result, several teams of anthropologists were able to study the tribe's first contact with civilization, and a film was made documenting the discovery of one of the last Stone Age cultures on earth.

With the Center's help, film crews obtained footage of a rare volcanic phenomenon, the explosive glowing gas clouds called *nuées ardentes,* when Mount Mayon erupted in the Philippines in 1968. Erupting at velocities of over 60 kilometers (37 miles) per hour with temperatures of from 1000 to 1200°C (1832 to 2192°F.), these unusual gas clouds can incinerate all living things over an area of several square kilometers in a matter of minutes.

The Center's International Correspondent Network includes more than 2,000 scientists and research institutions in 144 countries, covering every continent and ocean and every major scientific discipline. Among the institutions are the Japan Meteorological Agency in Tokyo; the Soviet Academy of Sciences in Moscow; the Council for Scientific and Industrial Research in Melbourne, Australia; Environment Canada, in Ottawa, Ontario; the U.S. Coast Guard National

Associate editor John Job prepares the final edited version of an event report. The Center may issue from 5 to 10 reports each week.

Response Center in Washington, D.C.; and the International Institute of Volcanology in Catania, Sicily.

Because it is independent of formal government channels, the correspondent network often reports events long before the news media and government agencies. For example, a detailed telegram describing the eruption of Bezmianny volcano near Kamchatka, Russia, on Nov. 3, 1969, arrived from Soviet correspondent Yuri M. Doubik of the Institute of Volcanology in Petropavlovsk immediately after the eruption. At about the same time, a telegram arrived from the United States Embassy in Moscow reporting that specific information about the eruption would not be made available. The Center quickly sent a copy of Doubik's telegram to the embassy and reported details of the eruption to scientists in other countries.

Sometimes, the Center is the only link between an event and the rest of the world. During the tragic earthquake in Managua, Nicaragua, which left about 10,000 persons dead, 20,000 injured, and 300,000 homeless in December 1972, the Center received firsthand information from an amateur radio operator in Managua who was broadcasting from a portable rig in his car. No other communication was possible because Managua had lost all electric power immediately after the quake. The Center's radio transmitter maintained contact with Managua and relayed information to the rest of the world, until normal communications channels could be reopened with the stricken city.

Staff dedication is a major asset to the Center as its workload and correspondent networks expand. In 1973, the Center established the first of several secondary reporting programs, the International Student Alert Network, in order to broaden its coverage in areas with few scientific centers. By 1975, hundreds of high school science teachers and more than 60,000 students in over 1,000 schools in 26 countries had joined the student network.

The students and teachers helped the Center to survey the United States in 1973 for madake, an unusual species of bamboo known to bloom only once every 120 years. The latest flowering cycle began in Japan in 1960 and peaked in 1969. The Center offered to help Smithsonian Institution botanists find out if there was a correlation between the flowering cycle of the Japanese madake and that of the madake introduced to the United States more than 80 years ago.

The Center sent pamphlets explaining the project and photographs of madake bamboo to several hundred high school classes throughout the United States and asked students to look for the flowering plants, take photographs if they saw one, describe it, and send a dried specimen to the Center. About 120 samples were collected from 16 states, and two-thirds of them were positively identified as madake. The students benefited from this opportunity to take part in a field project,

A science student collects a sample of madake bamboo during the Center's 1973 survey to determine if the U.S. madake bloom, which occurs only once in 120 years, is related to the Japanese bloom.

A Center correspondent studied and reported to the Center on a glacial ice cave near Juneau, Alaska, in August 1970. A small lake, called Lake Linda, drained through the ice cave and mysteriously disappeared.

A geologist collects samples of ash on a new volcanic island, born near Antarctica in 1968. The island's birth was the first event reported by the Center in its watch on the changing face of the earth.

and botanists learned from students that madake, previously known in only four Southern states, actually grows in several others and that its flowering cycles in Japan and the United States are indeed related.

Students and their teachers reported more than 200 events to the Center in 1974 and 1975, and one-third of these were then reported to the scientific community as significant short-lived phenomena. During 1976 and 1977, the Center plans to organize a student survey of tar accumulation on U.S. coastal beaches.

In 1974, the Center established the International Pollution Alert Network to trace long-term global pollution trends and help scientists assess the frequency, magnitude, and geographic distribution of significant pollution events. The network has more than 400 pollution-monitoring programs in 76 countries.

Since 1968, after Apollo astronauts observed and reported volcanic activity in Central America, the Center has also been deeply involved in U.S. space programs and the application of space age technology to the study of earth from space. After several large-scale natural events on earth were unexpectedly observed by unmanned

U.S. spacecraft, the National Aeronautics and Space Administration (NASA) in 1971 asked the Center to coordinate the systematic observation of such events by the Earth Resources Technology Satellite, now called Land-Sat. This one-year project involved identifying those earth phenomena detectable by the satellite and then scheduling observations as the satellite passed over the involved areas. The satellite's sensors observed volcanic eruptions, earthquakes, forest fires, defoliations, the formation of new islands, glacial surges, and oil spills. The data provided were analyzed in detail by NASA scientists and compared with on-the-spot ground observations of the events supplied by Center correspondents.

The Center coordinated similar projects during the three manned Skylab missions in 1973 and 1974, and the astronauts received daily reports of volcanic eruptions, earthquakes, insect infestations, forest fires, vegetation die-offs, and major pollution incidents that they might observe from space. On Dec. 11, 1973, the Skylab 4 astronauts photographed and described a volcanic eruption in the Galapagos Islands before the Center knew about it. It was one of several short-lived events reported first from space.

Now an independent, nonprofit organization, the Center has become a unique international clearing house for the collection, dissemination, and storage of information concerning short-lived events, a prototype for international monitoring systems of the future. It has shown that worldwide reporting networks, even when composed of volunteers, can make significant contributions to basic research and global environmental assessment.

As a result of the data accumulated on short-lived events during its first eight years, the Center is also beginning to develop the capacity for long-term environmental assessment. For example, it recently computerized data documenting most of the major oil spills that have occurred since 1968. This information can now be used to make assessments of global trends in marine pollution. Such assessments may be of enormous value to national and international environment agencies.

In its own way, the Center for Short-Lived Phenomena is helping to bring the earth into closer focus. For its support, it relies on subscribers, foundations, and government contracts. For its success, it relies on thousands of volunteer correspondents–from scientists to grade school students–and on the hard work and enthusiasm of its own small staff, qualities that were vividly shown during the two days after the Guatemala earthquake.

Shortly before noon on February 6, a few hours after the earthquake report had gone to the printer, Citron found Golob still in his office after his all-night stint and invited him to lunch with a visitor. Golob declined. He was waiting for one more message about the volcano in Alaska and then was going home to bed. As Citron and his guest went out the door, the crackling sound of static sent Golob to the short-wave radio. Something was coming in from Alaska.

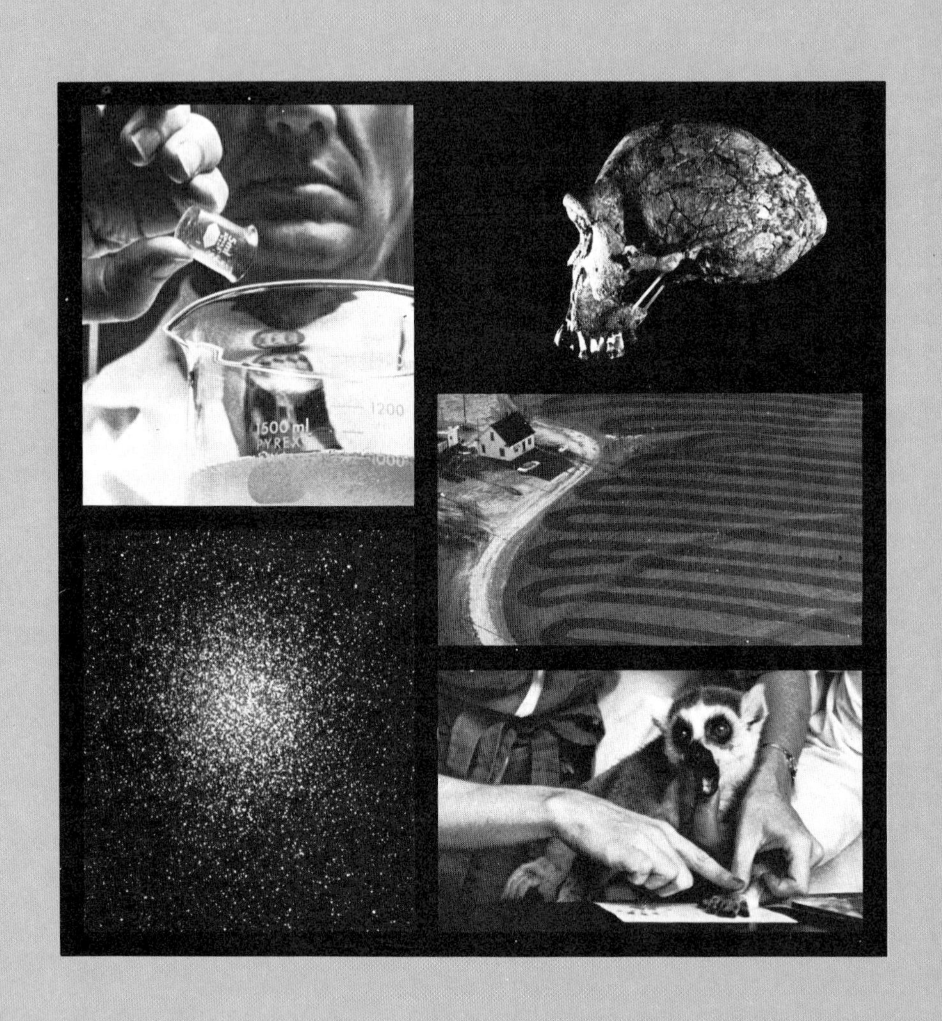
1500 ml
PYREX
1200

Science File

Science Year contributors report on the year's major developments in their respective fields. The articles in this section are arranged alphabetically by subject matter.

Agriculture

A long-awaited breakthrough in soybean yields was achieved in 1976 by John J. Hanway and Ramon L. Garcia of the Iowa State University Department of Agronomy in Ames. After two years of field experimentation, they found that spraying the plant leaves with liquid fertilizer while the seeds are growing in their pods increased soybean yields by from 4 to 8 bushels per hectare (10 to 20 bushels per acre), an increase of more than 50 per cent. The spray they used had about the same ratios of nitrogen, phosphorus, potassium, and sulfur as do soybean seeds.

The scientists reported that the first spray should be applied as soon as the beans can be felt in the top pods of the plants. They applied from one to three additional sprays at 10- to 14-day intervals, using 35 liters per hectare (19 gallons per acre) of solution.

Ozone protection. Donald J. Lisk of the pesticide residue laboratory at Cornell University in Ithaca, N.Y., and his co-workers reported in 1976 that diphenylamine shows promise in alleviating ozone injury to plants through altering the biochemistry of the plant leaves. Lisk has long been seeking chemical compounds to protect plants against air pollution absorbed through their leaves. The chief air pollutants for plants are ozone and compounds of sulfur and nitrogen. The heaviest levels of such pollutants are found in highly industrialized areas where they are released into the air from burning fuels.

Nitrogen efficiency. Cleve A. L. Goring of the Dow Chemical Company in Midland, Mich., developed Nitropyrin, a chemical to be put on the soil that increases root uptake of nitrogen fertilizer by crops. The chemical, also called N-Serve, was being tested extensively in 1976.

Nitrogen losses for soil-applied nitrogen fertilizer normally range from 75 per cent in the tropics to 50 per cent in temperate zones. *Nitrosomonas* bacteria oxidize much of the soil-applied nitrogen before the plants can absorb it. The oxidized nitrogen is then washed out of the soil and pollutes lakes and streams. Nitropyrin is poisonous to the *Nitrosomonas* bacteria. It also reduces denitri-

Fields of Bucklin, Kans., bear the unique pattern of chisel plowing, a new technique to keep soil from blowing away in drought-ravaged areas. The plows bring clumps of soil to the surface from depths of 15 to 20 centimeters (6 to 8 inches), and the clumps then help to keep the dry topsoil in place.

Agriculture

Continued

A mechanical currant harvester vibrates the currant bushes to shake off the berries. It can do the work of about 500 human pickers.

fication, the process by which soil-applied nitrogen is released into the atmosphere where it is useless to plants. Reduction of nitrification with this chemical can last up to three months.

Don M. Huber, associate professor of botany and plant pathology at Purdue University in Lafayette, Ind., has reported other advantages for Nitropyrin. Huber says it greatly reduces the incidence of some major crop diseases, such as bacterial root infections, suppresses the release of nitrous oxides and other nitrogen compounds into the atmosphere, and extends the season during which nitrogen fertilizer can be applied to plants.

Chemical regulators. Interest in chemical regulators for plant growth reached a new high in 1976, with 29 companies researching and developing such chemicals. The growth regulators for sugar cane and corn are among the most promising that have been developed. Chemical ripeners were used on 6,069 hectares (15,000 acres) of sugar cane in Hawaii during 1975. The ripeners suppress the vegetative growth of the sugar cane and cause carbohydrates to accumulate. This produces from 10 to 12 per cent more sugar per unit of sugar cane.

Agronomist Al J. Ohlrogge of Purdue University applied DNPB herbicide to corn as a growth regulator. The growth rate of the corn was slightly interrupted and its productive energy shifted earlier from its roots, leaves, and stalks to its reproductive parts – its ears. Ohlrogge said the growth regulator caused the plants to pollinate earlier and produce more and larger ears per plant, with a 5 to 10 per cent increase in grain yield.

Corn syrup. Another 1975 corn development was the production by hydrolysis of a high-fructose corn syrup. This is the first corn-derived sweetener comparable in sweetness to the sucrose in sugar cane and sugar beets. The high-fructose syrup could replace all imported raw sugar by 1980. This would create a new market for the nation's 5- to 6-billion-bushel crop.

New crop varieties. Two new varieties of rice – IR32 and IR34 – were de-

A puncture-strength test of field corn, *top,* by Terry Colbert of the Missouri Agricultural Research Station shows how well the stalks of new corn varieties will resist breaking and falling over in severe weather. In another test, Marcus Zuber saws a stalk, *left,* into pieces and removes the pith, *above,* to see how much strength this substance contributes.

veloped in 1975 by the International Rice Research Institute at Los Baños in the Philippines. Both varieties are resistant to many insects and diseases. IR32 is also moderately adapted and IR34 highly adapted to the deepwater growing conditions that are found in the Philippines and Southeast Asia.

The Agricultural Research Service and the Nebraska Agricultural Experiment Station at Lincoln jointly introduced Lancota, a disease-resistant, hard, red winter wheat. The new wheat – a high-protein variety – was developed through 20 years of cooperative effort.

Richard Lover, Conrad Miller, Ervin Humphries, and Samuel Jenkins of the North Carolina Agricultural Experiment Station at Raleigh developed three improved varieties of pickling cucumbers. They named them Addis, Sampson, and Calypso. The North Carolina research center also developed Boone, an easily threshed barley, and Saleem, a high-yielding oat.

The Virginia Truck Research Station at Norfolk originated Exmore, a southern pea that offers early maturity and good quality after freezing. Green Ruler, a snap bean with a strong beany flavor, was developed by the Michigan Agricultural Experiment Station at East Lansing.

A cooperative breeding program on experimental farms in Indiana, Illinois, and New Jersey developed a home-garden type of Golden Delicious apple, called Sir Prize. A pink-fruited market tomato called Traveller '76 was produced in Arkansas. Florida Belle, a very large strawberry, was developed in Florida, and Benton, another very large strawberry, was produced in Oregon.

New blueberry and plum varieties extend southward by 320 to 480 kilometers (200 to 300 miles) the fruit-producing boundaries for these plants. The blueberries, which can now be grown in Florida, are called Flordablue and Sharpblue. The plums, which can grow in Alabama, are called Crimson, Purple, and Homeside.

Hybrids. Lynn S. Bates of Kansas State University in Manhattan, trying to produce hybrid crosses between plants of different species, genera, and even families, used immunosuppressant chemicals such as E-amino caproic acid (EACA) on cereal grains and legumes to produce the hybrids. His aim is to show that it is possible to overcome the cross incompatibility between different species of plants. Bates developed hybrids of wheat and barley and barley and rye.

Animal health. Clyde K. Smith and his associates at Ohio State University in Columbus developed an orally administered vaccine for the treatment of colibacillosis, a disease caused by *Escherichia coli* bacteria that kills newborn calves. A new vaccine for virus-induced scours, or diarrhea, was also developed by scientists at the University of Nebraska in cooperation with the Department of Agriculture and private industry. This vaccine reduces calf deaths in a herd from 10 per cent to as little as 1 per cent.

Another technique for controlling diseases in calves came from Gabel Conner of the Michigan Agricultural Experiment Station. He perfected a method of vaccinating unborn calves against scours by injecting the vaccine through the flank of a pregnant cow.

No-plow farming. A 12-year study conducted by the Ohio Agricultural Research and Development Center at Wooster shows that growing corn on unplowed fields can significantly increase the yield in many types of soil. The no-plow system also substantially reduces soil erosion, increases retention of rain water, allows quicker planting, and saves on labor and tractor fuel.

A special planter that can operate on fields that are covered with stalks and roots must be used. It cuts a narrow furrow and lays the seed at the same time. A band of fertilizer is then laid along the furrow. Insecticide is applied later. The Ohio study showed that, on well-drained silt-loam soil, the yield of corn with no plowing equaled or surpassed that with conventional plowing.

A demonstration greenhouse at Becker, Minn., was erected in 1975 by the University of Minnesota Agricultural Experiment Station, the U.S. Environmental Protection Agency, and the Northern States Power Company of Minnesota. The greenhouse, largest of its type in the nation, covers 23,700 square meters (22,000 square feet) and is heated by water from an electric generating plant. [Sylvan H. Wittwer]

Anthropology

Discoveries related to the origins of human beings continued to come from Kenya, Tanzania, and Ethiopia in East Africa in 1975 and 1976. In March 1976, Richard E. Leakey of the National Museums of Kenya announced the discovery of a 1.5-million-year-old skull on the east side of Lake Turkana (formerly Lake Rudolf) in Kenya. It was dated by the potassium-argon radioactive decay process.

Leakey and his wife, Meave, found the skull buried up to its browridges in gray-brown soil. After its missing parts were reconstructed by Alan Walker of Harvard University Medical School, the skull closely resembled the *Homo erectus* skulls known as Peking man. These were discovered in China in the 1920s and, at that time, dated to about 375,000 years ago.

This discovery, Leakey says, "raises questions about the true age of Peking man," which has never been reliably dated. He suggested that the Peking man fossils may well be redated to 1-million years earlier after they have been re-examined on the basis of his new *Homo erectus* find.

Leakey also reported finding, in the same part of Kenya, another skull that is much older than the *Homo erectus* skull. This one, similar in appearance to a skull he found in 1972 that is known by its catalog number as 1470, is from 2.5 million to 3 million years old. It has remarkably human (nonaustralopithecine) features for a skull that can be dated to such an ancient age.

In October 1975, Mary N. Leakey, Richard's mother, found the jaws and teeth of eight adults and three children in the Laetolil lava deposits 40 kilometers (25 miles) south of Olduvai Gorge in Tanzania. The lava flow has been accurately dated by the potassium-argon method to from 3.35 million to 3.75 million years old, so these jaws and teeth are the oldest reliably dated remains of early man yet discovered.

In December 1975, C. Donald Johanson of Case Western Reserve University in Cleveland reported finding manlike remains at Hadar in the Afar region of central Ethiopia. He found about 150 bones and bone fragments from two infants and three to five adults that are at least 3 million years old. Because they were found together, Johanson believes these individuals were a social group that probably died together, perhaps in a flash flood.

Johanson took 35 of the bones and constructed a composite fossil hand that is about the same size as today's human hands and appears capable of as much dexterity. Because both juvenile and adult bones of the same types were found, Johanson believes anthropologists should be able to learn from them how certain bones grew and developed in early man.

Ramapithecus. As the anthropologists sifted through the fossil remains of the late Pliocene and early Pleistocene epochs in Africa to determine which were human and which australopithecine, others searched through early Pliocene Epoch deposits in other parts of the world for fossils representing the common ancestor of both these groups. The best candidate at present is *Ramapithecus,* whose partial jawbone was found in the Siwalik Hills of India in 1910 and who was thoroughly described in 1934 by G. Edward Lewis. Louis S. B. Leakey found other *Ramapithecus* fossils at Fort Ternan, Kenya, in 1962, and named them *Kenyapithecus* to distinguish them from the Indian find. Leakey dated his discovery to about 14 million years ago.

David R. Pilbeam of Yale University and Ibrahim Shah of the Geological Survey of Pakistan returned to the Siwalik Hills in an area now in Pakistan and reported in March 1976 that they had found additional *Ramapithecus* jaws and teeth, which they were able to date to between 8 million and 12 million years ago.

Ramapithecus-like fossils were also found in Hungary, Miklós Kretzoi of Budapest reported in October 1975. The cranial skull fragments of at least 20 individuals were discovered in lower Pliocene Epoch deposits at Rudábanya in northeastern Hungary in excavations for a new section of an iron mine between 1967 and 1975. Another 18 postcranial fragments had also been found in the same area. Although the Hungarian anthropologists recognized that these fossils are quite similar to the original *Ramapithecus* fossils from India, they, like Louis Leakey, preferred to assign new names to their specimens, *Rudapithecus* and *Bodvapithecus.*

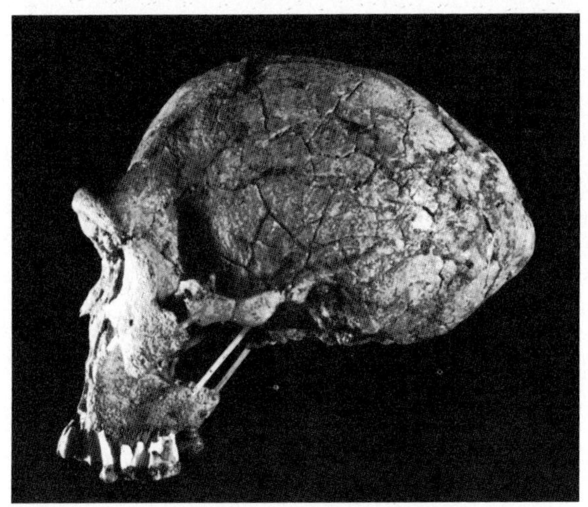

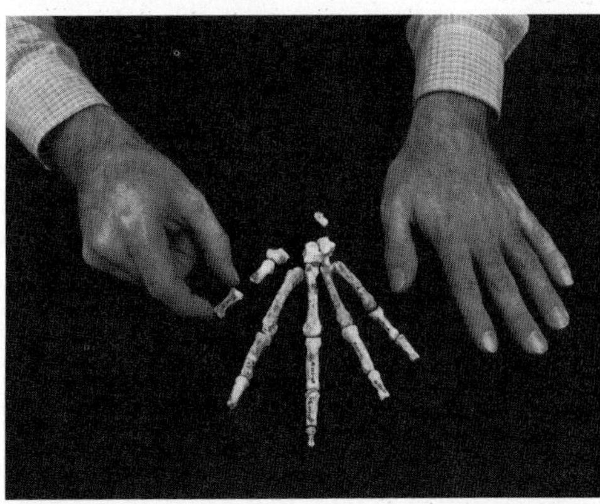

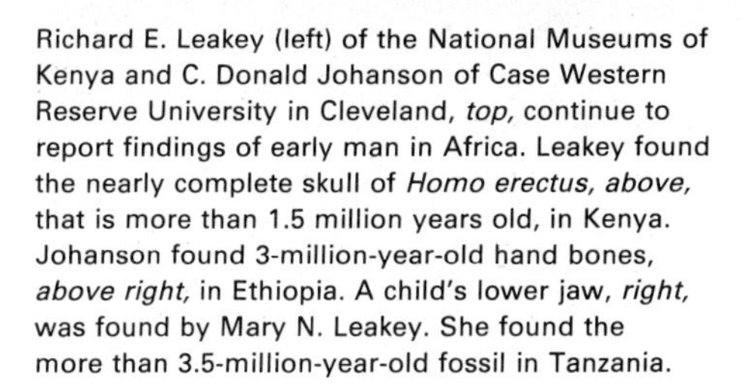

Richard E. Leakey (left) of the National Museums of Kenya and C. Donald Johanson of Case Western Reserve University in Cleveland, *top,* continue to report findings of early man in Africa. Leakey found the nearly complete skull of *Homo erectus, above,* that is more than 1.5 million years old, in Kenya. Johanson found 3-million-year-old hand bones, *above right,* in Ethiopia. A child's lower jaw, *right,* was found by Mary N. Leakey. She found the more than 3.5-million-year-old fossil in Tanzania.

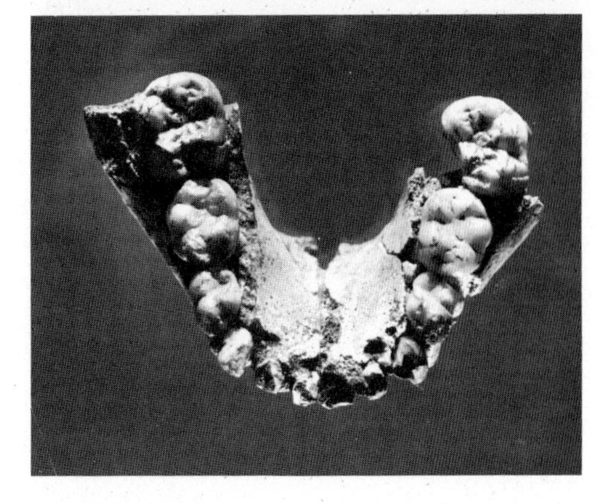

Anthropology

Continued

Miners' lungs. The mummified lungs of 22 Indian miners buried in a cemetery in northern Chile were studied in 1975 by scientists from the Medical College of Virginia at Richmond and the University of Santiago, Chile. The mummies date from A.D. 1550 to 1600, a time when the Spaniards forced the Indians to extract precious metals from the primitive mines.

The lungs of 12 individuals showed signs of extreme emphysema. Microscopic study of portions of the lungs showed evidence of severe coniosis, a disease caused by inhaling fine silica and ore dust from the mines. Chemical analysis showed large amounts of silver, copper, lead, and mercury in the lungs. The disease pattern observed is similar to that seen in some miners' and sandblasters' lungs today. See DEAD MEN DO TELL TALES.

Salem witches. Many explanations have been advanced for the witchcraft scare and trials of 1692 in Salem, Mass., including mass hysteria and fraud, but none have proved entirely satisfactory. Twenty people were executed there and two died in prison. The most recent attempt to explain the bizarre behavior at Salem was made by Linnda R. Caporael of the University of California, Santa Barbara. She suggested in April 1976 that a physiological factor, poisoning from the fungus ergot (*Claviceps purpura*), was the primary cause of the peculiar hallucinations associated with the witchcraft fear.

Ergot grows on many cereal grains, especially on rye, a staple of the Puritans. Caporael based her conclusions on the fact that some Salem children began suffering from the peculiar symptoms well before witchcraft was suggested as the cause, that the symptoms closely matched those observed in ergot poisoning, and that the hot, damp weather of 1692 was ideal for the growth of ergot fungus. This fungus produces a number of powerful alkaloids such as isoergine, which contains about 10 per cent of the hallucinogenic activity of LSD. Isoergine is also found in ololiuqui, the ritual hallucinogenic morning-glory seeds used by the Aztec Indians of Mexico. [Charles F. Merbs]

Archaeology

Old World. The most ancient shipwreck ever found was announced by George Papathanassopoulos, president of the Hellenic Institute of Marine Archaeology, in September 1975. He reported that a team of scuba divers led by Peter Throckmorton, an American photojournalist and expert on underwater archaeology, found an early Bronze Age ship, dating to about 2500 B.C. It was under about 20 meters (65 feet) of water off the small island of Dokos, between the island of Idhra (Hydra) and the Greek mainland.

Throckmorton reported that the first evidence of the wreck was two heaps of broken pottery lying on the seabed. A pile of stones found in shallower water may have been the ship's ballast, which could have fallen out when the vessel struck a rock and broke in half. Twenty-five shards from the cargo of pottery, brought up by the diving team, were identified as parts of cups, pitchers, and sauceboat-shaped pots. All were from the second period of the Bronze Age in the Cyclades, a string of Greek islands.

Throckmorton estimated that only about 5 per cent of the ship's timbers remain. He hopes that these may provide some idea of Cycladic shipbuilding techniques. Until now, the only knowledge of these techniques has come from representations of the ships in decorations on pottery from these islands. These illustrations show ships with 20 or 30 oars on each side and a curved prow, often with a fish hanging from it. The prow rises 2 or 3 meters (8 or 10 feet) from one end of the vessel. The other end of the boat terminates in a battering ram.

Spanish Armada. In addition to the technological information they may yield, shipwrecks are of archaeological interest because they provide undisturbed and precisely dated information. Nothing found on them can be more recent than the day on which the ship went down. In January 1976, archaeologists learned of the progress of several underwater investigations at the Seventh International Conference on Underwater Archaeology, at the University of Pennsylvania in Philadelphia.

Archaeology

Continued

Ancient clay tablets from the city-state of Elba were unearthed by Italian archaeologists south of Aleppo, Syria. The tablets dated to 2300 B.C., and were written in Sumerian and strange Semitic script.

The ships involved included King Henry VIII's flagship, the *Mary Rose,* which sank in 1545 while fighting French ships off Portsmouth on the south coast of England.

Archaeologist Colin Martin of the University of St. Andrews in Fife in Scotland reported on five ships of the Spanish Armada found at various points on the Scottish and Irish coasts where they were blown by storms. Martin stated that King Philip II of Spain used ships from many ports in Europe, ranging from Venice, Italy, to Rostock, Germany, on the Baltic Sea, to mount his 1588 seaborne attack on Queen Elizabeth's England.

Other shipwrecks. Off the island of Anglesey in northwest Wales, archaeologist Peter Davies of Liverpool University is supervising the recovery of King Charles II's royal yacht, *Mary.* The yacht, made for the king in 1660, was used by the Royal Navy to transport dignitaries. It capsized in the early 1670s while on a voyage from Dublin, Ireland, to England. Davies found the skeleton of a 20-year-old woman who was apparently trapped in the wreckage when the *Mary* capsized.

A famous shipwreck that occurred a century later involving John Paul Jones's 40-gun warship, the *Bonhomme Richard,* was being investigated in the summer of 1976 by underwater archaeologists from Great Britain and the United States. The ship sank in September 1779, off Flamborough Head in Yorkshire, England, after a fierce battle with the British ship *Serapis.* This was the last sea battle of the American Revolution in British waters.

New discoveries from China. Chinese archaeologists announced the discovery in 1975 of the oldest known evidence of human toolmaking. Chopping tools made of quartz or quartzite were found in the Nihowan Basin of northern China.

An accidental find in Shensi Province in the summer of 1975 corroborated legends concerning the Ch'in dynasty's first emperor, Shih Huang Ti (259?-210 B.C.). According to the tradition, Huang Ti conquered six kingdoms, including Korea and half of Indochina,

Archaeology
Continued

built the Great Wall, and standardized the Chinese language. He is also said to have abolished the feudal system, ruled through a civil service, and distributed land to the peasants. Huang Ti is believed to have begun building his own tomb near his summer palace in the Black Horse Hills, close to the modern city of Lin Tuang. According to legend, the tomb had a secret entrance and was covered with earth and trees so that it could not be seen. Huang Ti's harem is said to have been buried with him. His coffin was placed in a burial chamber at the center of the tomb and molten copper was then poured into the chamber to seal it.

The tomb was uncovered by a man digging a well. Chinese government archaeologists have found some 6,000 life-sized pottery horses and warriors in the tomb. Confucianism did not permit human sacrifice, which was common before Huang Ti's time. The emperor presumably substituted the pottery figures for human beings. Consequently, the harem, if it is found, may consist of pottery figures. [Judith Rodden]

New World. The remains of a 12-year-old child, buried about 5500 B.C., were found in L'Anse Amour burial mound and village on the Labrador side of the Strait of Belle Isle in 1975. James A. Tuck of the Memorial University of Newfoundland discovered it 1 meter (3.5 feet) below the surface.

Artifacts in the grave included a walrus tusk, some bone spearpoints, a bone whistle or flute, a decorated ivory toggle, and what is probably the earliest known toggling harpoon head. There also were six stone projectile points or knives, and nodules of graphite with red ocher stains. This excavation and others conducted earlier by Tuck in the same general area have established the fact that Indian groups moved north along the Atlantic Coast of Canada into Labrador from 7000 to 5000 B.C.

Charles W. McNett and his associates of the American University in Washington, D.C., uncovered a campsite north of Stroudsburg, Pa., that lies below three later levels representing early Archaic, late Archaic, and late

Life-sized warriors and horses made of pottery and buried for more than 2,000 years were found in an emperor's tomb in northern China.

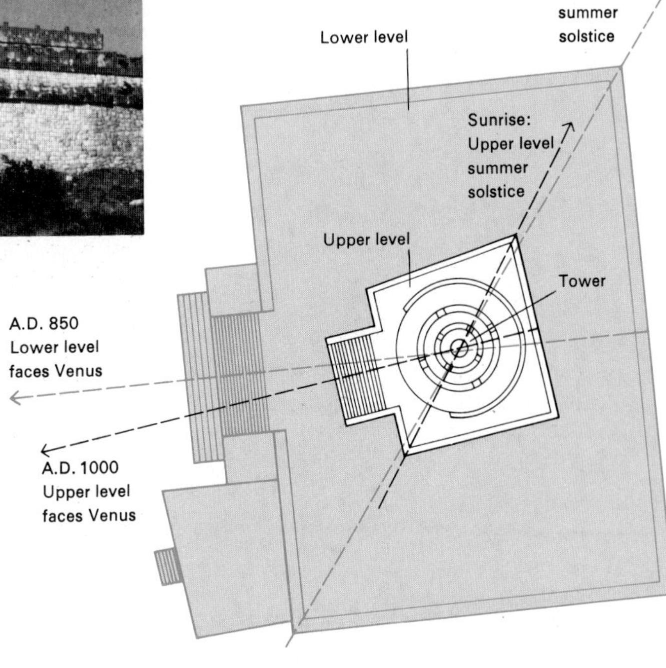

The Caracol Tower at Chichén Itzá, a Mayan city in Mexico, was probably used as an ancient astronomical observatory, according to United States and Mexican scientists. The building's upper level does not line up with its lower floor. This may have been done deliberately to compensate for changes in the earth's alignment with other planets and the sun between the periods when the top and bottom levels were built.

Archaeology
Continued

Woodland tribes. The earliest level was radiocarbon-dated at about 8672 B.C. It contained some stone spearpoints of the ancient fluted design.

Indiana flint. For many years, archaeologists in the Middle West have found tools and other artifacts made of a bluish-gray nodular flint or chert, often attributed to ancient stone quarries in Harrison County, Indiana. A 1975 study of this area by Mark F. Seeman of Indiana University in Bloomington has increased our understanding of how prehistoric Indian hunters used this material between 10,000 and 8000 B.C.

The flint nodules were extracted from limestone of the St. Louis rock formation after they had been exposed by weathering. Some nodules were as large as 129 square centimeters (20 square inches), though most were much smaller. This type of colored flint has also been found in Kentucky, Tennessee, and southwestern Illinois. Indian groups that date from about 10,000 to 6000 B.C. and from about 1000 B.C. to A.D. 500 made the implements in the Harrison County area. A large number of so-called turkey-tail spears or knives found in early Woodland period burials (beginning about 1000 B.C.) were made of this chert. So were more than 8,000 partially completed disks found in a small mound at the Hopewell site near Chillicothe, Ohio. Artifacts made from this flint between 1000 B.C. and A.D. 500 were traded among tribes from northern Lake Huron and Lake Michigan to Louisiana, and from Iowa to New York.

David A. Breternitz of the University of Colorado uncovered a great kiva, or ceremonial chamber, dating to the early 900s in Mesa Verde National Park in Colorado. The kiva's design represented two early stages of Pueblo prehistory. A Sinagua Indian village near Flagstaff, Ariz., was excavated in 1975 and 1976 by William J. Beeson and Howard P. Goldfried of California State University in Sacramento. They found nine houses with masonry rooms, some stone-lined rectangular pit houses, and an earlier pit house with clay-plaster walls. They also excavated sev-

Science Year Close-Up

Research In the Garbage Can

What do the words *New World archaeology* bring to mind? To some, it is the ancient cliff dwellings of the Anasazi Indians, tinted red by a setting Colorado sun. Or howler monkeys screaming in the jungle that shrouds the Mayan pyramids in the ruins of Tikal in Guatemala. But, for more than 200 University of Arizona students in Tucson, Ariz., New World archaeology is more recent and less romantic—yesterday's garbage, with some of the "freshest" artifacts ever examined.

These students are participating in the university's Garbage Project, a novel study begun in 1973 by Associate Professor of Anthropology William L. Rathje. The project's goals are similar to those of any traditional archaeological project. In an attempt to understand modern American society better, Rathje and his students, working in cooperation with the Tucson Sanitation Division, analyze the material that Tucson's 300,000 inhabitants discard. While making their rounds, Tucson Sanitation Division personnel bag and label a specified number of household pickups from preselected areas, which they deliver to Garbage Project student volunteers who sort the contents of each bag at the Sanitation Division maintenance yard. Wearing lab coats, rubber gloves, and surgical face masks, the students examine every item found in each bag. If it is available, they record such information as the original product weight and price as indicated on the container, the weight of discarded food, and the composition of the item and its container. Information identifying the source of the garbage is discarded.

A French title on the Tucson Garbage Project's symbol gives it a certain air.

The data is subjected to statistical analyses and examined for patterns and trends in consumer behavior. These must be interpreted in the light of official census statistics on income and family size, data on prices and consumption, and the state of the economy.

Many of the Garbage Project's findings are disturbing. For example, its figures show that the average Tucson family discards, in edible form, about 9 per cent of the food it purchases. This means that about 8,600 metric tons (9,500 short tons) of edible food worth from $9 million to $11 million is buried in Tucson's sanitary landfill each year. Middle-income areas waste more food than do lower-income neighborhoods. The smallest waste of edible food is where nearly half the families are below the poverty level.

Data analyses conducted in 1975 with the help of the National Science Foundation led to some interesting discoveries about how consumers react to such economic crises as food shortages. For example, when sugar was in short supply and prices increased markedly early in 1975, consumers threw away more edible sugar and sugar products than in 1974 or in 1976. An analysis of beef discards in the spring of 1973, the year of the much-publicized beef shortage, showed the same pattern of wasting large amounts of suddenly scarce and expensive foods. One interpretation of this is that people experiment with and discard cheaper forms of a scarce product in the face of rising prices.

The effects of general economic pressures, such as recession and inflation, also show up in the garbage can. The average Tucson family spent $26.05 per week in 1973 and $26.30 in 1974 for food. However, these figures mask the fact that food prices increased about 14 per cent from 1973 to 1974. Tucson families bought less food, especially less meat, during 1974. When food prices increased another 12 per cent in 1975, the average amount spent for a week's groceries rose to $29.19. Despite this, the garbage showed that consumers were still trying to cope with inflation. They tried to cut food costs by buying more of less expensive forms of protein and filler foods such as bread. Low-income families were most susceptible to this economic pressure. Their garbage showed the largest decrease in overall food purchases and in foods having high nutritive value.

Although the project's findings are useful, and often intriguing, the work can be hard and messy. But the students do not let this discourage them. They have formed volleyball, basketball, and softball teams and each year, wearing T-shirts stenciled "Le Projet du Garbage," they play the Tucson Sanitation Division personnel. Rathje and his students have also composed a garbage musical, and every Friday the lyrical strains of "I'm a Garbage Project Dandy" and "Duke of Slops" echo from a campus pub. [Sherry L. Jernigan]

Smashed beer mugs were found at the site of 70 houses that date to 2300 B.C. The houses were excavated near Chanduy, Ecuador, by Donald W. Lathrop of the University of Illinois at Urbana.

eral graves. The site was occupied three times after the volcanic eruption of Sunset Crater in what is now northern Arizona in the late A.D. 1000s.

Central America. Excavations at Abaj Takalik, near Retalhuleu in western Guatemala, by archaeologist John A. Graham and anthropologist Robert F. Heizer of the University of California, Berkeley, have uncovered a series of sculptured stone altars and stelae, or stone pillars. They are the earliest found so far in the area and are older than any from the lowland Maya area. One was inscribed with a date in the Mesoamerican calendric system that placed it in the early A.D. 100s. Another had a date, only partially preserved, that placed it sometime in the last three centuries B.C. About 50 large sculptured stone monuments have been identified as works created in the local style, and two others bear a stylistic resemblance to Olmec art of the southern Gulf Coast of Mexico.

These finds are undoubtedly part of the Isthmian complex, a cultural development during the middle and late formative period in Central America. Examples of this culture have been found extending from the Isthmus of Tehuantepec and the western and southern Chiapas highlands in Mexico into Guatemala.

South America. Continuing investigations in Ecuador confirm that the Valdivia culture was a complex development. Donald W. Lathrop of the University of Illinois in Urbana in collaboration with others has been excavating a town site of about 2300 B.C. near Chanduy. The town measured 400 by 500 meters (440 by 550 yards). Seventy houses have been located, and three have been excavated.

Earl C. Saxon of the University of Durham in England, who has been working for several years in the Mylodon cave in southern Argentina, reported new information. Saxon's 1976 excavations revealed that giant ground sloths occupied the cave 13,000 years ago. People occupied it 11,000 years ago, as evidenced by tools and guanaco bones from the food they presumably ate. [James B. Griffin]

Astronomy

Planetary Astronomy. Jupiter's magnetosphere may be a planetary nebula that truly deserves the name, according to a report by three Israeli astronomers in 1976.

A standard planetary nebula consists of a hot star surrounded by an immense luminous shell of gaseous hydrogen, sulfur, and other elements. It looks like a faint, round planet when seen from Earth, but it has no physical relationship to planets.

However, I. Kupo, Y. Meklar, and A. Eviatar of Tel Aviv University in Israel discovered glowing emissions characteristic of ionized sulfur in Jupiter's magnetosphere. These are presumably caused by the same physical processes that take place in a standard planetary nebula.

Robert Brown of Harvard University in Cambridge, Mass., analyzed the Israeli observations and reported in April that the ionized gas around Jupiter has an electron temperature of thousands of degrees and a density of 3,000 electrons per cubic centimeter (49,000 electrons per cubic inch). Many astronomers suspect the sulfur atoms may be knocked off the surface of Io, Jupiter's second closest satellite, by intense proton bombardment from Jupiter's magnetosphere. If so, Brown's estimates for other related quantities imply that the loss of sulfur from Io's surface in the last several billion years must have come to 2.8 kilograms per square centimeter (40 pounds per square inch).

Io's peculiar atmosphere. Sulfur is not the only substance knocked off Io by the energetic particles of Jupiter's magnetosphere. Sodium and hydrogen were the first atoms discovered escaping from Io. In April 1976, Laurence Trafton of the University of Texas at Austin and Guido Münch of the California Institute of Technology (Caltech) independently reported that they had found potassium escaping from Io.

The emerging picture of Io's atmosphere is complex and controversial, but a few general features are becoming clear. The Jovian magnetosphere stretches out in a thin disk that is inclined at a different angle from that of Io's orbit, so Io continually passes in

"We've almost mastered your language, but some of your adjectives are a little zlivruquok."

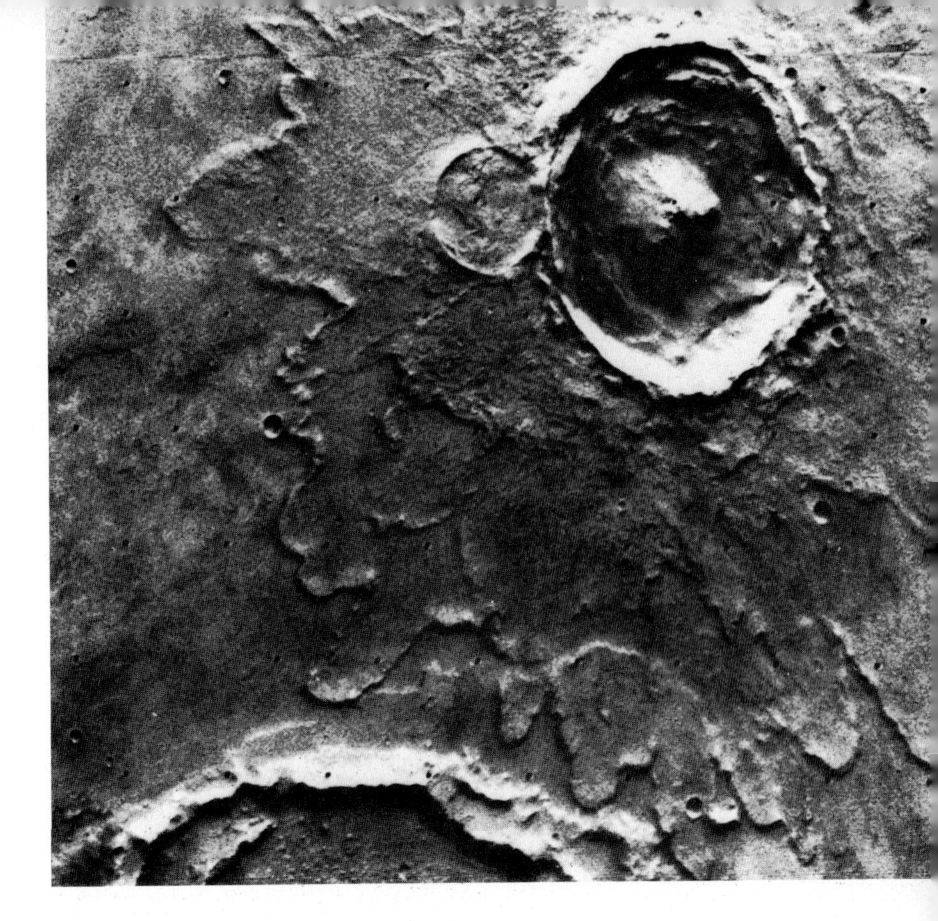

Rough Martian terrain photographed at the proposed landing site for Viking I caused postponement of the July 4, 1976, landing.

and out of the thickest part of the magnetosphere. Because of this, the number of atoms knocked off its surface varies. The escaped atoms, which glow by reflecting sunlight, move into a doughnut-shaped orbit around Jupiter after leaving Io. When they are eventually ionized by electron bombardment, they lose the ability to reflect sunlight, and they disappear from view.

The Great Red Spot has dominated the appearance of Jupiter for almost 100 years. This huge elliptical feature in Jupiter's south tropical zone is about 40,000 kilometers (25,000 miles) long and 13,000 kilometers (8,100 miles) wide. Early theories emphasized the apparent uniqueness of the Great Red Spot and its long-term stability. Most of these theories attributed the phenomenon to a disturbance originating deep in the atmosphere, perhaps at the planet's surface.

Modern theorists tend to discount the spot's apparent uniqueness because many smaller red spots have been found in recent years. Astronomers also discount the idea of a surface disturbance, because new theories about the internal structure of the giant planets indicate that they probably do not have a solid surface. In 1973, Andrew P. Ingersoll of Caltech described the Great Red Spot as merely an unusually large free vortex in the atmosphere that is maintained by the heat radiated by the clouds within it.

Tony Maxworthy and L. George Redekopp of the University of Southern California explored another possibility in 1975. They believe that the Great Red Spot, as well as many other features in Jupiter's atmosphere, may be a visible manifestation of a soliton, a special type of atmospheric wave that maintains its shape indefinitely.

In the kind of clockwise atmospheric flow that exists in Jupiter's south tropical zone, a soliton moves as a high-pressure area with an elliptical shape like the Jovian spots. In counterclockwise flows, it has a quite different shape that resembles a group of markings called the South Tropical Disturbance seen in Jupiter's south equatorial belt between 1900 and 1939.

These similarities alone would not substantiate a new theory of the Great Red Spot except for another striking factor. When one soliton overtakes another, the faster one appears to leapfrog over the other – it accelerates rapidly and reappears on the other side without changing shape and resumes its normal speed. Maxworthy and Redekopp report this is precisely what happened each of the nine times that the South Tropical Disturbance overtook Jupiter's Great Red Spot. The disturbance should have taken three months to pass through the Red Spot, but it actually did so in 14 days.

Icy Pluto? Astronomers know little about Pluto, the most distant planet from the Sun. Telescopic observations are exceedingly difficult because Pluto reflects so little sunlight, and telescopes cannot resolve the planet's reflected light into a disk. Until 1976, astronomers knew only that Pluto rotates every 6.4 earthdays, its diameter is less than 6,400 kilometers (3,977 miles), and its mass is about 18 per cent of Earth's at most. Because of Pluto's great distance from the Sun, we also assume that its surface must be extremely cold, with peak temperatures of no more than −212°C (−350°F.).

In April 1976, Dale P. Cruikshank, Carl B. Pilcher, and David Morrison reported new observations of Pluto, obtained with the 4-meter (158-inch) optical telescope at Kitt Peak National Observatory near Tucson, Ariz. Their observations indicate that solid methane ice may cover at least part of Pluto's surface.

Like most ices, methane ice should reflect light well. The brightness of an object at a known distance can be used to estimate its size. However, a small ice-covered object may reflect as much light as a much larger rocky object. If Pluto is indeed covered with methane ice, it may be only about 2,415 kilometers (1,500 miles) in diameter – not much larger than the Earth's moon. This means that the planet's mass must be much less than previously thought, or else its mean density would be unrealistically high. It may be closer to 2 per cent of the Earth's mass than the 18 per cent estimated earlier.

The Martian polar caps have been the center of controversy since they were discovered in 1666. Many scientists believe they are primarily water ice; others argue that they are primarily frozen carbon dioxide, or dry ice. If they are dry ice, it would have an important influence on climatic and atmospheric stability on Mars because the polar environment would then control the release of carbon dioxide, the main component of the atmosphere.

Pictures from Mariner 9, which orbited Mars for most of 1972, added fuel to the controversy. Daniel Dzurisin and Karl R. Blasius of Caltech reported in late 1975 that the Mariner photographs and other data showed that the southern polar cap is at a considerably higher elevation than the northern one. Both caps thin out toward the edges in a regular series of gentle waves.

Dzurisin and Blasius concluded that the southern polar cap must be under less atmospheric pressure because it is at a higher elevation than the northern cap. If the caps are carbon dioxide and their average temperature is the same, the southern polar cap should evaporate more easily because of the lower atmospheric pressure. Thus, the northern cap would grow, while the southern one would shrink. However, the southern cap has been dominant since it was discovered. Dzurisin and Blasius conclude therefore that the cap is water ice, not carbon dioxide.

Laboratory experiments reported in February 1976 by geologists Bruce R. Clark and Rosemary P. Mullin of the University of Michigan indicate that solid carbon dioxide would flow quite easily under a wide range of temperatures below its freezing point, much as glaciers do on Earth.

Clark and Mullin concluded that the carbon dioxide ice may not be as unstable as the arguments of Dzurisin and Blasius make it appear. They suggest that enormous amounts of frozen carbon dioxide may be trapped within the dusty icecap on Mars. The combination of overlying weight and low resistance to flow would force the carbon dioxide ice to flow outward from the center in waves. They estimated that an ice sheet with a radius of 1,000 kilometers (620 miles) would reach equilibrium when its central thickness was 3.1 kilometers (2 miles), close to what has been observed. [Michael J. S. Belton]

Stellar Astronomy. Astronomers began operating the world's largest optical telescope in 1975 and 1976. They also located both the nearest and the farthest known galaxies, detected extreme ultraviolet stellar radiation for the first time, observed the brightest *nova* (exploding star) in 33 years, and investigated some of the fundamental properties of the sun.

New instruments. The first photographs with the new 6-meter (237-inch) telescope on Mount Pastukhov in Russia – the world's largest optical telescope – were made late in December 1975, climaxing a 16-year development program. Regular observations began on Feb. 7, 1976, with quasars, binary systems, and magnetic stars among the first objects to be studied.

The National Aeronautics and Space Administration launched the largest of its eight Orbiting Solar Observatory satellites on June 21, 1975. The satellite carried French and American ultraviolet telescopes to investigate the solar chromosphere, a thin, hot layer of gas just above the sun's visible surface.

Extragalactic developments. Hyron Spinrad of the University of California, Berkeley, announced on June 30, 1975, that observations of the faint galaxy 3C123 revealed that it is about 8 billion light-years from the earth. This is the greatest distance yet measured for any galaxy. The spectrum of 3C123 may provide new clues about how galaxies evolve, because we observe this object at a much earlier stage in the life of the universe than any galaxy studied previously. See ASTRONOMY (Cosmology).

A new analysis of neutral hydrogen emission at the 21-centimeter wavelength was reported by S. Christian Simonson III of the University of Maryland in November 1975. His study revealed that our Milky Way Galaxy has a small companion galaxy that is only 55,000 light-years from the sun. This object, classified as a dwarf irregular galaxy, is the Milky Way's closest neighbor. It is only about one-third the distance to the Magellanic Clouds, previously thought to be the nearest galaxies. The newly discovered galaxy had gone unrecognized in optical sky pho-

Far-ultraviolet photos reveal new detail in the constellation Orion. A 30-second exposure, *below,* shows faint blue stars not normally seen in visible light. A 100-second exposure, *below right,* shows an extended nebula that is called Barnard's Loop.

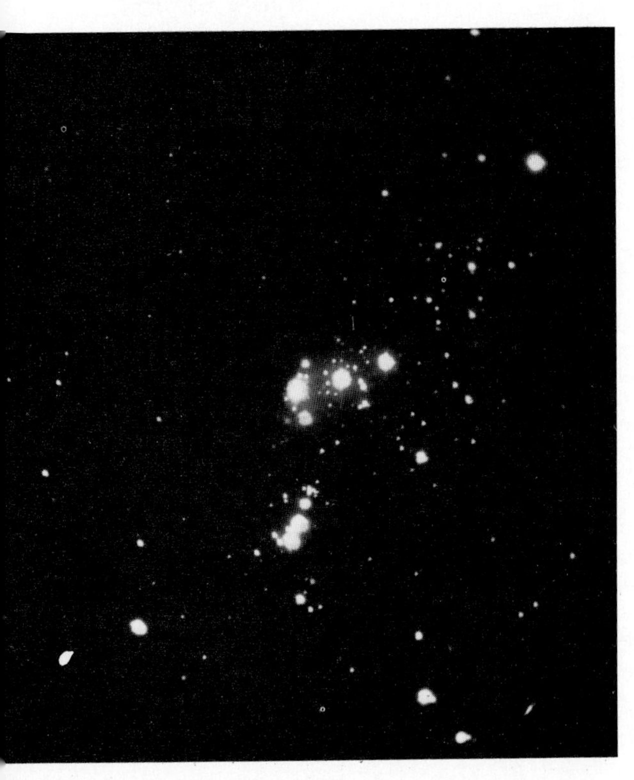

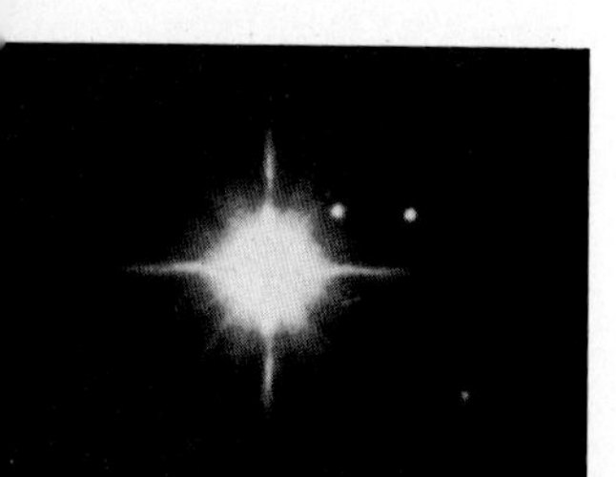

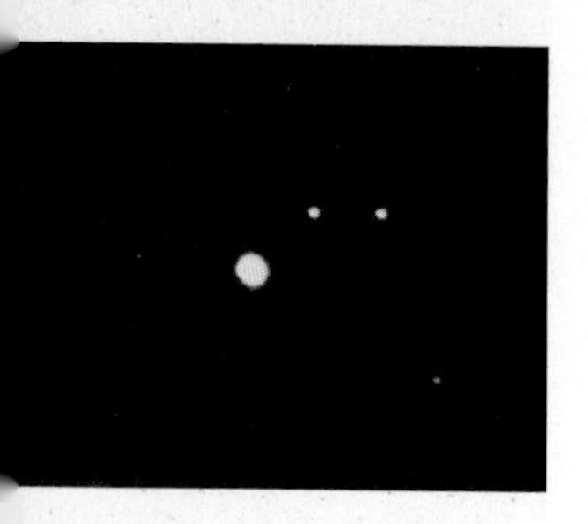

Nova Cygni, the brightest exploding star seen since 1942, exploded in August 1975, *top*. It faded gradually during the following months.

tos because it lies hidden behind a portion of the Milky Way's central disk.

A spinar model. F. Michael Flasar of Harvard University and Philip Morrison of the Massachusetts Institute of Technology, both in Cambridge, Mass., made an important contribution to the study of galaxies that emit radio waves. In March 1976, the astronomers proposed a spinar model for the radio emission from the galaxy Cygnus A. The emission comes from two widely separated regions on opposite sides of the galaxy. The scientists attribute the emission to a continuing source of energy within each region. The sources, called spinars, supposedly were formed when a massive object in the center of the galaxy broke apart and its two spinning halves flew off in opposite directions. Each of these spinars generates energetic particles and occasionally ejects a cloud of magnetized plasma, or ionized gas. These particles and clouds produce the radio emission.

Extreme ultraviolet objects. Observations by U.S. astronauts during the Apollo-Soyuz mission in July 1975 detected the first extreme ultraviolet radiation measured from astronomical objects beyond the solar system. Starlight at these wavelengths of the electromagnetic spectrum normally is blocked by clouds of neutral hydrogen in interstellar space. However, measurements with a telescope operating in the extreme ultraviolet developed by a team of astronomers led by Stuart Bowyer at the University of California, Berkeley, showed that some nearby hot stars can be detected in what was previously termed the "unobservable ultraviolet."

The white dwarf star HZ 43 was the strongest source observed. Such white dwarfs are collapsed stars that have exhausted their nuclear fuel and are cooling, slowly reddening, and fading away. The measurements indicate that HZ 43 is one of the hottest known stars of its type.

The extreme ultraviolet telescope also detected Proxima Centauri, the star nearest to the sun. This is surprising, because Proxima was thought to be too cool to produce detectable radiation in the extreme ultraviolet.

Measurements of another white dwarf, LP 380-5, were reported in March 1976 by Conard C. Dahn and Robert S. Harrington of the U.S. Naval Observatory in Washington, D.C. They found that LP 380-5 is the second reddest – hence presumably the second coolest – white dwarf known.

Novas. Nova Cygni 1975, the brightest exploding star seen since 1942, was discovered on Aug. 29, 1975, by Minoru Honda of Kurashiki, Japan. The eruption was unique because it occurred in two stages that produced the greatest rise in brightness yet observed for such a star.

On August 3, Martin Elvius of the University of Leicester in England and his colleagues had found another nova with the aid of the Ariel 5 satellite. This object, called Nova Monocerotis 1975 or A0620-00, gave off powerful X rays but did not become very bright optically. By contrast, no X rays were detected from Nova Cygni 1975, nor from any previous conventional nova. This suggests that the eruptions occurred in two different kinds of binary star systems. In a conventional nova, the outburst occurs in a layer of material drawn from a large red star to the hot surface of a companion white dwarf. But in an X-ray nova, the object that attracts the material may have a much more powerful gravitational field and could be a neutron star or a black hole.

In related theoretical work, reported in January 1976, Donald D. Clayton and Fred Hoyle of Rice University in Houston calculated the properties of dust grains that form in the gas ejected by a nova explosion. They reported that grains produced in this manner would have an unusual chemical composition. They also pointed out that this kind of material has been found in lunar soil and in some meteorites, suggesting that at least a small fraction of moon dust may actually be nova dust.

Solar research. Two independent groups reported in January 1976 that the entire sun pulsates like a beating heart every 2 hours and 40 minutes. The teams were led by Andrei B. Severny of the Crimean Astrophysical Observatory in Russia and J. R. Brookes of the University of Birmingham in England. This pulsing may enable astronomers to study the internal structure of the sun, just as seismic waves help to determine the earth's interior structure. [Stephen P. Maran]

High-Energy Astronomy. The X-ray bursts associated with globular clusters of stars became one of the most exciting topics in astronomy in late 1975 and 1976. Information on the X-ray bursts was reported at a meeting of the High Energy Astrophysics Division of the American Astronomical Society in January 1976. Data from the Uhuru X-ray satellite identified four globular clusters as possible optical counterparts of X-ray sources. A survey by one of the Orbiting Solar Observatory satellites, OSO-7, confirmed three of these identifications and discovered two more X-ray sources in globular clusters.

There are at least 150 globular clusters connected with our Milky Way Galaxy, each containing some 100,000 stars. The clusters are scattered at random in a spherical halo and show a marked concentration toward the center of the Galaxy. The clusters are almost as ancient as the Galaxy itself – 10 billion years.

Astronomical puzzle. The discovery of X-ray sources associated with globular clusters presents astronomers with a considerable puzzle, especially since the cluster sources show the same kind of energy and intensity variations found in binary, or double-star, X-ray sources within the Galaxy. But it seems unlikely that any binary systems could exist in the globular clusters.

Most binary sources contain an old, compact object such as a neutron star paired with a young, massive supergiant no more than 100 million years old. The X rays are produced when gas from the young, massive star falls on its companion. The problem is that, in the clusters, only stars that are about 20 per cent less massive than the sun are still following the normal sequence of stellar evolution. All those with greater masses evolved long ago to a point where they do not follow that sequence. This means that any massive binary system that could have evolved to form a neutron star could not still contain a massive nuclear burning companion. Such a companion would have burned out long ago, leaving the neutron star without a source of material to produce the X rays.

However, George W. Clark of the Massachusetts Institute of Technology (M.I.T.) in Cambridge suggested that recent binary systems might have been formed in the clusters when one star captured another. The remnants of early massive stars that evolved and collapsed long ago must still be in the clusters. Clark suggested that these remnants, with their powerful gravity, could capture passing stars which then supply them with the material that produces the X rays.

One difficulty with Clark's explanation is in trying to determine how often such captures might occur. No observational evidence yet supports the possibility, and none of the globular cluster sources behave like binary systems.

Another explanation. Several astrophysicists have suggested that X rays could be produced in globular clusters as matter falls into black holes that have masses equal to 1,000 suns. A black hole is a collapsed giant star so dense and with such great gravity that not even light can escape. P. James E. Peebles of Princeton University in New Jersey pointed out in 1972 that superdense cores could form as globular clusters evolved. When such central cores collapsed, they would form massive black holes. In most clusters, the gas emitted from stars escapes as fast as it is generated. But if the cluster had a massive black hole at its center, the gravitational pull would be strong enough to draw in the emitted gas, thus producing X rays.

The existence of black holes rests up to now entirely on observational evidence from only one object. Observations of Cygnus X-1 support the presence of a black hole that is equal to about 10 solar masses.

The peculiar behavior of the globular cluster X-ray sources also resembles the behavior expected for massive black holes. Herbert Gursky and Jonathan Grindlay of the Center for Astrophysics in Cambridge, Mass., and their colleagues in the Netherlands reported two giant X-ray bursts from the globular cluster NGC 6624 in November 1975. Soon after that announcement, Clark and his colleagues at M.I.T. and William Forman and Christine Forman-Jones at the Center for Astrophysics discovered additional burst sources. M.I.T.'s Small Astronomy Satellite-3, in particular, has revealed a wealth of details about the behavior

Omega Centauri, *above,* is similar to the globular clusters that emit strange X-ray bursts. Profile of a burst from NGC 6624, *right,* shows how X rays suddenly peak in half a second, then fade to background level again.

Profile of an X-ray Burst

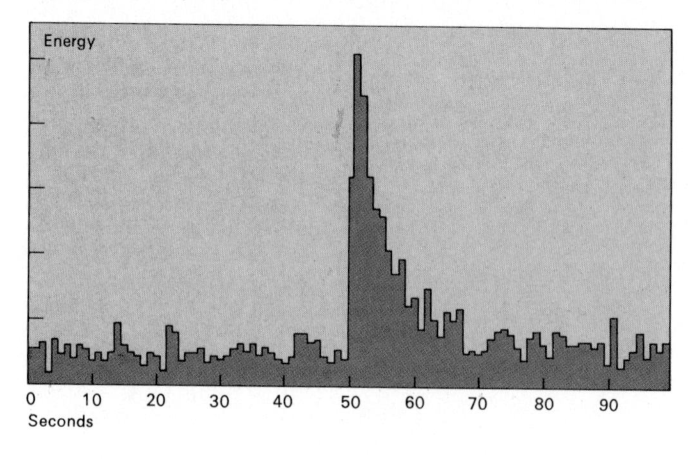

and spectral characteristics of these puzzling bursts.

During a burst, the source NGC 6624 brightens to an X-ray brightness about 30 times the normal level in half a second. Then it takes 10 seconds to fade back to the previous level. Each burst had an identical pattern. Grindlay and Gursky reasoned that the same pattern could not be repeated by chance. It either reflected some characteristic of the underlying source, or it was caused by some characteristic feature of the region around the source that modified the appearance of the bursts.

Grindlay and Gursky suggested that the bursts occur when the source generates an initial rapid pulse. Some of the radiation reaches an observer unchanged. But some could be scattered by the cloud of gas surrounding the source and reach the observer with varying delays from different regions. This would cause the pulse to fade slowly. The scattering of the X rays in the cloud modifies the spectrum in a way that can occur only if the gas in the cloud is very hot. The length of time it takes a burst to fade indicates how big the cloud is, and the spectral measurement gives its temperature. An observer can then calculate how massive the central object must be in order to keep the hot cloud bound to it. Grindlay and Gursky found that the central object in NGC 6624 should be about 1,000 solar masses.

Clark and his colleagues found that bursts from other sources had different characteristics and behavior. However, the peculiar features of the bursts from each source apparently remain constant, providing a characteristic fingerprint of that source.

What all this means is still rather obscure, but the excitement among astronomers is quite understandable. The precise and distinct behavior seen in the bursting X-ray sources has not been observed previously in any of the well-studied binary X-ray sources. The energy involved is so large that this behavior must be tied at some fundamental level to the basic energy processes of the source. Further studies of these sources may shed light on the existence of massive black holes and permit astronomers to determine their characteristics. [Riccardo Giacconi]

Cosmology. Hyron Spinrad of the University of California, Berkeley, announced on June 30, 1975, that he had measured the greatest red shift known for any galaxy. When a galaxy is moving away from us, characteristic lines in its spectrum are shifted toward the red side. By measuring the displacement of these lines, astronomers are able to determine how fast the source is moving and how far away it is.

Spinrad found that a faint galaxy – 3C123 – is moving away from us at one-third the speed of light, indicating that it is some 8 billion light-years away, the most distant known galaxy. The galaxy had been known as a source of radio waves, but it was so faint on photographs that it was difficult to obtain a sharp enough spectrum to measure the red shift. Spinrad overcame this problem by using a sophisticated image-scanning device at the Lick Observatory on Mount Hamilton, Calif., that stored light from the galaxy until a spectrum could be obtained.

Gobbling galaxies. Independent studies – by Simon White at Cambridge University in England in January 1976 and by Jeremiah Ostriker and Scott Tremayne at Princeton University in New Jersey in December 1975 – revealed that giant galaxies can absorb their dwarf companions. The gravitational drag on a dwarf galaxy as it moves through the diffuse outermost regions of a giant neighbor can cause the smaller galaxy to slowly spiral in toward the other, until it is swallowed up. The larger galaxy promptly becomes brighter.

This process has important implications for cosmology. If giant galaxies of a particular type have the same intrinsic brightness, differences in their apparent brightness would show their relative distances from us.

However, the new effect implies that giant galaxies brighten as they grow older, which means that the distant, younger galaxies are not as bright as previously assumed, and hence are closer. Consequently, the mutual gravitational attraction between distant galaxies is strengthened, decreasing the probability that the universe is open and will expand indefinitely.

Expanding universe. Astronomers generally believe that galaxies are more

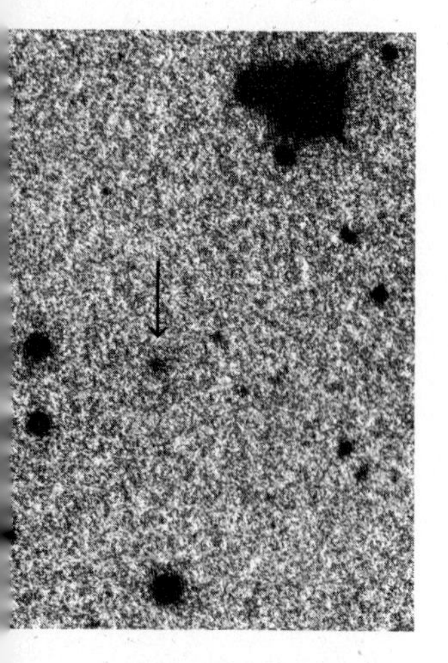
The most distant galaxy in the universe (arrow), 8 billion light-years from the earth, was identified in June 1975.

or less evenly distributed in the universe and moving away from each other at a speed that increases proportionately with distance. This is known as the Hubble expansion effect. However, distribution is not at all uniform on the large scale. Our Local Group of galaxies – containing the Milky Way, the Andromeda Nebula, and a number of smaller systems – is on the edge of a much larger concentration of galaxies known as the Virgo supercluster. The excess of mass in this region, measured by counting galaxies, would distort the uniformity of the Hubble effect.

If this distortion could be measured, it would be a sensitive test of whether the universe is open and will expand indefinitely, or whether it is closed and will someday contract and collapse. If the universe is closed, a large deviation in the evenness of the expansion would be expected as galaxies in our group decelerate and eventually fall back toward the Virgo cluster. On the other hand, in an open universe, there would be little deviation from a uniform expansion. Allan R. Sandage of the Hale Observatories in Pasadena, Calif., who made a study of this effect, reported in December 1975 that existing data tend to support – but cannot conclusively prove – the theory of an open universe. See THE FATE OF THE COSMOS.

Collapsed stars. Cosmologists have so far been unable to find enough material to provide the gravitational attraction needed to eventually reverse the universe's expansion. Some astronomers suggest that the missing mass might be hidden in black holes – giant stars that have collapsed to objects so dense that their gravity prevents even light from escaping.

The radiation produced as an early generation of massive stars collapsed into black holes should contribute to the background radiation in the night sky. However, John Thorstensen and Bruce Partridge of Haverford College in Pennsylvania reported in September 1975 that their observations of the radiation in the night sky indicated that the amount of matter that might be found in black holes is much less than needed to close the universe.

Quasar absorption lines. The source of the gas that shows up in absorption spectra in front of quasars has long been a subject of controversy. Some astronomers maintain that the gas, which is often moving much faster than the quasar, has been ejected from the quasar. Others argue that the absorption lines are produced by intergalactic clouds or halos of galaxies that lie along the line of sight to the quasar.

Using the 508-centimeter (200-inch) telescope on Mount Palomar in California, astronomers Alex Boksenberg and Wallace Sargent reported in May 1975 that they had split absorption red shifts from quasars into velocity pairs that were consistently separated by a range of about 100 kilometers (60 miles) per second. Physicist John N. Bahcall of the Institute for Advanced Study in Princeton, N.J., showed in August 1975 that inadequate resolution in the observations could strongly influence this result. He concluded that higher resolution would break up the velocity pairs into many smaller ones. He believes that the existing data favor an intergalactic origin.

The fundamental constants. Margaret Burbidge and her co-workers at the University of California, Berkeley, monitored the Lacertid AO 0235+164 during an intense optical outburst in December 1975. Lacertids, named after the quasi-stellar object BL Lacertae, are a class of faint optical counterparts of radio sources. They are highly variable and, unlike quasars, they normally have optical spectra without lines. But during the outburst, Burbidge and her co-workers discovered two sets of absorption lines with different red shifts that indicated Lacertid AO 0235+164 is extremely distant.

More dramatically, Morton Roberts and his colleagues at the National Radio Astronomy Observatory in Charlottesville, Va., found that the 21-centimeter radio wavelength of neutral hydrogen had also been greatly red-shifted and matched one of Burbidge's absorption sets. The frequencies of the hydrogen and optical structure lines observed in the spectrum are determined by certain fundamental atomic constants. They cannot have varied appreciably over the billions of years since Lacertid AO 0235+164 sent out the radiation we are now observing, or the optical and radio red shifts would not agree. [Joseph Silk]

Biochemistry

Research reported in 1975 and early 1976 provides a clue to the mechanism by which bacteria obtain and exchange certain genes, including some that make them resistant to various antibiotics. Bacterial resistance to any single antibiotic is determined by a gene that is part of a circular piece of deoxyribonucleic acid (DNA) called a plasmid. Scientists have known since 1974 that bacteria can easily exchange such genes among themselves, creating plasmids with a variety of resistance genes. These give various disease-causing bacteria the troublesome ability to survive attack by an assortment of antibiotics.

The new research done at universities in Geneva, Switzerland; Boston; Stanford, Calif.; and Seattle reveals that often each of these genes has a type of DNA known as an insertion sequence on either side of it. Further, the insertion sequence on one side is inverted with respect to the insertion sequence on the other side.

Insertion sequences, like all DNA, are chains of chemical units called bases. Several years ago, researchers in several countries discovered that certain spontaneous mutations in bacteria were caused by short lengths of DNA of unknown origin that had been inserted into the DNA of the bacteria's chromosome. These were appropriately called insertion sequences.

The insertion sequences found in the new research were detected when plasmid DNA that had been chemically treated was inspected with an electron microscope. DNA is usually double stranded – composed of two parallel chains, or strands, of bases. The strands are held together by the natural attraction of specific bases opposite each other all along the strands, but they were separated by alkali treatment. The resulting single strands of plasmid DNA were treated with acid to allow bases that attract one another to come together. When the strands were viewed, they had folded back on themselves in several spots, forming loops and double-stranded stems. The loops contain the resistance genes and the stems are the insertion sequences held together by their attracting bases.

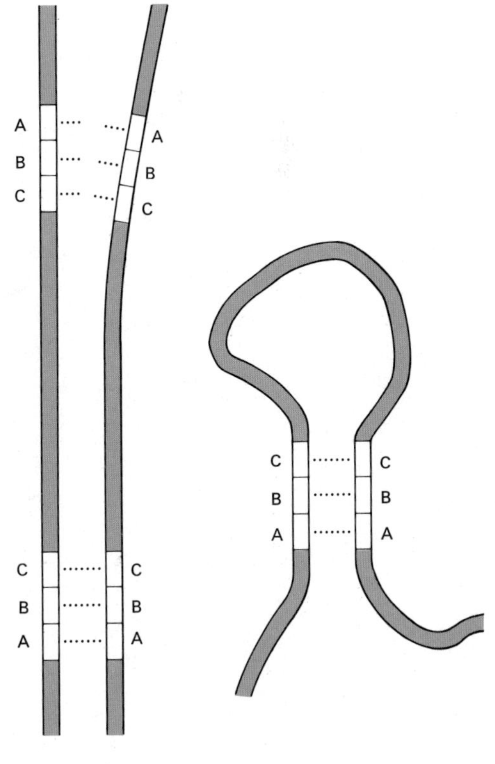

Loops and double-stranded stems on single strands of plasmid DNA, *above,* revealed that inverted insertion sequences—the stems—often flank antibiotic resistance genes—the loops—in bacteria. Details of experiment, *right,* show that scientists first split the normally double-stranded plasmid DNA, then observed the stem and loop indicating that the insertion sequences—A, B, C—were inverted in regard to their chemical components.

The insertion sequences in plasmids could be crucial to their ability to transfer resistance genes from one to another, although just how they might do so is not obvious. In any event, the discovery provides new avenues to explore in the hope of understanding and one day controlling drug resistance in bacteria.

Messenger RNA. Ribonucleic acid (RNA) is a substance much like DNA in several respects, including that it is composed of a chain, or strand, of bases. Over the past two years, scientists in many laboratories have found that messenger RNA (mRNA) strands from the cells of many living creatures, except bacteria and blue-green algae, have a chemical "cap," called a 7-methylguanosine cap, at one end.

A cell makes mRNA when a gene, which is a section of a DNA strand, is to be expressed. In most cases, genes express themselves by causing specific proteins to be made. Using the gene's structure as a template, or pattern, a cell forms an mRNA strand whose bases carry the gene's message for protein construction. The mRNA strand then interacts with a cell particle called a ribosome, which reads the message and helps assemble the protein.

Until 1975, little was known about the 7-methylguanosine cap beyond its chemical makeup and that it always caps the lead end of the mRNA strand – the end that passes through and is read by the ribosome first.

In March 1976, at a meeting in Keystone, Colo., several research groups that are studying the cap presented some of their findings. Some of the scientists were trying to determine when the cap was put on the mRNA strand. They used chemical inhibitors to interrupt the synthesis of mRNA strands at various stages, then checked to see if the cap was present. These experiments revealed that the cap is on the strand early in its synthesis.

Other scientists tried to determine the cap's purpose. G. W. Both, Amiya K. Banerjee, and Aaron J. Shatkin of the Roche Institute for Molecular Biology, Nutley, N.J., synthesized some mRNA strands with caps and others without caps. Then they added the

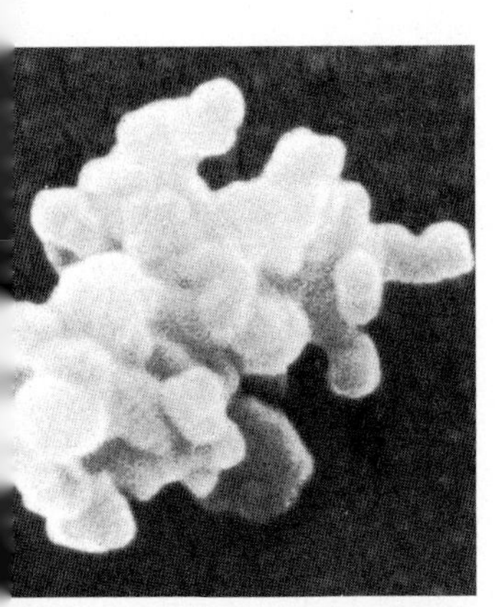

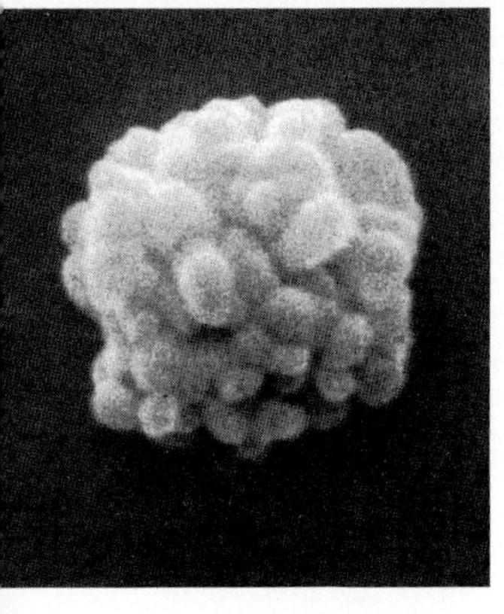

The membrane of a resting mitochondrion is branched with fingerlike extensions, *top,* but one of these cell structures performing its energy-producing function is much more compact, *above.* By studying the difference, scientists may learn more about how mitochondria work. Magnification is about 30,000 times.

capped strands to a mixture of ribosomes and all the other substances needed for protein synthesis. To an identical mixture, they added uncapped mRNA strands. After allowing time for protein synthesis, the scientists compared the amount of protein produced. The capped mRNA strands produced up to 20 times more protein than did the uncapped strands. This suggests that the cap plays an important role in starting the translation of the mRNA strand into a protein molecule. However, other researchers have found much less marked differences in the protein-synthesizing capacity of capped and uncapped mRNA strands.

Ribosomal RNA. Ribosomes are particles composed of proteins surrounding strands of RNA called ribosomal RNA (rRNA). Scientists have long wondered whether rRNA strands play a role in translating the mRNA or serve only as a skeleton around which the ribosomal proteins congregate. This year, support came for the theoretical function of rRNA.

The function was proposed in 1974 by John Shine and Lyn Dalgarno of the Australian National University in Canberra. It proposes that one end of a small rRNA strand plays some role in recognizing mRNA and starting the process through which the message it carries is translated into a protein.

The scientists based their theory on a careful analysis of certain bases on mRNA strands and the small rRNA strand. On the mRNA strand, they looked at the bases that are part of a sequence of bases called the ribosome binding site. If this sequence of bases is absent, the mRNA strand and a ribosome will not attach to each other. Several of these mRNA bases will pair with bases the scientists found at one end of the small rRNA strand. Shine and Dalgarno assumed that this pairing might be an essential initial step in mRNA-ribosome binding and subsequent protein synthesis.

The support for this theory came from experiments reported in December 1975 by Joan Steitz and Karen Jakes of Yale University. They added mRNA strands to ribosomes and then added enzymes that break down ribosomes and unpaired RNA bases. From the resulting materials, the scientists isolated complexes consisting of the end of the small rRNA strand paired to the bases of the ribosome binding site of the MRNA strand.

Protein analysis. Early in 1975, Patrick H. O'Farrell of the University of Colorado in Boulder reported he had developed a new technique for analyzing material containing proteins. The O'Farrell technique is a modification of an older one called polyacrylamide gel electrophoresis. It and other methods used in the past often could separate the proteins in a substance only to the extent that 10 or 20 protein types were isolated together.

O'Farrell separates proteins in polyacrylamide, a jellylike substance, on the basis of their molecular weight and their electrical charge. The technique is so sensitive that it isolates every protein in a cell. For example, O'Farrell isolated 1,000 different protein components from an extract of the cell of the bacterium *Escherichia coli.*

Perhaps the most exciting future use of this sensitive technique is in the study of how mutations, or changes in genes, change proteins. Scientists will be able to detect changes in the location of some proteins within the gel that are associated with various mutations. This will tell them which proteins play a role in the mutation. And, by collecting and analyzing the proteins, researchers can determine exactly what changes occur in them as a result of mutation.

A related experiment using the new technique was reported in January 1976 by Jane Peterson and Edwin H. McConkey, also of the University of Colorado in Boulder. By adding a chemical to mouse cells that do not normally produce hemoglobin, the scientists caused the cells to begin making it. Then, using the O'Farrell technique, they found 11 new proteins in the hemoglobin-producing cells. The scientists are now trying to determine the roles these proteins play in producing hemoglobin.

Chromosome structure. Researchers in several laboratories in Australia, England, and the United States are gradually uncovering the detailed molecular structure of the chromosome during interphase – the stage in which a living cell is not dividing. See GENETICS. [Julian Davies]

Books of Science

Here are 33 outstanding new science books suitable for general readers. The director of libraries of the Smithsonian Institution selected them from books published in 1975 and 1976.

Agriculture. *This Hungry World* by Ray Vicker. The booming population growth, worn-out land, depleted fish and wildlife stocks, adverse weather, urbanization of farmland, and other phenomena create a specter of coming starvation for much of the world. The author traveled widely to view the conditions he describes. He discusses many proposed solutions to these problems and the difficulties to be encountered in their implementation. (Scribner's, 1975. 270 pp. $9.95)

Archaeology. *Pyramid* by David Macaulay is a simple and effective account of how two of Egypt's pyramids might have been built. The large drawings almost tell the story themselves. They include sketches of pyramid-building tools and their use. (Houghton-Mifflin, 1975. 80 pp. illus. $7.95)

Astronomy. *Black Holes in Space* by Patrick Moore and Iain Nicolson. Evidence suggests that there are points called black holes in the universe with gravitational forces so strong that even light cannot get out. The authors fill in the background of astronomical facts from which the theory of these paradoxical black holes comes. The speculations about this phenomenon are presented in relatively simple terms. (Norton, 1976. 126 pp. illus. $7.95)

The Dark Night Sky: A Personal Adventure in Cosmology by Donald C. Clayton is the autobiography of a young astrophysicist written around key facts about the universe. The text covers the findings of great astronomers, power of the sun, discovery of the galaxies, decay of radioactive elements, relativity, expansion of the universe, and other fundamentals. (Quadrangle, 1975. 206 pp. illus. $9.95)

Eyes on the Universe: A History of the Telescope by Isaac Asimov. The noted science writer traces the development of the telescope and related equipment from the simple Dutch optic tube to the giant radio telescopes of today. He introduces the fundamentals of optics, mechanics, and electronics to give an understanding of the complexity of these devices and to explain how each development extended our knowledge of the universe. (Houghton-Mifflin, 1975. 274 pp. illus. $8.95)

The Universe: Its Beginning and End by Lloyd Motz shows how the universe evolved from the primordial materials at its beginning to stars, galaxies, atoms, molecules, and living things. Motz also speculates on the ways the universe might end and how the earth might escape destruction. (Scribner's, 1975. 343 pp. illus. $14.95)

Biology. *The Quest for Man,* edited by Vanne Goodall. Six scholars present essays on many aspects of the development of human beings. Included are theories of man's adaptation to society and the environment, his structural development through evolution, his cultural evolution, his behavioral development, and his attempts to understand his origins. (Praeger, 1975. 256 pp. illus. $17.50)

He and She: How Males and Females Behave by S. Carl Hirsch is a narrative about how males and females of various species of animals, fish, birds, and insects interact, particularly in the reproduction and care of offspring. The author explains how scientists and amateur observers have helped to clarify the varied and often strange behavior between the sexes. He points out that products of the human brain – logic and planning, language and education, and science and art – make humans unique in breaking free from the dominance of inborn behavioral patterns. (Lippincott, 1975. 160 pp. illus. $7.95)

Botany. *The Blossom on the Bough: A Book of Trees* by Anne Ophalie Dowden. A leading botanical illustrator describes the physiology of trees – how they grow, reproduce, and die. She cites the functions of the fruit, seeds, leaves, and other parts of the tree, and briefly notes the major features of trees in U.S. forests and points out the features of different kinds of trees. (T. Y. Crowell, 1975. 71 pp. illus. $7.50)

The Plant Hunters by B. J. Healey is a history of searches throughout the world by botanists since the 1700s to discover and collect samples of new plants and seeds. Many of these have become flower-garden and crop plants. (Scribner's, 1975. 214 pp. $8.95)

Ecology. *City and Suburb: Exploring an Ecosystem* by Laurence Pringle. This

Books of Science

Continued

book describes the effect of city and suburban development on plants, animals, and weather. The science of these effects is simply described, and the city emerges not as an insulated man-made environment, but as an ecosystem of its own, interacting with other natural systems. (Macmillan, 1975. 56 pp. $5.95)

Entomology. *Butterflies in My Stomach: or Insects in Human Nutrition* by Ronald Taylor. In this book, a physician examines a variety of aspects of insects for human consumption, including their value for survival in the wilderness and the prospects of their commercial availability. He includes a survey of ancient and existing practices throughout the world among many insect-eating peoples. (Woodbridge Press, 1975. 224 pp. illus. $8.95)

How Insects Communicate by Dorothy Hinshaw Patent. Insects communicate by sight, sound, feel, smell, and through message-carrying chemicals called pheromones. The author describes insect organs and mechanisms involved in such communication. In doing so, she covers the patterns of courtship, mating, searching for food, and defense against predators. (Holiday House, 1975. 127 pp. illus. $5.95)

Borne on the Wind: The Extraordinary World of Insect Flight by Stephen Dalton is a brief but highly articulate explanation of the flight mechanisms of ants, beetles, butterflies, wasps, and many other insects. The author notes the aerodynamic peculiarities of the skin surfaces, muscles, and motion of insects. This work is based on Dalton's unique high-speed photography techniques, and the book includes many spectacular photographs. (Reader's Digest Press, 1975. 160 pp. illus. $18.95)

Geology. *Lands Adrift: The Story of the Continental Drift* by Malcolm E. Weiss. This book is a highly simplified but carefully constructed version of the continental drift theory and the evidence that supports it. (Parents' Magazine Press, 1975. 64 pp. illus. $4.95)

Treasures from the Earth: The World of Rocks and Minerals by Benjamin M. Shaub is a basic mineralogy text for lay readers. Shaub describes the location, process of formation, identification, and testing of minerals, and how they are used. The book includes a brief glossary, a list of museums, and a list of sources of mineralogy materials. (Crown, 1975. 223 pp. illus. $15.95)

History. *Scientist Versus Society* by Vivian Werner contains brief accounts of the tribulations of six creative scientists whose work and lives were assailed or ignored by a world bound to traditional knowledge and social values. The scientists were victims of various forms of prejudice, but their contributions prevailed. (Hawthorne Books, 1975. 160 pp. $6.95)

Thinkers and Tinkers by Silvio Bedini describes the contributions of the mathematicians who surveyed and explored the United States. Bedini describes the instruments and mathematical tables that the surveyors and navigators used. His work is based on a study of documents, almanacs, atlases, and museum artifacts related to surveying, astronomy, and mapmaking. (Scribner's, 1975. 520 pp. illus. $17.50)

Medicine. *The Siege of Cancer* by June Goodfield is a literary account of the struggle to find out what causes cancer, presented as both an intellectual and a human problem. Goodfield's work is based on long periods spent with the researchers she sought to understand. She writes about the major discoveries in medicine, genetics, chemistry, and epidemiology that contributed to our knowledge about cancer. (Random House, 1975. 240 pp. $8.95)

Natural History. *Beautiful Swimmer: Watermen, Crabs and the Chesapeake Bay* by William Warner is a natural history of the Atlantic blue crab and the waters in which it is found. The description is blended with an account of the men who have developed the industry that markets this crab for food. The story covers a year's cycle in the Chesapeake environment. (Little, Brown, 1976. 304 pp. illus. $10)

The Mountain World by David Costello is a comprehensive description of the physical and natural elements of the United States high-country environment. The author tells how the mountains were formed, changed by rivers and climate, and explored by man. There are also chapters on forests, flowers, mammals, and birds. (T. Y. Crowell, 1975. 305 pp. illus. $7.95)

The Ends of the Earth: The Polar Regions of the World by Isaac Asimov ranges through the fundamentals of several fields of science that explain the polar regions and how they differ from each other. Worldwide weather, ice ages, continental drift, animal and plant life, electromagnetism and the weather, explorations, and other topics are woven into a comprehensive picture of these strange areas. (Weybright and Talley, 1975. 363 pp. illus. $15)

Oceanography. *The Sargasso Sea* by John and Mildred Teal. Basic physical and biological science are used to explain this strange sea within the North Atlantic Ocean. The authors blend vivid descriptions of the life and phenomena of the Sargasso Sea with its mythology. The region emerges as a unique ecological environment impacted finally by man and his wastes. (Little, Brown, 1975. 216 pp. illus. $10)

The Sea's Harvest: The Story of Aquaculture by Joseph E. Brown. Ocean-food businesses have grown from haphazard beginnings to organized industries, including sea farms. This brief book describes sea-farming methods and the sea farms' economic value. (Dodd-Mead, 1975. 96 pp. illus. $4.95)

Technology. *Wankel: The Curious Story Behind the Revolutionary Rotary Engine* by Nicholas Faith is a journalistic account of the development of the Wankel engine. The description of the technology is interspersed with the story of the people and the business enterprises involved in turning this invention into a commercial product. (Stein and Day, 1975. 249 pp. $10)

Windmills by Suzanne M. Beedell surveys the various types of windmills used throughout the world. Cutaway drawings and photographs show the intricacies of construction and of the machinery, gears, and power-chain systems of this source of power. (Scribner's, 1976. 143 pp. illus. $12)

Zoology. *Among the Elephants* by Ian and Oria Douglas Hamilton. A scientist and his wife report on their studies of the social life of African elephants. This book is an ecological study based on field observations of three elephant families in one large kinship group. The beasts, the land, and man are all viewed in their complex interrelationship. (Viking, 1975. 285 pp. illus. $14.95)

The Galapagos – The Enchanted Islands by Richard Hough. The Galapagos Islands are home for many species of familiar but seemingly displaced and variant animals. The isolation of these islands has preserved a living laboratory of evolution. Hough discusses the volcanic origin of the islands and explains the characteristics of their strange animals and how they adapted to this stark environment. (Addison-Wesley, 1975. 88 pp. illus. $5.50)

Golden Eagle Country by Richard Olendorff gives observations on the development and lives of eagles, owls, hawks, and falcons in the western Great Plains. The author emphasizes aerial courtship, nest building, mating, and hatching. He also describes the natural process of restoring farmland. (Knopf, 1975. 202 pp. illus. $12.95)

The Life and Lore of the Bird in Nature, Art, Myth and Literature by Edward A. Armstrong traces man's knowledge of birds from ancient times to the present. The book includes sections on mythological birds, birds in art, sports, literature, and fashion, as well as the science of ornithology and animal conservation. (Crown, 1975. 271 pp. illus. $15.95)

The Secret Life of Animals by Lorus and Margery Milne and Franklin Russell draws together observations of many scientists on the behavior of animals of all kinds in life functions, such as courtship, mating, caring for young, escaping predators, and adapting to environmental conditions. The mechanisms of the senses that guide the behavior are detailed, and there are many color photographs. (Dutton, 1975, 214 pp. illus. $29.95)

To the Brink of Extinction by Edward R. Ricciuti. By describing how seven types of animals lived and were driven to or near extinction, the author demonstrates the elements that lead to extinction. Techniques for reversing this trend are also featured. (Crown, 1975. 177 pp. illus. $15.95)

Wonders of Frogs and Toads by Wyatt Blassingame. Frogs and toads were the first animals with voices to appear on dry land 350 million years ago. Blassingame briefly but comprehensively covers the kinds and characteristics of this family of animals, its habitat, life cycle, and place in the environment. (Dodd-Mead, 1975. 80 pp. illus. $4.95) [Russell Shank]

Botany

Satellite data are providing range managers and ecologists with maps and tables giving standing-crop data for selected plant groups or topographical areas in 1976. Eugene L. Maxwell of the Colorado State University Department of Earth Resources in Fort Collins reported in January the successful use of computers to interpret data from Land-Sat 2, the Earth Resources Technology Satellite.

The satellite's remote-sensing devices gather high-altitude photographs of grasslands. When these are analyzed spectroscopically for chlorophyll-green colors, observers can detect electromagnetic energy reflected and given off by plants. From this, they construct computerized microfilm images of the test sites that show the type of vegetation and conditions of the range.

E. L. Fritz and Stanley P. Pennypacker of Pennsylvania State University's Department of Plant Pathology, University Park, reported in October 1975 that they employed Land-Sat's remote-sensing data to map vegetative changes caused by excessive zinc levels in the soil. Their maps show less vegetation damage and change as distance increases from a zinc smelter (see ECOLOGY). Use of the satellite to detect and monitor rangeland conditions, plant diseases, air-pollution damage to the vegetation, insect problems, and human activities has great promise.

Purple power. Of all the forms of life that directly convert the sun's energy into chemical energy, only one is known that does not use chlorophyll, the green pigment that is the basis of photosynthesis. The bacterium *Halobacterium halobium* uses bacteriorhodopsin, a purple-pigmented protein similar to rhodopsin, or visual purple, found in the eye. Cell biologist Walther Stoeckenius and his associates at the University of California, San Francisco, announced in March 1976 that they had duplicated *H. halobium*'s system for changing light into energy.

The scientists added the purple pigment to microscopic artificial vesicles, or small membranous sacs. When light was applied, the bacteriorhodopsin began pumping hydrogen ions, or pro-

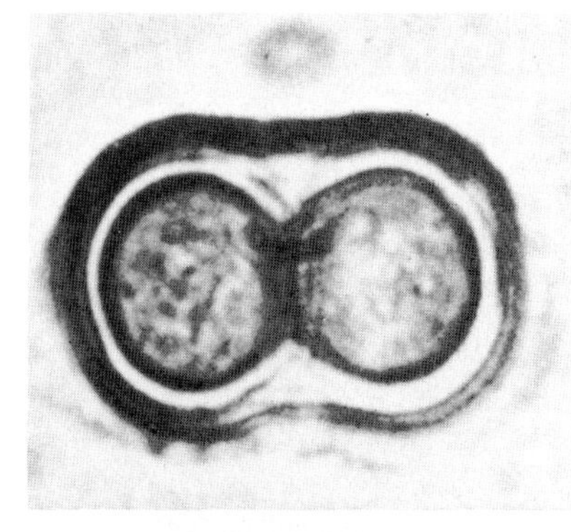
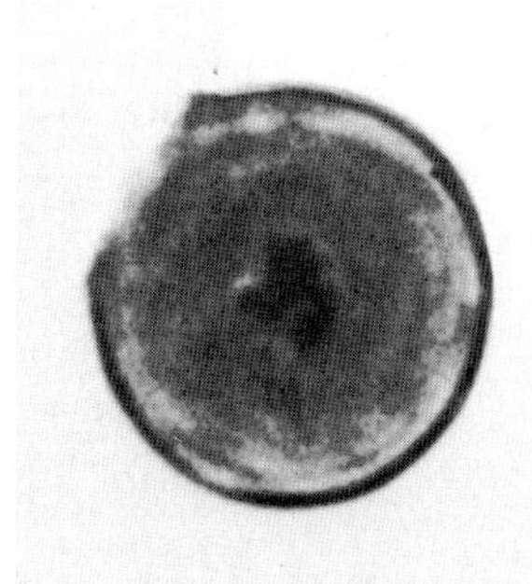
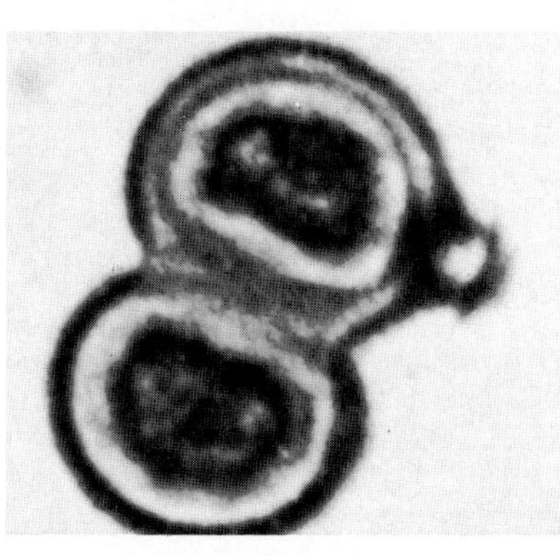
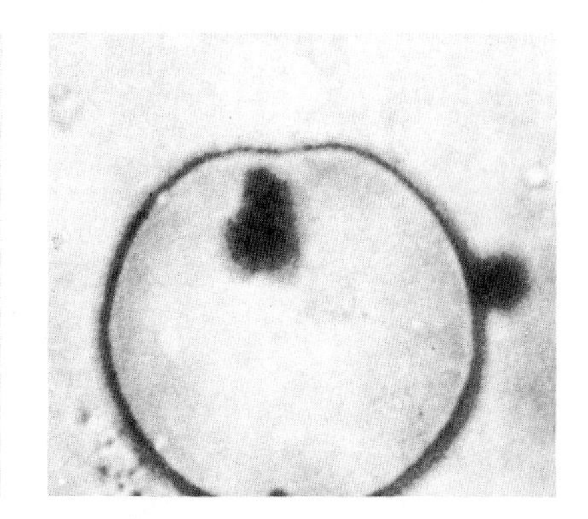

Modern nonnucleated algae cells, in varying stages of decay, *top row,* are similar in structure to fossil algae, *bottom row,* forcing scientists to reassess the fossils, which had been thought to be the oldest known nucleated cells. All the photomicrographs are magnified about 2,200X.

tons, out of the vesicles, thus creating an electrical potential.

They added ATPase, the enzyme that produces adenosine triphosphate (ATP), the universal energy-storage molecule. The artificial system then began to manufacture ATP. The system is believed to operate similarly to the way chlorophyll-based plants produce chemical energy from light.

Stoeckenius suggested that the chlorophyll-based and rhodopsin-based systems represent two important evolutionary pathways for the utilization of sunlight. Chlorophyll uses light to produce energy in plants; rhodopsin uses light to produce sensory information in the eyes. *H. halobium* retains what is probably an ancient rhodopsin-based method to produce energy. Because the pigment is very stable, scientists believe it may be useful in man-made solar-energy conversion systems.

A Precambrian misunderstanding. Paleontologist Elso S. Barghoorn and geologist Andrew H. Knoll of Harvard University in Cambridge, Mass., reported in October 1975 that fossil evidence of eucaryotic cells in the late Precambrian is incorrect. Billion-year-old fossils, discovered more than 10 years ago in Bitter Springs, Australia, were thought to be both procaryotic and eucaryotic.

There are basically two kinds of cells found in nature: eucaryotes, which have nuclei surrounded by protoplasm, and procaryotes which have only protoplasm. Molecular evolutionists believe that the procaryotes arose early in the earth's history, as life was evolving from primordial chemical solutions. Fossils of primitive procaryotes are found in the oldest Precambrian rocks. Some blue-green algae and bacteria fossils—both procaryotes—date back 3.4 billion years.

Barghoorn and Knoll subjected the living procaryotic algae species *Chroococcus turgidus* to a series of experiments in which they allowed the cells to decay partially. This produced stages in which the cells looked just like what were previously thought to be a dozen different groups, including six that were eucaryotic. Barghoorn and Knoll believe that what was thought to be nuclear material in the original fossil studies was actually degraded protoplasm. Thus, they concluded that the fossils represent only procaryotic algae in various stages of decay when they were fossilized.

The rise of the eucaryotic lines of cells from their procaryotic ancestors was a major leap in evolutionary history. Therefore, it is important to learn when and where this transition occurred. In light of this new interpretation, no Precambrian eucaryotic cells have yet been found.

Molds and mosquitoes. The probable life history of the water mold *Coelomomyces psorophorae* was discovered in 1975 by botanists Howard C. Whisler and Stephen L. Zebold of the University of Washington, Seattle, and Joseph A. Shemanchuk, Lethbridge Research Station, Alberta, Canada. At different stages in its 20-day life cycle, the fungus is a parasite of two animal forms, mosquito larvae and the aquatic copepod *Cyclops,* a small crustacean.

The researchers also discovered mating types and sexual reproduction in this fungus which will allow researchers to try to develop more vigorous strains of the organism. Seeding mosquito-breeding ponds with fungus-infected copepods may become an important biological control of mosquitoes, because the fungus kills the larvae.

Reveille for seeds. Plant scientists Sterling B. Hendricks and Raymond B. Taylorson of the U.S. Department of Agriculture's Research Service in Beltsville, Md., reported in 1975 that germination of dormant seeds can be increased and quickened by inhibiting catalase, an enzyme in seeds. They used sodium nitrite, hydroxylamine salts, and thiourea to enhance germination of the seeds. In experiments with dormant seeds of lettuce (*Lactuca sativa*) and pigweed (*Amaranthus albus*), they showed that these chemicals interfere with catalase which normally breaks down the plant's hydrogen peroxide. When catalase is inhibited, other naturally occurring enzymes that use hydrogen more effectively act on the hydrogen peroxide and hasten the germination of the seed.

One possible application of this research is weed control. Hastening germination of weed seeds would allow farmers to use herbicides before crops are planted. [Howard S. Irwin]

Chemical Technology

The major emphasis in chemical technology in 1975 and 1976 was on innovations in energy, pesticides, and plastics. There were also many individual accomplishments in specialized fields, such as industrial waste treatment.

Cleaning up coal has become a priority for chemical technologists. Electric power station operators expect coal to replace oil as a major fuel source. But most of the coal that can be burned economically to produce steam to turn generator turbines contains sulfur. This is released in gases in quantities that violate pollution standards. Desulfurizing these gases is difficult and expensive, so researchers developed ways to remove sulfur before burning the coal.

Sulfur in raw coal has two forms—organic, which is chemically bound to the coal; and pyritic, which consists of separate particles of pyrite, an iron and sulfur compound, mixed in with the coal. Techniques for physically separating the coal and pyrite frequently take advantage of the differences in specific gravity between the two. Processors use special solutions to float one away from the other. In another method, the coal-pyrite mixture is passed through a cyclone separator, which removes the heavier pyrite particles from the lighter ones by centrifugal force.

Hazen Research, Incorporated, of Golden, Colo., reported in early 1976 that it had developed a process using magnetism. The raw coal is exposed to traces of iron carbonyl vapor, which forms a thin magnetic coating on pyrite but not on coal. The coal then passes low-intensity magnetic separators that draw out the pyrite.

Unfortunately, physical separation methods work on only about 16 per cent to 17 per cent of U.S. coal. Most raw coal contains more organic sulfur than pyrite, and physical methods cannot remove organic sulfur. However, some chemical separation methods can remove both forms.

Chemical cleaning. TRW Incorporated scientists in California developed a chemical method for removing pyrite by treating pulverized coal with an iron sulfate solution. The solution reacts with the pyrite to form a liquid mixture

A polyurethane screen, designed for sizing such abrasive materials as gravel and sand, will last up to 30 times longer than screens made of wire.

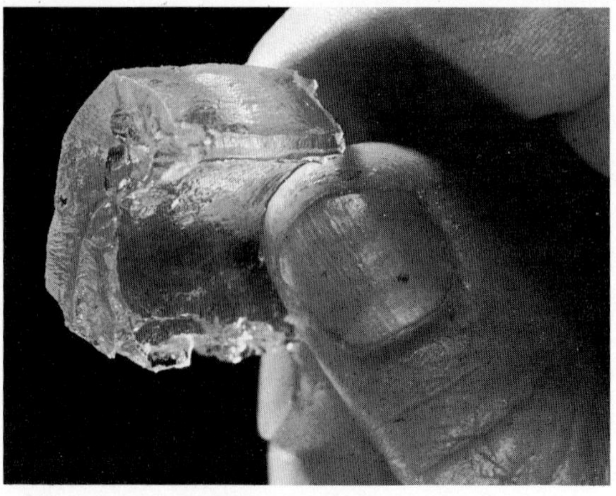

Two grams of super slurper, *left,* a highly absorbent starch graft copolymer developed by U.S. Department of Agriculture scientists, absorbs 1,000 grams of water. After taking up water, a flake of super slurper, *above,* looks like an ice cube, but feels warm and rubbery. Improved versions in powder, flakes, or sheets absorb up to 2,000 times their weight and may be used for such products as diapers.

of iron sulfate and sulfuric acid, which is poured off from the solid coal particles. TRW scheduled a test unit for completion in late 1976.

Battelle Memorial Institute of Columbus, Ohio, has evolved a similar technique that takes out almost 99 per cent of the pyritic sulfur and 70 per cent of the organic. Battelle uses sodium and calcium hydroxides, which dissolve both forms of sulfur.

The Edison Electric Institute and the Southern Company, an electric utility in Wilsonville, Ala., tested a revolutionary sulfur-removing process from early 1974 to mid-1975 and found that the process can reduce sulfur content to as little as 0.6 per cent, and boost the heating value of the refined coal. Called the solvent-refined-coal process, it converts crushed coal mixed with solvent to a liquid at high temperatures and pressures. This produces hydrocarbon gases, hydrogen sulfide, and three liquid layers of differing density. The lightest layer contains the valuable chemical by-products phenol and cresylic acids. The middle layer, the solvent, is recycled to dissolve more coal. The heaviest layer, the liquid refined coal, can either be burned in a power plant, or cooled down below 184.5°C (300°F.), at which point it turns into a solid that can be stored or shipped.

The pesticide industry is being reshaped because of stricter government regulations. Farmers and gardeners can no longer treat their crops with broad-based chemical pesticides in large quantities, and therefore, technologists concentrated on developing methods for dispersing existing pesticides more efficiently. One method that gained popularity during the year involved encapsulating – enclosing the pesticide in small capsules that release it slowly and in timed sequence. Pennwalt Corporation of Philadelphia began marketing a microencapsulated organophosphate, a substitute for DDT. These capsules range in diameter from 30 to 50 microns. Because the tiny capsules release a relatively small amount of pesticide at a steady rate over a given period of time, the crop area receives the proper level of pesticide, and there is no excess to be washed away into streams or other waterways.

Technologists have also begun encapsulating insect pheromones – sexual attractant substances. The encapsulated pheromones are used to reduce mating by confusing male insects that are trying to locate females. The relatively expensive pheromones are most economically effective when dispensed by slow-release techniques. Zoecon Corporation of Palo Alto, Calif., has taken this idea a step further by combining a pesticide with a pheromone for flies. The pellets are scattered on the ground, where the pheromone lures flies into contact with the deadly insecticide.

In August 1975, chemists at the University of Kansas reported the first synthesis of a cockroach pheromone. U.S. Department of Agriculture and Japanese scientists also synthesized the substance independently. The synthetic pheromone may prove useful in controlling the German cockroach, a species prevalent in the United States.

Scientists at the University of California, Riverside, also reported in August that they had developed a combination of chemicals to attract eye gnats. These minute, two-winged flies that swarm around the heads of people and animals have been very difficult to control. The new attractant might help solve this problem. The compound is made from decaying chicken-egg solids, a substance on which eye gnats like to feed. The attractant could be used along with a pesticide or some other insect-control method to rid an area of these troublesome flies.

Super materials. Several innovations in synthetic materials were introduced during the year. E. I du Pont de Nemours & Company of Wilmington, Del., in early 1976 unveiled a supertough nylon that it claims can withstand impact much better than other such materials. Du Pont also introduced Somel, a new polymer that possesses rubberlike properties but can still be processed on conventional plastics-forming equipment. Du Pont predicts it will be used for seals, tubing, hoses, and shoe soles.

PPG Industries of Pittsburgh developed a new sheet-molding compound, called XMC, that it claims is up to four times stronger than steel but weighs only half as much. The compound is produced by impregnating glass fibers with liquid resin, then crisscrossing them on a rotating drum to form an X-shaped reinforcement pattern. After being unrolled from the drum, the material is cut into sheets, and molded by compression into any desired shape. PPG foresees widespread use for XMC in automobiles and appliances.

The Coca-Cola Company, Incorporated, of Atlanta, Ga., and PepsiCo, Incorporated, of Purchase, N.Y., have test-marketed disposable plastic soft-drink bottles and have scheduled wider distribution for 1976. Plastic-bottle manufacturers had to develop material that would not leak carbon dioxide gas – the carbonating agent – to prevent the bottled soft drinks from going flat. Coca-Cola's bottles are made of an acrylonitrile-styrene copolymer, while Pepsi's are a polyester. *Modern Plastics* magazine estimated that about 1.8 million metric tons (2 million short tons) of plastic resins would be needed each year if plastic bottles replaced all the glass and metal containers for soda and beer in the United States.

Plastic bottles have several advantages, including lighter weight. Coca-Cola's new 0.946-liter (32-ounce) plastic bottle weighs only about one-eighth as much as a comparable glass bottle. The plastic bottles can be dropped from a typical table height without breaking. And if they should break, they break into a few large fragments, without sharp edges. Manufacturers also claim that slightly less energy is required to make the plastic containers than to make cans and refillable glass bottles.

Super sponges. Hercules, Incorporated, of Hopewell, Va., developed a superabsorbent fiber that can hold 45 times its weight in water. The fiber, called Aqualon, is a biodegradable and water-insoluble carboxymethylcellulose that has important applications for such products as diapers and medical dressings. Because the water absorbed is not released under pressure, each Aqualon product can only be used once. Hercules scheduled production to begin at the end of 1976.

U.S. Department of Agriculture scientists developed an even more highly absorbent material, which they call super slurper, made of starch graft copolymers. In the form of sheets, film, flakes, or powder, it can absorb 2,000 times its weight in distilled water. Its developers claim super slurper may replace cellulose as the major absorbent material, because the starch product will be cheaper to manufacture and is much more absorbent than cellulose.

Metallic "glass." United Technologies Corporation of East Hartford, Conn., developed metallic "glass," a strong, corrosion-resistant layer for metal surfaces ranging from razor blades to industrial turbine blades. A laser beam scans a metallic surface, heating it to more than 2000°C. This causes a very thin layer of the metal to melt. Then the surface is suddenly cooled, leaving a glassy surface on the solid metal. The quick cooling takes place in an inert gas, which prevents the molten metal from oxidizing.

Shrimp versus sludge. Dow Chemical's Texas Division came up with a plan in 1975 to use brine shrimp in an effort to harness nature for pollution control. Dow's system would use algae and brine shrimp to consume bacterial wastes generated by its 26-hectare (65-acre) petrochemical waste-treatment facility in Freeport. The company would not only solve its sludge disposal problem, but the treatment would also result in the daily production of about 272 kilograms (600 pounds) of brine shrimp, a potential animal food.

To treat the wastes conventionally, Dow would have to use daily 3.6 metric tons (4 short tons) of chemicals, 151,400 liters (40,000 gallons) of fresh water, and 100,000 kilowatt hours of electricity. Operating costs for the shrimp system are only one-third of the conventional system, and brine shrimp thrive in the Dow plant's salty wastes.

New deodorizers. Monsanto Flavor Essence, Incorporated, of St. Louis reported in October finding a new type of room deodorizer. Most so-called air fresheners are strong perfumes that mask unpleasant odors. However, the new deodorizers cause a selective loss of smell. They block a person's perception of unpleasant odors without interfering with his perception of pleasant smells, such as cologne. [Frederick C. Price]

Chemistry

Theoretical chemists joined experimentalists in 1975 and 1976 to attack the pressing problems of dwindling supplies of energy, food, and materials. They searched for better understanding of reduction reactions in which atoms or groups of atoms gain electrons or lose oxygen, or both.

Chemists focused on developing better catalysts – substances that speed up chemical reactions – for reduction of such small, abundant molecules as carbon monoxide (CO), carbon dioxide (CO_2), nitrogen (N_2), and oxygen (O_2). These reactions are important in producing synthetic fuels, making ammonia fertilizers and other key chemicals, and converting chemical energy to electricity.

Catalyzing coal conversion. One particularly important problem concerns the conversion of coal to such pumpable, energy-rich fuels as methanol (CH_3OH) and methane (CH_4), which are easily transported and stored. Although the conversion process is complex, the basic chemical problem centers on finding an efficient means of reducing carbon monoxide by hydrogen to give methane. Nickel (Ni) is often used as a catalyst.

Chemist William A. Goddard III of the California Institute of Technology (Caltech) in Pasadena extended the methods of molecular quantum mechanics in 1976 to calculate accurately the chemisorption, or bonding, of such small molecules as CO to solid surfaces. These calculations are important because researchers cannot determine many properties of adsorbed molecules experimentally. Goddard studied such intermediate molecular fragments as methylene (CH_2) and formyl (HCO) that occur in concentrations too small to detect. His calculations were accurate for those fragments that have been studied in experiments. For example, bond energies for the Ni-H bond and the Ni-CO bond agree closely with experimental data.

Goddard proposed a reaction mechanism that involves the carbon atom of CO bonded to two Ni atoms. Addition of H_2 to one Ni-CO bond leads to chemisorbed HCO and H fragments

A chemist at Argonne National Laboratory holds a synthetic "leaf" that uses chlorophyll to convert light into energy in a kind of photosynthetic process.

and releases energy. A CO molecule can then insert itself into the Ni-H bond, releasing more energy and producing another HCO fragment. The HCO fragments accumulate at the nickel surface. Further addition of H_2 to chemisorbed HCO may then occur in steps to yield methane and water, which are released from the surface.

Goddard's proposed mechanism is consistent with experimental data reported in 1976 by M. Albert Vannice of the Corporate Research Laboratories, Exxon Research and Engineering Company in Linden, N.J., but predictions of various reaction rates are needed to check the theory.

In the critically important area of chemical catalysis, Goddard and other theorists may soon be able to study how atomic changes affect the nature of catalytic sites. They might then be able to specify ideal catalyst shapes and compositions to enhance particular reaction pathways.

Meanwhile, experimentalists are also making great strides in their studies of the reduction of carbon monoxide and its close relative, molecular nitrogen. Both are among the most inert molecules known – their chemical conversion to useful products normally requires enormous energies. Hence, high temperatures are needed for reasonable conversion rates.

But in March 1976, Earl L. Muetterties of Cornell University in Ithaca, N.Y., reported that clusters, or discrete molecular groupings, of osmium and iridium atoms catalyze the production of methane from CO and H_2 in solutions. These solution systems may or may not use the same mechanistic steps Goddard suggested for the reaction at a nickel surface. But researchers know that the three metal atoms in the osmium cluster are arranged in a triangle, and that the four metals in the iridium cluster form a tetrahedron. Muetterties believes that such discrete metal-atom clusters are reasonable models for metal surfaces in describing processes such as chemisorption.

Recent advances in the photochemistry of transition metal cluster complexes by Mark S. Wrighton of Massachusetts Institute of Technology (M.I.T.) in Cambridge and Thomas J. Meyer of the University of North Carolina make possible the study of a new type of catalytic process. It could be important for the synthesis of gas fuels from coal-based raw materials and such intermediate chemicals as olefins and acetylene.

The active catalyst is formed by irradiating a transition-metal atom cluster in the presence of a specially designed polymer. Transition metals, such as iron and cobalt, have an incomplete inner electron shell. Irradiation produces small molecular fragments from the cluster that are trapped and stabilized by groups attached to the specially designed polymer.

Alan Rembaum and Ami Gupta of the Jet Propulsion Laboratory in Pasadena, Calif., irradiated dicobalt octacarbonyl, $Co_2(CO)_8$, with short wavelength ultraviolet light in the presence of polyvinylpyridine (PVP) to make Co-PVP, a light-gray powder.

Co-PVP catalyzed the addition of H_2 and CO to olefins or to acetylene to give aldehydes at room temperature and atmospheric pressure. When tiny PVP spheres were used as the support, yields were high. Tests have shown that Co-PVP, which is called a photocatalyst because it is produced by light, is active for long periods of time, usually more than 1,000 hours.

Fixing nitrogen. Practical considerations motivate experimentalists seeking better catalysts to reduce nitrogen to ammonia (NH_3). Plants require nitrogen that has been fixed, or combined chemically with hydrogen or oxygen, in order to grow. Most ammonia now used in nitrogen fertilizer is made in the Haber-Bosch process, a complex industrial method that requires much energy. A metal catalyst allows the reduction of nitrogen at about 550°C (1022°F.) and at about 200 times atmospheric pressure. The resulting ammonia makes many other nitrogen-containing chemicals in addition to fertilizer.

Chemists are looking for better catalysts to reduce nitrogen to ammonia. They also hope to find low-energy routes to hydrazine (N_2H_4), a potential fuel and a generally useful chemical.

In 1975 and 1976, John Bercaw of Caltech, Joseph Chatt of the University of Sussex in England, and Eugene E. van Tamelen of Stanford University in Stanford, Calif., made important discoveries relating to chemical reduction of N_2. Bercaw used a zirconium-based system to reduce N_2 to hydrazine, while Chatt and Van Tamelen independently used molybdenum complexes to reduce N_2 to ammonia under very mild conditions. Chatt also converted related nitrogen-tungsten compounds to ammonia and other products.

Fuel cells. Finding an efficient and economical oxygen electrode would remove the largest single obstacle to making a practical fuel cell – a device that produces electricity directly from the chemical reaction of oxygen and a fuel such as hydrogen. An efficient oxygen electrode must be able to reduce molecular oxygen rapidly and directly to water in a fuel cell.

Two major technical problems must be solved before such an electrode can be built, however. The more important and difficult problem is finding catalysts that will allow rapid transfer of the four electrons and four protons needed to reduce O_2 to water in an acid medium in the cell. The other problem is providing for adequate physical transfer of reactants, products, electrons, and heat to and from the catalyst sites. The catalyst sites must be incorporated in an electrically conductive solid that has wetted internal pores a few atomic diameters wide.

A team of investigators including Howard Tennent of Hercules, Incorporated, Wilmington, Del.; James P. Collman, Michel Boudart, and Henry Taube of Stanford University; and Fred C. Anson and Robert Gagné of Caltech proposed in 1975 that a catalytic oxygen electrode could be built by attaching transition-metal complexes as discrete active sites to otherwise inert, but conducting, electrode surfaces. The design and synthesis of suitable catalysts for the active sites appear to be formidable tasks. However, research on the enzyme laccase produced strong clues as to what is needed to reduce oxygen to water rapidly.

Laccase is a copper-containing protein of high molecular weight that may be isolated from the latex of the Japanese lacquer tree *Rhus vernicifera* or from the apple-rot fungus *Polyporus*

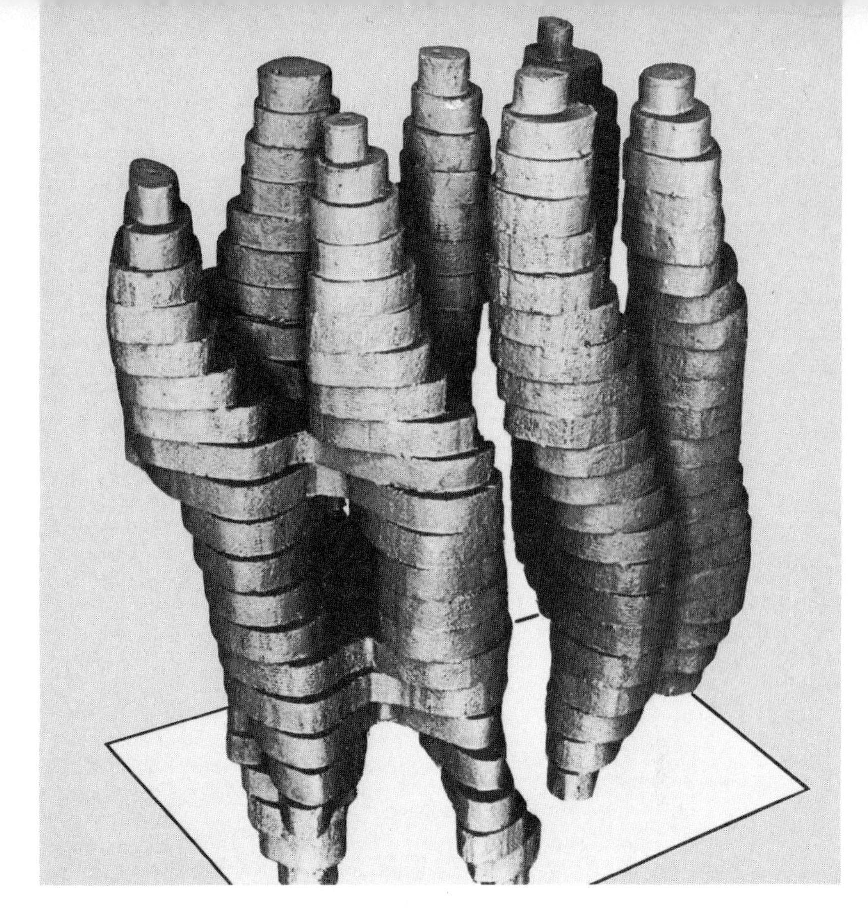

A three-dimensional model of the protein molecule found in the cell membrane of *Halobacterium halobium* was made during 1975. The protein plays a role in the bacteria's conversion of light to chemical energy without the use of chlorophyll.

versicolor. Laccase contains four copper atoms, each capable of accepting an electron. Fully reduced laccase reacts directly and rapidly with oxygen to yield water. Because oxygen reduction by laccase is so rapid, chemists are tempted to base designs of synthetic active sites on what they believe to be the structure of the four-copper unit in the enzyme. Research by Harry B. Gray of Caltech and Bo G. Malmstrom of the University of Göteborg in Sweden during 1975 and 1976 suggested that two of the four copper atoms act as a binuclear unit holding oxygen in place until all four electrons needed for reduction can be delivered to it, producing water directly.

Magnetic measurements on laccase by Edward I. Solomon of M.I.T. and Massimo Cerdonio of the University of Rome produced a great deal of new information about the structure of this binuclear oxygen-fixing unit in 1975. Solomon and Cerdonio determined that the two coppers in the binuclear unit are so close to each other in the protein that an electron on one interacts with an electron on the other, thereby drastically affecting the magnetism of the molecule. This close interaction of the coppers may be quite important in providing a site for four-electron O_2 reduction.

Solar energy is another field where chemists made important contributions in 1975 and 1976. Although it has been known for over 100 years that irradiation of electrodes in electrochemical cells results in photocurrents, several fundamental problems remain to be solved before such effects can be used to convert solar energy efficiently.

In a typical photoelectrochemical (PEC) cell, two electrodes are placed in an electrolyte. One electrode is in darkness. The other, the photoelectrode, is exposed to light. When light strikes the photoelectrode, oxygen is produced by oxidation of water, and electrons flow to the dark electrode where hydrogen bubbles off. In effect, a PEC cell splits H_2O just as may be done in electrolysis by electrical current from a battery or other energy source. But in this case, the energy comes from sunlight.

Chemistry

Continued

Titanium dioxide (TiO_2), a semiconductor, is often used as the photoelectrode in PEC cells. But TiO_2 responds only to short-wavelength (ultraviolet) light. It is not a good photoelectrode for solar-energy conversion because most available sunlight is in the visible region. Two photoelectrodes that can collect visible light may be fabricated from cadmium sulfide (CdS) and cadmium selenide (CdSe). However, another serious problem arises with CdS and CdSe. They deteriorate upon irradiation.

Chemists Arthur B. Ellis, Steven W. Kaiser, and Wrighton of M.I.T. reported a possible solution to the electrode deterioration – a new type of electrolyte. They built PEC cells using CdS and CdSe semiconductor electrodes immersed in a polysulfide electrolyte in water solution. This type of PEC cell does not produce H_2 and O_2 from H_2O, but it converts light to electricity with more than 5 per cent efficiency. The M.I.T. team's breakthrough was the discovery that the polysulfide medium stabilizes the CdS and CdSe photoelectrodes. They found no measurable deterioration of the electrodes over reasonable time periods. These results indicate that it is likely that PEC cells with stable photoelectrodes that will split H_2O to H_2 and O_2 using visible light will be built within the next few years.

Superheavy natural elements? Scientists reported evidence in June 1976 for two natural elements that have 116 and 126 protons in their nuclei. The heaviest confirmed natural element, uranium, has only 92 protons. Scientists have created elements with as many as 106 protons in accelerators, but these quickly break up into lighter elements.

Thomas A. Cahill and Robert C. Flocchini of the University of California, Davis, Robert V. Gentry of Oak Ridge (Tenn.) National Laboratory, and researchers at Florida State University in Tallahassee based their claim on unusually large X-ray "halos" formed by bombarding ancient African mica with protons. Nuclear chemists and physicists are attempting to confirm their findings. [Harry B. Gray]

Communications

Transmitting communications by light beams guided through glass fibers became a practical reality during 1976. By midyear, several companies were selling newly developed optical-fiber communications systems, which have many advantages over copper-wire systems. Being thinner, optical fibers require less space and weigh less. They also have less cross talk between channels and suffer less interference from lightning or radio transmissions.

In optical communications, signals similar to radio waves are transmitted on a beam of light, which focuses on the end of a glass fiber only 0.05 to 0.15 millimeter in diameter. But it may be hundreds of meters long. Light cannot "leak" from the fiber because the glass material is formed in such a way that light traveling at an angle toward the edge of the fiber will bounce back toward the center.

Component costs for optical systems have decreased, and they will eventually be economical for many applications, including telephone links. Present components already are sufficiently high in quality and low enough in cost for use in special-purpose systems and prototype test systems. For example, the U.S. Department of Defense announced in April 1976 that 1,450 meters (4,757 feet) of copper wire in the A-7 attack plane was being replaced by 78 meters (256 feet) of optical-fiber cable. The optical-fiber cable weighs only 1.6 kilograms (3.6 pounds) compared to the 14.5 kilograms (32 pounds) of copper cable it replaced.

The Dorset County Police in Bournemouth, England, installed a fiber optics system in late 1975 after lightning disabled its communications system. In April 1976, Harris Electronics Systems Division of Cleveland began marketing a general-purpose, optical computer hookup system. The system has 32 channels, each capable of carrying up to 16 million bits (0's and 1's) of data per second.

Most of the new systems generate the light signal with a gallium arsenide light-emitting diode (LED) – similar to, but smaller than, the diodes used for the readouts on digital wrist watches

Technicians prepare Intelsat IV-A, a telecommunications satellite designed to relay transatlantic telephone calls, for launch Sept. 25, 1975.

and pocket calculators. The signal to be transmitted varies the current as it passes through the LED, causing corresponding variations in the brightness of the light. A receiver at the other end of the fibers detects the variations in the light signals and translates them into copies of the original message.

In early 1976, Bell Laboratories in Atlanta, Ga., began evaluating a system that uses 600 meters (1,969 feet) of cable containing 100 optical fibers. Bell technicians have simulated a light pathway longer than 600 meters by connecting some of the fibers at their ends and sending light signals back and forth. For example, if two cables are connected at one end, the light would travel 600 meters down one and 600 meters up the other, a total of 1,200 meters (3,937 feet). Bell technicians are evaluating cable made by the Western Electric Company and by Corning Glass Works, as well as different types of light-signal sources.

The farther a beam travels, the more strength, or intensity, it loses. A light beam traveling along experimental fibers now being tested can go about 750 meters (2,461 feet) before losing half its original strength. Meanwhile, many of the new commercial fiber optical systems are using a cable made by Corning Glass that contains six glass fibers. Each can carry a light beam 150 meters (492 feet) before the beam loses half its strength. Corning's cable costs $13.50 per meter (about $4.11 per foot). When carrying 100 million bits of data per second, this cable is comparable to high-quality copper coaxial cable.

Digital data communications. More than half the computers in the U.S. are now linked to other computers by telephone lines. Regular telephone circuits also connect many computer terminals with their central computer. The data they exchange range from travel reservations to credit checks and inventory-control information.

Until recently, most of this data was converted to tones that were transmitted in the same manner as voice signals. But in March 1976, the Bell System began operating switched digital service for computer exchanges between

Communications
Continued

18 metropolitan areas and planned to extend the service to nine more areas later in the year. In a digital system, the data is converted to a stream of bits rather than tones. Digital service transmits data more efficiently because it can carry 56,000 bits of data (equivalent to about 7,000 typewritten characters) per second. Tones can only transmit about 120 characters per second. At this high rate, the computers will be linked together for shorter periods. Consequently, transmission costs will go down.

Microcomputers. Small computers known as microprocessors are becoming more attractive as components of communications systems. Silicon integrated circuits can have 10,000 transistors interconnected on one thin 5-millimeter-square silicon wafer. This allows small computerlike circuits to be manufactured as a single component. Prices for these microprocessors fell below $30 early in 1976. See INVASION OF THE MICROCOMPUTERS.

The Bell System Transaction II telephone, introduced in 1976, contains a microprocessor. The new telephone can be used by businesses to check a customer's credit-card standing instantly. The telephone has a visual data display similar to the readout on a pocket calculator and is connected to a computer. A store clerk inserts a credit card and uses the touch-tone telephone buttons to add other information, such as the amount of the purchase. The microprocessor in the phone sends this information to the central computer as soon as a line is available. The microprocessor also checks the data for errors. Then the computer sends back credit information, which appears on the telephone's visual display.

The New York City Police Department installed a message-handling system in early 1976 that uses four microprocessors working with a common memory system. Two of the microprocessors receive messages and store them in the computer memory. The other two constantly check both the memory and transmission channels. As soon as a channel is open, they retrieve the messages and send them on to their destination. [John A. Copeland]

Drugs

Medical researchers reported on their investigations of levamisole at a conference held in Bethesda, Md., in December 1975. Their findings on the drug's action and its effects on cancer and other diseases associated with defective immunity were encouraging.

Two microbiologists, Gerard E. and Micheline D. Renoux of the University of Tours in France, found in 1972 that this compound, then used as a deworming agent in humans and other animals, stimulated the immune system. This attracted the attention of cancer researchers who were trying to develop compounds that stimulate the body's immunity to cancer.

Many researchers believe that one of the body's two immune mechanisms, humoral immunity, which produces antibodies, may actually stimulate tumor growth. But the other system, cell-mediated immunity, apparently retards cancerous growth. By a yet-undefined mechanism, levamisole increases cellular immunity when it is below normal. Normal cell-mediated immunity is not increased by the drug.

In the treatment of cancer with the drug, the researchers' findings were mixed. J. Leonard Lichtenfeld of the Baltimore Cancer Research Center found it had no effect on cancer patients. But Alexandro Rojas of the Angel H. Roffo Instituto de Oncologia in Buenos Aires, Argentina, reported that levamisole increased the survival rate of breast-cancer victims.

Laboratory studies by Elizabeth W. Doller at the Milton S. Hershey Medical Center in Hershey, Pa., indicate that metastasis – the spread of tumors to distant parts of the body – can be prevented in hamsters if levamisole is administered before metastasis begins.

Since levamisole enhances immunity only when that immunity is deficient, it may be especially useful when used in combination with cancer-fighting drugs that depress immunity. Cancer researcher Michael A. Chirigos of the National Cancer Institute found that one combination increased the number of cures in animals with leukemia.

Levamisole also shows promise in the treatment of other diseases. Two reports at the Bethesda meeting described improvement in patients with herpes virus infections, and complete remission of symptoms in a woman with systemic lupus erythematosus, an inflammatory disease of the skin, joints, and blood vessels in which the immune system is abnormal.

Levamisole's side effects are loss of appetite, nausea, diarrhea, nervousness, and fatigue. The drug is toxic in large doses, but its side effects are less severe in general than those of most anticancer drugs. If further studies confirm and define its efficacy, levamisole may become an important weapon against many serious and fatal diseases.

New antibiotic. Chemist Frederick M. Kahan and his associates at Merck Sharp and Dohme Research Laboratories in West Point, Pa., reported in September 1975 that they have developed a new antibiotic drug that has been effective against every bacterial strain on which it has been used. The new drug is a combination of PCS, a form of an antibiotic known for 20 years, and DFA, a chemically altered form of D-alanine, a protein that bacteria require in their cell walls. PCS is rarely used because it has serious side effects. In small amounts, DFA prevents bacteria from building cell walls, thus killing them. Paradoxically, bacteria thrive on larger doses. A small amount of PCS prevents this.

The drug is so successful against bacteria that some scientists fear that it might kill off the important bacterial flora in the lower intestine. But in mice, at least, the drug is absorbed into the bloodstream before it reaches the lower intestine. The antibiotic, as yet unnamed, is being tested on monkeys.

New warning to users of oral contraceptives was sounded with publication of two British studies in 1975. William H. W. Inman, principal medical officer of the Committee on the Safety of Medicines in London, and medical researcher Martin P. Vessey, now at the University of Oxford, compared the incidence of heart attacks in women who took birth-control pills with those who did not. They found that women from 30 to 39 years old who used the

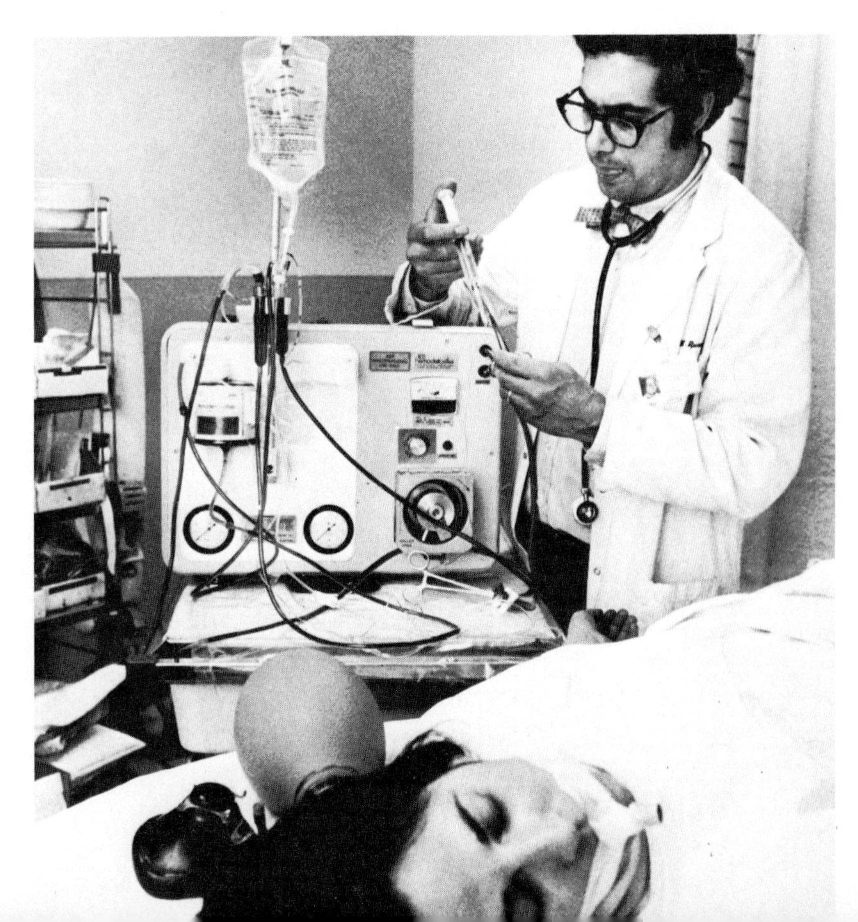

A new filtering machine uses charcoal particles to filter toxic substances from blood of poison and drug-overdose victims.

"pill" suffered more than 2½ times as many heart attacks as nonusers. In women from 40 to 44, the incidence was more than five times higher.

Scientists already know that those who take oral contraceptives regularly are more likely to develop blood clots, with possible severe damage to the brain, eyes, and lungs. Potential hazards would be detailed in proposed new brochures to accompany these drugs.

Vitamin preparations. On April 12, 1976, Congress curbed the power of the Food and Drug Administration (FDA) to regulate certain vitamin preparations. It passed legislation that prevents the federal agency from limiting the potency or combination of ingredients of vitamin and mineral products. Certain substances, such as vitamins A, D, K, and folic acid, where misuse can be hazardous, are excepted. The new law also affirmed an earlier court decision which prohibited FDA classification of vitamins and minerals as drugs when the dosage is greater than the recommended dietary allowance set by the National Research Council.

New drugs introduced in the United States in 1976 included:

- Bricanyl; Brethine (terbutaline), a drug for the treatment of bronchial asthma and reversible bronchospasm associated with bronchitis and emphysema. The compound relaxes the muscles in the bronchial passages but stimulates the heart less than other drugs.
- Catapres (clonidine), a compound to lower blood pressure in persons with hypertension. This drug apparently affects brain centers that control how much the body's small arteries constrict. It is particularly effective in some patients when used with diuretics, which eliminate excess water from the body. However, the new compound may produce drowsiness and lead to a dangerous rebound in blood pressure if its use is stopped abruptly.
- Nebcin (tobramycin), a new aminoglycoside-type antibiotic drug for use in infections caused by the bacteria *Pseudomonas aeruginosa.* This microorganism seems susceptible to the new antibiotic's action even when resistant to other drugs. [Arthur H. Hayes, Jr.]

Ecology

Research on the effects of plutonium (Pu) in the environment intensified in 1975 and 1976. Plutonium isotopes are produced when a uranium atom is split by nuclear fission. The isotope Pu-239 can also be split by fission. Because it may serve as fuel for future reactors, it is important to understand plutonium's effect on the environment.

Ecologists Roger C. Dahlman, Ernest A. Bondietti, and L. Dean Eyman of the Oak Ridge National Laboratory in Tennessee reported in late 1975 on how plants absorb plutonium. They conducted their investigations in a drained Tennessee flood plain that had low levels of plutonium isotopes, probably from water discharged from Oak Ridge nuclear facilities some 30 years ago.

The ecologists' findings suggest that plant roots take up only from 0.0001 to 0.001 of the plutonium concentrated in soil, but the amounts that individual plants will absorb vary greatly. One plant can take up 10 times as much as another of the same species.

However, radioecological research on how Pu-238 and Pu-239 are absorbed from soil is complicated by gradually settling plutonium isotopes that entered the stratosphere during nuclear-test explosions that began in the mid-1950s. Plutonium particles dropped on the outside of a plant make it difficult for scientists to measure the amount of radioactive isotopes inside. But continued research should greatly increase our knowledge of plutonium radioecology within a few years.

Plants and pollution. An ecologist, Marilyn J. Jordan of Rutgers University, reported in the winter of 1975 the effects of emissions from two zinc smelters near Lehigh Water Gap at Palmerton, Pa., on the surrounding forest.

Jordan calculated that daily emissions since 1960 probably ranged from 6,000 to 9,000 kilograms (13,228 to 19,842 pounds) of zinc particles; 70 to 90 kilograms (154 to 198 pounds) of cadmium; and less than 90 kilograms each of copper and lead. She tested the effects on seed germination, root growth, and seedlings in soil collected 2 kilometers (1.2 miles) and 40 kilometers (25 miles) from the smelters.

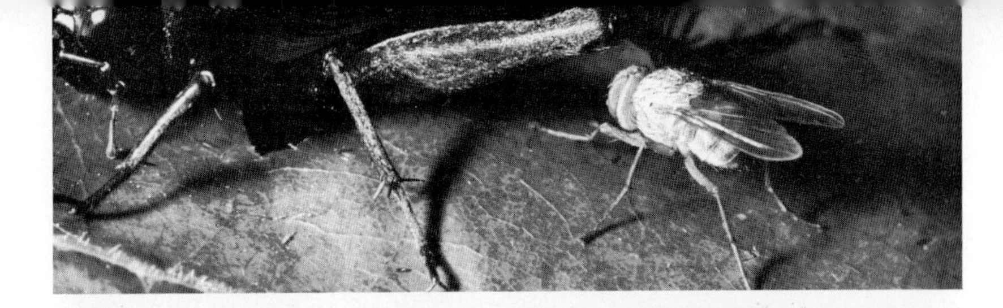

A male cricket's mating song attracts parasitic fly, *top*. The fly then deposits larvae on the cricket, *center and bottom*. Larvae will feed off the host's flesh. Scientists believe this may be why some male crickets do not sing but find indirect ways to meet females during the mating season.

Jordan grew white pine and red oak seedlings in pots containing top soil collected from the test sites and found significant differences in their growth. After two growing seasons, red oak seedlings grown in soil taken from near the smelters had significantly less new twig growth and fewer leaves than those grown in soil taken farther from the smelters. White pine seedlings grown in the two soil types showed similar results. Also, those in soil taken from near the smelter became chlorotic, or yellowed, and had few roots.

Jordan also tested the effects of various concentrations of zinc and cadmium on red oak seedlings grown in washed sand. She prepared five concentrations of zinc sulfate in solution – 0.1, 1, 10, 100, and 1,000 parts per million (ppm) – and three concentrations of cadmium sulfate – 0.1, 1, and 10 ppm. She added the different metal concentrations to various groups of plants. Emerging red oak seedlings watered with 1,000 ppm zinc died within a week. Seedlings given 100 ppm zinc, 10 ppm cadmium, or both produced only one spurt of new growth instead of the usual three. They also had significantly shorter than normal stems and leaves, and became chlorotic and died in from three to five months. After five months, Jordan took the remaining seedlings to the laboratory for analysis.

The seedlings given 10 ppm zinc became chlorotic in much the same way as trees growing near the smelters. The mean weights of roots, stems, and leaves were significantly lower in plants watered with either 10 ppm zinc or cadmium than those watered with 0.1 ppm zinc or 1 ppm cadmium. Growth was more retarded in plants watered with concentrations of both metals.

These experiments indicate that high levels of zinc in the soil cause noticeable damage to vegetation. Nevertheless, if high zinc levels in the soil had been the only stress on the forest ecosystem for the past 50 years, the Lehigh Water Gap area would probably be much more densely vegetated than it is.

Smelting operations have been carried on there since 1898, and normal amounts of vegetation have failed to

Ecology

Continued

A semicircular marsh in the Mississippi Delta, made of silt dredged from the river bottom by the U.S. Army Corps of Engineers, helps replace Louisiana's vanishing wetlands.

grow back following repeated forest fires during the past 50 years. The problem is apparently a combination of high zinc levels plus repeated fires. High zinc levels in the soil apparently inhibit reproduction in the area's plants, preventing rapid growth of new vegetation after a fire. Without new vegetation, the burned slopes begin to erode, and eroded areas are even less likely to produce new growth.

Zinc and cadmium do not wash out of the soil. However, adding limestone helps to reduce the amounts of the metals available to the plants. Limestone reacts chemically with zinc and cadmium to form compounds that roots have difficulty absorbing, but ecologists do not know its long-range effectiveness. Nevertheless, without some action to offset the damaging effects of metal in the soil, the burned-out areas may remain barren for centuries.

Playing dead. Ecologists Alan B. Sargeant and Lester E. Eberhardt of the Northern Prairie Wildlife Research Center in Jamestown, N. Dak., reported in July 1975 on how adult ducks play dead when confronted by red foxes. Playing dead appears to be a highly developed survival behavior of ducks.

Sargeant and Eberhardt conducted their tests in completely enclosed outdoor pens containing underground fox dens. They used five mated pairs of 1-year-old red foxes and two litters of 5-month-old red fox pups. The ducks included pen-reared mallards and pintails and wild blue-winged teal, pintails, and wood ducks. Depending on the size of the pen, the ecologists released up to 10 ducks and observed them from nearby blinds. The adult foxes were tested during the evening, when they are normally active, and the pups were tested primarily during the day. The scientists often did not feed the foxes for one or two days before testing, to make them hungry.

Sargeant and Eberhardt watched the foxes catch about 50 ducks. Most of the birds tried to escape, but all of them became immobile as soon as the foxes seized them. The birds extended their heads, kept their eyes open, and held their wings tightly against their bodies for from 20 seconds to 14 minutes. The immobile ducks appeared to be alert and aware of escape opportunities.

The birds played dead each time they were seized by foxes. The scientists observed 29 birds survive capture this way. The others also appeared to have played dead, but they were nevertheless killed by the foxes before the researchers could rescue them.

Seed feed. Ecologists H. Ronald Pulliam and Mariner Riley Brand of the University of Arizona at Tucson reported in the summer of 1975 on the relationship between the production of grassland seeds and seed-eating animals. They studied an area of 3,200 hectares (7,907 acres) about 16 kilometers (10 miles) southeast of Elgin, Ariz., that contains three major seed predators – harvester ants, grasshopper sparrows, and rodents. They were trying to determine if these animals compete for food; if they eat different kinds of seeds; and if so, how this has affected evolution of seeds.

In southeastern Arizona grasslands, perennial grasses produce seeds after the summer rains, when the harvester ants and rodents are becoming inactive and the first sparrows are arriving to spend the winter. Annual forbs, or herbs, produce seeds after the winter rains, when the sparrows leave and the ants and rodents again become active.

The scientists set out 15 square, wood-framed traps to catch falling seeds. They also collected soil samples and examined them in the laboratory, removing and counting all seeds. Every two weeks they counted the sparrows and watched them to see what they were eating. They also trapped rodents every two weeks and examined the contents of their cheek pouches and waste matter. In addition, they observed six colonies of harvester ants.

Pulliam and Brand divided the grassland seeds into two broad classes – seeds with smooth coats and seeds with hairs or other projections. They estimated that 52 million grass seeds per hectare (2.5 acres) were produced after the summer rains and 328 million forb seeds per hectare after the winter rains.

The ants, foraging mainly in spring and summer, removed about 6.5 million hairy seeds per hectare during the year. Sparrows, most active in autumn, ate about 4 million smooth seeds. Rodents ate about 3 million smooth seeds per hectare, mostly in the spring.

Pulliam and Brand found that there was little competition or overlap in eating patterns among the three animal species because of the differences in times of feeding and their choice of seed types. The two types of seeds were produced at times when the animals that preferred them were not feeding. Since the animals did not pose a threat, a sufficient number of seeds were able to germinate and produce new forbs and grasses. So these data support the idea that feeding pressures have been important in evolving both the times seeds are produced and the external structures of their coats.

Intentional land burning in Southeastern states promotes the growth of legumes – plants that bear seeds in pods. The seeds are eaten by quail and turkey. To improve the management of wildlife habitats, scientists need to learn more about possible relationships between land burning and seed germination. In late 1975, ecologists Robert E. Martin, Robert L. Miller, and Charles T. Cushwa of the Virginia Polytechnic Institute and State University reported on how heat affects the seed germination of 18 species of legumes.

They subjected the seeds to both moist- and dry-heat treatments, at temperatures ranging from 30°C (86°F.) to 110°C (238°F.), reasoning that in a forest fire, combustion produces a certain amount of moisture. They placed seeds in wire baskets and exposed them to heat for four minutes – about the average time that seeds on a forest floor would be exposed to a fire. After heating the seeds in either an enclosure at 100 per cent humidity or in an oven, the scientists then placed them between moistened blotters in shallow dishes.

Plants treated with moist heat germinated at temperatures ranging from 45°C (113°F.) to 80°C (176°F.); those treated with dry heat, from 45°C to 90°C. The most effective temperature was 70°C for moist heat and 90°C for dry. Temperatures over 90°C appeared to kill seeds. Moist heat increased germination in eight species; dry heat, in seven. This supports the value of burning as an aid to bird-food plant production. [Stanley I. Auerbach]

Electronics

A measure of the pace of development in large computer memories was seen in early 1976 when several semiconductor firms introduced the so-called 16-K (16-kilobit) memories. The new models were announced shortly after the first full-scale production of their predecessors, the 4-K memories.

Over the last five years, semiconductor memories have grown rapidly in capacity and performance. The improvements have come through large-scale integration (LSI) – the manufacturing of thousands of transistors simultaneously on one tiny piece of silicon, and interconnecting them to form complex electronic subsystems for a wide variety of applications.

Since 1970, when Intel Corporation introduced the read-write semiconductor memory with 1,024 bits, or binary digits, it has become possible to build very large and fast memory systems that can compete economically with magnetic-core memories. By 1974, LSI techniques had produced the 4-K system, 4,096-bit memory chips with even greater economies in size and cost.

The 16-K and CRAY-1 memories. The 16-K memory has 16,384 bits. Each bit is represented by a single transistor "cell." In addition to the 16,384 cells, each chip contains sense amplifiers, decoders, input/output circuits, and interconnecting lines. All of this is on a piece of silicon approximately 0.51 by 0.41 centimeter (0.2 by 0.16 inch). The cycle time for these memories – the time required to write or find a bit in the memory – is about 450 nanoseconds (450 billionths of a second).

Most computer activity is now concentrated on microcomputers and small and medium-sized machines (see INVASION OF THE MICROCOMPUTERS). But Seymour R. Cray of Cray Research Incorporated is gambling that there is still a demand for very large-capacity, high-performing computers.

By the end of 1975, he had completed assembly of the first CRAY-1 – a very fast advanced scientific computer. The CRAY-1 computer is intended to solve complex scientific problems at speeds significantly higher than those of any other existing computer. Thus it will be

"Nope; no chess problems until you've finished your quadratic equations!"

Science Year
Close-Up

Calculators In the Classroom

The students were totally absorbed; no one looked up as we entered the classroom. They were working with small electronic calculators and were obviously deeply interested. "This has been true ever since we've had the calculators," the teacher said. "On the day before Christmas vacation, when the school chorus and band came through the halls caroling, only four students even looked up to see the activity. The rest continued doing their math, too absorbed to notice what was going on around them."

This is typical of teachers' comments in all junior and senior high school classrooms in the Denver area where calculators are being used in a mathematics program developed by the University of Denver Mathematics Laboratory. The teachers say that students have to know more, rather than less, mathematics to use a calculator. This refutes the common objection that students will become dependent on calculators and will not master facts and basic computational skills.

"My students were intrigued with signed [+ and −] numbers, and discovered their laws from calculator usage," a seventh-grade teacher declared. "I asked them to use the calculator for checking only, in the beginning, but I am now convinced it can be used for discovery and concept building."

An algebra teacher reported that by using calculators, his students easily learned how algebra worked in a way he had not been able to make clear to them before. In evaluating algebraic expressions, his students were able to do so many problems in such a short time that they quickly recognized how important the structure of the problem can be. They realized, he said, that failure to observe mathematical principles, not sloppy computation, caused many wrong answers.

Other teachers said that some students have discovered for themselves the rules for placing decimal points when dividing decimal fractions. A girl who could not memorize the multiplication tables mastered them when she and the calculator "communicated" without pressure or outside interference. Some students mastered interest, discount, and commission problems in half the time when they could concentrate on the meaning and interrelationship of these processes, leaving the arithmetic computations to the computer. Remedial students who had not mastered their long division suddenly grasped the process when they used calculators to guess at separate digits of the quotient, check them out, modify, and check again until they got the correct answer.

Students studying decimal equivalents of common fractions first got unit fraction values and then predicted nonunit fraction values and checked them quickly on the calculator. They discussed repeating decimals and, with the calculator doing the computation, looked at many examples. Without the tedium of long division hand computations to distract them, they talked knowingly of periods and patterns and analyzed the concepts.

Hand calculators are a fact of our everyday life – they are available to almost everyone. As far back as 1974, calculator manufacturers estimated that 1 out of 4 households in the United States had a calculator. The choice may be whether students will use them surreptitiously – and then perhaps incorrectly – at home, or whether teachers will allow their use as a powerful tool for learning in a guided manner in the classroom. Objections that the calculator will become a crutch and will widen the gap between high and low achievers have been disproved.

Teachers who use the calculators in advanced mathematics courses such as trigonometry and advanced algebra are also unanimous in their enthusiasm. Arithmetic computation becomes only a means to an end in such high school courses, and the calculator performs these computations accurately and rapidly. This allows the student to concentrate on such concepts as mean, mode, and standard deviation; trigonometric functions and their characteristics; roots of polynomial equations; and the limiting values of sequences.

Even teachers who were at first dubious agreed that the University of Denver program proved that the calculator can be a tool for learning, not for by-passing, mathematics. The parents of one group became equally enthusiastic after an evening session with the calculators. [Ruth I. Hoffman]

A modified television set, a telephone, and an electronic key pad are all the equipment needed to participate in Viewdata. The system, being tested by the British Post Office, permits a subscriber to tap into information storage banks for items such as news, weather, sports, or advertising, at his convenience.

able to handle more complex problems, such as weather forecasting.

A principal feature of the CRAY-1 is a very large semiconductor memory. It has a capacity of up to 64 million bits, with a cycle time of 50 nanoseconds.

Computer X rays. Radiologists are mating the computer to the X-ray machine. A new technique known as computerized axial tomography can produce highly detailed X-ray pictures of cross sections of the body. Benefits include faster, more accurate X-ray examinations than by conventional systems, and less exposure to potentially harmful radiation for the patient.

The principle was first developed by researchers at EMI Limited in England for scanning the brain. In 1976, it is being extended to the whole body by that company and by General Electric Company, Ohio Nuclear Incorporated of Solon, and Digital Information Sciences Corporation of Silver Spring, Md., among others.

In the typical scanner, a patient lies on a couch that extends through an opening in a framework that contains the X-ray source and an array of detectors. As the X-ray source is precisely step-rotated around the patient's body, it emits pulsed radiation. Varying densities of bone and tissue absorb varying amounts of radiation, which the detectors pick up. A computer converts this data into a television picture that can be displayed on a screen or stored on a magnetic disk.

The EMI scanner, known as the CT 5000, uses 30 detectors. A 20-second scanning sequence produces details of all types of tissue contained within a 13-millimeter-thick slice of the body. A General Electric version considerably shortens the scanning time. It uses a 30-degree pulsed X-ray beam that rotates around the patient in just 4.8 seconds. The absorption information is collected by 320 detectors. This fast scanning eliminates blurring due to involuntary movement by the patient.

Electronic vision. An improvement has been made in efforts to provide artificial vision for the blind. In 1974, a team of electronic and medical researchers at the University of Utah in

Salt Lake City and the University of Western Ontario in Canada demonstrated that electrical stimulation of the cortex of the brain could produce light pulses, known as phosphenes, which might be the basis of artificial vision for the blind. In those experiments, electrodes were implanted in the visual cortexes of volunteer blind patients. Using a computer, the researchers determined the pattern of phosphenes as a function of where they placed the electrodes and mapped this pattern for each patient. With this technique, the patients were able to perceive rudimentary letters and patterns.

In 1976, this work was carried a step further. A man who had been blind for 10 years has been able to read sentences and phrases in braille by "seeing" them. The experiment was designed by William H. Dobelle and Michael G. Mladjovsky (who developed the earlier system) and Jerald R. Evans of the Institute for Biomedical Engineering, University of Utah; T. S. Roberts of the Department of Surgery at the University of Utah Medical Center; and J. P. Girvin of the Department of Clinical Neurological Sciences, University of Western Ontario.

The researchers implanted 64 electrodes in the patient's skull and brought them out through a connector mounted behind his ear. The electrodes were assigned to braille dots in accordance with a computer-determined phosphene map and, when electrically stimulated, the patient "saw" the corresponding braille letter pattern. With this technique, he was able to read much faster than he was able to with his fingers.

Laser acupuncture. The growth of interest in acupuncture in treating various disorders has led to development of electronic equipment that may improve on the ancient technique of inserting needles into the skin to alleviate pain and cure ailments. Walter Kroy of Messerschmitt-Bolkow-Blohm GmbH (MBB) in Munich, West Germany, led a team of researchers in developing a system in which the classical acupuncture needles are replaced by painless laser beams.

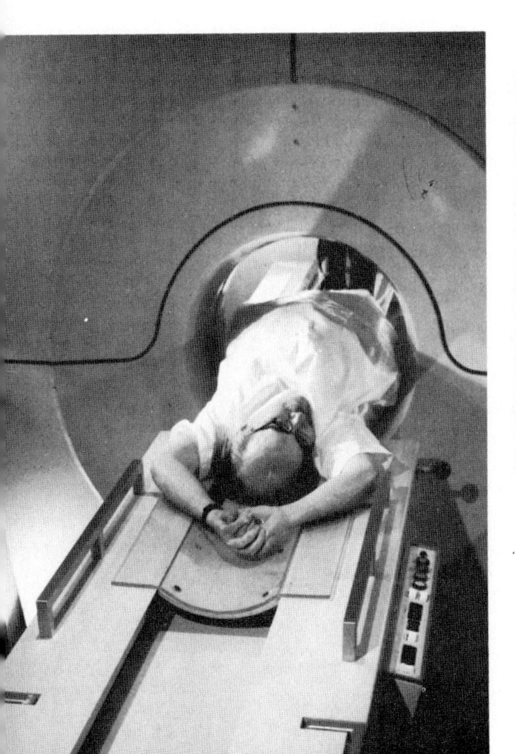

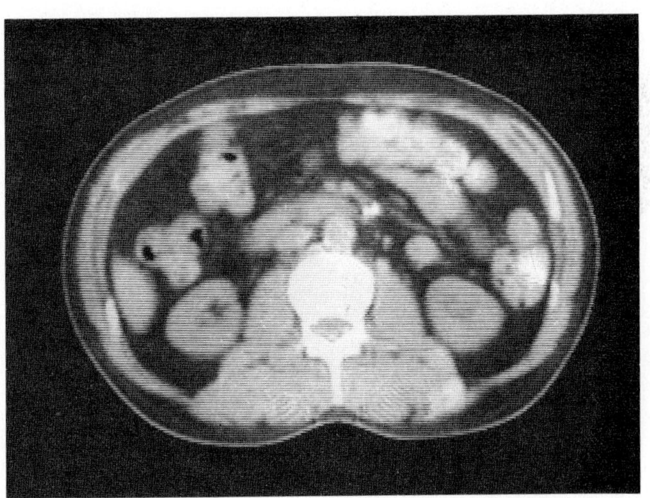

High-speed computerized body scanner, *left,* takes X-ray pictures in less than five seconds. A 13-millimeter thick scan through the patient's abdomen, *above,* clearly reveals his kidneys on each side of his spine.

Electronics

Continued

The MBB equipment, called Akuplas, uses a helium neon gas laser that emits a 1-milliwatt beam of light at a wavelength of 0.632 micrometer (0.0000249 inch). The beam is applied to the patient's acupuncture points by a probe connected to the light source by a glass-fiber cable. The probe handle functions as an electrode that measures skin resistance on a meter. A reduction in skin resistance indicates an acupuncture point. The beam penetrates the tissue to a depth of 10 to 15 millimeters, the same depth to which conventional needles are inserted.

News on demand. In early 1976, an imaginative new service for British television viewers began experimental trials. Called Teletext, it permits home viewers to call up "pages" of information such as news, sports, and weather maps for display on TV sets equipped with special encoders. The service is being broadcast by the British Broadcasting Corporation and the independent Broadcasting Authority.

In Teletext, the signals ride piggyback on the regular television signal. They are transmitted during a portion of the blank period after each 625-line frame composing the regular picture is completed. The digital signals are transmitted at a rate of one row of 40 characters every 50 microseconds. A full page of information consists of 24 rows. A viewer can call up each of about 800 pages by punching a description on a calculator-type keyboard.

The British Post Office plans to offer a similar service called Viewdata. It will have the same type of information as Teletext, but also will permit interaction between the viewer and the system, and even between the viewers themselves. Viewdata links a television receiver equipped with a decoder to a central data bank via telephone lines. Through a decoder keyboard, the viewer has access to virtually unlimited pages of information. It will permit him, for example, to work out mathematics problems at home on a remote computer. Other services to be offered include wire service news, stock quotations, and educational and training courses. [Samuel Weber]

Energy

Exploiting the vast coal deposits in the United States to free the nation from dependence on foreign oil was a major focus of energy research during 1975 and 1976. Researchers also concentrated on refining solar-power technology, improving nuclear safety, and developing unconventional energy sources.

Fossil fuels. Mobil Oil Corporation scientists in February announced a unique process for converting coal into high-octane synthetic gasoline. The process adds one step to the sequence of converting coal to methanol.

The final step is based on principles of molecular engineering, using zeolites, crystals of aluminosilicate, that are commonly used at oil refineries in the cracking process. The Mobil researchers developed special zeolite crystal structures to break down the methanol molecules into a limited range of hydrocarbons and water. Although 25 per cent of these hydrocarbons are gases, most of the rest form liquid gasoline.

Since methanol is the essential raw material in the process, gasoline can be made from any material that can be converted to methanol. Such materials include natural gas, wood, and paper.

In October 1975, the Environmental Protection Agency (EPA) successfully completed tests of its new facility for experimenting with fluidized-bed combustion, a clean, efficient way to burn coal. The plant was built for the EPA in Linden, N.J., by the Exxon Research and Engineering Company.

The coal is ground up, mixed with granules of a noncombustible substance, such as limestone, and laid on a plate in the combustion chamber. Air is then forced through holes in the plate, suspending the coal and granules in the streams of forced air. The limestone reacts chemically to remove sulfur dioxide. Because this process operates at lower temperatures than conventional furnaces, nitrogen oxide emissions are also reduced.

Several new methods for removing sulfur from coal were announced during the year. These ranged from a magnetic process to chemical methods employing acid or hydroxides. See CHEMICAL TECHNOLOGY.

Rows of solar-collector panels on the roof of a school in Atlanta, Ga., *top,* trap the sun's energy to power the building's heating and cooling system in a practical experiment with solar energy. Other solar-energy schemes include a car, *above,* covered with photovoltaic cells that convert sunlight to electricity and black venetian blinds between windowpanes, *right,* that use heat from the sun's rays to warm forced air of a heating system.

Nuclear safety. In October, scientists at the Argonne National Laboratory near Chicago announced the development of a new device for detecting dangerously high temperatures in the Liquid Metal Fast Breeder Reactor (LMFBR). This controversial type of reactor theoretically will produce more fuel than it consumes. An LMFBR demonstration plant is scheduled to begin operating in 1983. See ECOLOGY.

In the LMFBR, liquid sodium will circulate as a coolant around the fuel core and transfer the heat from the atomic reactions to a steam generator. Liquid sodium is ideal for handling the high temperatures inside the LMFBR because its boiling point is about 870°C (1600°F.). However, if anything prevents the liquid sodium from circulating, the temperature of the core would rise unchecked, causing the sodium to boil. Eventually, the nuclear fuel and its container would melt, releasing dangerous radioactive materials.

The Argonne scientists have developed temperature-resistant pressure sensors, containing plates of lithium niobate crystals, that can be placed in the liquid sodium to detect pressure changes when sodium begins to boil. Such changes produce electrical charges on the surface of the crystals, and this electrical signal is passed on to monitoring instruments. If the sodium starts to boil, the reactor would be shut down. This pressure-detection system will supplement existing temperature sensors.

Solar developments. The largest solar heating and cooling system in the United States went into operation at a school in Atlanta, Ga., in the fall of 1975. The experimental system fills 60 per cent of the heating and cooling requirements for the building.

On the roof, 576 solar-collector panels heat water that is then stored in two underground tanks which hold 56,781 liters (15,000 gallons) each. In winter, the hot water is used to heat the building. In summer, one of the tanks can be filled with cold water to cool the building at night and on sunless days. On sunny summer days, an air conditioner, which is powered by heat energy rather than electricity, cools the building.

Solar-energy researchers are monitoring all aspects of the experimental system to determine how to make it more efficient and thus make such systems economically attractive.

In April 1976, John C. C. Fan, a researcher at the Massachusetts Institute of Technology's Lincoln Laboratory in Lexington, described new materials to improve solar heating. Absorbers for trapping light and retaining heat are usually made of nickel-black or chromium-black. However, these materials can operate efficiently only at temperatures below 200°C (392°F.). New materials, made with gold and manganese dioxide, have resisted temperatures up to 400°C (752°F.).

Fan also described newly developed, transparent metallic films that can be applied to windows like wallpaper, replacing storm windows. These films are heat mirrors, which allow light to enter the house but prevent radiated heat from escaping. The experimental films consist of extremely thin layers of silver and titanium oxide or tin and indium oxide bonded to plastic sheets.

Boeing Aerospace Company scientists concluded in late 1975 that it is technically and economically possible to build huge satellites that would produce electricity from the sun's energy. The satellites would then transmit the power to the earth in the form of highly concentrated microwave beams. One solar-power satellite could produce up to 10,000 megawatts.

Ice-cube power. Westinghouse Research Laboratories in Pittsburgh announced in February that it was testing a heating and cooling system powered by the energy released when water changes to ice. The system consists of a heat pump and a block of ice, covering 1.6 square meters (20 square feet) and 1.2 meters (4 feet) thick, buried behind an experimental house.

In summer, as the ice melts, it cools the house. In winter, a heat pump extracts heat energy from the water, causing it to freeze slowly. The system is best suited for areas where winter and summer are about equally long.

Harnessing the oceans. U.S. Navy and Carnegie-Mellon University scientists reported in July 1975 that it is now technologically possible to use the temperature differences between the surface and depths of the oceans to produce electricity, an idea first conceived by a French physicist in the 1880s.

Science Year Close-Up

The Firewood Crisis

Dwindling petroleum reserves and restrictions on its distribution make headlines. Yet, for at least one-third of the human race, the energy crisis is a daily scramble to find wood for cooking dinner. This search for firewood, once a simple chore but now a day's labor in places where forests have receded, has been strangely neglected by governments and economists as well as by the general press.

While chemists devise ever more sophisticated uses for wood, at least half of the timber cut in the world is still used as fuel for cooking and heating. In most poor countries, 9 out of every 10 people depend on firewood. And all too often, the growth in population is outstripping the growth of new trees.

The firewood crisis is most acute in the densely populated Indian subcontinent and in the semiarid stretches of central Africa along the Sahara.

One morning on the outskirts of Kathmandu, the capital of Nepal, I watched a steady line of people trudge into the city with neatly chopped and stacked loads of wood on their backs. I asked my taxi driver how much their loads would sell for. "Oh wood, a very expensive item!" he exclaimed without hesitation. "That load costs 20 rupees [about $2] now. Two years ago it sold for six or seven rupees."

Firewood prices are still climbing throughout Africa, Asia, and Latin America. Those who can, pay the price; many forego other essentials. They accept wood as one of the major expenses of living. In Ouagadougou, Upper Volta, the average family spends over one-fourth of its income on firewood. Those who cannot pay forage in the surrounding countryside if there are enough trees within walking distance. Others gather twigs, bark, garbage, or anything else that can be burned.

The firewood crisis has provoked little world attention, because most of those affected by it are illiterate and the shortage is localized. A shortage in Mali or Nepal is of no immediate consequence to someone building a campfire in the United States.

Unfortunately, the economic burden placed on the local poor is not the only consequence of the firewood shortage. The accelerating of the deforestation process throughout Africa, Asia, and Latin America, caused in part by fuel gathering, results in soil erosion and expanding deserts that reduce world food productivity.

The single most destructive result of the shortage on the Indian subcontinent is probably not denuded land and the erosion and floods that follow. It is the alternative fuel most of the people have been forced to use – cow dung. In many areas, hand-molded dung patties have been the only source of fuel for generations, and their use is spreading, robbing the soil of badly needed nutrients and organic matter. Dung is also used for cooking rather than for fertilizer in the Sahelian region of Africa, in Ethiopia and Iraq, and also in the nearly treeless Andean areas of Bolivia and Peru.

The deterioration of land resources can be reversed only through concentrated national efforts and strong leadership. Preserving the earth's productive capacity calls for greater use of kerosene and natural gas in the homes of the poor. Unfortunately, this would require great reductions in the prices of these fuels, perhaps through government subsidy. And it would be feasible only if the waste and comparatively frivolous uses of energy in the industrial countries, which are depleting petroleum reserves so quickly, were very sharply reduced.

Alternative energy sources now being considered in industrial nations need efficient, inexpensive models for the developing nations. For example, an economical device that could cook dinner with solar energy collected earlier in the day would be ideal. Water power, where it is available, is another alternative.

Indian scientists have experimented for decades with a device that breaks down manure and other organic wastes into methane gas for cooking and leaves a rich compost for the soil. The device works well, but it costs too much for the rural villages where the fuel problem is mushrooming.

Most of all, the inexorably growing need for firewood demands more massive tree-planting programs than most government officials have contemplated. Somehow, the suicidal deforestation of Africa, Asia, and Latin America must be reversed. [Erik P. Eckholm]

Cooling Fuel to Clean It

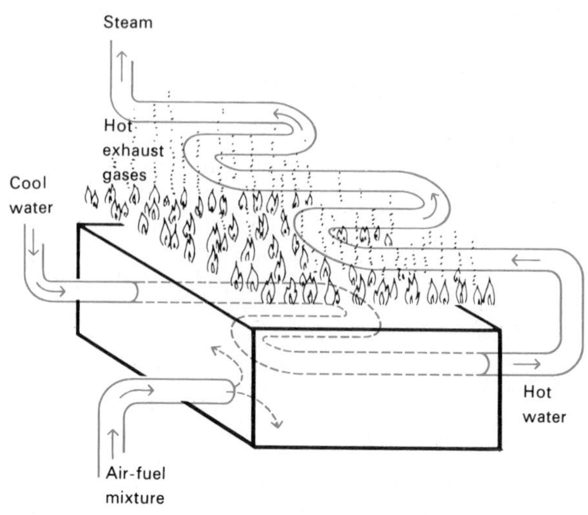

A transpiration burner's flat flames emit few pollutants because they are cooled by water that is piped through the fuel-air mixture beneath them. The warmed water is heated further in coils passing through exhaust gases above the flames to produce steam for turning a turbine.

On a floating platform, warm surface water would be used to vaporize a liquid with a low boiling point, such as ammonia. The resulting "steam" would then turn a turbine. The generated power would be transferred to shore by cables. Some of the power would also be used to pump cold seawater up from the depths to condense the vapor for recycling.

Certain problems still exist, including finding economical building materials that would not corrode in salt water and protecting the platforms from storms. Nevertheless, TRW, Incorporated, of Cleveland, which conducted a one-year study of the concept, urged Congress to fund a $1-billion program for building an experimental platform that would begin generating power by 1980.

A variation of this concept was patented in December 1975 by chemical engineer Donald F. Othmer of the Polytechnic Institute of New York. Othmer's design used solar panels to heat surface seawater to boiling, thus producing steam to turn turbines. The steam is condensed at low pressure by cold water pumped up from the depths.

Tornado in a tower. A Grumman Aerospace Corporation researcher's scheme to generate electrical power from artificial tornadoes was reported in October 1975. The tremendous power of tornadoes results from high winds and low atmospheric pressure at the center of the swirling wind column. The low atmospheric pressure gives tornadoes the power to lift automobiles and uproot trees.

Grumman researcher James Yen designed a circular tower for accelerating the speed of ordinary winds to tornado force. In theory, the wind enters through vertical vents near the top of the tower – but only on the side from which the wind is blowing. All other vents are closed. The wind then swirls inside, gathering speed, and low pressure develops at the center of the tower. The low pressure sucks outside air in at the bottom. As it is pulled toward the top of the tower, the air turns a turbine generator's blades at the bottom of the tower. [Darlene R. Stille]

Environment

A three-year study by the Ford Foundation, released on Feb. 27, 1976, concluded that far more of the public's general ill health is caused by injuries or exposure to hazardous materials on the job than had been thought.

An earlier study conducted in 1975 by University of Washington researchers for the National Institute for Occupational Safety and Health reported that a survey of 908 workers in two states revealed 346 cases of probable occupational disorders. These included respiratory conditions, skin disorders, toxic symptoms, and eye diseases.

But federal efforts to identify hazardous materials in industry bogged down. The National Cancer Institute (NCI) admitted on March 29, 1976, that it had been testing 150 chemicals suspected of causing cancer on the job. However, the institute said its staff was unable to complete the testing and release the results. Other federal agencies responsible for enforcing occupational and environmental health standards have relied in the past on the NCI's findings to ban a number of chemicals from industrial use or impose strict limits on exposure.

Charging that NCI bureaucrats had starved his program of funds and staff and brought it to the point of collapse, Umberto Saffiotti, head of the NCI effort to identify potential cancer-causing agents, resigned in May 1976.

One chemical, identified by NCI on April 8, 1976, as a cancer-causing compound in tests with both rats and mice, was Kepone, a pesticide used in 40 household products to control ants, roaches, and other pests. Kepone was manufactured at a small plant in Hopewell, Va., mostly for export. The plant was closed in July 1975 when doctors found that workers exposed to high levels of Kepone developed tremors, blurred vision, loss of memory, and other disabling symptoms.

A grim bicentennial passed almost unnoticed during the year. The Rachel Carson Trust noted that Sir Percival Pott first established in 1775 that an environmental agent, soot, caused cancer among English chimney sweeps. In 1975, the cancer death rate reached

Wildlife expert Gerald Craig places young prairie falcons in a nesting place on a Colorado cliff. The birds, bred in captivity, will help rebuild the dwindling population.

record levels in the United States, and it was increasing at a rate of about 3 per cent per year. According to statistics released by the federal government on March 1, 1976, the growth rate is at its highest since World War II. Scientists do not clearly understand the rapid rise in cancer mortality, but they believe that most cancers are caused by such agents as diet, cigarette smoke, and industrial chemicals.

On Dec. 8, 1975, the three winners of the 1975 Nobel prize in physiology or medicine blamed environmental factors for causing up to 80 per cent of human cancer. Cancer researchers David Baltimore of the Massachusetts Institute of Technology, Renato Dulbecco of the Imperial Cancer Research Fund Laboratory in London, and Howard M. Temin of the University of Wisconsin believe that about half of these factors can be eliminated.

Good-by red dye. The Food and Drug Administration (FDA) banned the use of amaranth, called Red Dye No. 2, on Feb. 12, 1976, because its safety has not been proved. The dye, used to color foods such as soft drinks, candy, ice cream, and some meats, is one of the most widely used coloring agents. The FDA ban came after experiments seemed to indicate that the dye, in high doses, caused cancer in laboratory animals.

In 1970, two reports from Russia claimed that amaranth caused cancer and harm to reproductive systems in animals. There have since been a number of studies attempting to settle the question of Red Dye No. 2's safety. After reviewing the available evidence, the FDA concluded that, although the evidence that Red Dye No. 2 causes cancer in female rats is still uncertain, the safety of the chemical had not been proved. Existing products with the coloring may be sold, but the dye may no longer be added to foods, drugs, or cosmetics in the United States.

Plutonium stand-off. The Nuclear Regulatory Commission (NRC) has not yet decided whether plutonium-fueled reactors will be permitted to operate. However, the Energy Research and Development Administra-

An experimental plastic dam 5 meters (16 feet) high and 60 meters (197 feet) wide blocks a channel in Italy's Po River Delta. It can be inflated or deflated to control the water level. Engineers hope to install three of these dams to block the high tides and floods that endanger Venice.

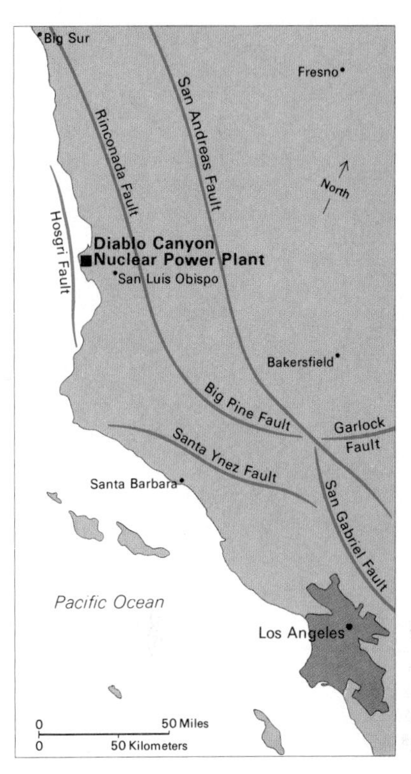

The Diablo Canyon nuclear power plant on the California coast, *above*, was designed to withstand strong tremors on the Rinconada and San Andreas faults. But the Hosgri Fault, recently discovered 2½ miles away, may prove a greater hazard.

tion (ERDA) at the end of 1975 approved an overall plan for developing the plutonium-fueled Liquid Metal Fast Breeder Reactor (LMFBR) as the centerpiece of U.S. energy development.

ERDA approved a final environmental impact statement on Dec. 31, 1975, describing the LMFBR program as environmentally acceptable. The program will cost at least $10 billion of federal funds. See SCIENCE POLICY.

The Natural Resources Defense Council, which has opposed the use of plutonium and filed the lawsuit that compelled the production of ERDA's impact statement, went to court to block construction of the prototype LMFBR planned for Clinch River, Tennessee. The environmental group released an internal NRC memorandum on Jan. 19, 1976, conceding that present safeguards against the theft of plutonium by criminal or terrorist groups were inadequate.

The nuclear power fight emerged as a major political issue in several parts of the United States in 1975 and 1976. In California, a coalition of citizens' groups forced a statewide vote on nuclear power on June 8, 1976. California is one of 22 states that allow voters to decide for or against specific pieces of legislation, and a petition drive undertaken in 1975 put Proposition 15 before the voters. This complex measure would have cut the output of California's three operating nuclear reactors and banned all new ones until the state legislature decided that their safety systems were effective. The legislature was also to judge on the effectiveness of radioactive-waste disposal.

Proposition 15 might have effectively halted nuclear power development in the state. Efforts to pass it were led by David Pesonen, an attorney who directed earlier campaigns against specific power plants. Pesonen's groups had previously prevented the construction of two nuclear power plants in areas threatened by earthquakes. However, Proposition 15 was defeated in the statewide referendum by a 2 to 1 margin. Similar initiatives on nuclear safety were on the ballot in Oregon and Colorado in November. [Sheldon Novick]

Genetics

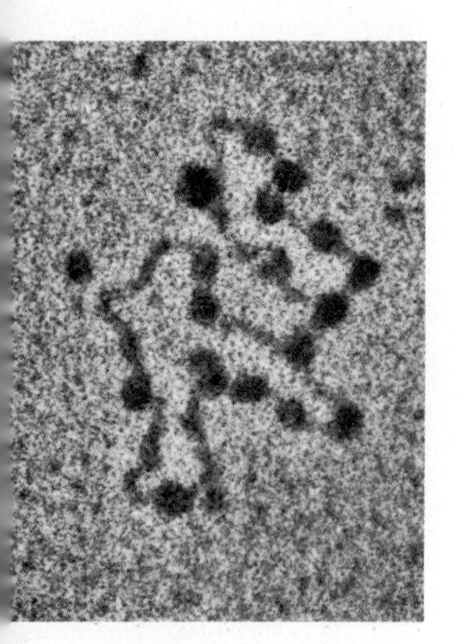

Minichromosomes from a virus that infects pigs, magnified 320,000X, look like strings of beads. Scientists think that all chromosomes assume this form when the cell is not dividing.

Biochemist Jack D. Griffith of Stanford University in California developed important evidence during 1975 supporting a recently proposed model of the molecular structure of the interphase chromosome. This chromosome form, called chromatin, is highly extended and threadlike. It is the form a chromosome takes in the nucleus of a cell when the cell is not dividing.

According to the model proposed in 1974 by Roger D. Kornberg of the Medical Research Council Laboratory in Cambridge, England, chromatin is organized like beads on a string. Each bead consists of a tiny bleb, or bubble, of a class of protein molecules called histones. They are associated with a short, coiled segment of the enormously long molecule deoxyribonucleic acid (DNA), which carries hereditary information in the cell. Small stretches of DNA between the coiled segments in the beads form the string that holds the beads together.

All chromosomes contain DNA, the long, double-stranded molecule whose celebrated double helix resembles a twisted ladder. The base pairs, whose sequence determines the genetic information in that molecule, form the crosspieces of the ladder. DNA generally has a negative electrical charge in the cell. The histones, composed of amino acid subunits, have an overall positive electrical charge because they contain large amounts of the positively charged amino acids lysine and arginine. The positively charged histones and the negatively charged DNA molecules seem to attract one another and form a tight DNA-histone complex.

There are five types of histones – lysine-rich H1, intermediates H2A and H2B, and arginine-rich H3 and H4. Chromosomes have about as much histone as DNA and, except for histone H1, the histones are all present in equal amounts. All histones except H1 seem to be involved in forming the chromatin beads of the Kornberg model. Kornberg proposed that each bead consists of eight histone molecules – two each of H2A, H2B, H3, and H4 – surrounded by about 200 base pairs of DNA. The length of the DNA string between adjacent beads is still unclear. Kornberg believes it is so short that the beads practically touch one another. Other investigators suggest that the bridge may be 40 or more base pairs long. Kornberg calculated each bead to be about 100 angstrom (A) units (0.00001 millimeter) in diameter, too small to be visible with a light microscope but large enough to be observed in an electron microscope.

The new evidence supporting the Kornberg model came from visual observations in the electron microscope and biochemical studies. Working with the electron microscope, Griffith examined the DNA of a monkey virus known as SV40. This viral DNA is circular, unlike the spiral DNA in higher organisms, but it is associated with the same kinds and amounts of histones as typical chromatin.

Observing these viral "minichromosomes" in the electron microscope, Griffith discovered that their structure is unmistakably like beads on a string. His measurements indicate that each bead is about 100 A in diameter, as Kornberg predicted, and contains about 170 base pairs of DNA, close to Kornberg's suggestion of 200. Adjacent beads are separated by about 40 base pairs of DNA.

Biochemical tests have been made by geneticists Dean R. Hewish and Leigh A. Burgoyne of Flinders University in South Australia. Hewish and Burgoyne extracted chromatin by chemical methods and then exposed the chromatin to an endonuclease enzyme. Endonuclease cuts DNA molecules at numerous places and digests the fragments into small pieces, except where the DNA molecule is bound with protein and thereby protected from the enzyme. The undigested pieces of DNA formed a regular series in such lengths as about 200, 400, 600, and 800 base pairs long. Thus, the 200 base pairs of DNA within each chromatin bead evidently resist endonuclease digestion, and the increasing lengths of undigested DNA found by Hewish and Burgoyne correspond to the length of DNA in a single bead, or in two or more adjacent beads.

The beadlike structures have been observed by other researchers as well. Ada L. and Donald E. Olins of the Oak Ridge National Laboratory in Tennessee found them in prepared chromatin. C. L. F. Woodcock of the University of

Lemur paw prints are being taken and studied by scientists at Duke University. The prints will be examined for significant patterns in a genetic study on primates. Researchers hope to correlate the print patterns with chromosomal variations.

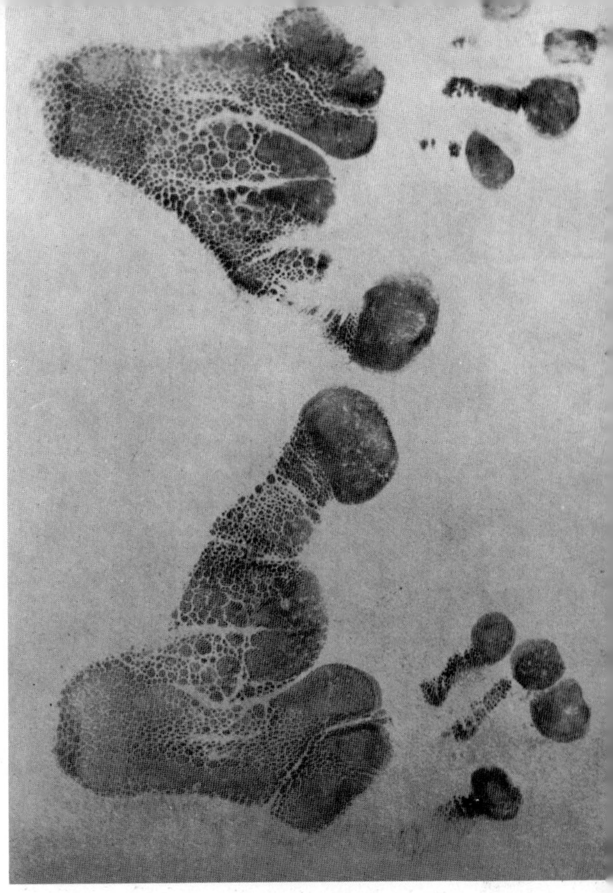

Massachusetts in Amherst calls them "ν-bodies." The term is a pun; ν is the Greek letter *nu,* pronounced *new.*

Other important biochemical evidence comes from test-tube studies of isolated histones. Biochemist Joseph A. D'Anna, Jr., and biophysicist Irvin Isenberg of Oregon State University in Corvallis studied the interactions between pairs of isolated histones. They found that the pairs of histones H3 and H4, H2A and H2B, and H2B and H4 form stable complexes. Similar studies by Kornberg and Jean O. Thomas at Cambridge in 1975 yielded comparable results. Geneticists now generally agree that histones H2A, H2B, H3, and H4 combine to form the chromatin bead with DNA.

The finding that histone H1 does not seem to form a complex with the others is very important. Indeed, histone H1 is unlike the other histones in a number of ways. Cells do not contain the same relative amounts of it as of the other histones, for example, and the histone H1 molecule has about twice as many amino acids as the other histones.

All these observations strongly support the Kornberg model, but that model is only for the structure of chromatin. Chromosomes change in structure throughout the life cycle of cells. In the earliest stages of cell division, the chromosomes begin to coil or fold into a progressively shorter and thicker form that gradually becomes visible in the light microscope. Just before their distribution into the new cells formed by division, the chromosomes are tightly coiled. This is the stage in which chromosomes are most easily studied with the light microscope. How the interphase chromosome becomes transformed into the chromosome seen in later stages of cell division is not known, but current opinion favors histone H1 as the agent responsible for the transformation.

The discoveries of the past year on chromosome structure have given scientists confidence that they are on the right track in their long attempt to discover what the chromosome – a structure first observed by microscopists in 1888 – really is. [Daniel L. Hartl]

Geoscience

Moon rock 76535, taken from Taurus Mountains landing site by Apollo 17 astronauts, was found to be at least 4.53 billion years old. It adds to the evidence of the solar system's age.

Geochemistry. A long-standing goal of geochemists – to find rocks that provide evidence corroborating the age of the solar system – may have been reached in 1976. Researchers working with rock sample 76535, taken from the Apollo 17 landing site in the Taurus Mountains on the moon, have found this evidence. The rock weighs 155.5 grams (5 ounces) and is a little larger than a ping-pong ball.

Up to now, geochemists have used only chondritic, or stone, meteorites to determine the age of the solar system, and an age of 4.55 billion years has been reliably established with them. The rocks of the earth have undergone many extensive chemical changes that erased their primitive records. But moon rocks have not. Although the moon has undergone chemical changes somewhat like those on earth, its history is much simpler.

Rock 76535 is a troctolite that consists primarily of olivine and plagioclase feldspar. The best evidence for its extreme age comes from the work of Dimitri Papanastassiou and Gerald Wasserburg of the California Institute of Technology (Caltech), based on the decay of rubidium into strontium.

In this system, the rubidium isotope of relative mass 87 decays with a half-life of 49 billion years to a strontium isotope of relative mass 87. If all the minerals in a rock crystallized at the same time and retained all their initial rubidium and strontium, the ratio of strontium 87 (from rubidium decay) to rubidium 87 should be identical in all of them. Another significant measurement is the ratio of strontium 87 to strontium 86 at the time the rock crystallized. This is because all the strontium 87 does not come from decay of rubidium, and strontium 86 is a stable isotope not involved in the decay.

The rubidium-strontium relations in four olivine and four plagioclase samples from rock 76535 indicate that the minerals crystallized about 4.53 billion years ago, close to the age established for stone meteorites.

Once we know the age of troctolite 76535, we can subtract the strontium 87 produced from rubidium decay and determine this rock's initial ratio of strontium 87 to strontium 86. That calculation indicates that the ratio in the troctolite minerals is nearly identical to the ratio previously established for the two isotopes of strontium in stone meteorites.

Because such a ratio has not been found in the chemically changed rocks on earth, and only rarely in those from the moon, the ratio provides further assurance of the great age of troctolite 76535. A fifth olivine sample in the troctolite did not show this ratio, which indicates a slight disturbance of the rubidium-strontium ratios in some of the rock's minerals.

Geochemists at the University of California, San Diego, made similar measurements on the same troctolite using the decay of radioactive samarium 147 to neodymium 143, and determined an age of about 4.53 billion years. They presented data for only the total rock sample, however, and determined the age by assuming that the neodymium would have had the same initial ratio of neodymium 143 to neodymium 144 that existed 4.55 billion years ago in stone meteorites.

Other data show clearly that the troctolite has undergone some change over its 4.53 billion to 4.55 billion years. Three laboratories – the State University of New York in Stony Brook, the National Aeronautics and Space Administration laboratory in Houston, and Caltech – reported the potassium-argon age is 4.3 billion years. Argon 40, the product of potassium decay, is a gas, so some of it could conceivably be lost under conditions that would not dissipate other, less volatile, elements.

Uranium-lead dating of the troctolite by Fouad Tera of Caltech indicated that the uranium-lead ratios in the rock were disturbed 4 billion years ago.

The record preserved in the troctolite is not ideal, but it is the best evidence yet found for lunar rock of an age that approximates that of meteorites. The findings of the age tests are not likely to stem from the random effects of chemical disturbances. We can assume that the uranium-lead changes did not seriously disturb the other decay systems. Scientists are still looking for other examples of primitive lunar rock, but troctolite 76535 may be the best specimen that will ever be found.

Oxygen isotope studies. Careful measurements by University of Chicago geochemists Robert Clayton, Naoki Onuma, and Toshiko Mayeda of oxygen isotopes in rock show that terrestrial, lunar, and some meteoritic rocks form a common rock family. Oxygen consists mainly of an isotope of mass 16, but contains 0.2 per cent of isotopes of mass 17 and 18. These two isotopes are important in spite of their scarcity because chemical processes tend to produce small changes in the amounts of all three isotopes. The percentage change in the ratio of oxygen 18 to oxygen 16 is twice that in the ratio of oxygen 17 to oxygen 16.

The Chicago scientists report that data indicates enstatite chondrites—stone meteorites with pure particles of iron—are in the same family as the terrestrial and lunar rocks. On the other hand, oxygen in the rare carbonaceous chondrites—stone meteorites that are rich in carbon, and whose iron is entirely in silicate minerals—differs markedly from oxygen in the earth and moon rocks. The differences indicate that carbonaceous chondrites have been enriched with pure oxygen 16.

Newer data for oxygen isotopes from other chondrites, the "ordinary" ones that constitute over 90 per cent of all stone meteorites, reveal still further differences. These stone meteorites are divided into two groups—H for higher total iron content and L for lower total iron content. The H group contains a larger percentage of iron in the metallic state than does the L group. Oxygen in the H and L chondrites contains less of the pure oxygen 16 component than the earth-moon oxygen. The H and L chondrites also differ slightly in the amount of pure oxygen 16 that they naturally contain.

These findings will undoubtedly change our concepts of the origin of the planets and other objects of the solar system in important ways. On the basis of chemical data alone, most scientists have maintained that the various kinds of meteorites could be explained by starting with matter like that in carbonaceous chondrites and chemically reducing various fractions of the iron to metal until all iron was in the metal phase in enstatite chondrites.

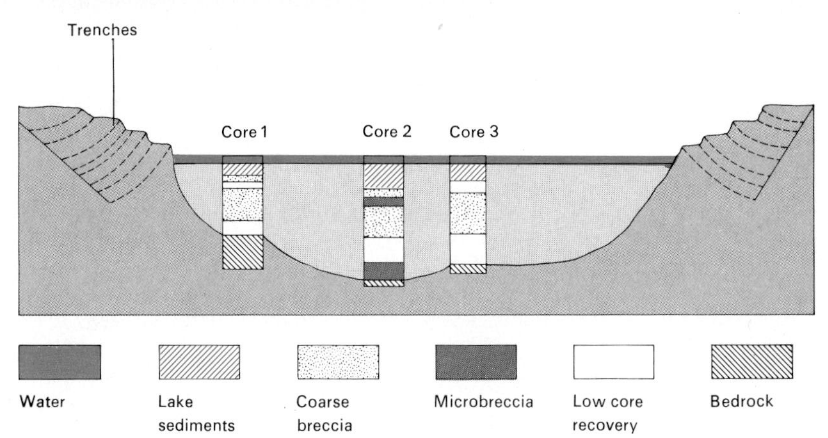

Lonar meteorite crater in state of Maharashtra, India, was studied by a joint Indian-American research team because it is the only major crater on earth comparable to impact craters on the moon. They excavated trenches around the rim and bored cores into the rock below the crater's lake to find evidence that a meteorite caused the crater.

The new oxygen-isotope data show that the overall process was not that simple. Although chemical reduction, or removal of oxygen and removal of iron, undoubtedly helped to produce the different kinds of stone meteorites, they were not derived by progressive reduction of silicate material in objects with the same oxygen isotopes. The earth and moon must have come from sources whose oxygen isotopic composition differed from those of meteorites.

Allende chondrite aluminum. Aluminum as we know it consists of a stable isotope of mass 27. When the solar system began, there was probably a second aluminum isotope of mass 26, but that isotope is unstable and decays with a half-life of 0.7 million years. So any aluminum 26 that might have been present when meteorites formed 4.55-billion years ago has long since decayed.

Aluminum 26 decays to a stable isotope of magnesium, also of mass 26. Geochemists assume that the unusual amounts of magnesium 26 in meteoritic matter could provide evidence of aluminum 26 when the material formed.

Caltech's Typhoon Lee and Papanastassiou reported such an abundance of magnesium 26 in the Allende carbonaceous chondrite that fell in 1969 in northern Mexico. Lee, Papanastassiou, and Wasserburg have now reported new data suggesting that aluminum 26 causes the magnesium abundance.

The Caltech scientists separated various silicate fractions from the chondrite and compared the amount of the magnesium 26 (using magnesium 24 as a reference isotope) to the aluminum-magnesium ratios of the silicates. Although the aluminum now consists entirely of aluminum 27, any aluminum 26 initially present would have behaved chemically exactly like the heavier isotope. The data positively correlates the abnormal amount of magnesium 26 and the aluminum-magnesium ratios of the samples, good evidence that the silicates once contained aluminum 26.

This is also an indication of the primitive age of the Allende silicates. The aluminum 26 could have been an important source of heat in the parent body because of the heat radioactivity supplied. Such a heat source could explain some chemical changes caused by heat in meteorites. [George R. Tilton]

Geology. A new look at uranium ore reserves in the United States has caused a drastic reduction in their estimates. In April 1976, Michael Lieberman of the Energy Resources Group at the University of California, Berkeley, applied methods developed for predicting oil resources to determine uranium resources in the United States. To the surprise of most geologists, he found that 87 per cent of the high-grade uranium that can be mined has already been discovered.

Uranium makes up only about 0.00016 per cent of the earth's crust. Natural processes have concentrated this element in a few ore deposits that are a thousand times richer than they would be if the ore was spread more evenly throughout the crust. These deposits provide most of the uranium for U.S. nuclear reactors and other needs.

By 1958, uranium exploration in the Western United States had been so successful that the price of uranium oxide was once as low as $3 per kilogram ($6 per pound). But the Middle East oil embargo of 1973 boosted the importance of uranium as an alternative energy resource and prices rose rapidly. The price is now nearly $20 per kilogram ($40 per pound). As one result, the Westinghouse Electric Corporation announced in 1975 that it could no longer fulfill long-term commitments for uranium to fuel Westinghouse-built reactors.

Evaluating the supplies. There are two different methods for evaluating the uranium potential, and scientists using one or the other are getting radically different answers. Vincent E. McKelvey, now director of the U.S. Geological Survey, developed the first technique. He divides the resources into deposits that are feasible to mine at present-day metal prices and lower-grade deposits that might be economically mined if the price increases. McKelvey further divides deposits into those that have been well explored, those that probably will be discovered, and those that may possibly be discovered in the future.

Using this method, a figure for the reserves is arrived at by asking a number of experts how much ore they believe is in each category, adding up their estimates, and dividing by the

Volcanic Mount Baker in Washington, in 1976, spewed from 10 to 100 times its normal volume of steam. Scientists studying the volcano hope to determine the cause of the activity and if there is any danger of an eruption.

number of experts. The Geological Survey has estimated the U.S. uranium supply in this way. The Energy Research and Development Administration has produced similar, though more cost-oriented, estimates that depend in part on the informed opinions of engineers and scientists actively working on uranium resources.

M. King Hubbert of the Geological Survey developed the second method for estimating resources. In 1956, he predicted that oil production on land in the continental 48 states would reach a peak about 1968 and decline thereafter. Hubbert used the historical trend of oil-finding success in the industry to make that startling prediction, and his methods were severely criticized. However, his predictions proved to be accurate. Hubbert later improved his technique, relying increasingly on historical trends instead of expert opinions. The Geological Survey abandoned the optimistic predictions made by earlier methods during the last year and now has published estimates that are close to Hubbert's.

Lieberman's analysis. Lieberman used Hubbert's methods to analyze uranium resources, taking the time uranium ores were discovered as the leading indicator. Because most surface uranium deposits were discovered in the 1940s and early 1950s, most uranium discoveries since that time have been made by drilling wells and lowering radiation counters into them. Lieberman therefore made the number of feet drilled his second indicator.

He measured the success rate by the pounds of economically minable uranium oxide discovered per foot of borehole drilled. As drilling has gone on, that success rate has fallen, because the most likely places have already been covered and the search turns to less likely locations. The dropping success rate looks mathematically much like the decay rate of radioactive atoms – every 15 million meters (50 million feet) of exploration cuts the success rate in half. If we assume that the same process will continue, it is easy to estimate how much uranium will yet be discovered by extending drilling plans into the future at a diminishing success rate. Lieberman showed that only 13 per cent of the total supply of minable uranium is yet to be discovered. An independent analysis by Ian MacGregor, a geochemist at the University of California, Davis, who used quite different data, has confirmed Lieberman's conclusions.

As a result of these studies, several leading mining companies have shifted their exploration in the United States to the search for low-grade uranium deposits, which will undoubtedly be mined in the future if the price of uranium continues to rise.

Fluid inclusions. Lincoln Hollister, professor of geology at Princeton University, made a special study in 1975 and 1976 of fluid inclusions – bubbles filled with fluids – in a variety of rocks that were once deeply buried. Mineralogists first noticed fluid inclusions more than a century ago in certain rock crystals. Improved microscopes later showed that fluid inclusions are quite complicated, containing liquid, gas, and several kinds of solid crystals.

Jewelers distinguish gemstones from one another and from synthetic gemstones by examining the fluid inclusions. Many natural emeralds, for example, contain complex fluid inclusions that identify individual stones.

One of the first geologists to use the information in fluid inclusions was Edwin Roedder of the U.S. Geological Survey. He examined the crystals in metal ore deposits. Presumably the fluid in the inclusion is a sample of the "juice" that created the ore deposit. Unfortunately, most inclusions are so tiny they can be seen only through the lens of a microscope.

Indirect techniques have been developed to study inclusions. Under a microscope, the sample is heated and cooled to determine the freezing temperatures of the fluid in the inclusion or the temperatures at which a gas bubble appears or disappears. For instance, salt lowers the temperature at which ice freezes and so the amount of salt in a water-filled inclusion can be estimated by its freezing temperature.

Roedder's work on metal ore deposits showed that the inclusions often contained salt water. Other laboratory studies showed that metal chlorides, such as zinc chloride and lead chloride, were soluble enough at high temperatures to account for the high concentration of metals in ore deposits. In other words, the salt in the fluid inclusions was telling us that it was furnishing the chloride carrier that transported the metal into an ore deposit.

The temperature at which a crystal formed can also be determined from fluid inclusions. If an all-water inclusion is formed at high temperature, the water shrinks slightly as the crystal cools and the fluid inclusion then consists of a small vapor bubble and liquid. Skilled observers have become so proficient that they can look briefly at a sample through the microscope and estimate the temperature with reasonable accuracy.

Hollister and his students have been studying the fluid inclusions in ordinary rocks rather than those in the crystals of ore deposits. They hope to learn more about the origin of the rocks, but this is difficult work. The crystals in ordinary rocks are much smaller and less perfect. Also, most of them have been cracked many times during their long history, producing new fluid inclusions as the cracks healed. But by searching patiently, Hollister has found a few isolated fluid inclusions that date from the time the rock was formed.

These isolated inclusions tell an interesting story. In rocks that have been buried at high temperatures and pressures deep in the continental crust, many of the fluid inclusions are rich in carbon dioxide instead of water. No one had suspected that carbon dioxide was so common at great depths, and geologists are now reconsidering theories of the conditions that existed when the rocks formed.

The work with fluid inclusions goes rather slowly. It takes time to heat and to cool each inclusion under the microscope while observing it carefully. The amount of material that has been studied is also quite small. The contents of all the fluid inclusions studied in this way by all the scientists working on them since the microscope was invented would probably add up to less than a teaspoonful of liquid. Yet the studies have produced important information about the formation of rocks in the continental crust. [Kenneth S. Deffeyes]

Geophysics. The nature of the earth's interior received the increasing attention of geophysicists in 1975 and 1976. They concentrated particularly on the asthenosphere, the region of hot, soft rock on which the lithospheric plates that form the earth's crust move about. The lithospheric plates are presumed to vary in thickness from a few kilometers beneath the axis of the mid-ocean ridge system to several hundred kilometers under the continents.

Francis Boyd of the Carnegie Institution of Washington's Geophysical Laboratory and Peter Nixon of the Geological Survey of Lesotho in Africa produced a model of the lithosphere in 1975. The model was based on an analysis of rock fragments from deep within the earth that were blasted to the surface through the Kimberlite pipes. These natural subsurface columns of rock fragments contain most of the South African diamonds.

The scientists' model suggests that the upper zone of the lithosphere has no basaltic material while the lower zone is filled with basaltic material. The base has a region of sheared lherzolites, or pyroxene rock, and large rock masses containing garnet and pyroxene are beneath this. The sheared zone appears to correspond with temperature changes that Boyd and Nixon estimated would be caused by the shearing process in the underground rocks.

The Boyd-Nixon model depends on accurate methods of determining the temperature and pressure of the underground mineral assemblages. At an International Conference on Geothermometry and Geobarometry in October 1975 at the State University of Pennsylvania, some geophysicists raised doubts about the temperature estimates. Conference participants concluded that the Boyd and Nixon methods of determining the temperature were not accurate enough to pinpoint the temperature changes they suggested. However, the structure and types of rock levels in their model seemed reasonable.

The asthenosphere boundary. The Pennsylvania conference was followed in November by another in Vail, Colo., sponsored by the Geological Society of

Continental Collisions Formed Asia

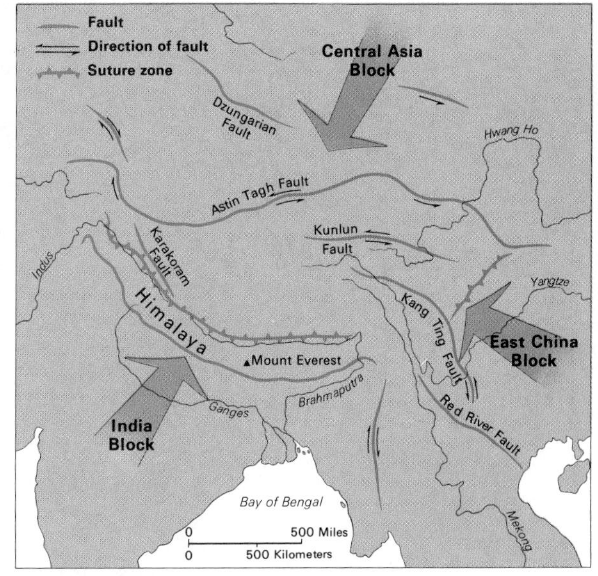

Source: P. Molnar and P. Tapponier, Science, © 1975 American Association for the Advancement of Science.

Himalayan range, upper left to lower right, seen from Apollo 7 spacecraft, is one result of the collision of three continental blocks that, according to other geophysical evidence, formed the continent of Asia.

America. This conference focused on the unique nature of the lithosphere-asthenosphere boundary – the interface between the surface plates and the material on which the plates move.

The problem is how to define this boundary. The asthenosphere has been defined as the region in which rock is so hot that it will flow. But scientists cannot determine how deep in the earth it is when it becomes this viscous.

In a simple model of plate tectonics, geophysicists often assumed that the boundary was at the region in the earth's mantle where the velocity of seismic waves temporarily slow down. They assumed that low velocity was associated with low viscosity, perhaps even with partial melting. But theoretical and experimental data now suggest that low velocity and low viscosity are not necessarily synonymous.

Further complications arose from studies of earthquake shear waves that are reflected from the core of the earth. Geophysicist Thomas Jordan of the Scripps Institution of Oceanography in San Diego, Calif., reported that the delay times for such waves are significantly great when compared with the times that are estimated for a homogeneous earth. Jordan's data also suggest that variations in the structure of the crustal material under continents and under oceans extend at least to depths of 450 kilometers (280 miles) and perhaps 650 kilometers (400 miles).

Another complication is the presence of rock material at depths of 50 to 80 kilometers (30 to 50 miles) in the lithospheric plates under the oceans that permits seismic waves to travel at extremely high velocities. Some observers suggest these high velocities may be due to anisotrophy – an orientation of the crystals of minerals such as olivine. Others believe the lower lithosphere may be greatly enriched in garnet. If either theory is true, current notions of the chemical and physical evolution of the lithosphere must be revised.

Seismic reflection experiment. The Consortium for Continental Reflection Profiling (COCORP) conducted a seismic reflection study in 1975 on the structure of the underground Rio Grande graben, or rift valley, south of Albuquerque, N. Mex. The seismic waves projected underground were reflected from depths of 40 to 50 kilometers (25 to 30 miles). Preliminary examination of the data obtained suggests that the graben's boundary faults can be traced even at great depths. Since the energy returns at these depths were so strong, it was suggested that the seismic waves must have reached a magma chamber, or source of molten rock. Such a magma chamber had been proposed on the basis of other data by Alan Sanford of the New Mexico School of Mines. These data may yield not only interesting details of geologic history and structure but also information on mineral and energy resources.

Deep rock measurements. The U.S. Geodynamics Committee in 1976 was supervising the measurements being taken in a deep well that the Amoco Production Company and Shell Oil Company were drilling in Gratiot County, Michigan, in the rocks of the Michigan Basin. In their search for oil, the drillers were trying to reach the crystalline basement rocks beneath the Precambrian rocks.

The measurements may furnish information about the dynamics of vertical movements in the basin rocks and their relationship to major horizontal movements of the lithospheric plate of which they are a part. The gravity, thermal, stress, and other geologic data from this study should also contribute to understanding the basin dynamics.

Ocean-floor studies. The U.S. submersible *Alvin,* which participated in the 1974 French-American Mid-Ocean Undersea Study (FAMOUS) exploration of the Mid-Atlantic ridge, conducted a series of dives in February 1976 in a spreading center of the Cayman Trough in the Caribbean Sea. It made 15 dives along two 1,800-meter (6,000-foot) scarps on either side of the ridge center.

The crew examined these steep slopes and collected samples at 70 points. The uppermost rocks were pillow lavas, representing the uppermost rocks of the oceanic crust, and beneath these were gabbros, or coarse-grained igneous rocks, showing the effects of heat and movement over undisturbed gabbros. Serpentine, peridotite, and dunite that were found on the western side may represent the types of rocks of the oceanic mantle. [Charles L. Drake]

Geoscience

Continued

Paleontology. Studies of parts of the skeleton of an ancient mammal found in 1975 in Mongolia have advanced our knowledge of the evolution of birth mechanisms among early mammals. Zofia Kielan-Zaworowska of the Polish Academy of Sciences discovered the fossil bones of the ancient mammal *Barunlestes butleri* in the Barun Guyot Formation of rocks in the Gobi Desert. These rocks date to the late Cretaceous Period, about 70 million years ago.

Paleontologists determined that this mammal, like most modern mammals, was a placental animal – its young developed completely inside the mother's body before birth. Certain characteristics of the animal's teeth and skull are found only among placental animals. However, the pelvic region showed that *B. butleri* also had prepubic bones, a pair of struts projecting forward from the pubis, which are found today only in egg-laying mammals such as the platypus and spiny anteaters, and in marsupials such as the kangaroo. No living placental animals have such prepubic bones.

The prepubic bones in *B. butleri* suggest that the evolution of reproduction processes and skeletons were proceeding at different rates in early mammals. Some Cretaceous species appear to have developed the teeth and skull of placental animals while retaining more primitive reproductive methods. By the early Tertiary Period, 60 to 65 million years ago, placental mammals no longer had prepubic bones.

Trilobite eyes. Recent studies by Euan N. Clarkson at the University of Edinburgh in Scotland and Ricardo Levi-Setti of the Enrico Fermi Institute at the University of Chicago cast doubt on the belief that some trilobites could not see sharp images.

Trilobites were small prehistoric sea animals with segmented bodies encased in a horny exoskeleton, or shell. Scientists have collected fossils of them dating from the early Cambrian Period, some 600 million years ago, to the late Permian Period, about 225 million years ago, when they became extinct. The trilobite's eyes are the most ancient visual system known.

"Some of these youngsters have come up with a terrific new idea—feathers."

A 100-million-year-old fossil angiosperm leaf was found in a central Kansas rock. The leaf's well-preserved details gave paleontologists much new information about the growth of the fruit-bearing plants during that period.

Most of the trilobites had eyes that were a multifaceted series of lenses – compound eyes similar to those of today's flies and other insects. They contained up to 15,000 small, closely packed prisms covered by a single cornea. Through such eyes, trilobites probably received a mosaic image as light fell on the underlying retina. However, several species of trilobites had schizochroal eyes, which had large, thick biconvex lenses under individual corneas. Until these studies, scientists thought schizochroal eyes could only detect the movement of large objects, and then see them only indistinctly.

But the microscopic studies of the schizochroal eyes of trilobites by Clarkson and Levi-Setti show that these biconvex lenses were double-element lenses. The outer portion was made of calcite, but the composition of the inner element has yet to be determined. The complex interface between the two elements formed a distortion-free lens similar to the Cartesian oval optical lens designed by French scientist René Descartes in 1637. It was first constructed by Dutch astronomer Christian Huygens in 1690 to eliminate spherical aberration in optical lenses.

Clarkson and Levi-Setti believe that the trilobites' schizochroal eye structure enabled them to form sharp, undistorted images, and also to concentrate light in the dark sea. Far from being a poorly adapted structure, the schizochroal eye may have been an early experiment of nature in converting the external multifaceted visual system of these animals into one in which a more complete image was projected onto the retina. If so, it represents the first step in the independent evolution of a type of eye found today in higher animals.

Fossil leaf-mining. A fossil example of leaf-mining was discovered in August 1975 by Leo J. Hickey of the Smithsonian Institution and Ronald W. Hodges of the United States Department of Agriculture. Leaf-mining is the process by which insect larvae bore tunnels between the upper and lower skins of leaves as they eat the inside of the leaf. The paleontologists found the fossil in the Sheridan Pass area of western Wyoming, near Dubois. The rock layers it came from were originally silt deposits in a swamp formed 55 to 40 million years ago during the Eocene Epoch.

The fossil leaf was of the genus *Cedrela,* a member of the family Rosidae. The tunnel in the leaf contained evidence of loosely packed material excreted by the burrowing larva. The tunnel led to a terminal chamber where the larva presumably entered its pupal stage. The structure of the leaf mine was similar to that produced today by the insect genus *Phyllocristus.* This insect, found in the subtropical areas of Central and South America, mines the leaves of the *Cedrela.*

Fossils and plate tectonics. For more than 100 years, paleontologists and zoogeographers have assumed that land connections existed between western Europe and North America at various times during the Cenozoic Era, which began 65 million years ago and extends to the present. The theories were based on the close resemblance among mammal fossils on the two continents during these periods, particularly during the early Eocene Epoch some 50 to 55 million years ago.

Scientists assumed that an overland route must have extended from western Europe across Asia and eastward to North America by way of what is now the Bering Strait. However, many of them stressed that there were closer affinities between the animals of America and those of western Europe than existed between either group and Asian animals during the early Eocene.

Paleontologist Malcolm C. McKenna of the American Museum of Natural History in New York City has now postulated a direct land connection between America and Europe across the North Atlantic during the Eocene Epoch. In a recent review of the fossil and geologic data, McKenna applied plate tectonic theories of continental reconstruction to support his theory that a trans-Arctic land mass existed between what are now Greenland and Norway. When the Atlantic Ocean rift extended northward about 45 to 50 million years ago, it interrupted this corridor and created a major barrier to continued animal migration. At this time, the fossil record shows distinct differences began developing in the animal species because of their isolation from each other. [Vincent J. Maglio]

Immunology

The potential value of a particular type of anti-antibody surfaced in 1975 and 1976. An antibody is a protein produced by a type of lymphocyte – a key cell in immunological reactions – when a virus, bacteria, toxin, or other intruder enters the system. Antibodies are very specific – that is, they have an active site that recognizes and attacks only the intruder to which the lymphocyte originally responded. The specificity of the active site is determined by its shape, which fits surface molecules called antigens on the intruder just as a key fits a lock.

We have known for some time that the active site of antibodies is itself an antigen to animals other than the one in which the antibody is formed, and any animal injected with another's antibodies will eventually form antibodies against their active sites. These anti-antibodies that recognize only the active-site portion of an antibody molecule are called anti-idiotype antibodies.

Anti-idiotype antibodies. Hans Binz and Hans Wigzell of the Uppsala University Medical School in Sweden reported in November 1975 that they used anti-idiotype antibodies to obliterate certain specific immunological reactions. The scientists produced these anti-antibodies by repeatedly injecting lymphocytes from one strain of rats (strain L) into hybrids of that and another strain (strain L × strain DA).

Because they are genetically half strain L, the hybrid rats' immune system found no foreign antigens on the injected lymphocytes, so they did not react against them. On the other hand, the injected lymphocytes reacted immunologically against DA components, that part of the hybrid rats' cells that they recognized as foreign, and some of these lymphocytes produced anti-DA antibodies. This is part of what is known appropriately as a graft-versus-host reaction. The immunological systems of the injected rats mounted a counterattack, forming antibodies against the specific anti-DA antibodies.

Binz and Wigzell went on to purify these anti-idiotype antibodies from the hybrid rats and used them in a series of fascinating experiments. For example, they chemically attached the anti-antibodies to plastic beads and packed the beads in columns into which they poured a liquid containing lymphocytes from strain-L rats. The lymphocytes that passed through the columns were tested for anti-DA activity and found to have none. Any lymphocytes with anti-DA activity had been retained in the columns by the anti-antibodies, which recognized the lymphocytes' specific anti-DA capacity. The passed lymphocytes could mount reactions against all but DA antigens.

Binz and Wigzell are now trying to induce strain-L rats to produce anti-idiotype antibodies against DA antibodies to see if this will abolish the rats' capacity to recognize and attack DA tissue. The work has obvious possible future implications for patients undergoing graft and transplant operations.

In a related experiment, Binz and Wigzell revealed an important fact about two types of lymphocytes – B cells, which produce antibody, and T cells, which react to antigens in at least two important protective ways. The scientists isolated rat B cells and T cells to which the same specific anti-antibodies reacted. Next, they found that the B cells absorbed all the anti-antibodies that could react with the T cells and vice versa. This implies that both types of lymphocytes recognize a DA component in the same way, a finding that is extremely relevant.

While there is solid evidence that B cells recognize a specific antigen by way of antibodies in their outer membrane, the nature of the T-cell receptor, or antigen-recognition molecule, is not clear. Now it seems possible that some T cells use the same recognition unit as do B cells, but it may be part of a molecule that is not a typical antibody.

This hypothesis was supported by experiments performed by Klaus Eichmann, Klaus Rajewsky, and their colleagues at the University of Cologne in West Germany. These investigators reported in April 1976 that they injected into mice a component of *Streptococcus* bacteria as an antigen, causing the mice to generate specific antibody. Then the scientists isolated this antibody and injected it into guinea pigs, which responded by making anti-idiotype antibody. When the researchers isolated this anti-antibody and injected it into normal mice, it stimulated the formation of antibody to the specif-

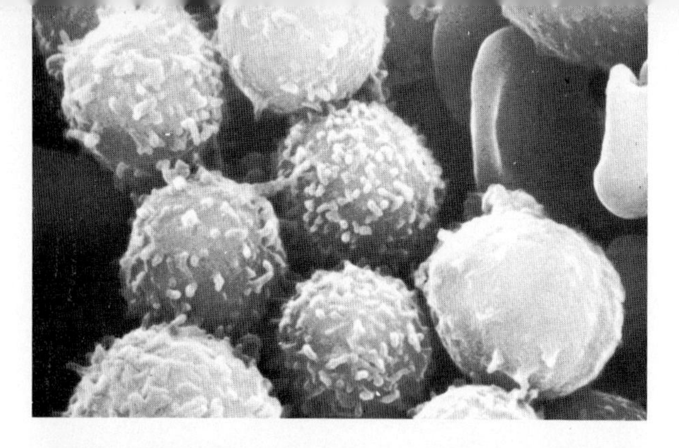

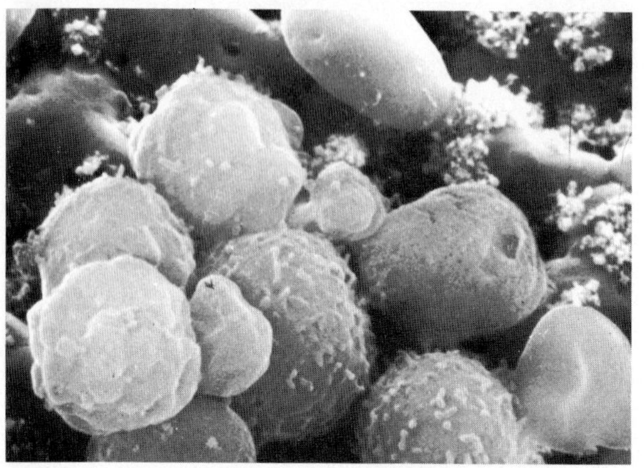

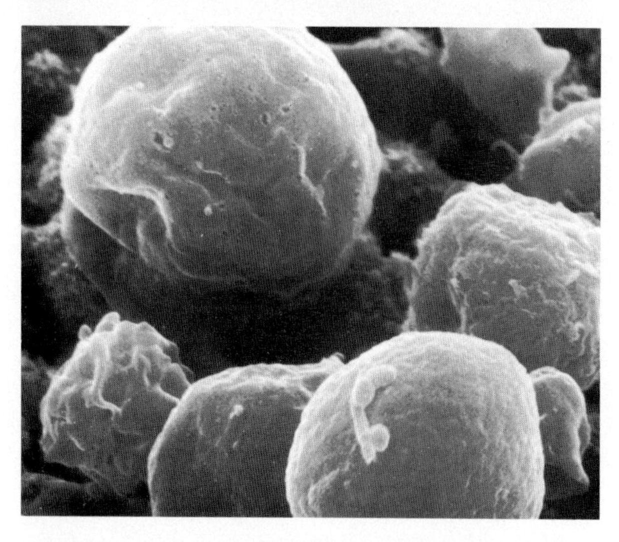

Lymphocytes, key cells in the immune response, *top,* include many with tentaclelike villi. These cells, from a normal mouse, are probably B cells, which make antibodies. Ten days after a mouse is injected with a leukemia virus, most of its lymphocytes have a relatively smooth surface, *center.* After 30 days, *above,* nearly all of them are grossly abnormal, with holes and a spongy texture. These changes may be related to markedly lowered immune responses that accompany leukemia.

ic component of *Streptococcus* much like immunization with the bacteria themselves. In other words, specific lymphocytes in the normal mouse could be triggered to form antibody against the bacterial component either because the lymphocyte receptors recognized antigen or because the anti-antibodies recognized their receptors. Furthermore, the Cologne researchers found that the same anti-antibodies not only induced B cells to produce antibody, but also induced T cells to react.

T-cell subclasses. It became clear during the year that two and possibly three subclasses of T cells can be identified by different antigens present on their cell surfaces as well as by their different immunological functions. Edward A. Boyse of the Sloan-Kettering Institute for Cancer Research in New York City and Harvey Cantor of the Harvard Medical School in Boston reported in June 1975 that they have clearly linked genetically determined T-cell surface molecules with specific T-cell functions in mice. Boyse and his colleagues had showed these antigens to be present only on lymphocytes – hence their name, Ly antigens. Of the eight or nine Ly antigens that have been identified, Ly 1, Ly 2, and Ly 3 are present only on T cells, but all three seem to occur simultaneously only on more primitive T cells. As T cells differentiate in the maturing mouse, some of them apparently begin to express either Ly 1 or both Ly 2 and 3.

According to Cantor and Boyse, those T cells that express only the Ly 1 antigen can function either as helper T cells – so named because their presence helps B cells produce antibodies – or in certain types of delayed allergic reactions. On the other hand, T cells that express both Ly 2 and Ly 3 antigens can become killer cells, lymphocytes that attack foreign tissues or foreign organisms directly. Other T cells with Ly 2 and Ly 3 function as suppressor lymphocytes, retarding the production of antibodies. This function may be important in regulating immunity.

More recent information suggests that suppressor T cells may actually be a subclass of T cells distinct from killer T cells. Clearly, we are merely at the beginning of lymphocyte subclassification. [Jacques M. Chiller]

Medicine

Dentistry. Researchers in Finland reported in August 1975 that xylitol, a chemical sometimes used as a sugar substitute, seems to inhibit the growth of bacteria on teeth.

In tests at the University of Turku, Arje Sheinin and his associates compared the results of using chewing gum containing xylitol with those of using chewing gum containing sugar.

Subjects who chewed gum containing sugar developed the most dental plaque – the thin film of food particles and bacteria that leads to tooth decay – and those using xylitol gum developed the least. A control group that chewed no gum had an intermediate amount of plaque. This indicated that xylitol has a direct effect on the bacteria causing plaque.

In another study, the same researchers used chewing gum to study the effects of xylitol on tooth decay. They randomly assigned 102 dental and medical students who averaged 22 years of age to xylitol and sugar groups. The students chewed between three and seven sticks of gum per day.

Both groups were checked for caries after six months, and again 12 months after the study began. The results clearly showed that sugar causes dental decay, as expected. However, the xylitol group had far less decay than expected.

Bacteria and tooth loss. Michael G. Newman and Sigmund S. Socransky of the Forsyth Dental Center and the Harvard School of Dental Medicine in Boston have implicated bacteria in periodontosis, a gum disease that destroys tooth-supporting bone in adolescence, in research reported by the National Institute of Dental Research in October 1975. Because little dental plaque or inflammation is associated with this disease, scientists had not previously suspected bacteria.

The two dental researchers recovered and cultured bacteria from the sites of periodontosis that could not be found in healthy areas of the patient's mouth. They later infected healthy rats with these organisms and produced severe loss of tooth-supporting bone. Researchers hope for an antibiotic treatment for the disease. [Paul Goldhaber]

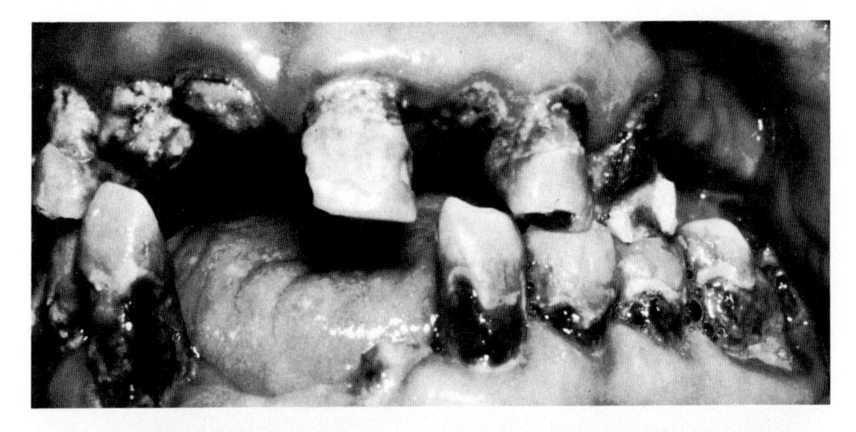

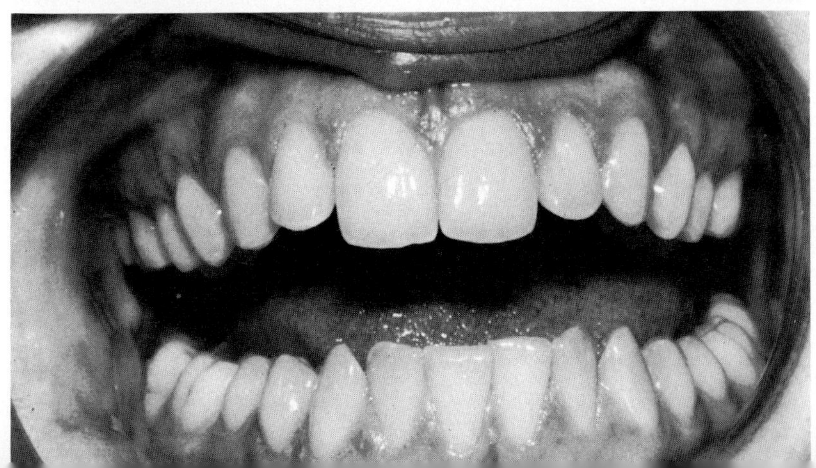

Rampant tooth decay, *top right,* often follows radiation therapy for head and neck cancer. But a new prevention program, using fluorides and other chemicals, keeps a victim's teeth decay-free, *bottom right.*

Internal Medicine. A virus or virus-like agent associated with multiple sclerosis (MS) was reported by Werner and Gertrude Henle and their associates at Children's Hospital in Philadelphia in December 1975. The Henles demonstrated that the blood of MS patients and most of the people who had close contact with them contained antibodies to this infectious agent.

MS is a disabling disease that frequently causes death. The disease first destroys the tissue surrounding the nerves and then the nerves themselves. It produces weakness, loss of muscle control, and speech disturbances. The variable and changing course of MS has frustrated efforts of investigators seeking its cause and an effective treatment. During the past year, however, a line of experimental research begun by Dr. Richard I. Carp at the Institute for Research on Mental Retardation in Staten Island, N.Y., has provided strong evidence that an infectious agent causes the disorder.

Carp first reported in 1972 that material taken from nine MS patients seemed to have a profound impact on the white blood cells of mice inoculated with the material. Extracts of brain tissue from three patients who had died of the disease and tissue from the spleen of another, as well as blood and cerebrospinal fluid from these patients, had been injected into the brains or abdominal cavities of the mice.

The number of polymorphonuclear neutrophils – a kind of white blood cell – declined remarkably while the percentage of lymphocytes, another kind of white cell, increased within 16 to 48 hours. These changes in the otherwise healthy mice persisted for 11 months. Perhaps even more impressive, the effects on white blood cell production could be repeatedly reproduced by injecting material obtained from the inoculated mice into other animals. Similar changes in white blood cells did not appear with injections of normal human brain or spleen tissue. The implication of these observations is that MS patients harbor an infectious agent with viral characteristics, such as the ability to replicate itself in animal cells.

"You've got whatever it is that's going around."

Medicine

Continued

An experimental sonar device that sends out inaudible sound waves which come back as audible echoes may help Dennis Daughters, 8 months old and blind since birth, to "see" as a bat does.

Carp expanded his observations by working with cell cultures and showed that the agent that depressed cell growth in test tubes could be transmitted from one cell culture to another just as it had been transmitted in the mice. Furthermore, when material from the fluid bath of these cells was injected into mice, the polymorphonuclear neutrophils were reduced as expected. These studies demonstrated that 80 per cent of tissue material taken from 71 MS patients contained the infectious agent. No material taken from healthy persons produced such results.

The Henles reproduced all of Carp's work and extended it. For instance, they found antibodies in the relatives of MS victims which halted the neutrophil reduction when injected into mice. This and other evidence led them to identify the agent as a virus or viruslike particle. These observations have convincingly demonstrated MS to be infectious.

Although the agent has not been directly proved to be the cause of MS, its association with the disease has been conclusively proved.

The Henles have suggested that this virus may be a common, relatively harmless childhood infection which, if it is acquired in adolescence or early adulthood, produces far more severe results. Some researchers suspect that perhaps the immune system of genetically susceptible people may be unable to contain the infection, thus permitting MS to develop over a long period of time.

Breast cancer treatment remains a subject of intense debate within the medical community. Cancer researcher Gianni Bonadonna and associates at the Instituto Nazionale Tumori in Milan, Italy, in collaboration with an international team of investigators supported by the United States Public Health Service, reported in February 1976 that drugs used as an adjunct to surgery for breast cancer produced encouraging results.

Underlying Bonadonna's study was the knowledge that lymph nodes in the axilla, or armpit, are often tumorous at the time of breast surgery. Cancer is more likely to recur in patients with

Science Year
Close-Up

New Leads On Diabetes

Important research advances in 1975 and 1976 greatly added to our understanding of diabetes mellitus, a disease in which the body cannot store or use sugars properly. Until now, scientists had believed that diabetes mellitus was one disease with two forms – juvenile-onset diabetes, which generally strikes children, and mature-onset diabetes, which usually strikes adults. But researchers have found that these are distinctly different diseases, apparently set off by completely different biological processes. Mature-onset diabetes is most likely passed on genetically from parent to child. Juvenile-onset diabetes may be caused by a virus.

In juvenile-onset diabetes, the pancreas, a gland behind the stomach, does not produce enough insulin, a hormone that helps the body regulate the storage and use of sugars. In mature-onset diabetes, the pancreas produces insulin, but its target tissues – such as liver, fat, and muscle – do not respond properly.

In 1975, three teams of researchers found that the poor response of the target tissues is caused by a defect on the surfaces, or membranes, of their cells. In the early 1970s, researchers at the National Institutes of Health in Bethesda, Md., and at Stanford University in California had laid the groundwork in experiments with insulin and leucocytes, a type of white blood cell. Although leucocytes are not the normal target cells for insulin, the scientists believe their response to insulin is the same as that of the cells in the target tissues. These researchers found that insulin does not attach as tightly to leucocytes from individuals with mature-onset diabetes as it does to leucocytes from healthy individuals.

Then, biochemist Melvin Blecher and his colleagues at Georgetown University Medical Center and Veterans Administration Hospital in Washington, D.C., reported in January 1975 that leucocyte membranes have specific receptors to which insulin and glucagon, another hormone involved in sugar regulation, attach. The receptors are molecules into which the hormones fit like a key in a lock.

Blecher developed a test to determine whether this attaching, or binding, ability varied between healthy and diabetic individuals. He found that only about one-third as much insulin and glucagon attached to leucocytes from mature-onset diabetics as to leucocytes from healthy persons. He also found this weak binding ability in apparently healthy young adults from families in which one or both parents had the mature-onset form of diabetes.

But the test is still too complicated and too expensive to be widely used.

Blecher's findings suggest that mature-onset diabetes has a strong genetic link. And doctors may someday be able to determine the insulin-binding ability of cells from children of families with a history of this disease. Those most likely to develop the disease should then be able to control their diet and weight to delay its onset and reduce its severity.

Endocrinologist Stefan S. Fajans of the University of Michigan at Ann Arbor in 1975 provided further evidence that mature-onset diabetes is genetically linked by studying two groups of young diabetics – those with mature-onset symptoms and those with juvenile-onset symptoms. He found that 85 per cent of the first group had a diabetic parent, while only 11 per cent of the second group did. He also found a history of diabetes through three generations in 46 per cent of the first group, but in only 6 per cent of the second.

Meanwhile, Robert B. Tattersall and D. A. Pyke of King's College Hospital in London studied pairs of identical twins in which at least one had diabetes. They found that if the diabetic twin was less than 40 years old when diabetes first appeared, then in 59 per cent of the cases, the other twin also contracted it. But if the diabetic twin was over 40 when the disease appeared, then in 92 per cent of the cases, the other twin contracted it. Although both percentages were high, this study again indicates strongly that mature-onset diabetes is tied more directly to genetic factors than is the juvenile-onset form.

In fact, juvenile-onset diabetes probably results from a viral infection that severely damages or destroys insulin-producing cells in the pancreas. For years, reports in medical journals have suggested some link between juvenile-onset diabetes and such viral infections as mumps, German measles, and upper respiratory infections. Epidemiologist

Harry A. Sultz of the State University of New York at Buffalo studied the medical records of Erie County, New York, for a 25-year period. He reported in 1974 that the number of mumps cases rose and fell three times during this period, with peak outbreaks about every seven years. He also found that outbreaks of juvenile-onset diabetes followed mumps outbreaks by an average of 3.8 years. Sultz suspects this lag might represent the time required for the mumps virus to permanently damage the pancreas.

Despite such evidence, scientists have had difficulty proving that viruses can damage the pancreas in laboratory animals. But in the late 1960s, virologist John E. Craighead of the University of Vermont produced diabetes in mice by infecting them with an animal virus. However, this virus apparently does not infect human beings.

Then, in January 1976, virologist Sidney Kibrick and his colleagues at the Boston University School of Medicine and the Medical College of Virginia reported that they had produced diabetes in mice with a virus known to infect humans, the Coxsackie virus. This confirmed similar experiments done in 1973 by virologists D. Robert Gamble and K. W. Taylor of England, who had long investigated a possible diabetes link with this virus.

However, strains of mice that developed diabetes after viral infections had a genetic defect that made their pancreas susceptible to viral damage. Mice without this defect did not develop diabetes. Therefore, it seems likely that a genetic factor is involved in juvenile-onset diabetes, but it differs from the genetic link in mature-onset diabetes.

It will be many years before any of these developments directly affect the treatment of diabetes. Researchers must first confirm the effectiveness of Blecher's test for the binding ability of insulin and glucagon. And the test must be simplified. In addition, scientists must establish the link between Coxsackie viruses and diabetes more firmly before they attempt such preventive measures as vaccination. Nevertheless, these developments provide strong leads for future research on diabetes, and they suggest that it someday could be prevented. [Thomas H. Maugh II]

that complication. Efforts to improve their chances for recovery through postoperative radiation have been fruitless. The failure of localized radiation therapy is thought to mean that cancer in the lymph nodes indicates that the disease is already widespread and cannot be cured by local X ray.

Bonadonna reported on 391 patients between June 1973 and September 1975 in whom cancer had spread to axillary lymph nodes by the time they had breasts removed. Half of these patients received no active antitumor drugs, while others were treated with cycles of three different tumor-killing drugs. The antitumor drug therapy has been very well tolerated so far, and results have been very favorable. The disease recurred in 24 per cent of the first group of patients and in only 5.3 per cent of those treated with drugs. Patients with fewer tumorous lymph nodes achieved greater benefit. The treatment was equally effective for young and older women. More than 75 per cent of the recurrences were at sites far from the breast and armpit, supporting the thesis that tumorous lymph nodes often indicate that the disease is already widespread. This makes treatment with drugs a better approach than radiation, which has only a localized effect.

The researchers cautioned that they cannot yet conclude whether suppression of recurrence at this early time will be associated with prolonged survival. However, this integrated approach to treatment will probably provide important benefits for those who are afflicted with breast cancer.

Diverticular disease. Drs. A. J. M. Brodribb and Daphne M. Humphreys added substantial clinical support in February 1976 to the theory that dietary deficiency may cause diverticular disease. They reported results of a study at the Radcliffe Infirmary at the University of Oxford in England. Diverticular disease affects the colon, a part of the large intestine. It is perhaps the most common bowel disease among adults in the industrialized world. Despite advances in diagnosis and treatment, its incidence appears to be increasing. Its victims slowly develop weak areas in the intestinal wall that eventually become pouches which trap

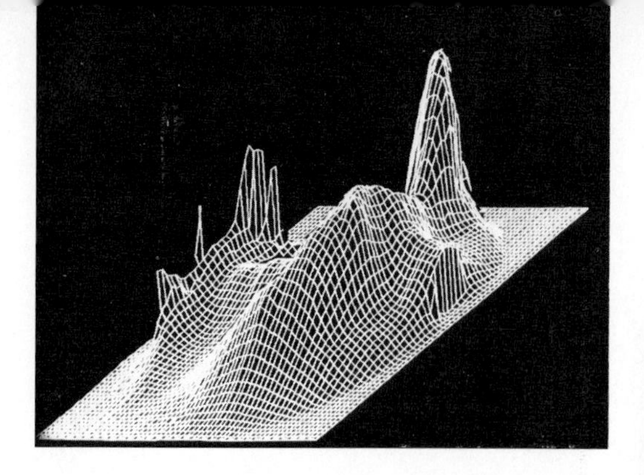

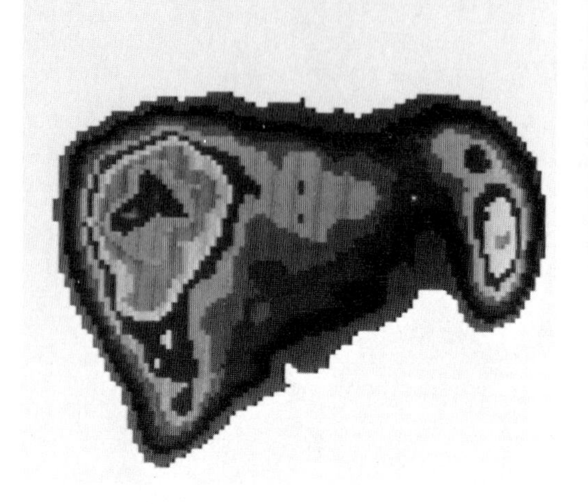

Information-storage and video-display units and an electrostatic printer, *above,* are linked by computer to a gamma-ray scanner to provide doctors with more detailed information for diagnosing disease. The isometric view of a cirrhotic liver, *top left,* and the cross-section scan of another liver, *left,* are both products of the flexible new system.

intestinal contents. The infections that result can cause pain, fever, and improper intestinal function.

In 1972, London surgeon Neil S. Painter and Denis P. Burkitt of England's Medical Research Council suggested that the disease may be caused by a deficiency of dietary fiber. Cereals containing bran appear to be the principal sources of this fiber, and epidemiologic studies indicate that diverticular disease and its inflammatory complication, diverticulitis, are more common in communities with low bran intake than in regions of high bran intake.

By checking the crude-fiber consumption of patients with diverticular disease and of healthy people, Brodribb and Humphreys showed that the diets of those with the disease contained half the amount of fiber seen in diets of normal people. There was no evidence that patients had ever changed their dietary habits, and the older ones may have always had fiber-deficient diets.

To further test the importance of fiber deficiency, the investigators treated 40 patients who had clinical and X-ray evidence of diverticular disease. They ate supplemental dietary fiber in the form of 24 grams (0.85 ounce) of wheat bran per day for six months. Treatment was rated extremely beneficial by 83 per cent of the patients, and all experienced some improvement. The patients reported sharp decreases in nausea, distention, and pain, as well as in diarrhea and constipation, which are alternating symptoms in diverticular disease.

Physiological studies showed that the amount of time it took for food to travel through the intestines was normal in all cases, and there was less intestinal pressure, which probably causes the pain that is a feature of the illness. Biochemical studies failed to reveal any harmful side effect of this treatment.

These studies suggest that dietary fiber, long thought to be biologically inert, may have important biological functions. This apparently safe and effective dietary modification not only holds real promise of improving the well-being of those with diverticular disease, but also might prevent the development of the disease in others. See NUTRITION, Close-Up.

Balkan kidney disease. Studies carried out jointly by virologists in Great Britain and Yugoslavia appear to have demonstrated in November 1975 that a virus is responsible for Balkan nephropathy. A kidney disease, Balkan nephropathy may affect about 20,000 residents of adjacent areas in Bulgaria, Romania, and Yugoslavia. It begins unobtrusively in young adulthood, and is characterized by proteinuria, an excess of serum proteins in the urine, and eventually leads to kidney failure. It is associated with mild anemia and high blood pressure. Balkan nephropathy is usually fatal. The localization of the disease in certain villages and its tendency to occur in family clusters have encouraged speculation that the disease has a primarily genetic basis.

K. Apostolov and P. Spasic, currently at the Royal Postgraduate Medical School in London, and N. Bojanik of the Vojno Medical Academy in Belgrade, Yugoslavia, examined biopsy tissue from the kidneys of seven patients known to have the disease. They noted changes indicating that kidney tissue was slowly being destroyed. In addition, swollen cells lining the tubules in the kidneys were found to be degenerating, and examination by electron microscope revealed that these cells contained particles characteristic of virus infection. Strikingly, similar particles were found in biopsy specimens obtained from seemingly unaffected children in families where one case had appeared. The existence of these particles in children who had no kidney disease is further evidence that a viral agent causes the disease. Antibodies were also found, attached to kidney cells in these children.

The findings strongly suggest that a coronavirus is involved. This virus family is known to have a preference for kidney tissue, most commonly in domestic animals. It can be transmitted from swine intestine to humans. The fact that most of those afflicted with Balkan nephropathy live in rural areas where pig farming is a common occupation, lends strong epidemiological support to the researchers' clinical observations. [Michael H. Alderman]

Surgery. A Boston surgeon is using bougies, small bullet-shaped pieces of metal, to repair closed esophagus, a birth defect that makes it impossible for babies to swallow because the esophageal tube is incomplete. A gap occurs between the upper and lower ends of the esophagus, which are themselves closed off. W. Hardy Hendren of Massachusetts General Hospital described his new method of dealing with this problem in September 1975.

Hendren placed the bougies in the separated halves of the esophagus. The upper bougie was put in place through the baby's mouth while the lower one had to be pushed up from an opening made in the stomach.

Hendren worked with engineers at the Francis Bitter National Magnet Laboratory at the Massachusetts Institute of Technology to devise an electromagnetic system that gradually draws the bougies together and reduces the gap separating the esophagus ends.

Hendren placed the infants in the center of an intermittent electromagnetic field – on for 60 seconds and off for 90 seconds – that is designed to attract the bougies. This treatment continued around the clock. A constant force cannot be used because it would injure the esophageal wall. The first two patients required 58 and 30 days of treatment, respectively, before the bougies – and the two ends of the separated esophagus – were judged to be close enough for the surgeon to operate on the esophagus.

There was some concern that the magnetic field might have an adverse effect on the infants. However, attending doctors could find no evidence of damage. Hendren believes that intermittent magnetically activated bougies may be used in other ways. For example, he plans to use the technique in infants with an imperforate anus, a similar birth defect at the lower end of the digestive tract.

Vasectomy reversal. A successful technique to reverse vasectomies was announced at the annual meeting of the American College of Surgeons in San Francisco in October 1975. Surgeons Sherman J. Silber of the Veterans Administration Hospital in San Francisco and Earl Owen of Sydney Australia, reported on 20 patients they had operated on to restore the continuity of the vas deferens, a duct or tube that carries sperm from the testicles to the penis.

The vas deferens in all these patients had been tied off as a means of sterilization. Many men have hesitated to have vasectomies because little could be done to restore fertility later if the patient changed his mind. The vasectomy-reversal success rate was extremely low, largely because the vas deferens has a diameter of only 0.3 millimeter.

Silber and Owen applied microsurgical techniques. They employed a 40-power, foot-operated zoom microscope and techniques developed in three years of experimental microsurgery on rats (see SURGERY IN MINIATURE). They sewed two layers of sutures, or stitches, to join the two ends of the vas deferens. The inner sutures join the inner lining of the vas deferens and the outer stitches join the muscular wall. The outer sutures are important because uninterrupted muscular activity is necessary in order to move the sperm through the tube. Within a year, the doctors reported, 15 of the first 20 patients had impregnated their wives.

An estimated 4 million men in the United States have had vasectomies. Now that this procedure is no longer irrevocable, many more may seriously consider this means of family planning.

Artificial joints. Dr. John P. Albright of Iowa City, Iowa, reported in 1975 on 50 patients who have had total knee-joint replacement, often due to arthritis or injury. He reported good to excellent results in more than 80 per cent of the patients. Despite these encouraging developments, there is still some disagreement on the best type of artificial knee to use. There are about 35 different artificial knees available.

The ankle joint is an even more complex problem than the knee joint. The ankle functions as a simple hinge during walking, and it must also be able to compensate for uneven terrain. Furthermore, the thin bones in the ankle make it difficult to anchor an artificial joint.

Despite these problems, two Brooklyn, N.Y., doctors in September 1975

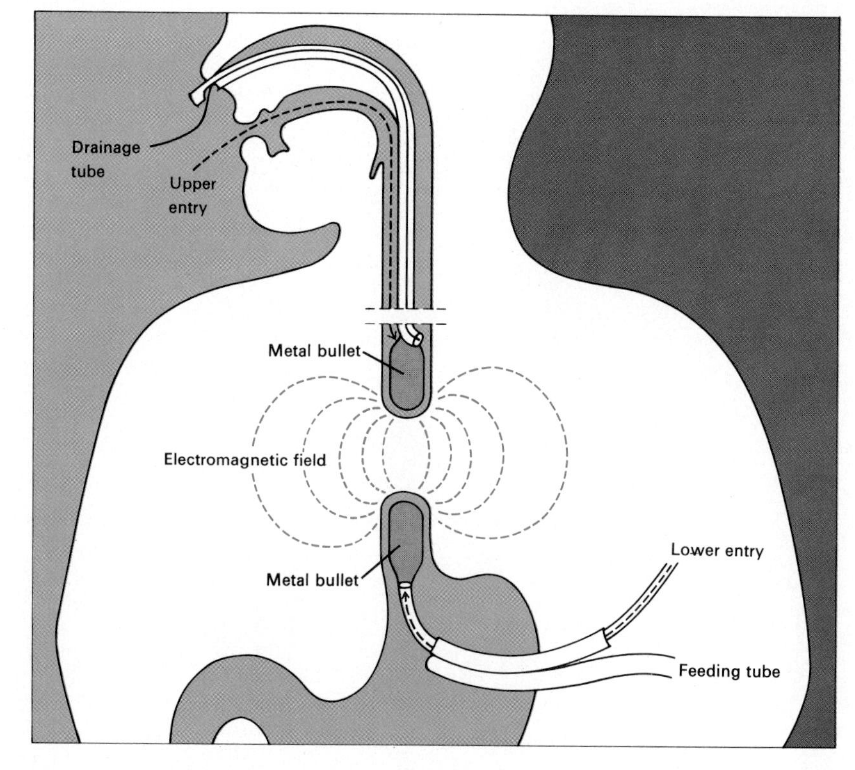

An electromagnetic field pulls two metal "bullets" together to help doctors link the separated ends of a closed esophagus. When the two ends have been pulled close enough, they can be surgically opened and connected to make a normal esophagus.

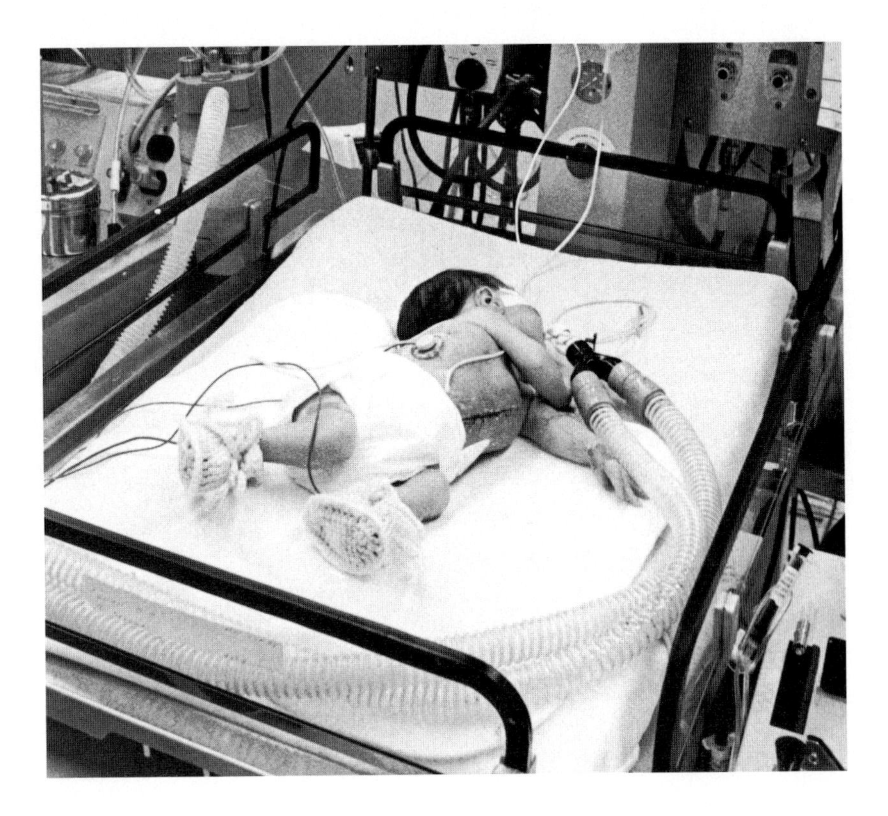

Christopher Wall, Jr., rests after surgeons covered his exposed heart with flaps of skin. The boy was born with his heart protruding through a hole in his chest. He was the first to survive this defect.

successfully inserted artificial ankle joints into a 23-year-old mailman whose bone tissue had been adversely affected by immunosuppressive drugs he had been taking after a kidney transplant. Surgeons Eduardo Alvarez and Harvey Manes removed the diseased portions of the tibia – the thicker leg bone – and of the thinner fibula, and replaced them with a block-shaped piece of polyethylene. Then they removed the talus, or anklebone, and replaced it with a metal device shaped to fit the lower end of the block.

The surgeons were pleased with the results of this unusual operation, but they stressed that it was still an experimental procedure and it would take time to judge how well the artificial joint would hold up. Replacement joints in the lower limb are subject to special stresses because they have to support the weight of the body. Factors such as obesity, which may have contributed to the original destruction of the joint, also complicate the patient's recovery after an artificial joint has been implanted.

Obesity and surgery. Dr. Edward E. Mason of Iowa City has developed an operation to reduce the size of the stomach. The new operation is a variation on the popular intestinal by-pass operation to combat morbid obesity, in which the patient is more than twice normal weight.

The original procedure rearranges the intestinal tract so that food by-passes most of the small intestine. The large intestine is left undisturbed because it exists primarily to absorb water. Patients with uncontrolled appetites can then eat all they want and still lose weight because most of the food cannot be absorbed in the large intestine. The by-passed intestine is generally left in place so the surgeon can restore the normal digestive route if weight loss becomes too severe.

A number of problems associated with this operation, such as severe diarrhea, uncontrolled loss of potassium or calcium, liver failure resulting in jaundice, and a variety of vitamin deficiencies have led Mason and his associates to try reducing the stomach.

A new balloon pump used during open-heart surgery converts the continuous flow of blood from a heart-lung machine to a pulsating flow that emulates that from the healthy heart. When synchronized with the re-established heartbeat after the surgery, the device can help the heart heal.

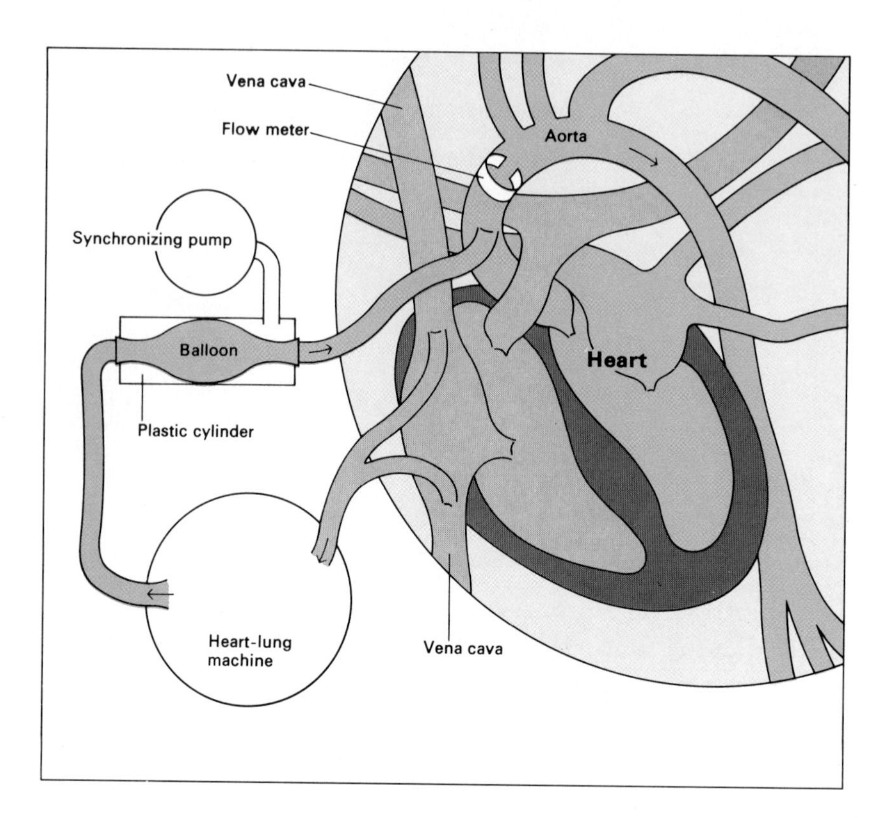

The Iowa doctors surgically reduce the stomach to 10 per cent of its normal size, leaving only a small pouch to receive food. The rest of the stomach is left in place and can be restored should the need arise. The very small stomach pouch makes it uncomfortable for the patient to eat large meals.

The operation is technically difficult because the stomach is harder to reach in the obese patient than the small intestine, and Mason reports a 3.8 per cent mortality rate. However, this is only about half the death rate that occurs in small-intestinal by-pass procedures, so it provides a safer road to weight reduction. The new stomach pouch may expand in time, and the long-term effect of this new technique is not yet certain.

Balloon pumping. The balloon pump, inserted into the aorta to assist patients with failing hearts during and after heart surgery, has been in use since 1967. In 1975, Dr. David Bregman of Columbia-Presbyterian Medical Center in New York City reported having developed an external balloon pump which avoids the risks inherent in placing the balloon in the aorta.

Bregman's device, used in conjunction with a heart-lung machine, consists of a balloon inside a rigid cylinder installed in the line between the machine and the patient. The cylinder is connected by a hose to a standard balloon pump which synchronizes the heart-lung machine blood flow to the patient's own heart rate. As a result, the body is better supplied with blood while the patient is on the pump. Most important of all, the new system helps during the critical period when the patient is being taken off the pump, and the heart has to resume the full burden of pumping.

Bregman used the device on 22 patients who were having surgery for coronary artery disease or for aortic valve disease. About half of them had been severely disabled by this disease but all did well after surgery. Without the new devices, Bregman said, at least three or four patients would have had severe cardiac failure during or after surgery. [Frank E. Gump]

Meteorology

Meteorologists gave major attention to climate change, air pollution and its threat to the stratospheric ozone shield, and weather modification in 1975 and 1976. They also made use of advances in remote-sensing techniques and satellite instruments to gather data.

Climate change. A number of climatologists believe that the increasing amount of carbon dioxide in the atmosphere will cause a worldwide rise in temperature. Such a rise, even if only 1 or 2 degrees, might cause the polar icecaps to melt partially, raising the level of the oceans and flooding many coastal areas.

Syukuro Manabe and Richard Wetherald of the Geophysical Fluid Dynamics Laboratory of the National Oceanic and Atmospheric Administration (NOAA) in Princeton, N.J., reported in December on a computer model of the climate showing that the effects of increased carbon dioxide would be much more complex than simple worldwide warming. They estimated that if the burning of fossil fuel continues at the projected rates, the amount of carbon dioxide in the earth's atmosphere will increase by 20 per cent by the year 2000. They then constructed an atmospheric model to see what would happen if the amount of carbon dioxide doubled.

They found that the rise in temperature would vary geographically. The temperature of the surface air at the poles would rise by as much as 10°C (18°F.), while over the globe as a whole, surface temperatures would increase by only 2.9°C (5.2°F.). At the same time, the stratosphere would cool by 6°C (11°F.). In addition, they found that humidity would increase, and evaporation and precipitation rates would also rise.

Atmospheric pollution. The use of fluorocarbons in pressurized aerosol cans is another source of pollution that may cause climate changes. The U.S. Government Interagency Task Force on Inadvertent Modification of the Stratosphere reported in June 1975 that continued use of the gas as a propellant may reduce the amount of ultraviolet-absorbing ozone in the stratosphere by

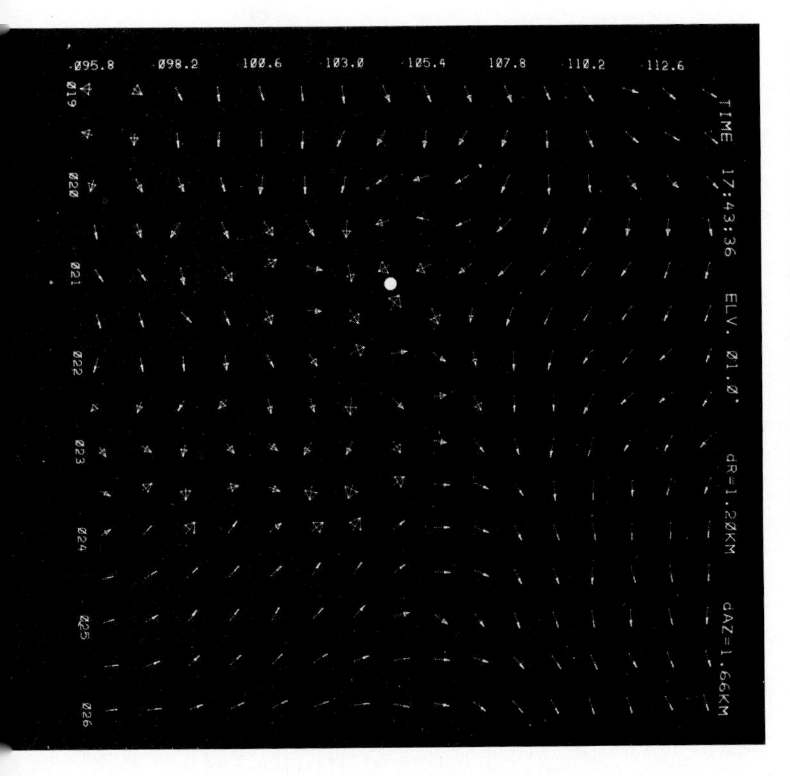

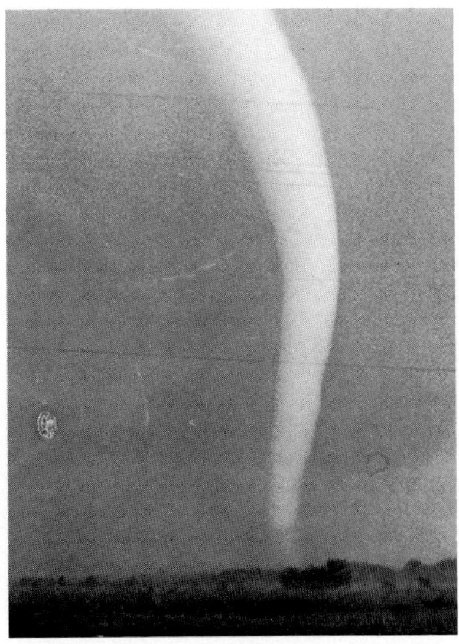

A computerized radar display, *left,* reveals telltale storm pattern that preceded a tornado in Oklahoma in June 1975, *above.* Arrows show wind direction. Arrow length and arrowhead size indicate amount of rain and turbulence, respectively.

Photos from weather satellites help meteorologists keep close track of storms. Hurricane Katrina, *above,* swirled off the coast of Baja California on Sept. 3, 1975. By the end of the month, cloud patterns over the Pacific Ocean appeared relatively calm, *right.* Hurricane Olivia formed near the west coast of Mexico in October, *opposite page.*

up to 7 per cent. That reduction could cause an annual increase in skin cancer as more of the sun's ultraviolet radiation penetrates the atmosphere and reaches the earth's surface.

Estimates of the danger to the ozone shield vary widely and, as a result, several programs have been started to study the ozone layer, using aircraft and satellites as well as observations from the surface. As part of the Global Air Sampling Program, commercial aircraft are carrying instruments to measure ozone content and other meteorological variables along regular air routes. The purpose of the program is to determine how aircraft flights affect ozone, carbon monoxide, dust, water vapor, and temperature.

Weather modification. Steven R. Hanna and Frank Gifford of NOAA's Atmospheric Turbulence and Diffusion Laboratory in Oak Ridge, Tenn., predicted in October 1975 that the proliferation of electricity-generating power parks could inadvertently cause changes in regional weather. According to Hanna and Gifford, each park releases as much heat, moisture, and dust into the air as come from such natural sources as volcanoes, thunderstorms, and giant bushfires. Large cumulus clouds and atmospheric whirlwinds frequently form as a result, and areas as far as 100 kilometers (62 miles) downwind may experience increased fogs.

Sensing severe storms. Meteorologists are taking advantage of many advances in remote-sensing devices to analyze and determine the characteristics of severe storms. Perhaps the most noteworthy advance in surface instrumentation was the refinement of color-display Doppler radar, first used by the National Hail Research Experiment in Grover, Colo.

Radar meteorology has advanced considerably since its introduction in the 1950s. By 1962, display systems could use four or five shades of gray to clarify information. Later advances permitted display patterns that indicated wind speed and direction. The new technique uses 16 colors to highlight meteorological information, much of which might be lost when only gray

Meteorology

Continued

shades are used. Greens and blues show less intense areas of a storm, while yellows and reds indicate areas of greater turbulence. The resulting picture is displayed on a color-television screen.

In separate reports during the summer of 1975, Donald W. Burgess of the National Severe Storms Laboratory in Norman, Okla., and G. R. Gray of the National Center for Atmospheric Research in Boulder, Colo., reported that the technique makes it possible to identify tornado conditions by distinguishing between different flow patterns in storms. Analysis of these patterns can detect large, slow rotation about halfway up in a parent storm long before the visible portion of a tornado funnel can be sighted.

The radar technique also can play a part in weather modification. By locating major inflow and updraft regions, it can show where cloud seeding should be concentrated to prevent severe storms from developing.

International cooperation among meteorologists and climatologists continued. Data are now being processed from the largest single meteorological data-gathering project in history, the Global Atlantic Tropical Experiment (GATE), conducted under the Global Atmospheric Research Program (GARP). GATE was set up to study tropical weather systems, which have important effects on global circulation.

GATE concluded its initial data-gathering phase in September 1974. A data-transmission system now being developed will begin distributing the information gathered by mid-1977. The data will flow from processing centers in 10 countries to two world data centers in Washington, D.C., and Moscow, then will be distributed to the scientific community for use.

GARP is also studying the Southern Hemisphere. Its Tropical Wind, Energy Conversion, and Reference Level Experiment began with the launch of the Nimbus 6 satellite in June 1975. The project is to measure winds, temperature, and atmospheric pressure in conjunction with a constant-level balloon project. [David L. Jones]

Microbiology

Two particularly significant and interesting developments occurred in microbiology in 1975 and 1976. A new method using bacteria to check substances for their ability to cause cancer gained growing support, and a new response of bacteria was discovered.

Bacteria-based test. Many common chemicals normally encountered in the environment and many more used to preserve, flavor, or color foods may be carcinogens, or agents that cause cancer. In standard tests used to detect these carcinogens, scientists expose laboratory animals such as rats to a suspected substance to see if cancer develops. This is not only a very long, but also a very costly process because large numbers of animals must be tested to obtain statistically meaningful results. Worse, these tests may not always be reliable.

But a much simpler, faster, and less expensive process, the work of microbial biochemist Bruce N. Ames and his colleagues at the University of California, Berkeley, showed increasing promise. The Ames group recognized that at least 85 per cent of known carcinogens are also mutagens. This means that they cause permanent changes called mutations in the genes of cells. Moreover, they cause mutations in all forms of life, including bacteria.

This behavior of carcinogens in bacteria is the basis for Ames's new screening procedure. About 1 million cells of the bacterium *Salmonella typhimurium* are spread over the surface of a solid nutrient material. The bacteria are a special strain with a genetic defect that makes them unable to synthesize the amino acid histidine. And, since the nutrient material contains no histidine, the bacteria will not grow on it.

Next, the scientists place a small amount of the suspected carcinogen on the nutrient material. If the chemical causes mutations, some of the bacteria will acquire the ability to synthesize histidine and grow into colonies on the nutrient material. This growth is easily detected within a day or two and indicates that the suspected substance is a mutagen, implying that it is also a carcinogen.

Normal-looking human sperm, *below,* and sperm coated and distorted by microorganisms called T-mycoplasmas are both from infertile men. They were cultured to encourage growth of T-mycoplasmas. Results support a theory that these microorganisms cause some infertility.

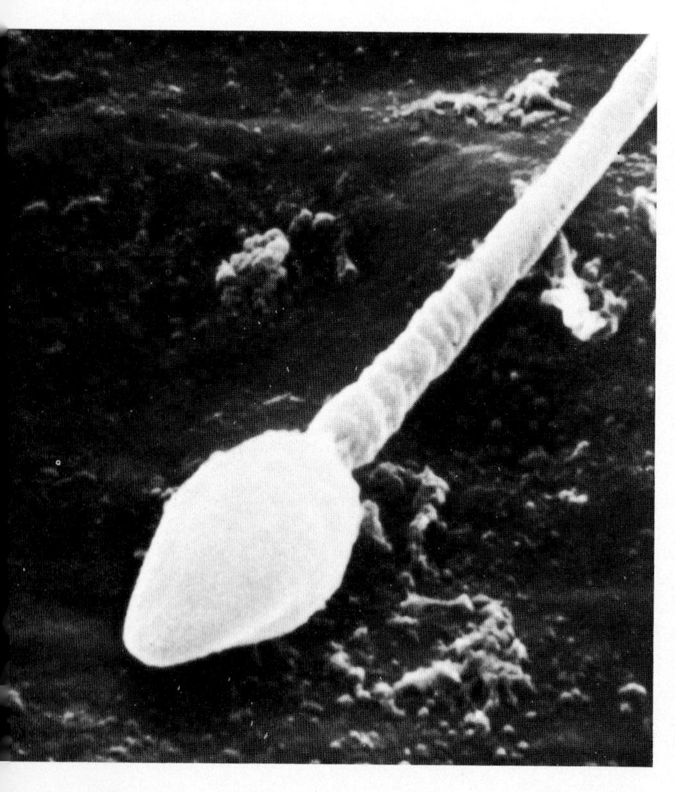

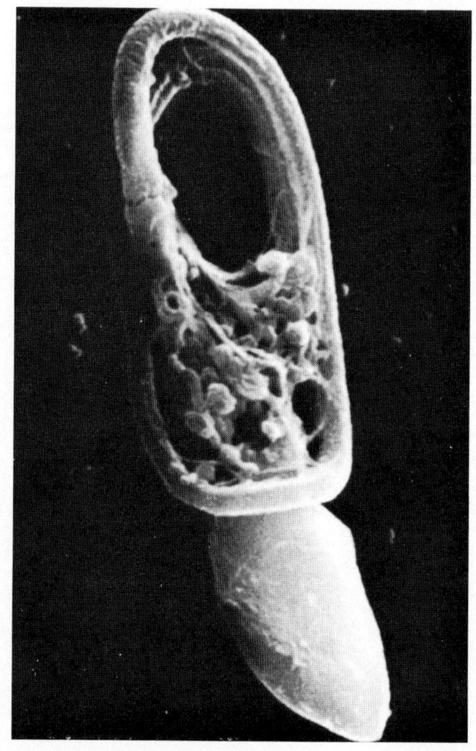

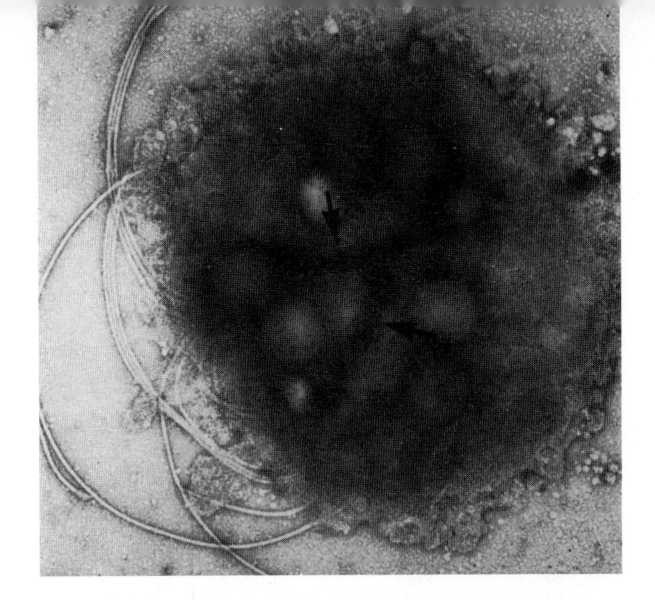

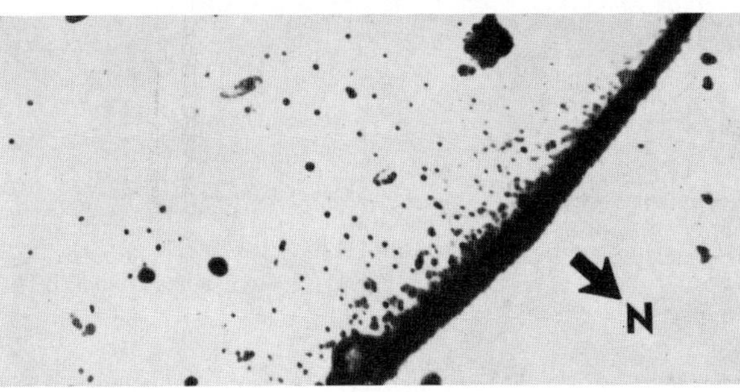

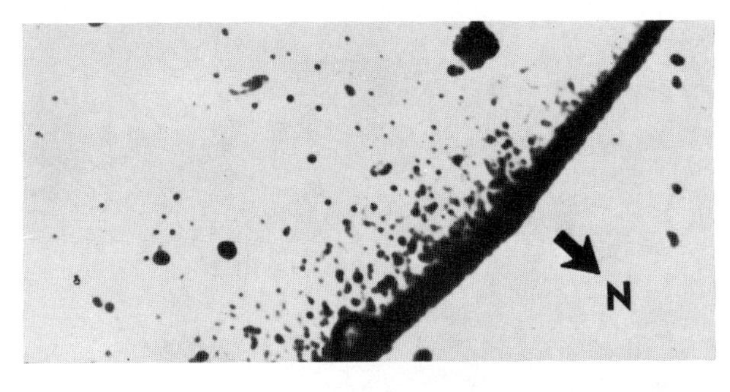

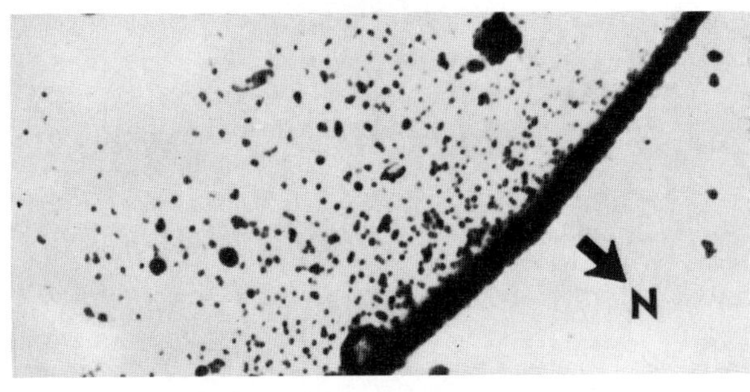

Iron-rich beads, arrows, *top,* probably play a role in magnetotaxis in bacteria. The bacteria crowd to one edge of a waterdrop under the influence of the earth's magnetic field, *second top,* then migrate in opposite direction when a magnet is placed there, *second bottom and bottom.*

While testing the procedure, Ames and his colleagues found that some known carcinogens were not causing mutations in the bacteria because the materials could not penetrate the bacteria's cell wall. To solve this problem, the scientists isolated and began to use a strain of the bacteria whose cell walls did not block the carcinogens.

They confronted a still greater problem in that some substances are not themselves carcinogens but are converted into carcinogens in the body. This is caused by enzymes in microsomes, tiny particles located in liver cells that normally detoxify some chemicals by slightly modifying them. Because experimental animals also have microsomes in their liver cells, tests with these animals identify such carcinogens. But bacteria have no microsomes. To compensate for this problem, Ames isolated the microsomes from rat liver tissue and included them in the test.

The new procedure, now called the *Salmonella*/microsome test, has been studied extensively in the past year in several other laboratories, most notably those of the U.S. Food and Drug Administration.

Ames published some of his results in December 1975 and April 1976, and they showed that his new procedure is remarkably accurate. For example, his first report listed 174 known carcinogens that he and his associates had tested. An impressive 156 of them caused mutations in the bacteria. In other tests, many chemicals not suspected of being carcinogenic did not cause mutations in the bacteria.

Magnetotaxis, a new response of bacteria, was reported in December 1975. Bacteria respond to a number of stimuli by migrating toward or away from them. For example, some bacteria swim toward molecules of food, a process called chemotaxis. Some that require oxygen for growth migrate toward oxygen (aerotaxis), while bacteria to which oxygen is toxic migrate away from it (aerophobia).

Bacteriologist Richard Blakemore, working at the Woods Hole Oceanographic Institution in Massachusetts, discovered a response he called magnetotaxis while studying bacteria in marine and marsh muds. While observing drops of mud with a high-powered

Science Year Close-Up

The Wayward Cells

A major laboratory resource has come under suspicion with the discovery that hundreds of cultured cell lines are not what they were thought to be.

Cells from humans and other animals are invaluable in laboratory research. Such cells are grown in the laboratory for use in studies of cell biology and reproduction, biochemistry and nutrition, toxicity, sensitivity to radiation, kinetics, and structural components. They are also used to study a vast spectrum of individual diseases.

But the increase in the use of varied cell lines for research has also increased the chance of contamination. One cell left on an improperly sterilized test tube or culture dish can multiply in the next cell culture it contacts and destroy its research value. If an experimenter is not aware of the contamination, he may be basing his entire research on the wrong cell line.

The problems of cell contamination, which are only now becoming widely recognized, can be traced back to at least 1952 when George E. Gey of Johns Hopkins University School of Medicine in Baltimore announced he had succeeded in growing cells from a human tumor in the laboratory. This was the first time human cells had been successfully cultured. The peculiar nature of the tumor was not to become known until 1971 when Howard W. Jones, also of Johns Hopkins, discovered that Gey's successful cell culture was not a common cancer, but a rare, very aggressive type. It had killed the donor, Henrietta Lacks, a 31-year-old Negro woman, within eight months. Gey had named the cell culture HeLa after her.

Gey's HeLa cells were rapidly distributed to other research centers and, within a short time, researchers began growing and exchanging more cell cultures from other sources. The rapid increase in the use of these varied cell lines, often by inexperienced workers, made cross contamination inevitable. The faster-growing cells – those best adapted and longest in cultivation – would clearly win out in a mix-up and overgrow the culture.

Routine microscopic examination of cells in culture vessels cannot reveal the original species, nor can it distinguish between cell lines from different individuals of the same species or between male and female cells of the species.

However, other tests can distinguish between species. One, called complement fixation, identifies an unknown cell if it joins with an antibody previously produced against cells of a known species. Joseph L. Melnick of the Baylor College of Medicine in Houston used this test in 1958 to discover that three separate cultures, thought to be rabbit cells, were actually HeLa or other human cells.

David Franks of Cambridge University in England and his collaborators at the National Cancer Institute in Bethesda, Md., detected additional cell contamination in 1962 by using the mixed agglutination reaction. In this test, different cells clump together when they are of the same species.

I now believe that this was an early clue to widespread HeLa cross contamination. Stanley M. Gartler of the University of Washington in Seattle found further evidence of contamination in 1966. He discovered that 17 other human cell lines were producing the same enzymes as were HeLa cells. One enzyme, a form of glucose-6-phosphate dehydrogenase (G6PD, type A), is found in 20 to 25 per cent of the world's Negro people, but not in Caucasians. He also found another relatively rare enzyme, phosphoglucomutase (PGM, type 1), in HeLa and all the other presumably different cell lines. Gartler concluded that all the cell lines must be HeLa because it was highly improbable that so many separate donors could have both these relatively rare traits.

Then, in the late 1960s, Torbjorn O. Caspersson and his associates at the Karolinska Institute in Stockholm, Sweden, learned how to identify individual chromosomes by longitudinal differentiation, the banding of chromosomes by various fluorescent dyes. For example, cells from males normally reveal an X and a Y chromosome, while female cells show two X chromosomes. Applying this method to the suspect cell lines in 1972, Ward D. Peterson, Jr., of the Child Research Center of Michigan in Detroit discovered that all of the cells examined lacked the male Y chromosome, even though at least some were supposed to have come from males. Most important in identifying HeLa cell contamination was the work by Orlando J. Miller of Columbia University in New York City who applied the Caspersson technique to HeLa cells and discovered a whole array of unique banding patterns distinct from those found in normal chromosomes.

Thus, by 1973, researchers had accumulated the following criteria for the ready identification of HeLa cells: The cells had G6PD, type A, and PGM, type 1; they lacked a Y chromosome; and they possessed an array of unique chromosomal markers.

My co-workers and I used all of these criteria to unmask the real nature of many cell lines in common use. In January 1973, Paula K. Hawthorne (Berry), Robert F. Flandermeyer, and I began to study six cancer cell lines from laboratories in Moscow. All were supposed to be human cells but were producing an uncommon virus previously seen only in rhesus monkeys.

By conventional staining, we found the chromosomes in all the cell lines were human. Then, in collaboration with Peterson, we found that all six cultures produced the rare G6PD, type A enzyme. We could not believe that chance had permitted six individuals, each with this rare genetic enzyme found almost exclusively in Negroes, to give cell specimens in Moscow. The riddle was solved when chromosome banding located many of the HeLa marker chromosomes.

We published the data on the Russian cell cultures and six contaminated cell lines from U.S. laboratories in 1974. A year later, we summarized the other lines we had found to be HeLa contaminants. Combining our data with all known reports of suspected cases of HeLa contamination, we listed 102 separate instances of contamination and more than 40 erroneous identifications. The list should aid researchers.

Our most recent studies reveal what we have suspected for some time. While HeLa cell contamination is widespread, other human and animal cell lines are also contaminating one another. Although contamination probably can never be completely eliminated, techniques for monitoring cell purity must be applied to reduce it and the problems it presents to biologists throughout the world. [Walter A. Nelson-Rees]

light microscope, he noticed that many of the bacteria migrated rapidly to one side of the drops. This occurred repeatedly and the bacteria always traveled in the same direction.

Blakemore was puzzled by this behavior and thought at first it was a response to light. However, when he tested the bacteria in the dark, they still migrated as before. Then he noticed that some small magnets were lying next to the microscope on the side toward which the bacteria swam. When he moved the magnets to the other side of the microscope, the bacteria moved in that direction. When he rotated the magnets around the slide, the bacteria followed like a compass needle.

Blakemore could not isolate the magnetotactic bacteria in pure culture, a necessary step before making a comprehensive study of them. However, he obtained nearly pure cultures by taking advantage of the magnetotactic behavior of the bacteria. He placed a drop of mud containing the organisms on a microscope slide and a drop of seawater at its northern edge. The magnetotactic bacteria, as sensitive to the earth's magnetic pole as a compass needle, swam into the waterdrop. Examining these bacteria with an electron microscope, he noted bundles of flagella, the long, hairlike appendages that enable bacteria to swim. But he also saw chains of beads of a dense, crystalline material in the cytoplasm of the bacteria. Such crystalline beads have not been seen in cells of any other bacteria, so Blakemore reasoned that they might cause the magnetotactic behavior.

Blakemore then examined the beads by a technique that reveals the metal content of very tiny particles. The metal present in highest concentration in the bacterial beads was iron. This suggests that these tiny beads are attracted to magnets, but how the beads direct the flagella, and thus control the direction in which the bacteria swim, is still a mystery.

The ecological significance of magnetotaxis is an intriguing question. Blakemore pointed out that the magnetotactic bacteria prefer to live where there is very little or no oxygen. Magnetotaxis might direct the bacteria downward, to the oxygen-poor sediments. [Jerald C. Ensign]

Neurology

Scientists in 1975 and 1976 have developed methods to study, for the first time, the activity of brain cells that seem to be involved in the command, or selective-attention, mechanism.

Vernon B. Mountcastle, James C. Lynch, Tom C. T. Yin, and others at Johns Hopkins University School of Medicine in Baltimore reported on their continuing studies of the electrical activity in the brains of monkeys during trained behavior involving coordinated eye and hand movements. Their findings indicate that there are populations of neurons, or nerve cells, in the parietal lobe of the cerebral cortex that control selective attention and coordinate the eye and hand movements by which monkeys – and probably human beings – control their surroundings.

The scientists use an ingenious experiment to monitor the electrical responses of hundreds of individual neurons, one at a time. They insert a microelectrode painlessly into a part of the parietal lobe of the monkey's brain to record the electrical activity while the animal is performing a learned task.

In one experiment, the monkey is trained to watch a moving light and to reach for it when the light's intensity changes. In another, the signal to reach out is pressure on the monkey's paw rather than a change in intensity of the light. A correct response earns the monkey a drink of juice through a tube.

After studying the responses of hundreds of neurons in the brains of many monkeys, the researchers reported that a significant number of neurons in the parietal area discharged electrically from 60 to 80 milliseconds before the monkey made any physical response. Thus, these neurons – only a small proportion of all the cells in this part of the brain – seem to be involved in ordering the coordinated eye-hand response.

Furthermore, these nerve cells did not react to any other sensory influence such as other visual signals, nor were they a direct part of the visual-motor muscle activity involved in the behavioral act. They also seemed to be affected by the animal's motivation; the monkey responded at high rates only if it wanted food or drink.

A frog's eye, transplanted to the ear position in an embryonic stage, does not function because the optic nerve did not connect to the brain's visual center. This result suggests that the optic nerve, as it develops, follows preprogrammed guidance clues that do not change even when the position of the eye is changed.

These continuing experiments are providing new ways to study the higher functions of the brain – the thinking functions – as well as helping us understand the problems of those who suffer damage in this part of the brain.

Frogs and seizures. Neuroscientists Frank Morrell, Noriko Tsuru, and their co-workers at Rush University of Rush-Presbyterian-St. Luke's Medical Center in Chicago, induced epilepticlike convulsions in bullfrogs in studies that may aid our understanding of epilepsy in humans. Graham V. Goddard of Canada's Dalhousie University in Halifax, N.S., discovered in 1967 that brief, low-intensity electrical shocks in various parts of a rat's brain produced an epilepticlike seizure when repeated once a day over a period of several days.

The stimulations produced no effect at first. But the cumulative effect of the repeated shocks somehow changed the electrical discharge in the brain, and finally produced seizure and convulsion. It was almost as if the rat stored each stimulation to produce a large electrical discharge when some threshold level was reached. After that point was reached, each brief stimulation would produce a full-blown convulsion. Goddard and others have also observed this "kindling effect," as Goddard calls it, in several other animals, including cats, mice, rabbits, and monkeys.

The Chicago researchers, finding that the kindling effect can be produced more quickly in bullfrog brains – in hours rather than days – are now experimenting with bullfrogs because a frog brain is simpler in some ways than that of a mammal. Anatomical and chemical studies as well as electrical measurements can thus be more readily made. They are continuing to unravel the bioelectrical-chemical interactions among the affected brain areas and cells in an attempt to explain how accumulation of the electrical shocks brings about the changes in the electrical discharge patterns that are comparable to discharges in epileptic seizures. Better understanding of human epilepsy may well result. The research may also help to explain mechanisms of electrical activity in the brain associated with how we store information.

Sugar in the brain. A new method to find the sites of actual activity in the brain when the body is carrying out a specific movement or function was reported in 1975 by neuroscientists Louis Sokoloff and Charles Kennedy of the National Institute of Mental Health (NIMH) in Bethesda, Md., Martin Reivich of the University of Pennsylvania Medical School in Philadelphia, and others.

The method depends on the fact that there is a greater energy, or metabolic, output in the brain areas that are performing a particular function than in areas that are not directly involved. For example, the visual centers of the brain are likely to be using relatively less energy than the auditory areas when a person is listening to music while the eyes are closed.

The scientists inject a radioactive sugar called 2-deoxyglucose into an animal's bloodstream and it readily enters brain tissues and cells, especially those that are active and converting glucose to energy at a high rate. Because it is radioactive, the glucose can be tracked.

Those cells involved in functional activity can be seen to "light up" on a photographic plate, and their interconnecting pathways can be readily traced. For example, in one experiment, the Sokoloff group exposed rats that had one ear closed to normal laboratory noise. Using their measuring method, the scientists demonstrated a distinct difference in activity between those parts of the rats' brains that received and processed the sounds and those parts that did not. The scientists also measured other sensory activities.

The method enables scientists to survey different structures of the brain simultaneously and to observe and map functional interactions for the first time. The Sokoloff technique was also used to study the olfactory system of rats. Gordon Shepherd of Yale University School of Medicine in New Haven, Conn., and his colleagues John S. Kauer and Frank R. Sharp at NIMH reported in November 1975 that various odors – including amyl acetate, cheese, and camphor – were handled by different cells in the parts of the brain that process olfactory information. [George Adelman]

Nutrition

Researchers reported in late 1975 that we may be literally overeating ourselves to death. Their studies were part of an effort to understand the causes of such chronic diseases as arthritis, cancer, heart disease, and stroke.

In developed Western nations, these illnesses have largely replaced infectious diseases as the most common killers. Because the number of people suffering from them often varies widely from nation to nation, some scientists have tried to relate the differences in disease rates to cultural patterns.

Food is one of the great differences among cultures. Nutritionists who have studied diets in various countries have correlated certain diseases with certain parts of the diet. For example, a low incidence of cancer of the colon has been associated with a high-fiber diet (see Close-Up). Unfortunately, this is not proof of a cause-and-effect relationship between the dietary component and the disease. But the data from such studies give clues to the diet variables that should be studied in laboratory experiments.

Diet and longevity. Evidence that diet influences life span was reported in October 1975 by Morris H. Ross of the Fox Chase Cancer Center in Philadelphia and Gerrit Bras of the Rijks University in Utrecht, the Netherlands. They studied a strain of white rats that in previous experiments had selected dietary components that predisposed them to develop tumors and other diseases associated with aging. They allowed 121 of the rats to choose their own diets after 21 days of life and then monitored them until the rats died. The rats could choose among three diets that were always available and differed only in the amounts of protein and carbohydrate.

Rats eating the largest total amount of food (24 grams or 0.85 ounces) had the shortest life span (540 days) and those eating least (18 grams or 0.63 ounces) lived longest (690 days). Surprisingly, life span correlated best with the amount of food eaten during the rat's early life (100 to 199 days). For every gram of food eaten above 18 grams, a rat lived 26 fewer days.

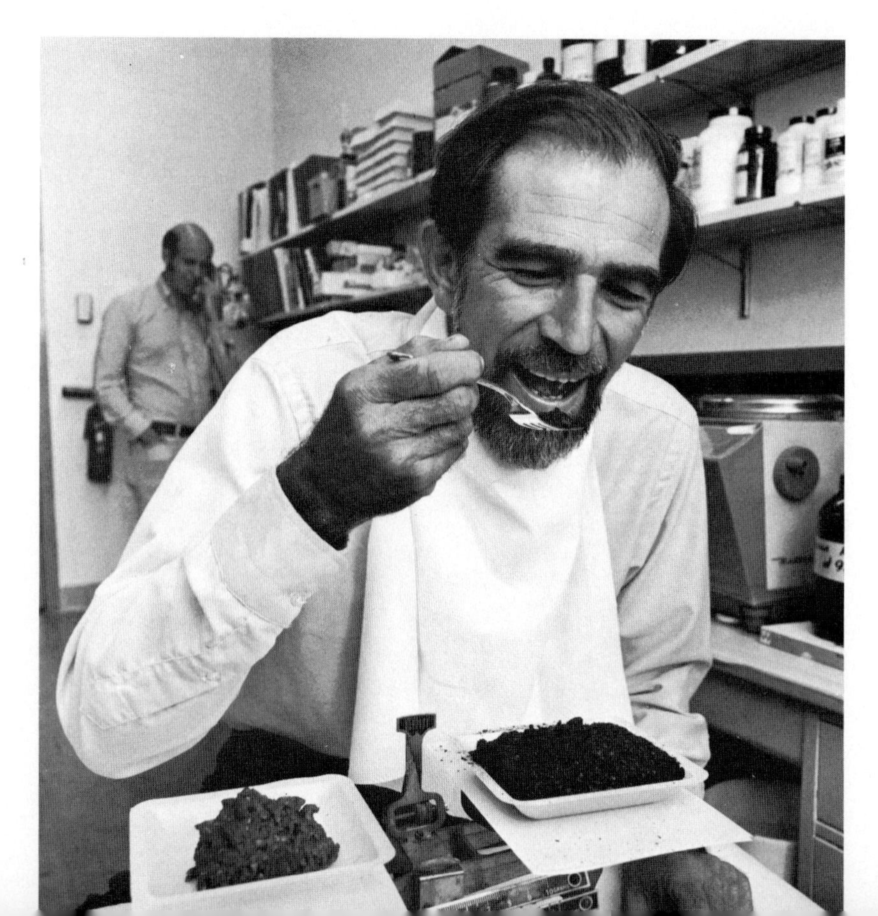

An investigator samples dried, flaky algae—one possible substitute for meat. The algae contain about half protein by weight, whereas meat has about 20 per cent.

Science Year Close-Up

Focus On Fiber

Stories of a new nutritional cure-all – dietary fiber, or roughage – overflowed newspapers and magazines in 1975 and 1976. People were encouraged to eat potatoes, wheat bran, and other foods that stimulate the movement of material through the intestines. The goal was large, frequent bowel movements and the promise was freedom from many diseases, including appendicitis, heart disease, and cancer of the colon.

These stories were based on the work of physicians and epidemiologists who studied the diet and diseases of African and other groups of people who ate foods rich in fiber, and were remarkably free from these diseases. On the other hand, such diseases are common in the Western world, where people generally eat foods containing far less fiber. No cause-and-effect relationship was ever proved, but some investigators concluded that fiber was the key factor.

Those of us who have been studying the role of fiber in the diets of laboratory animals continue our research with renewed urgency. We are trying to determine precisely what fiber is, what its effects are, and how it achieves them.

We have a reasonable answer to the first question: Dietary fiber is anything we eat that our own digestive systems, or those of bacteria in our systems, cannot break down. Fiber is composed primarily of plant cell-wall material, which, in turn, is mainly cellulose interlinked with hemicellulose, pectin, and lignin (see PLANTS: THE RENEWABLE RESOURCE). Other important constituents are cutins, found in the surface layer of leaves, fruits, and seeds, and the mucilages found in their outer cell walls. Although all are types of fiber, each acts differently in the digestive tracts of various animals.

The differences emerged in the 1950s and 1960s when reports linked diets high in saturated fat to atherosclerosis – fat and cholesterol deposits in the arteries. Rabbits fed diets that are rich in saturated fats developed high blood-cholesterol levels and atherosclerosis in from 6 to 12 months even though their diets contained no cholesterol. But when we fed rabbits the same amount of the same saturated fat for similar periods in 1964, their blood-cholesterol levels did not rise significantly. We found that the two diets differed most in the source of fiber. The commercial feed we used in 1964 contained a mixture of natural plant fiber while the diet that led to atherosclerosis in the earlier experiments was made by adding purified cellulose to semipurified sources of protein, carbohydrate, and fat.

Our tests and those of others showed that fiber somehow interacts with saturated fat to prevent blood-cholesterol build-ups. But pure cellulose did not act the same as the mixed fiber found in natural foods. For example, in 1974 tests on baboons and monkeys, we compared high-saturated-fat diets of the semipurified type to a mixed diet of fruit, bread, and vegetables for the baboons, and to commercial feed for the monkeys. In both cases, the animals fed the semipurified diet containing pure cellulose had more fat deposited in their arteries than did the animals fed the mixed-fiber diets.

Data from these primate experiments suggested that the type of diet affected the synthesis and composition of bile acids, the principal products of cholesterol metabolism in the liver. Bile acids are required for proper absorption of fat in the small intestine.

We then developed a theory of how fiber and bile acids interact. Fiber ties up the bile acids, leading to their excretion. The binding would have two effects. First, less fat would be absorbed. Second, because more of the cholesterol would have to be metabolized to make up for the excreted bile acids, less cholesterol would be available to produce the fatty deposits of atherosclerosis.

We then found that the bile acids and their natural derivatives, bile salts, could be tied up by different types of fiber. Our tests showed that alfalfa, wheat bran, cellulose, lignin, and several other materials bound bile acids and bile salts to some extent, but the amounts varied widely. For instance, the binding capacity of lignin is 20 times that of cellulose. The binding theory may be correct, but it is more complex than we originally thought.

Our experiments and others continue to determine exactly how the constituents of fiber affect heart disease and colon cancer. Diet plays an important role in these diseases and perhaps others, but fiber is only one component of a balanced diet. [David Kritchevsky]

Termite Pilaf

1 cup long grain rice
1/4 cup sesame seeds
1/4 cup butter
1/4 cup termites
1 chopped onion
2 cups beef bouillon

Brown rice and sesame seeds in butter until golden. Mix in chopped onion and termites. Transfer this mixture to a casserole dish and add beef bouillon. Cover and bake in preheated oven at 350° F. for 45 minutes.

Carol Miller of California Polytechnic State University pours grasshoppers into a blender, *left.* She uses the chopped insects, which contain about 60 per cent protein, in baking "Jiminy bread." Another of her recipes, *above,* calls for termites.

Protein consumption was important during the first 50 days of a rat's life. Those rats eating the largest amount of protein during this period lived longer, and they continued to eat about the same amount of protein throughout their lives. In contrast, the short-lived rats started off by choosing a diet relatively low in protein, then progressively changed to a high-protein, low-carbohydrate diet.

These findings may have great importance for those who plan human diets, especially diets for infants and children. But the experiment does not indicate whether the diet pattern caused the differences in longevity or the rats chose their diet pattern because of some underlying differences in their metabolism – the breakdown of food to energy and usable molecules. Also, the rats could not make choices on the levels of nutrients other than carbohydrate and protein.

Vitamin protection? Concern about the role of diet in noninfectious diseases extends to vitamins. Vitamin A is needed for good night vision and by the cells that make up the skin and linings of body passages such as the intestinal and excretory systems. Now it appears that vitamin A may protect the body against viral infection and cancer.

Physician Eli Seifter of the Albert Einstein College of Medicine in New York City reported in August 1975 that very high levels of vitamin A given to mice significantly decreased their vulnerability to smallpox virus. In a group of mice given 5 to 10 times the normal recommended amount of vitamin A and then infected with the virus, fewer became ill. Those that did were sick for shorter periods than groups without vitamin A. Seifter had shown previously that vitamin A increased the survival time of mice inoculated with tumor cells by from 50 to 70 per cent and even decreased the number of tumors that appeared.

Vitamin A is not recommended for disease therapy because overdoses can be toxic, producing headaches and pains in joints. Carotene, which is found in carrots, yellow corn, and other foods, is broken down into vitamin A in

Nutrition

Continued

the body. But too much of it turns the skin yellow. So scientists are looking for a substance with the beneficial qualities of vitamin A but without any side effects. In the meantime, nutritionists are worried that these findings may start a new dietary fad.

Vitamin E overrated. Taking large amounts of vitamin E is another dietary trend that concerns some nutritionists. Spurred by claims that vitamin E slows aging and increases sexual prowess, some individuals routinely take large dosages. But Phillip M. Farrell and John G. Bieri of the National Institutes of Health in Bethesda, Md., reported in 1975 that they could find neither beneficial nor toxic effects from large doses.

Farrell and Bieri compared a control group of people with an experimental group consuming from 100 to 800 international units per day – an average of 30 times the recommended allowance. Blood tests showed no differences between the groups except vitamin E levels in the blood were higher in the experimental group. The investigators also failed to find any significant benefit from the large vitamin E supplements when they examined volunteers' answers to a questionnaire.

Ill-advised food fads. Evidence has been accumulating that people who alter their diets because of widely publicized preliminary data may be doing themselves more harm than good. John Y. Reinhold and his colleagues at Pahlavic University in Iran studied the effects of increased dietary fiber on the metabolism of various minerals. They found a significant increase in the amount of calcium, magnesium, zinc, and phosphorus excreted by the test subjects. The increase was so great that the subjects were losing more of these minerals than their bodies were taking in. Continued loss of minerals beyond the 20-day test period would have led to mineral-deficiency diseases. Exact interpretation of the results for Americans is difficult because the fiber was added to the diet by switching the subjects from low-fiber white bread to a high-fiber, whole-wheat Iranian bread that is not widely available in the United States. [Paul E. Araujo]

Oceanography

Scientists in 1975 and 1976 began to report the results of large-scale ocean research projects started in the early and mid-1970s. Their experiments include re-creating the global climate of the last Ice Age, reconstructing how the ocean basins and edges of continents evolved, and determining the role oceans play in shaping global weather and climate.

Ancient oceans. In March 1976, scientists in the Climate: Long-Range Investigation, Mapping, and Prediction (CLIMAP) project described some details of what the earth was like during a typical August 18,000 years ago. By analyzing fossils of microscopic plants and animals recovered from deep-sea sediment cores, they determined that thick ice sheets on land stretched as far south as New York City. Ocean surfaces were an average of 2.3°C (3.7°F.) cooler than they are today, but in some equatorial areas they were as much as 6°C (9.6°F.) cooler. Northern and southern oceans were covered with ice, and sea level was at least 85 meters (279 feet) lower.

The Gulf Stream flowed far south of its present path. It ran almost west to east across the Atlantic Ocean, in contrast to its present northeasterly route from the southwestern Atlantic past Great Britain.

Computer scientists used this information to make mathematical models simulating the world climate of 18,000 years ago. The changes in sea surface temperature as determined by the computer models agree reasonably well with those derived by analyzing fossil pollen and data about glaciers on land. CLIMAP investigators are now assembling data that will indicate what both August and February temperatures were 18,000 years ago. This information will be used in new computer models to simulate seasonal changes in Ice Age climate.

The deep seabed. The Deep Sea Drilling Project entered a new stage in 1976, called the International Phase of Ocean Drilling. Earlier phases, from 1968 to 1975, outlined the ages of the major ocean basins and how they developed. This new phase tackles some basic scientific questions about the oceans and their evolution. These questions involve the nature and origin of the earth's crust beneath the ocean; the edges of continents, including sites of earthquakes and volcanoes; and what the ocean environment was like in past ages. Drilling for the international phase began in the North Atlantic in December 1975.

In another sea-floor study in January and February 1976, the U.S. submersible *Alvin* probed the Caribbean Sea's Cayman Trough, a gouge in the sea floor 3 to 4 times deeper than the Grand Canyon. The trough is located where the North American and Caribbean tectonic plates meet. It contains rocks that may provide evidence that the two plates are pulling apart and that forces inside the earth are pushing up molten rock to form new crust in the gap.

This expedition aimed to build on data collected in the Mid-Atlantic Ridge during the French-American Mid-Ocean Undersea Study (FAMOUS) in 1974. Project FAMOUS scientists found that lava forms new crust where two plates apparently separate along the ridge.

Polar regions. In April 1976, scientists in the International Southern Oceans Study completed a three-month survey of the Antarctic's Circumpolar Current in the Scotia Sea and the Drake Passage. They moored ocean-current meters across the Drake Passage early in 1975 and used the data from these readings to describe the unique characteristics and changes in the Circumpolar Current as it moves through the Drake Passage.

For example, they found that the current is not one massive body of water like the Gulf Stream. Instead, it seems to be made up of smaller currents flowing at different speeds. The Circumpolar Current is important because it controls most of the water exchanged between the South Atlantic Ocean, South Pacific Ocean, and the Indian Ocean. The Antarctic is the source of the deep, cold water for major parts of all oceans, and the Circumpolar Current distributes this water, which takes about 600 years to reach the tropics.

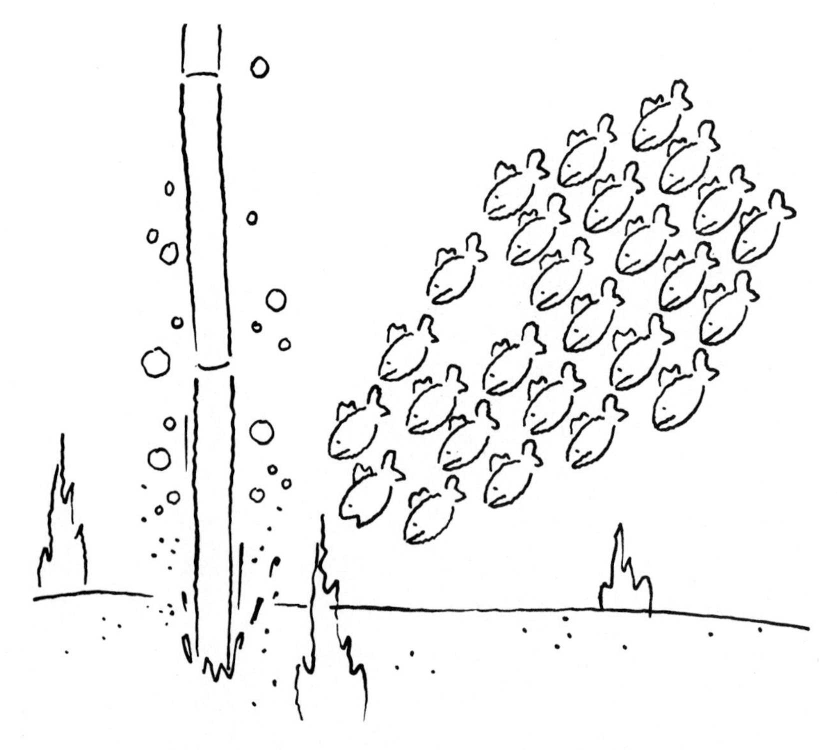

"They're taking a core sample, but they also took Herbie."

The amount of cold and warm water exchanged appears to play a key role in ocean-water circulation and in determining weather and climate.

In May 1976, scientists in the Arctic Ice Dynamics Joint Experiment (AIDJEX) completed field studies attempting to measure stresses on ice caused by winds in the Beaufort Sea. Their goal was to improve forecasting of Arctic weather and ice conditions. Computer scientists used the AIDJEX data to make a computer model designed to predict stress caused by ice on such offshore structures as oil rigs and the roughness of the bottom of ice sheets. Ice roughness distorts underwater sound, a problem for submarines navigating by sonar. The scientists are also trying to predict how far ice will extend and how fast it will be produced.

El Niño. Data gathered by oceanographers in mid-1975 may explain why nutrient-rich cold waters, which normally flow from the south past the coast of Peru, are occasionally replaced by warm waters from the north. Peruvians call this phenomenon *El Niño,* Spanish for *the child,* because it often starts about Christmas. El Niño disrupts populations of fish and of birds that depend on them for food. Scientists believe the fish either dive down to deeper cold water or leave the area.

In 1972, El Niño proved disastrous for the Peruvian fishing industry. The small Peruvian catch caused world production of fish meal for the year to drop to less than half the average.

Oceanographers could neither explain what causes El Niño nor predict when it might occur. However, Klaus Wyrtki, a German oceanographer with the U.S. North Pacific Experiment (NORPAX), proposed in 1974 that the effect of strong equatorial winds on the Pacific Ocean might cause El Niño. He theorized that southeast trade winds blowing strongly for more than a year speed up the westerly flow of the warm South Equatorial Current. This piles up warm water along the equator in the western Pacific Ocean, building up a large east-west slope. When the velocity of the trade winds slows down, the warm water flows back toward the eastern side of the Pacific.

Wyrtki suggested that trade wind conditions in late 1974 were right for producing an El Niño early in 1975. So NORPAX organized two cruises to analyze how an El Niño develops and to get a clearer picture of the water movements involved.

During the first cruise, from February 11 to March 31, the oceanographers observed a massive eastward invasion of warm water flowing along the equator to the Central American coast and then south to the coast of Peru. This indicated the start of an El Niño. During the second cruise, from April 17 to May 27, the warm water was already dissipating and ocean conditions off Peru were returning to normal.

The trade winds did not blow as long or as strongly as they have in the past, and the El Niño was not severe. Nevertheless, these studies apparently confirmed Wyrtki's theory.

By refining this technique of making predictions about ocean phenomena based on statistical relationships and then going out to observe them, we may be able to predict accurately the El Niño and other large-scale events in the oceans and atmosphere.

Ocean pollution. Most people think of ocean pollution in terms of oil slicks, sea birds covered with tar, or hundreds of poisoned fish. But some marine scientists are concerned that much less obvious pollutants will upset the balance of nature in the oceans by harming one of the key components of the food chain. This might involve such low concentrations of pollutants that it would have no immediate visible effect on large organisms, such as fish. For example, such pollution might harm phytoplankton, the microscopic "grass" of the sea and the food source for small animals, such as baby shrimp and fish larvae.

Oceanographers cannot study the effects of pollutants directly in the open oceans, because the environment is so large. On the other hand, they cannot study pollutant effects in a laboratory, because it is impossible to raise all the natural ocean plants and animals in such confined quarters. So U.S., Canadian, and British oceanographers of the Controlled Ecosystem Pollution Experiment (CEPEX) compromised by using large plastic bags to trap biological communities in the sea. The bags are 10 meters (33 feet) in diameter, 30 meters (99 feet) deep, and open at the top.

Collecting Sea Secrets

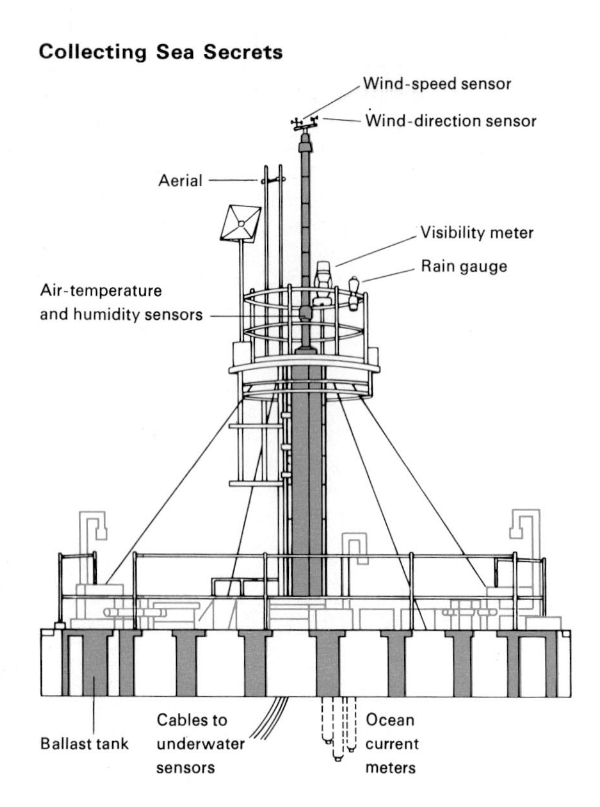

Oceanography

Continued

Great Britain's new data buoy has an array of sophisticated sensors to monitor weather and oceanographic conditions in the North Sea. It is the first in a planned European network of such data buoys.

In May 1976, the scientists began an experiment using three of these bags in a sheltered inlet of Vancouver Island, off the Pacific Coast of Canada. They filled the bags by pulling them up from the bottom, thus capturing vertical columns of water and all of the organisms living in them. They then added small quantities of pollutants to one bag and compared the organisms with those in the two pollutant-free bags.

The CEPEX scientists found indications that phytoplankton are among the most sensitive organisms. Of these, the centrate diatoms, the major food source for fish larvae, are the first killed by the pollutants. When centrate diatoms vanish, they are replaced by the kind of organisms that young jellyfish prefer to eat. If this happened in a large area of the sea, the fish population would eventually decline and the area would be overrun by jellyfish.

Nuclear waste disposal. Eight oceanographic institutions, commissioned by the Energy Research and Development Administration to study the possibility of burying radioactive waste beneath the sea floor, issued a progress report in March 1976. A total of 1,815 metric tons (2,000 short tons) of radioactive waste has been generated since nuclear power stations began operating in the United States in the late 1950s. By 1985, with a predicted 200 or more nuclear plants operating, the annual waste will reach 3,176 metric tons and will increase 20 per cent each year.

So far, the institutions have agreed that radioactive wastes could be enclosed in canisters and placed in holes drilled in the earth's crust under the sea. The holes would then be filled with concrete. In looking for suitable locations, the scientists are studying the properties of underlying rock and how bottom sediments mix with water. Should a buried container leak, they want to know how long it would take contaminated material to rise to the surface. They believe that offshore oil-drilling technology, coupled with what has been learned during the Deep Sea Drilling Project, would allow engineers to design a safe system for burying nuclear wastes. [Feenan D. Jennings]

Physics

Atomic and Molecular Physics. Researchers used lasers in 1975 and 1976 to study the intermediate- and high-energy levels of atoms and to form tiny particles in a laser beam. They also demonstrated a new method of amplifying light, which may serve as the basis for future lasers.

Giant atoms. When an atom absorbs energy, one or more of its orbital electrons moves to larger orbits, resulting in an excited atom that can be as much as 1,000 times larger than an unexcited one. Given enough energy, the electron can get so far away that it becomes free from the binding force of the atom's nucleus and leaves an ion behind. The minimum energy needed to remove the electron is called the ionization energy.

The laws of quantum mechanics limit energy changes in atoms to a series of discrete energy states that lie increasingly closer together as an atom approaches its ionization energy. A quantum number labels each state.

Atomic physicists have thoroughly studied the lower energy states of lightweight atoms, those with quantum numbers up to about 5. But physicists have been unable in the past to study higher energy states because they differ so little in energy. Also, highly excited atoms are so large that they often collide. The collisions usually supply the small amount of energy needed to ionize the excited atom. Consequently, unless they are in a dilute gas where collisions are infrequent, highly excited large atoms have short lifetimes.

Daniel Kleppner and his co-workers at the Massachusetts Institute of Technology (M.I.T.) in Cambridge reported in late 1975 that they had excited sodium atoms to energy states having quantum numbers between 26 and 37 in a low-density gaseous beam. They excited them with light from a pair of dye lasers, tuned to deliver precisely the energy difference between the states.

One laser supplied yellow light that boosted an absorbing sodium atom to a state with quantum number 3. By slightly changing the energy, or frequency, of a blue laser, the scientists pushed the partially excited atoms from state 3 to states between 26 and 37.

Researchers adjust a pair of dye lasers. Only one beam is visible. The lasers are used to excite sodium atoms in the vacuum chamber at rear to probe the atoms' energy-level structure.

Physics

Continued

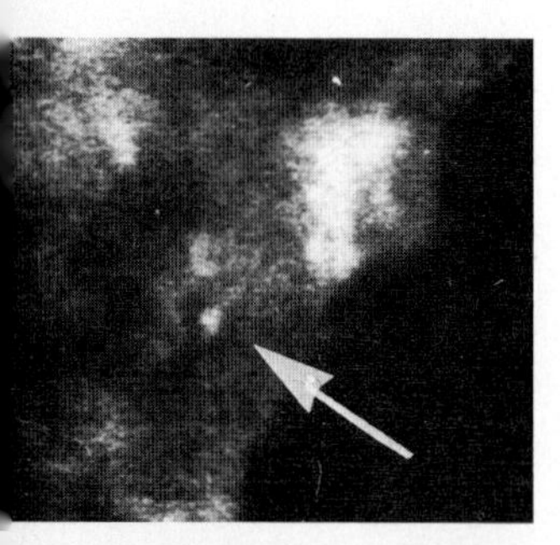

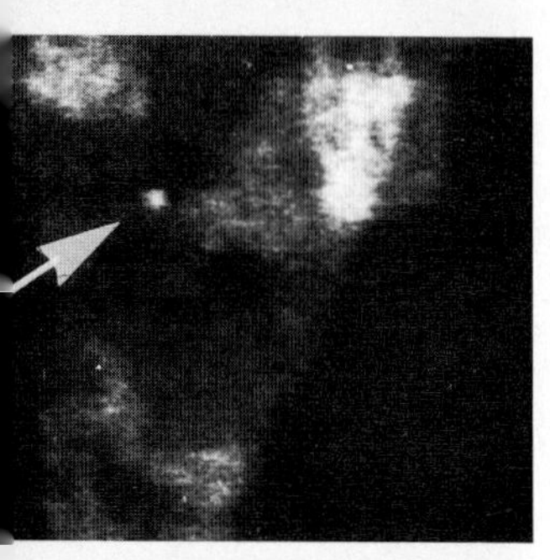

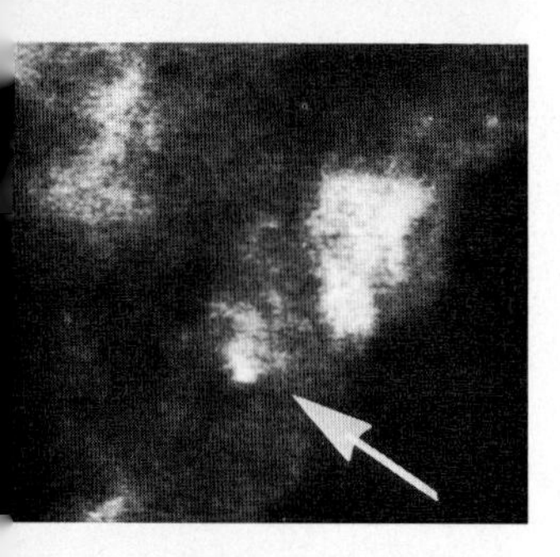

Arrows point to a mercury atom, magnified 4.5 million X, moving about on a thin carbon film. Micrographs were taken with a scanning transmission electron microscope. Scientists pieced such micrographs together in 1976 to make the first "motion picture" of the actions of individual atoms.

The M.I.T. group probed the structure of these giant atoms by applying an electric field across the gaseous beam. When that field supplied enough energy to overcome the binding pull of the atomic nuclei, the electrons became free and the remaining ions could be detected. Then, by increasing the field strength in discrete steps, the group counted the electrons in the various high-energy states.

Their method may have practical applications for astronomers. Kleppner has suggested that highly excited atoms might be used to detect photons of light in the infrared to microwave regions of the electromagnetic spectrum.

In related experiments at the Stanford Research Institute (SRI) in Menlo Park, Calif., Thomas F. Gallagher, Stephen A. Edelstein, and Robert M. Hill studied the sodium energy states with quantum numbers between 5 and 10. They irradiated sodium vapor in a pyrex cell with a pair of short-pulse dye lasers. They observed the effect of atomic collisions on the lifetimes of the excited states – the average time between an atom's excitation and subsequent return to a lower energy state by emitting a photon.

For quantum state 8, the SRI researchers measured a lifetime of 0.5 microsecond (millionth of a second) – a rather long time by atomic standards. They then introduced small amounts of inert helium or neon gas into the pyrex cell. They expected to see a more rapid decay caused by more frequent collisions. But collisions with neon atoms increased the lifetime of the 8th state to 1.39 microseconds.

The researchers attributed this puzzling effect to the mixing of various excited states by the collisions. Normally, the laser light drives a sodium atom to quantum state 8 and a photon is emitted when the atom returns to quantum state 3. However, a collision with an inert gas atom can very easily drive the sodium atom from its 8th state to a slightly more energetic 8th state having higher angular momentum. The laws of quantum mechanics forbid a direct relaxation from this state to quantum state 3. The atom remains in the higher state until a collision returns it to its original 8th state from which it then can relax to state 3. The time spent in the higher angular momentum state accounted for its prolonged lifetime.

Particle formation. While performing experiments in which laser light passes through cesium vapor in 1973, Columbia University researchers noticed that very small particles formed in the beam of light. Temporarily putting their original experiment aside, Andrew C. Tam, George Moe, and William Happer performed further tests to confirm the observations, and reported their findings in December 1975.

The scientists passed blue light from an argon ion laser through a glass cell that they kept heated to about 300°C (572°F.) to produce cesium vapor. To produce the particles, the cell had to hold enough hydrogen gas for random collisions to supply the small energy differences between certain laser frequencies and cesium energy states. Excited cesium atoms then formed cesium hydride plus a hydrogen atom upon colliding with hydrogen molecules. The cesium hydride molecules clumped together to form the particles.

Free-electron laser. Luis Elias, William Fairbank, John Madey, H. Alan Schwettman, and Todd Smith at Stanford University in Palo Alto, Calif., reported in December 1975 that they had demonstrated a new laser mechanism. They focused a 10.6-micron laser beam on a stream of electrons with an energy of 24 million electron volts traveling through a magnetic field. The laser beam was amplified 7 per cent per pass along the electron's line of flight.

A special double helix winding produced the magnetic field at right angles to the length of a 5.2-meter (17-foot)-long tube. The direction of the field rotated in screwlike fashion, making a complete revolution every 3.2 centimeters (1.25 inches) of tube length.

As the high-energy electrons crossed this field they could absorb or emit radiation. If both absorption and emission were equally likely, no light amplification would be expected. However, because the photons of laser light impart some momentum to the electrons when they are absorbed or emitted, the wavelength of the absorption photons differs slightly from the wavelength of the emission photons. This slight difference makes laser light amplification possible. [Karl G. Kessler]

Elementary Particles. Startling discoveries that open new areas of research are the sweetest fruits of science, and particle physics was blessed with a bumper crop in 1975 and 1976. Advances were expected in the wake of the discovery of the first psi particle on Nov. 10, 1974, but events far outran expectations. Although the results are still too new for physicists to be sure of their exact significance, it is clear that yesterday's speculation can quickly become today's experimental fact.

According to current theories, all matter can be built up from two kinds of simple subunits, called leptons and quarks, that have no internal structure. Based on recent experiments, physicists view the subunits as point concentrations of mass and electric charge.

Two families are founded on these subunits. Electrons belong to the lepton family, as do heavier negative particles called muons and two neutral particles known as neutrinos. Protons, neutrons, and about 300 unstable particles that make up the hadron family are combinations of quarks. The main difference between leptons and quarks is that the latter are never found alone. Quarks are always observed bound to other quarks, or to their antimatter companions, antiquarks. Leptons are unaffected by the force that produces these permanent bonds.

Explaining why quarks form such tenacious bonds has been the major loose end in the quark theory of hadrons. But the discovery that the psi particle exists in a variety of forms, each at a different level of energy, gave physicists a way to test the theory. Because the psi is believed to be a new type of quark bound to its antiquark, these levels must represent different patterns of motion by the quark and antiquark.

Ten such levels had been observed by May 1976. They were found at the SPEAR electron-positron storage ring of the Stanford Linear Accelerator Center (SLAC) in Palo Alto, Calif., and at DORIS, a similar ring at the DESY laboratory near Hamburg, West Germany. The SPEAR data came from the same team that discovered the psi – a collaboration of SLAC and University of California Lawrence Berkeley Laboratory (LBL) researchers.

All three groups were observing particles produced in head-on collisions between electrons and their antiparticles, positrons. In such collisions, the two particles can annihilate one another, producing an intense concentration of energy in a small space. This energy must go somewhere, and the formation of a quark-antiquark pair is one likely outcome.

Particle physicists have long lamented that nature had not seen fit to provide them with anything as simple and useful in their field as the energy-level pattern of the hydrogen atom was in atomic and molecular physics. This pattern was the key clue that led Danish theorist Niels Bohr to formulate his quantum theory of atomic structure in 1913.

Bohr was dealing with a well-understood force, the electrical attraction between hydrogen's lone electron and its nucleus. Armed with this knowledge, he invented rules that restricted the electron's motion to orbits with exactly the observed energies. More general versions of these quantum rules are now well established. Particle physicists hope to reverse Bohr's reasoning process and use the energy levels of the psi to understand the force binding quark to antiquark.

A constant force? Before this can be done, physicists must establish quantum numbers that describe the exact pattern of quark-antiquark motion for each level. For example, a particle can carry 0, 1, 2, or more units of angular momentum or some other characteristic, and these numbers determine the energy levels. Fewer than half of the psi's energy levels have yielded quantum numbers, but one feature already looks promising. The pattern may well be the result of a force that does not depend on how far apart the quark and antiquark are. And the force is immense, more than 10 tons. This is revealed by the spacing of the psi's energy levels – 100 million times greater than those of hydrogen.

A force of this type guarantees that quarks can never completely separate. No matter how hard they are hit, this constant force always reels them back in. Such a force could arise in a simple fashion. If the field that transmits the force is confined to the line joining the

Psi: An Atom for Elementary Particle Physics

Ten energy levels for the psi particle, *right,* had been observed by May 1976. Each level represents a different pattern of motion for the quark and antiquark believed to make up the psi. The energy levels are spaced 100 million times farther apart than those of a hydrogen atom, *far right,* which represent the quantized motion of an electron around a proton. Blue lines represent changes between energy levels.

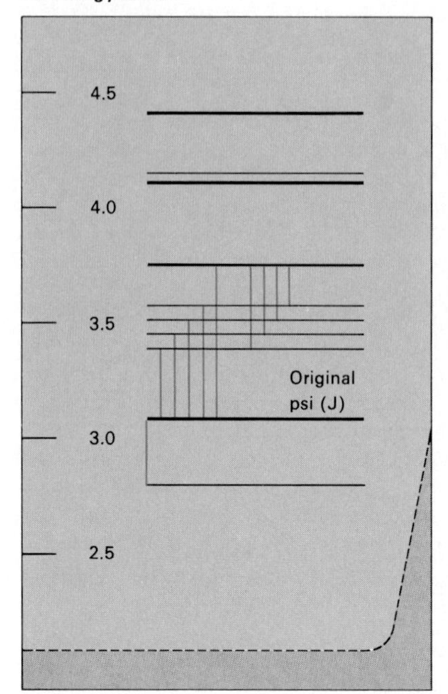

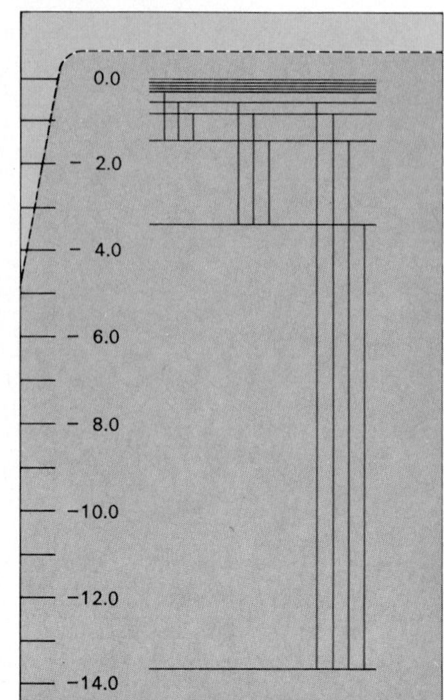

Physics

Continued

particles, it would not diminish with distance. An electrical or gravitational field spreads out equally in all directions from its source and weakens as the distance from the source increases. As a result, objects bound by these forces can eventually get free.

How many quarks? Although the quark model is now on a sounder footing, the number of different types of quarks required to explain all of the known hadrons has become an open question. Ordinary matter requires only two quarks, the basic building blocks of protons and neutrons. By 1964, when Murray Gell-Mann and George Zweig of the California Institute of Technology in Pasadena formulated the quark theory, a third quark was required. It was dubbed a strange quark because the particles that contained it acted in a strange way. The existence of the psi particle requires at least a fourth quark. And another feature of the SLAC-LBL data suggests that there may be more.

The SLAC-LBL physicists found that when positrons and electrons collide at energies slightly greater than those required to form a psi, the chance of forming hadrons rather than leptons increases dramatically. In the simple quark picture, all combinations of particle and antiparticle have an equal chance of emerging from an electron-positron annihilation. Thus, the only way to increase the chances that hadrons will be produced is to increase the number of kinds of quarks from which they can be built. The data suggest that as many as six kinds of quarks may be required.

Do quarks have charm? By far the greatest confusion surrounds experimental attempts to settle a question originally regarded as an easy one. Is the new quark that is responsible for the psi the same as the charmed quark invented in 1964 by Sheldon L. Glashow of Harvard University and James Bjorken of SLAC? Like the strange quark, the charmed quark is another variety. In the charmed-quark theory, later refined by Glashow and various co-workers, leptons and quarks are regarded as close relatives. The weak interaction, a process which can transform one kind of quark or lepton into another, is thought to operate almost identically on both particles.

When the charmed-quark theory was proposed, there were four known leptons and only three quarks. For balance, the theory required a fourth quark. When the psi was discovered, many theorists reasoned that it was likely composed of a charmed quark-antiquark pair. The crucial test of this theory was to find a charmed particle, one with net charm. It would be made of a charmed quark bound to another kind of quark with no charm.

Charmed quarks can be identified because they tend to transform themselves into strange quarks through a weak interaction. This should happen too quickly to observe a charmed hadron directly, so physicists must infer its presence by examining the products of its transformation. The signature of a charmed hadron is a weak interaction producing a strange hadron.

Events with exactly the right signature were announced in January 1976 by two groups observing neutrino reactions in bubble chambers, devices that produce photographs of the tracks of moving charged particles. One group, a multinational team at the European Center for Nuclear Research (CERN) in Geneva, Switzerland, used a bubble chamber named *Gargamelle.* The other group, a collaboration of physicists from the universities of California, Hawaii, and Wisconsin, and CERN, used a bubble chamber at Fermilab near Chicago.

Both teams observed events in which a neutrino collided with a nucleus in the bubble-chamber liquid and produced a muon, a positron, and a strange hadron. The presence of the muon implied that the neutrino had interacted and been transformed from an uncharged, and therefore trackless, particle to a negatively charged muon. In order to balance the negative electric charge picked up by the neutrino in this process, a quark inside a nucleus in the bubble-chamber liquid must also have been transformed. Because a strange hadron was produced, theorists

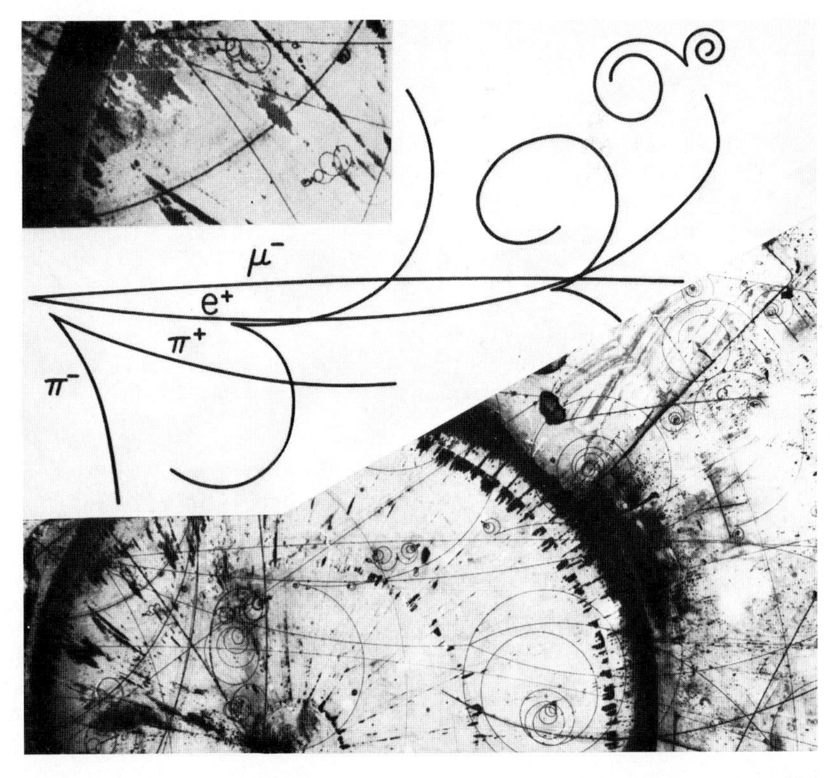

Tracks of fast-moving charged particles in a bubble chamber suggest that charmed particles exist. A neutrino (not seen), moving from left to right, collides with a nucleus in the bubble-chamber liquid. The collision produces a muon (μ^-), a positron (e^+), and a strange particle—a neutral kaon (not seen)—which breaks up into two pi mesons ($\pi^+$and π^-).

infer that it is the result of a charmed hadron undergoing transformation by the weak interaction. The positron, a lepton, was a signal that the charmed hadron was indeed transformed by a weak interaction. Thus, the charm signature was complete, and it seemed likely that the Glashow theory had been confirmed.

But the Fermilab photos had revealed 14 examples of the process by April, and an embarrassing fact became evident – the charm signature occurred too often. A typical event contained not just one but an average of four strange hadrons. While Glashow's theory might be stretched to accommodate an average of two, no existing theory works with as many as four strange hadrons.

Early in June, however, the SLAC-LBL team found events that produce a strange hadron and either one or three ordinary hadrons. They are all electrically charged, and thus observable. As a result, the SLAC-LBL researchers could measure the mass of the short-lived charmed hadron particle that transformed into the event products. Its mass is 1.86 GeV (billion electron volts), about twice the mass of a proton. This value fits the Glashow theory well.

How the particle observed at SLAC is related to the events at Fermilab and CERN is an open question. Several experiments suggest that more than one new phenomenon is being observed. A subgroup of the SLAC-LBL team of SPEAR has seen the same signature, minus the strange particles. A team from Harvard, Pennsylvania, and Wisconsin universities and Fermilab has also seen the lepton part of the signature with an electronic neutrino detector that cannot tell whether strange particles are present. But the pattern of lepton energies does not match that found in the bubble chamber.

One intriguing possibility is that, in addition to new quarks, new leptons may be involved in these phenomena. The SLAC-LBL data seem to fit this explanation better than any other. If it proves necessary to have six quarks, six leptons would be desirable to preserve the quark-lepton kinship of Glashow's theory. Better experimental data are needed before further theoretical progress can be made. [Robert H. March]

Nuclear Physics. Researchers perfected a technique during 1975 and 1976 that lets them measure the electromagnetic moments of unstable, excited nuclei, even those that last for only a picosecond (one-trillionth of a second) following the collisions that produce them. Electric and magnetic moments arise from the distribution and motions of charged particles in a nucleus. Measurements of nuclear electromagnetic moments have played key roles in the development of theories that describe the structure of nuclei.

The new technique refines one in which researchers polarize, or align, the spins of excited nuclei, then expose them to electric or magnetic fields. Physicists can calculate the moments by observing how such fields change the nuclei's emissions of characteristic radiations when the nuclei return to their original states. Unfortunately, random effects can destroy the polarization before the measurements are made. But Kunio Sugamoto and his colleagues at Kyoto University in Japan showed in 1968 that the polarization remains for long periods if the aligned nuclei are in special crystals that have highly symmetrical internal fields.

Stanley S. Hanna and his colleagues at Stanford University in Palo Alto, Calif., refined this method, and reported in 1975 that they could maintain the polarization of nuclei for as long as 10 seconds by using carefully chosen crystals. They used a polarized beam of particles from an accelerator to induce a nuclear reaction creating a desired excited nuclear state in a crystalline target including a suitable target nucleus. The recoiling products of the reaction reflected the polarization of the projectile beam and were trapped in the crystal lattice where their lifetimes were measured. This new technique promises to provide a wealth of new electromagnetic moment measurements that will pose stringent tests for theories about all atomic nuclei.

Superdense nuclear matter. Physicists used superhigh-energy, heavy-ion projectiles from the Lawrence Berkeley Laboratory's Bevalac in California and the Synchrophasatron at the Joint Institutes for Nuclear Research in Dubna, Russia, to search for a new type of superdense nuclear matter. T. D. Lee

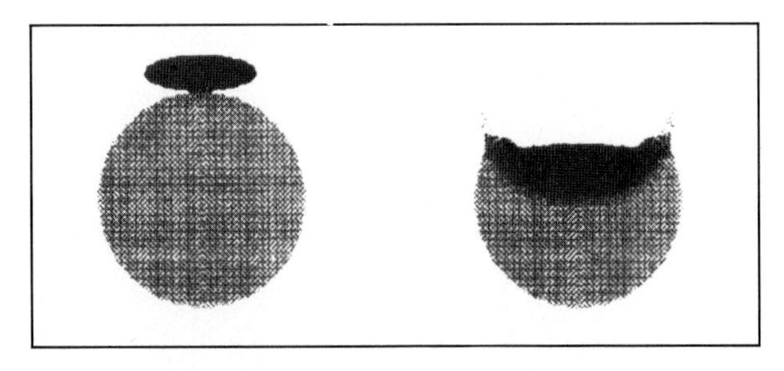

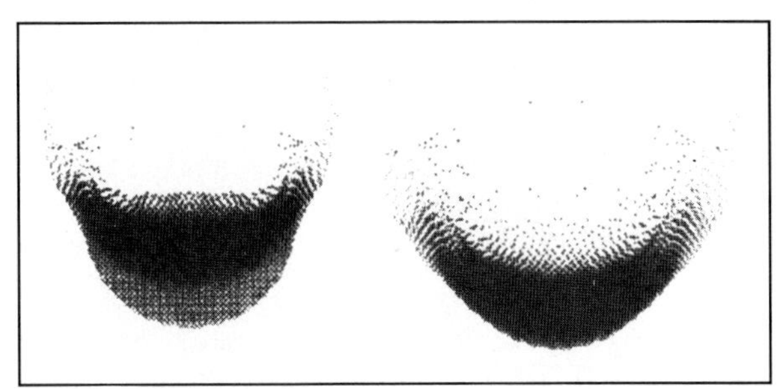

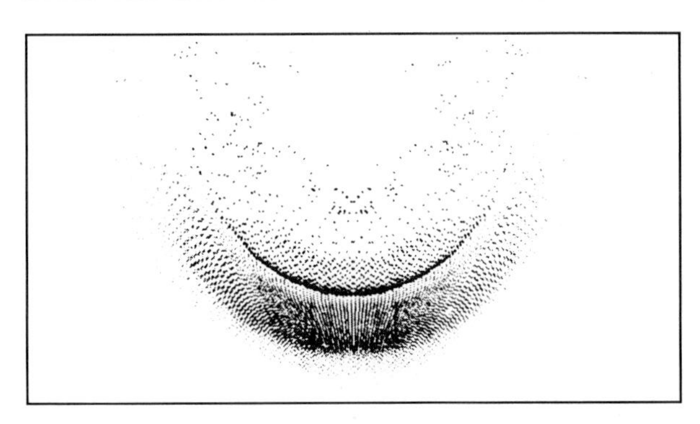

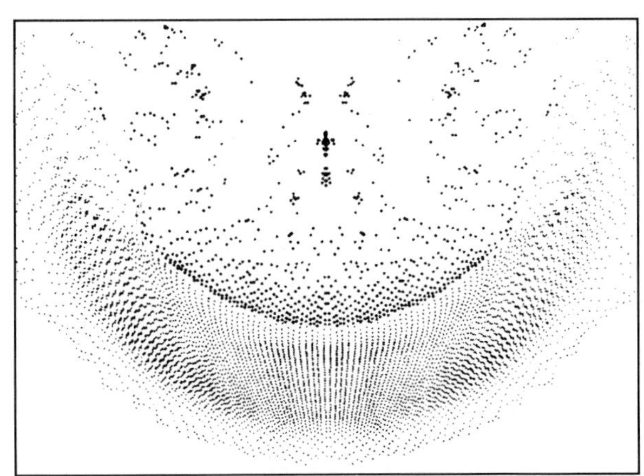

Computer drawings show what may happen when an energetic oxygen-16 nucleus (moving from top to bottom) hits and merges with a stationary silver-107 nucleus. A shock wave compresses the nuclear matter, which flies off in the direction of the shock-wave front. The nuclei are treated as drops of fluid for the computer calculation.

and G. C. Wick suggested in 1973 that superdense matter might exist. Walter Greiner and his associates at the Institute for Theoretical Physics in Frankfurt, West Germany, calculated that superdense nuclear matter might form and remain briefly stable if the density of the matter can be increased to several times its normal value.

Researchers look for shock waves created when a small, heavy ion such as oxygen hits a heavier and larger nucleus such as silver or uranium. Like the shock waves that aircraft cause in air when they exceed the speed of sound, the nuclear shock waves would form at an angle determined by the ratio of the incident projectile's velocity to the velocity of sound in the larger nucleus. The nuclei disintegrate and fragments fly off in the wave-front direction.

Erwin Schopper and his colleagues at the University of Frankfurt reported preliminary evidence of shock waves from the Bevalac early in 1975. They measured greater fragment yields at a particular angle with respect to the beam direction and took this to be the angle at which the wave front was moving. But more detailed measurements at Berkeley by two independent teams headed by Harry H. Heckmann and Andrew M. Poskanzer neither have such a simple interpretation nor show clear evidence of shock waves.

Meanwhile, Schopper's team began using alpha particle beams from the Dubna Synchrophasatron. Measurements over a wide range of alpha particle energies up to 6 billion electron volts per nucleon in 1976 again suggest shock waves and nuclear densities several times greater than normal.

Recent theoretical calculations of the superdense matter are as contradictory as experimental results. These calculations have tended to fall into two classes. One group considers the two nuclei to be fluids and uses the normal techniques of hydrodynamics. The other group recognizes the discrete nucleon structure of the interacting nuclei and performs a microscopic, collision-by-collision, nucleon-by-nucleon calculation using large digital computers.

Raymond Nix and his co-workers at the Los Alamos Scientific Laboratory in New Mexico used the hydrodynamic method and predicted significant densi-

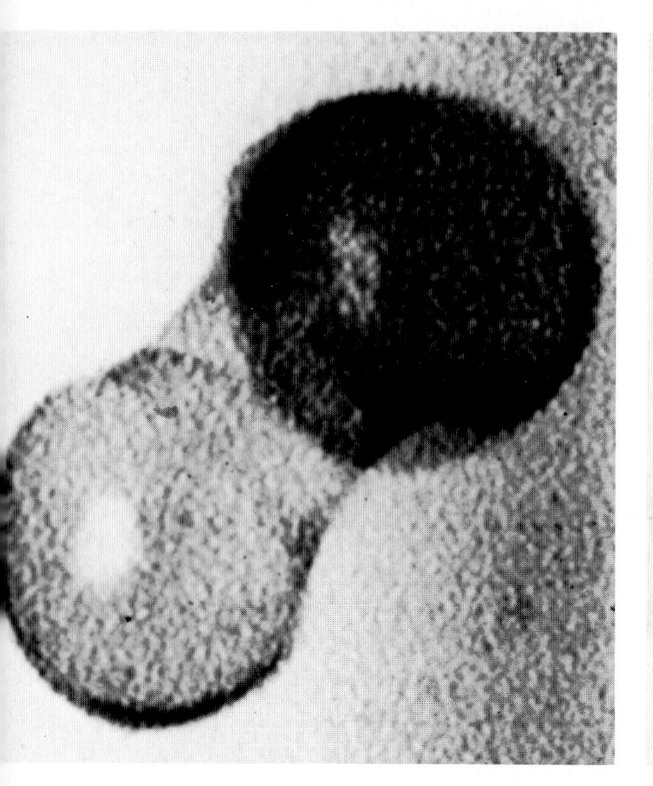

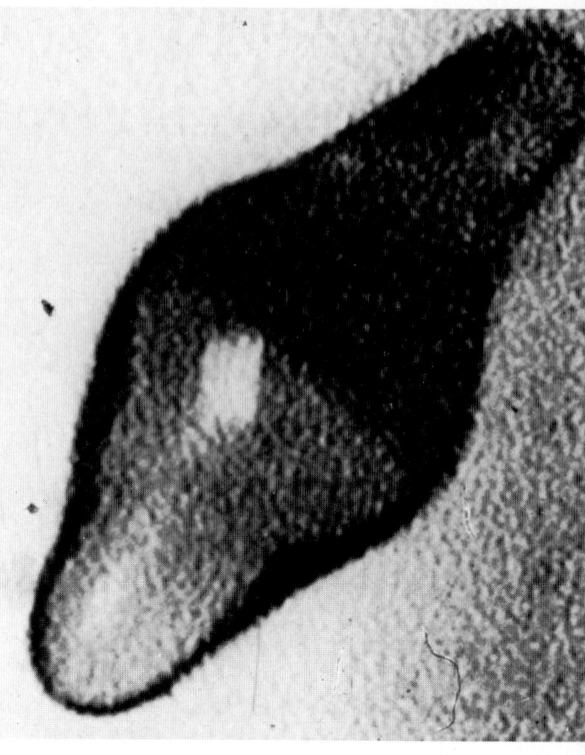

Physics

Continued

Two floating water drops, weightless in space, collide, *above*, and merge, *above right*. The collision is a pictorial representation of how certain heavy atomic nuclei collide.

ty increases. But John Negele and his colleagues at the Massachusetts Institute of Technology (M.I.T.) in Cambridge, using the microscopic method, found that the density should not increase substantially.

Hypernuclei. In a 1975 experiment at the European Center for Nuclear Research (CERN), Geneva, Switzerland, Bogdan Povh and his co-workers from the University of Heidelberg in West Germany found evidence for a "strange" kind of nucleus predicted more than 10 years earlier. Normal atomic nuclei are composed of two kinds of nucleons, neutrons and protons. Both belong to the family of elementary particles called baryons. A third kind of baryon, the lambda particle, differs from the proton and neutron only in possessing a quality called strangeness (see PHYSICS [Elementary Particles]).

The lambda particle is not normally found in nature because it lives for less than a billionth of a second. But by nuclear standards this is long enough for a lambda particle to replace a neutron or proton in a normal atomic nucleus to create a hypernucleus.

A simple hypernucleus was originally proposed independently in 1966 by Herman Feshbach and Arthur K. Kerman of M.I.T. and in 1965 by Harry J. Lipkin of the Weizmann Institute in Israel. It would result from a collision in which one of the nucleons in a normal nucleus changes into a lambda particle while all other aspects of the target nucleus remain unchanged.

Povh directed a beam of strange particles called K-mesons at such nuclei as beryllium, carbon, and oxygen and observed nonstrange particles called pi-mesons as the reaction products. Lipkin and John P. Schiffer of the Argonne National Laboratory near Chicago interpreted these measurements as involving the transfer of strangeness from a K-meson to a nucleon, converting it to a lambda particle. Excess momentum was carried off by the pi-meson, precisely as predicted by Feshbach, Kerman, and Lipkin. Such experiments reveal particle distributions in nuclei. [D. Allan Bromley]

Plasma Physics. Researchers conducted or planned several large experiments in 1975 and 1976 to determine the scientific feasibility of controlled thermonuclear fusion reactors. They hope to know by the early 1980s whether fusion, the process that produces heat in the sun and other stars, can be harnessed on earth to tap the vast supply of fusion fuel in the oceans – the hydrogen isotopes tritium and deuterium.

A successful reactor must heat the fuel to more than 100 million °C, hot enough so that tritium and deuterium atoms fuse and form helium, releasing useful energy in the process. The atoms are stripped of their electrons at these high temperatures, producing a hot gas of positive ions and electrons called a plasma. The plasma must be confined in such a way that its density (expressed in ions per cubic centimeter) times the confinement duration (in seconds) is greater than 10^{14}. Finally, the density of such heavy impurity ions as molybdenum must be kept below about 0.5 per cent of the electron density.

Magnetic confinement. In this approach to thermonuclear fusion, a strong magnetic field confines the superhot plasma. Most research is concentrated on tokamak devices, which were developed by Russian scientists in the mid-1960s. In them, a field about 100,000 times the earth's magnetic field confines the plasma in a toroidal, or doughnut, shape. Such a field exerts a containing pressure on the plasma equal to 100 times the earth's atmospheric pressure. A current flowing through the plasma heats it in the same manner as current flowing in a wire.

Bruno Coppi, Ronald Parker, and their co-workers at the Francis Bitter National Magnet Laboratory at the Massachusetts Institute of Technology in Cambridge reported in November 1975 that they had achieved a fivefold improvement in magnetic confinement. Using a tokamak called Alcator, they produced a plasma for which the product of density and containment time was 10^{13}. The plasma reached a temperature of only 8 million °C, but was essentially free of unwanted impurities.

The Tokamak-10 at Moscow's Kurchatov Institute began to operate in late 1975.

Researchers are examining ways to increase the plasma heat. The most promising technique is to inject a high-velocity beam of neutral atoms into the plasma. The atoms are unaffected by the magnetic field and deposit their kinetic energy in the plasma during collisions. Neutral-beam heating has been successfully demonstrated in such tokamaks as the ATC at Princeton University's Plasma Physics Laboratory in New Jersey, the ORMAK at Oak Ridge National Laboratory in Tennessee, and the TFR at Fontenay-aux-Roses, France. Plasma ions were heated from about 4 million °C to 13 million °C in these experiments.

Magnetic mirrors. Researchers at the University of California's Lawrence Livermore Laboratory (LLL) used neutral-beam heating to produce temperatures of 130 million °C. A team headed by Frederic Coensgen used LLL's 2XII-B magnetic mirror device, so named because strong magnetic fields at the ends of the device reflect, or mirror, the plasma particles back into the system.

The LLL researchers reported their high-temperature results in July 1975, and by March 1976 had reached a density-confinement product of 10^{11} while maintaining a plasma temperature above 100 million °C. These results indicate that mirror-confinement devices, as well as tokamaks, are approaching the reactor requirements.

The confinement and heating results are being extended by two larger tokamaks, the Russian Tokamak-10 (T-10) at the Kurchatov Institute in Moscow and the Princeton Large Torus (PLT) at Princeton. These devices began operating in 1975 at plasma-heating currents of about 1 million amperes. A tokamak reactor will probably require plasma-heating currents of from 5 to 10 million amperes.

Initial operation of both T-10 and PLT are very encouraging – plasmas produced by resistance heating have temperatures of 20 million °C, densities of 5×10^{13} ions per cubic centimeter, and confinement times of 0.05 second. The Princeton experimenters plan to inject several million watts of power into the PLT late in 1976 using neutral beams to reach temperatures of 40 million to 50 million °C. [Dale M. Meade]

Solid State Physics. Physicists made important advances during 1975 and 1976 bouncing neutrons off liquids and solids to study the internal motions of the atoms or molecules in those substances. The technique is much like a game of pool played with incredibly small balls – the neutron is the cue ball and the material's atoms are the object balls. Knowing the velocity and direction of a neutron before and after a collision, physicists can calculate the recoil velocity and direction of the atoms and molecules of a material. They can thus determine how atoms and molecules can be set in motion in materials.

Researchers need intense beams of neutrons for such studies. A team of scientists, headed by John M. Carpenter at Argonne National Laboratory (ANL) near Chicago demonstrated a new concept in neutron production that promises to be cheaper and better than the previous sources. Hydrogen ions (H^-) – atoms with an extra electron – are produced and accelerated to an energy of 750,000 electron volts (eV) in a Cockcroft-Walton accelerator. The H^- beam is then passed through two additional accelerators. A linear accelerator increases the energy to 50 million eV, then a proton synchrotron boosts the energy to 2 billion eV. When the beam passes through a thin plastic foil between these two accelerators, all electrons are stripped from the H^- ions. The resulting high-energy proton beam leaves the synchrotron and hits a lead target, producing great numbers of high-energy neutrons. A polyethylene moderator slows the neutrons to velocities useful for experiments.

Helium-3 waves. Using the CP-5 reactor at ANL between February and May 1976, Kurt Skold, Charles Pelizzari, Robert Kleb, and George Ostrowski completed a study of neutron scattering in liquid helium. Helium is the only material that exists as a liquid near absolute zero (−273°C or −459.7°F.). Two kinds of helium atoms exist in nature. Helium-3 (^{3}He) has two protons and one neutron in its nucleus, and the much more abundant helium-4 (^{4}He) has an additional neutron. The properties of ^{3}He and ^{4}He are notably different. At 2.2°C above absolute zero, ^{4}He becomes a superfluid able to flow through small tubes with absolutely no

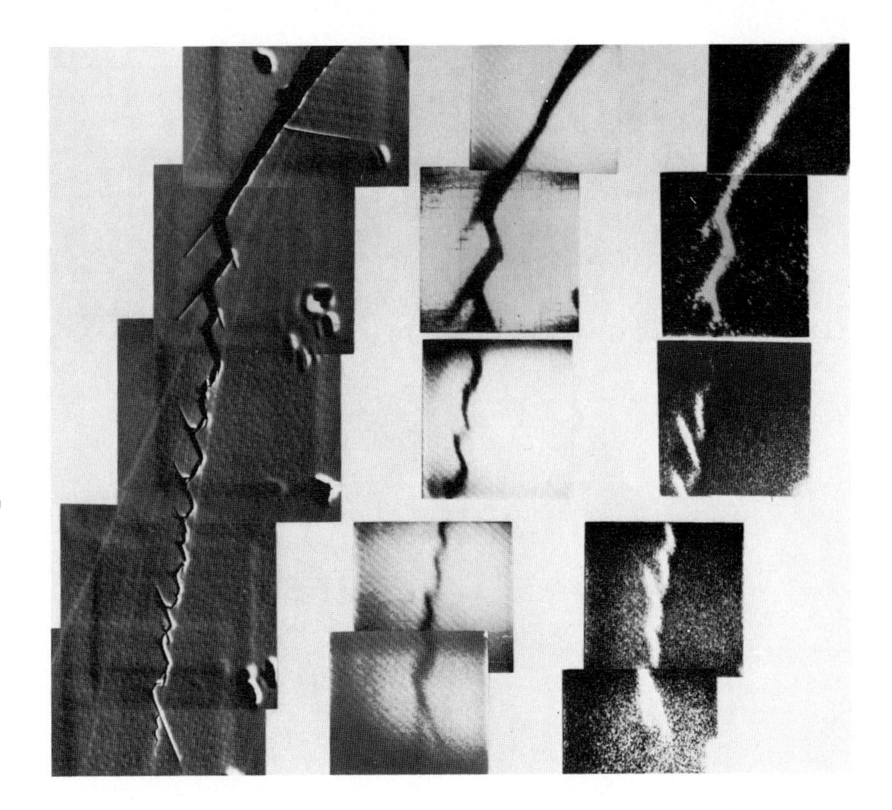

Cracks in niobium from exposure to hydrogen are revealed, left, by a scanning electron microscope. An ion microprobe, which is sensitive to chemical composition, produces photographs of the possible cause. Niobium ions appear in the center photograph and niobium hydride ions, which are thought to play a key role in the development of cracks, appear at right.

resistance. But ^{3}He increasingly resists flow as the temperature falls. Then suddenly it, too, becomes a superfluid at about 0.002°C above absolute zero.

The unusual behavior of ^{3}He allows it to transmit a unique sound wave called zero sound that cannot exist in ordinary fluids. University of Illinois physicists in Urbana produced zero sound in ^{3}He in 1966 using vibrating quartz plates. Because Skold's group used neutrons to excite the liquid helium, they showed that zero sound waves having wavelengths that are 1,000 to 10,000 times shorter than those previously studied can be excited in ^{3}He.

The researchers first passed a steady stream of neutrons through a shutter device that broke the stream into bursts of neutrons. The neutron bursts recoiled after colliding with ^{3}He atoms near the surface of the supercold target, and a neutron detector sensed their arrival. By correlating the arrival time with the time the neutrons passed through the shutter, the physicists could calculate the velocities of the recoiling neutrons. The amount of energy and momentum given to the sample indicated that the neutrons had set up zero sound waves in the ^{3}He.

In January and February 1976, Pat R. Roach of ANL and John B. Ketterson of Northwestern University in Evanston, Ill., verified another predicted quality of ^{3}He – the ability to transmit transverse waves. Sound waves in ordinary liquids are longitudinal, causing the molecules to move back and forth in the same direction that the wave moves. With transverse waves, the molecules move back and forth at right angles to the wave motion like a vibrating string.

We directed a sound wave with a frequency of 12 million hertz at a sample of liquid ^{3}He that was 0.0025 centimeter (0.001 inch) thick and measured both the reflected and transmitted waves. The experiments were run at temperatures as low as 0.002°C above absolute zero. By studying the reflected and transmitted waves we found clear evidence for transverse waves.

Weighing a neutron. In the quantum mechanical view of matter, all particles have a wave nature, and a

Physics

particle's wavelength is inversely proportional to its momentum. Suppose it were possible to split a beam of neutrons into two parts, one of which traversed a higher path than the other. Neutrons taking the higher path would have a smaller average momentum because the earth's gravitational pull would slow them. According to quantum mechanics, they would have a slightly larger average wavelength. Reflected and recombined, the two parts would interfere in much the same way that light waves do.

Roberto Colella and Albert W. Overhauser of Purdue University in Lafayette, Ind., and Samuel A. Werner, now at the University of Missouri, used a nearly perfect silicon crystal to split, reflect, and recombine a neutron wave. They used the interference pattern to calculate the weight of a neutron. Their experiment is important because it combines the effects of gravity and quantum mechanics.

Brittle niobium. Niobium may be used to make the walls of nuclear fusion reactors because it has certain desirable nuclear properties and a high melting temperature. But in the reactor the metal would be exposed to tritium, an isotope of hydrogen, and many metals, including niobium, become brittle and tend to crack when exposed to hydrogen at high temperatures.

Peter Williams, Charles A. Evans, Martin L. Grossbeck, and Howard K. Birnbaum of the University of Illinois in Urbana used a scanning electron microscope and an ion microprobe to study cracks in niobium. In the ion microprobe, a beam of high-energy oxygen ions struck the metal and produced either niobium ions or niobium hydride ions that formed an image.

Hydrogen is thought to move to those regions of the largest tensile, or stretching, stress and either weakens the bonds between niobium atoms or forms the brittle chemical compound niobium hydride. One microprobe showed niobium hydride concentrated at the crack and in the high-tensile stress region where the crack ends. The researchers hope to develop ways to control the cracking. [John B. Ketterson]

Psychology

The possibility that early visual experience may irrevocably affect the development of the visual cortex of the brain was reconsidered in July 1975 by Colin Blakemore and Richard Van Sluyters of Cambridge University in England.

Psychologists believe that both genetic factors and experience are important in determining how the developing brain learns to use information received from the senses. In one widely used technique for studying the separate contributions of these two factors, researchers raise experimental animals in controlled environments where the amount and kind of visual stimulation can be varied for different groups. Behavioral and physiological tests then reveal how differences in early visual experience affect the development of the visual cortex, where visual information is processed.

Pioneering work with kittens in the 1960s showed that even temporary deprivation of light to one or both eyes dramatically altered both brain functioning and behavior. An experimental kitten that had one eye covered behaved like a normally reared kitten in later tests only if the light-receiving eye was open. With only the light-deprived eye open, the kitten behaved as if it were blind. These effects continued even when the kitten was exposed to normal visual experiences. This suggested that the functioning of the adult visual cortex might be completely and irrevocably set by an animal's early visual experiences.

Blakemore and Van Sluyters set out to investigate this problem further. They exposed experimental kittens to various kinds of selected visual experiences by allowing them to look at restricted patterns of lines with one eye covered. They then compared recordings from cells in the visual cortices of the experimental kittens with those of kittens raised in a normal visual environment. They found that nerve cells in the visual cortex developed normally only if both eyes had been exposed at the same time to the various patterns.

They also compared their data to the results of similar experiments with rabbits and reported that rabbits appear

An overturned snail, *top,* uses small pebbles to help right itself. The snail extends its foot over the edge of its shell to probe for stones, then moves the stones to the back of the foot, *center pictures.* When enough weight is accumulated, the shell tips and the snail can right itself again, *bottom.*

less affected than cats by early manipulations in the visual environment. They suggest that early experiences may profoundly affect the developing visual systems only in species with binocular vision, such as cats, monkeys, and humans, which use both eyes to produce three-dimensional vision.

Visual illusions. Certain arrangements of lines and figures on a background may lead a viewer to believe he sees surfaces and outlines that are not there, Gaetano Kanizsa of the University of Trieste reported in April 1976.

The common factor in the displays is the presence of certain elements that appear to be incomplete representations of a familiar form or figure. For example, Kanizsa found that viewers tended to see the outlines of three angles as a completed triangle. The breaks in the joining lines led them to visualize another figure hiding the missing parts of the first figure. If the broken lines are joined, the second figure disappears.

Masculine-feminine scale. Sandra Bem of Stanford University in California in September 1975 challenged the traditional view of masculine and feminine as opposite end points on a single scale or dimension.

Bem has developed a sex-role inventory containing separate lists of masculine and feminine characteristics. She tested the lists on 1,500 male and female Stanford undergraduates. Asked to check those traits that they felt best described themselves, about half of the students chose only traits considered appropriate for their sex. Another 15 per cent checked only traits appropriate to the opposite sex. But the remaining 35 per cent rated themselves as possessing about equal amounts of both masculine and feminine traits.

Bem then asked the students to choose which of two activities they would engage in for pay. Feminine women and masculine men consistently avoided choosing an activity inappropriate to their view of their sex role, even when they would be paid more for that choice.

She also put the students into situations where she could measure such behavior traits as independence, assertiveness, playfulness, and willingness to respond to another creature or person.

Koko, a 4-year-old gorilla, talks to Penny Patterson of Stanford University with sign language. Koko is the first gorilla known to learn to talk with signs.

She reported that women who score high in femininity and low in masculinity on her inventory tend to be less independent and assertive. Men with high masculine and low feminine scores were less likely to engage in playful activities and were less responsive to other people.

Bem also considered data from other studies that show an association between high femininity in females or high masculinity in males with high anxiety levels, low self-esteem, and low self-acceptance. She concluded that persons who score roughly equal on masculine and feminine traits are better adjusted and more adaptable than highly feminine women or highly masculine men.

Childhood altruism. Parents can help their children learn to be responsible and to help others, according to a study published in May 1976. Martin L. Hoffman of the University of Michigan asked fifth-grade children in a Detroit school district to name three girls and three boys in the class who were most likely to follow the rules even when the teacher was not there, to try not to hurt the feelings of other children, to defend a child being teased by others, and to accept responsibility if they did something wrong.

Hoffman then studied the parents of the 40 girls and 40 boys who were named most often by their classmates. He found that most of the children with these traits have a parent of the same sex who communicates these values, and thus serves as a model. Most of the children also have a parent of the opposite sex whose discipline of the child included pointing out the harmful consequences of unsocial behavior and suggesting that the child apologize or otherwise make up for behaving badly.

Social evolution. The publication in late 1975 of Edward O. Wilson's book, *Sociobiology: The New Synthesis* has stirred continuing discussion and controversy. Wilson argues that genetic adaptation to environmental conditions is not the only variable that determines the survival and evolution of species. He maintains that social behavior also may have a genetic basis. He reached his conclusions by reviewing various

Psychology

Continued

animal societies, and drawing analogies with humans.

Many researchers have regarded changes in environmental conditions as the primary variables influencing the evolutionary selection of certain types of behavior that increase both individual survival and reproductive success in passing one's genes on to the next generation. Wilson points out, however, that the two goals do not always work together. For example, when males compete for a mate, the winning male increases the probability of passing on his genes. However, the chances of his own survival may be decreased by injuries sustained during the competition.

According to Wilson, the general tendency of animals to live in groups is directly related to the importance of social behavior in determining species survival. Group living increases the likelihood of individual survival because groups can provide protection from predators and can pool information about food and water sources. However, living in groups tends to involve dominance hierarchies. Because dominant animals mate more than others, only some members of the group are likely to reproduce.

Wilson gives two reasons why the subordinate members stay with the group, even if they are unsuccessful in reproducing. The traditional explanation is individual survival. But Wilson argues that staying with the group also promotes species survival. By staying with a group of close relatives, a subordinate animal increases the likelihood that genes similar to its own will be passed on, even if its own are not. Social behaviors that promote aiding one's kin therefore have survival value and are selected in evolutionary processes.

Some of Wilson's critics protest against his drawing analogies from the animal world and applying them to humans. Others argue that trying to sustain a genetic basis for human social behavior could lead to unwelcome political repercussions. Wilson denies any political motivation for his work. Rather, he says, sociobiology can help in better understanding the basics of human behavior. [Sally E. Sperling]

Public Health

The United States Public Health Service was directed in April 1976 to undertake one of the most ambitious public health efforts in history. The goal is to immunize the entire U.S. adult population against the swine influenza virus. At the request of President Gerald R. Ford, Congress appropriated $135-million to produce and distribute a vaccine to combat this potentially fatal disease. This unprecedented immunization program required 200 million doses of vaccine.

The identification of a new type of influenza virus stimulated this drive. The virus caused one death during a flu epidemic in February at Fort Dix, a U.S. Army base in New Jersey.

The new flu strain was significantly different from the Victoria influenza A virus, the last variant of the Hong Kong flu virus to threaten an epidemic. The Center for Disease Control in Atlanta, Ga., said the new virus had the same identifying protein coat as a virus that has occurred primarily in pigs for the past 50 years. This strain was responsible for the worldwide epidemic of 1918 in which 20 million persons, including 548,000 Americans, died.

Four major drug companies combined resources to produce and test the new vaccine in order to find the lowest possible dosage that will provide sufficient resistance to the virus.

Critics claim that this massive effort is a gamble based on the questionable assumption that a virus isolated in a single military community will cause an epidemic. They question whether it is reasonable to anticipate a recurrence of the enormous 1918 death toll, and argue that even if the swine virus does run rampant, sophisticated antibiotics are now available to control the complications that led to so many deaths.

Those in favor of the massive effort call the $135-million cost a bargain compared to the $3.2-billion to $4-billion cost of the 1968 Hong Kong flu epidemic, which killed thousands.

Unnecessary surgery became the subject of congressional debate and got front-page newspaper attention in 1976. A study begun in 1971 under physician Eugene McCarthy of Cornell

N-nitroso Compounds: Pervasive Poisons?

"Although there is no firm evidence that these compounds are a cause of cancer in man," food chemist Steven R. Tannenbaum of the Massachusetts Institute of Technology said, "there is no published report of any animal species that is resistant . . . and therefore, no reason to believe that man should not be susceptible." Tannenbaum, speaking at an American Cancer Society meeting in March 1976, was referring to N-nitroso compounds, combinations of nitrites with other nitrogen-containing compounds.

About 75 of these substances are now known to cause cancer in animals. The best known are the nitrosamines, compounds of nitrites and amines. Each of these compounds consistently induces cancer in a particular organ in specific animal species. And the target organ is the same whether the compound is administered by feeding or by injection. For example, according to William Lijinsky, director of the chemical carcinogenesis program at the National Cancer Institute's research center in Frederick, Md., nitrosoheptamethyleneimine induces cancer of the lung and esophagus in rats and cancer in the hamster's digestive system.

"No one yet knows why these target organs differ from species to species," says Lijinsky. Cancer may also be induced by separately feeding the precursors, nitrites and amines, which combine to make N-nitroso compounds.

Many scientists believe that humans encounter the precursors more often than the N-nitroso compounds themselves. People come into contact with nitrites primarily in foods that have been chemically cured or preserved, such as ham or fish, or in foods and water that contain nitrate. Vegetables, for example, particularly the root and leafy vegetables, contain large quantities of nitrates. These are reduced to nitrites by bacteria in the digestive tract. Tannenbaum showed in 1975 that certain vegetables containing above-average amounts of nitrate can lead to higher levels of nitrite in saliva than would be permitted in any food product. Whether this is something to worry about is unclear. In any event, according to Tannenbaum, "it would be essentially impossible for man to . . . avoid nitrate in a normal diet."

Amines, too, occur in foods, and also in tobacco smoke, drugs, and other commonly encountered substances. Lijinsky believes that there is some possibility of nitrosamines forming in the stomach.

Nitrates and nitrites are added to meat products to prevent botulism and to produce a fresh, red meat color. This has provoked much controversy. When its fat is not removed, for example, fried bacon forms a cancer-causing nitrosamine from its nitrite and an amine.

In the summer of 1976, bowing to scientific evidence, the U.S. Department of Agriculture began enforcing stricter standards for addition of nitrates and nitrites to meat products.

N-nitroso compounds are hard to detect. But chemist David Fine of the Thermo-Electron Corporation in Waltham, Mass., has developed a thermal energy analyzer that is far more sensitive than previous detection methods.

Fine's device detected a highly carcinogenic nitrosamine, dimethylnitrosamine (DMN), in the air at a factory using nitrosamines to manufacture fuel for missiles. The plant was shut down in April 1976 when the U.S. Air Force canceled its contract. DMN was later found around a DuPont Company plant in Charleston, W. Va. Nitrosamines were also discovered in the air in three Eastern industrial areas.

Fine and environmental toxicologist Samuel S. Epstein, at Case Western Reserve University in Cleveland, believe that people are likely to take in more N-nitroso compounds, or their precursors, from air than from food. Others disagree, saying that food and water pose the biggest risk.

DMN was first found to be toxic to humans almost 40 years ago. Hugo Freund, a Detroit physician, reported in 1937 that two industrial chemists who had inhaled fumes of DMN developed the same acute liver disease, and one of them died. "It's now come round full circle," says Peter Magee, director of the Fels Research Institute at Temple University School of Medicine in Philadelphia. "In large concentrations, DMN was clearly a hazard to those men. Are the traces of nitrosamines that have been found recently in the environment also a hazard?" Scientists do not yet know. [Gail McBride]

University Medical College in New York City produced some definitive, but controversial, data about unnecessary elective surgery in a report released in May 1976.

In McCarthy's study, members of two New York City labor unions were offered the opportunity to obtain an opinion from a second certified surgeon before undergoing surgery that had been advised by the patient's initial surgeon. In some instances, the second opinion was mandatory, and in others, voluntary, but none of the patients had to follow either opinion.

The second opinion failed to confirm the need for surgery in 35 per cent of cases (824 out of 2,373) in the voluntary program and in 16 per cent of the cases (161 out of 1,011) in the mandatory program. About 70 per cent of those in the mandatory and 85 per cent in the voluntary program followed the advice of the second surgeon, and did not undergo the surgery. From 1 to 4 years later, only 10 per cent of those who did not originally choose surgery reported that they later underwent the operation. Most of those who did not have surgery are well and receiving no medical care for the original symptoms.

These results suggest that peer review may reduce the amount of surgery. In addition, there is still no evidence of adverse effect on patients, although observers agree that more follow-up will be required to ensure that the recommended surgery was in fact unnecessary.

Cigarette smoking. Recent studies suggest an increasing use of cigarettes by adolescents in the United States. Epidemiologists Saul R. Kelson, James L. Pullella, and Anders Otterland of the Ohio Advisory Board on Smoking and Health reported in September 1975 on the prevalence of cigarette smoking among students in grades 7 through 12. They compared findings among 25,000 students in 1964 with 29,000 in 1971 in the same Toledo and Lucas County schools in Ohio.

The percentage of smokers increased substantially in every grade for both sexes over the seven-year period. The study showed a dramatic increase

"Has a kid come running past here, screaming 'I won't be a victim of Big Government'?"

Highest Risk Cancer Areas

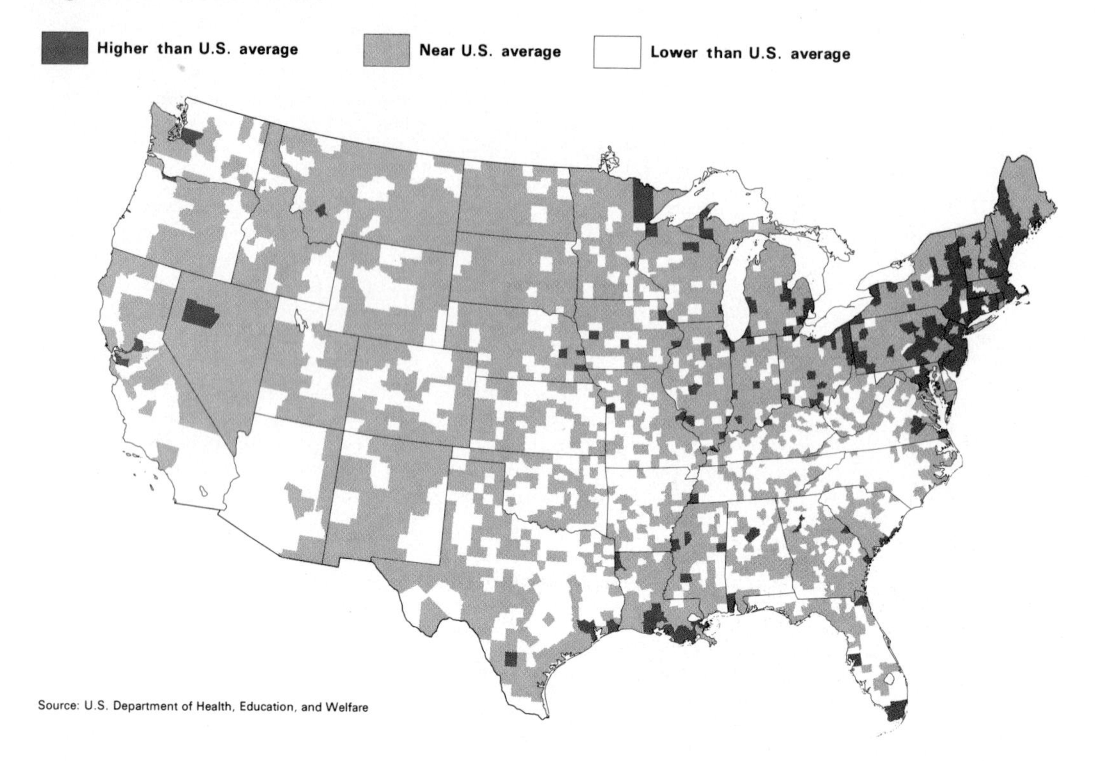

Public Health
Continued

A county-by-county survey of U.S. cancer deaths over a 20-year period pinpoints areas of high and low risk.

among the youngest group of girls; in fact, the rate of female smokers exceeded that of males in the eighth and ninth grades. In the 12th grade, 44.9 per cent of the boys and 37.9 per cent of the girls acknowledged that they smoked. Perhaps the only encouraging finding was that 45.6 per cent of the boys and 42.9 per cent of the girls surveyed in 1971 who previously smoked had stopped.

The findings suggest that widespread tobacco use continues, despite general recognition of the harm associated with cigarettes. And greater cigarette consumption by females, particularly by adolescent females, has reached the proportions of an epidemic.

Cancer epidemiologist Nicholas S. Wald of the University of Oxford, England, has recently reported that cigarette consumption in Great Britain continues to be high among men (4,000 cigarettes per adult per year) and is steadily increasing among women. Wald reported convincing evidence on Jan. 17, 1976, that correlates the change in smoking from unfiltered to filtered cigarettes to changes in the incidence of lung cancer and coronary heart disease in cigarette smokers.

Filter-tipped cigarettes were introduced in Great Britain in the late 1950s. By the early 1960s, they had largely replaced unfiltered cigarettes. Since that time, lung cancer deaths have dropped sharply, although the rate of heart disease deaths continues to rise.

Wald believes these data are consistent with the theory that the tar in tobacco smoke is the principal cause of lung cancer in smokers while carbon monoxide (CO) and other gases in the smoke may lead to coronary heart disease. Filters reduce the tar content of inhaled smoke, but the CO content of smoke is greater in filtered cigarettes.

CO is diluted in unfiltered cigarettes when air passes through the porous cigarette paper. However, the thicker paper on filters limits gas exchange, thus maintaining a high concentration of CO in inhaled smoke. Studies are underway to determine whether filtered cigarettes create a greater risk of coronary heart disease than unfiltered cigarettes. [Michael H. Alderman]

Science Policy

Scientists experienced a year of mixed political fortunes. On May 11, 1976, President Gerald R. Ford signed a bill re-establishing a science policy office in the White House. But in other areas, such as funding, scientists did not fare as well. Although research and development funds increased substantially in 1975 and 1976, much of the increase was offset by inflation. Also, scientists argued with their conscience, and frequently with each other, about how to control potentially dangerous genetic-engineering experiments.

Science and the White House. The bill Ford signed to re-establish a small Office of Science and Technology Policy in the White House, headed by a science adviser, essentially reinstated the science advisory apparatus abolished in January 1973 by former President Richard M. Nixon. Nixon shifted the focus of science policy development from the White House to the National Science Foundation (NSF).

Before reinstating the White House science office, President Ford established two top-level committees in November 1975 to study the federal government's science policies. One was the Advisory Panel on Contributions of the Technology to Economic Strength, and the other was the Advisory Panel on Anticipated Advances in Science and Technology. Many prominent scientists had urged Ford to re-establish the White House science policy apparatus, and many of them attended Ford's signing of the measure.

Science budget increases. Science and technology programs fared surprisingly well in the fiscal 1977 budget that Ford presented to Congress in January 1976. Because the start of the fiscal year shifted from July 1 to October 1, the budget included funds for the three-month transitional period. The Administration requested $24.7 billion for research and development and construction of facilities. Ford proposed an 11 per cent increase over fiscal 1976, which ended June 30, 1976, for basic research, including a 20 per cent hike for the NSF's basic research programs.

As in the previous two years, a large part of the proposed increases was slated for the Department of Defense and the Energy Research and Development Administration (ERDA). Defense research and development alone was to climb 13 per cent to $11.2 billion.

The proposed budget was much less generous toward the National Aeronautics and Space Administration (NASA), which was forced to delay starts on several space science projects, including the Large Space Telescope, a Pioneer spacecraft mission to Jupiter, and a Mariner spacecraft rendezvous with Uranus. However, NASA's $3.7-billion budget included about $1.3 billion for the space shuttle, and no major projects in progress had to be abandoned for lack of funds. The Administration also proposed modest increases for biomedical research and agricultural science and technology.

Energy funds up. Direct support for energy research and development was set to increase by 37 per cent, from $1.9-billion in fiscal 1976 to $2.6 billion in fiscal 1977, with another $600 million for related basic research and environmental studies. Most of the money will be channeled through ERDA.

Nuclear energy programs account for much of the proposed increases, particularly research on such technical problems as waste disposal and reactor safety (see OCEANOGRAPHY). The 1977 budget also provided about $655 million for the Liquid Metal Fast Breeder Reactor (LMFBR) program. This controversial effort, expected to cost $2 billion, is aimed at producing a plutonium-generating reactor for commercial use in the 1990s (see ECOLOGY).

Arguments continued over ERDA's heavy emphasis on nuclear energy. But congressional opponents of nuclear power were easily defeated in key votes on the LMFBR budget and on federal insurance for the nuclear industry.

Inflation and research. Considering the high inflation rates of the early 1970s, NSF Director H. Guyford Stever noted that the aim of Ford's proposals was "to counteract the gradual decrease of federal support for basic research, which has declined by about 23 per cent in terms of constant dollars since 1968." An NSF study published in July 1975 estimated that fiscal 1975 government outlays on basic research were about 8 per cent below the 1974 level in terms of purchasing power. Two months later, another NSF study predicted that the increases proposed

President Gerald R. Ford on Sept. 18, 1975, presented the prestigious National Medal of Science to the renowned chemist Linus C. Pauling at a White House ceremony.

for fiscal 1976 would "be more than offset by inflation." Ford's fiscal 1977 budget proposals would thus barely restore the inflationary losses of the previous two years.

While inflation nibbled at U.S. research budgets, it took savage bites out of science support in other countries. Great Britain, which experienced a 26 per cent rate of inflation in 1975, was the most severely affected. To counteract the inflationary trend, the government in February 1976 announced sharp cuts in expenditures for the next five years. This included closing Britain's two particle accelerators and making large reductions in other areas of so-called "big science."

A decline in U.S. leadership. In March 1976, the National Science Board, governing council of the NSF, published a report indicating that U.S. leadership in science and technology is gradually being eroded. Observers counted this, along with inflation, as a major factor in stimulating the research-budget increase. The report showed that – in terms of personnel and dollars – science and technology grew more slowly in the United States than in France, Japan, Russia, and West Germany during the early 1970s.

Although the U.S. is still the world's largest producer of scientific literature, the study showed its worldwide share of published research papers is declining. Also, productivity rates and the number of patent awards and innovations are increasing more rapidly in some other countries. But the study showed that U.S. research papers are the most often cited, implying that the quality of U.S. research is still highly regarded.

Energy planning. In a report sent to Congress in July 1975, ERDA indicated it is unlikely that the United States will be self-sufficient in energy supplies until about 1990. To ease energy shortages in the near future, the agency suggested that efforts to expand the use of coal, produce more electricity by nuclear reactors, and recover more oil and gas get the highest priority.

Next, researchers should develop geothermal energy technologies, ways of producing synthetic oil and gas from coal and oil shale, and methods for collecting solar energy. For the longer term, the ERDA report recommended concentrating on the LMFBR, thermonuclear fusion, and production of electricity from solar energy.

However, the Congressional Office of Technology Assessment sharply criticized the plan. It accused ERDA of concentrating on sophisticated technology while neglecting such areas as conservation. ERDA issued an updated plan in April 1976 with more emphasis on increasing efficiency in energy use.

Solar energy arguments. A major dispute erupted over the Ford Administration's solar energy program. The Administration proposed increased spending on solar energy projects—from $86 million in fiscal 1976 to $116-million in fiscal 1977. However, John M. Teem, who resigned as chief of ERDA's solar energy division in February, revealed that ERDA originally proposed a $275-million budget for solar energy. The Administration refused such a large increase, but Congress moved to restore much of the cut.

Teem also revealed that the Administration had pressured ERDA to move cautiously with its plans for a Solar Energy Research Institute (SERI), a research facility approved by Congress late in 1974. The National Academy of Sciences (NAS) had recommended a large national laboratory be established with a staff of about 1,500 and a budget of $50 million. However, ERDA planned a small SERI facility to begin operations by January 1977, with a budget of between $4 million and $6-million and a staff of 50 to 75 persons.

NSF under attack. Some of the year's most bitter exchanges occurred between critics and supporters of the NSF. Part of the NSF's function is to award federal grants to science projects that it approves. NSF's problems began in 1974, when several projects it supported were ridiculed by Senator William Proxmire (D., Wis.) as a waste of taxpayers' money. As a result of the criticism's publicity, the House of Representatives early in 1975 passed a measure giving Congress power to examine NSF grants before funding them and to

"All right. Now convert the whole thing to metric."

veto grants it believed unworthy of support. The Senate did not approve the measure, but the move caused great apprehension in the scientific community. NAS President Philip Handler called it "tantamount to book burning."

Another problem arose in 1975 from controversy over school science courses that had been developed with NSF support. Congressman John B. Conlan (R., Ariz.) attacked some of these courses as undermining American values. For example, one course touched frankly on such aspects of Eskimo life as sexual practices and leaving old people to die because Eskimos cannot support them in the harsh Arctic climate.

Both controversies led the House Committee on Science and Technology to conduct hearings in June and July 1975 on NSF procedures for deciding which grant proposals to fund. A number of scientists criticized the NSF's secrecy and charged that funds are concentrated at a few élite institutes.

NSF officials responded with proof that each grant application is reviewed by an average of more than six independent scientists. Their analysis showed that the distribution of NSF grants among various states represents a compromise between such factors as population and concentration of scientific resources. NSF also announced that copies of reviewers' reports would hereafter be available to grant applicants upon request.

The congressional committee in January 1976 declared that it was satisfied with NSF's system and ordered no major changes.

Biomedical research policy. The budgetary and personnel problems that have plagued the National Institutes of Health (NIH) in Bethesda, Md., for several years came under the close scrutiny of a seven-member President's Biomedical Research Panel. NIH's problems included the dismissal of two NIH directors, scientists' charges that overemphasis on cancer research had diverted funds from equally deserving research areas, and hints that politicians rather than scientists are shaping NIH research policy.

The panel issued its report in April 1976, giving the NIH essentially a clean bill of health. It recommended that NIH should continue to be controlled by the scientific community and should have a more stable funding base. The panel also argued that the privileged status of cancer research should continue.

However, there was a move in the Senate to reduce proposed budget increases for the National Cancer Institute and to redistribute some of the savings to other research fields. But the measure was defeated on Sept. 26, 1975, by a vote of 62 to 19.

Genetic-engineering rules. Perhaps the most important single issue in the area of biomedical research policy was the continuing effort to set controls on a revolutionary new technique by which scientists can transplant genes from any organism into a virus or bacterium. The technique promises advances in biomedical research, but it could also pose a serious health hazard if virulent organisms escaped from a laboratory.

Genetic scientists placed a voluntary moratorium on some of this work in July 1974, and an international group of geneticists met in February 1975 to discuss what controls, if any, should be placed on use of the technique. Then an NIH committee worked during 1975 to draft guidelines for governing NIH's support of such studies.

The committee drafted guidelines in December 1975 that would outlaw some of the more hazardous experiments and require that others be performed under conditions of strict physical safety. It recommended that for some experiments researchers use specially modified viruses or bacteria virtually incapable of surviving outside an artificial laboratory environment.

Some critics charged the committee's guidelines were too lax and others said they were too stringent. Nevertheless, NIH Director Donald S. Fredrickson issued a final version embodying most of the committee proposals in June 1976. See GENETICS.

Another ethical controversy erupted about whether relatives of a terminally ill patient in a coma can instruct doctors to halt treatment. The controversy centered around Karen Ann Quinlan, who was kept alive by machines as her parents fought a legal battle for permission to have the life-support systems stopped (see THE CHOICE TO DIE). [Colin Norman]

Space Exploration

One of the most important space events of 1976 was the U.S. Viking mission to Mars. The $1-billion mission had 750 persons on its flight team to guide the most complex unmanned craft ever flown.

Two Viking space vehicles were launched toward Mars on Aug. 20 and Sept. 9, 1975. The first reached Mars orbit on June 19, 1976, and began to search for a landing site for the first U.S. spacecraft on Mars. The second Viking was to reach Mars orbit on August 7, with a landing expected a few weeks later. Each landing craft carried cameras, such devices as a soil-sampling scoop, and biology experiments designed to determine whether life exists on Mars. They were the first craft to carry instruments designed specifically to look for life on another planet.

Apollo-Soyuz. After 14 years of manned space missions, United States astronauts and Russian cosmonauts finally met in space on July 17, 1975. Cosmonauts Alexei A. Leonov and Valery N. Kubasov manned the Soviet Soyuz spacecraft. In the Apollo were astronauts Thomas P. Stafford, Vance D. Brand, and Donald K. Slayton.

The rendezvous, two days after the July 15 liftoff, was the main event in the long-planned Apollo-Soyuz Test Project. With their spacecraft locked together by a tunnellike coupler, the five spacemen spent almost two days together performing joint scientific experiments and exchanging memorabilia. The scientific experiments ranged from studies of the earth to astronomical searches of the sky.

Preparations for the complex mission took much cooperation. Russia and the United States exchanged technical information to develop the docking adapter and the specialized communications systems used to link the two launch sites, two control centers, numerous tracking stations, and the two spacecraft. Each space crew learned the other's language.

After completing the mission, both crews landed safely. However, the three astronauts suffered temporary respiratory irritation when some propellant gas entered their spacecraft through a

The Russian Soyuz spacecraft, *below,* and the U.S. Apollo, *below right,* photographed each other before they docked together during their joint mission in July 1975.

vent during its descent to the Pacific Ocean on July 24.

In a less dramatic show of U.S.-Soviet space cooperation, an unmanned Russian Vostok spacecraft, launched into earth orbit on November 25, carried four U.S. biological experiments.

Salyut 4 dockings. The Russian Soyuz 18 space mission, launched on May 24, carried cosmonauts Pyotr I. Klimuk and Vitaly I. Sevastyanov to the orbiting Salyut 4 space station. The Soyuz 18 crew, the second to occupy Salyut 4, returned to earth on July 26. Their 63-day stay in space was the longest for Russian cosmonauts.

In November, an unmanned spacecraft, Soyuz 20, docked with Salyut 4. Observers believed the Russians used the unmanned mission to study methods of resupplying space stations.

Space-shuttle tests. The National Aeronautics and Space Administration (NASA) announced in October 1975 that it was abolishing the Apollo Spacecraft Program Office. The move was part of a reorganization to prepare for the coming of the space shuttle. The space shuttle's first orbital flight was scheduled for 1979, and NASA began preparing for the first flight tests of unpowered shuttle landings scheduled to begin in March 1977. In these tests, shuttle craft will be dropped from a Boeing 747 jet flying at 7,620 meters (25,000 feet), and test-pilot astronauts will guide them to the earth.

NASA prepared during 1976 to recruit the first shuttle-era astronauts, who will go on duty in 1978. The space agency particularly emphasized the opportunities for women and members of minority groups. NASA planned on a heavy recruitment program for the 30 positions to be made available.

The Johnson Space Center near Houston conducted the first mission simulation with a Spacelab mock-up connected to a simulated space-shuttle orbiter in February 1976. The space shuttle will eventually carry the Spacelab aloft. Astronaut F. Story Musgrave, physiologist Charles F. Sawin, and nuclear chemist Robert S. Clark were sealed in the Spacelab for seven days, testing biological, physiological, and cosmic-ray experiments.

Space scientists showed growing interest in the concept of self-sustaining space colonies, although such colonies could not be built until long after the space shuttle goes into operation. Nevertheless, NASA contracted during the year for several detailed studies on large space stations in earth orbit.

Interplanetary missions. Two unmanned Russian probes, Venera 9 and 10, provided the first photographs ever transmitted from the surface of Venus. Two Venus landers penetrated the planet's hot, crushing atmosphere on Oct. 22 and Oct. 25, 1975, and transmitted data for 53 and 65 minutes, respectively. The 118 minutes that they operated was more than triple the total time of previous Venus landers.

Meanwhile, the spacecraft that carried the landers circled the planet gathering data on the temperature variations in Venus' atmosphere and other conditions. See UNVEILING VENUS.

The sun also came under close scrutiny during the year. On Jan. 15, 1976, West Germany launched its second Helios probe. On April 16, it passed 43.4-million kilometers (27 million miles) from the solar surface – almost 3 million kilometers (1.9 million miles) closer to the sun than its predecessor Helios I.

The United States launched an Orbiting Solar Observatory satellite on June 21, 1975, to study the sun from an orbit around the earth and also study such astronomical targets as X-ray sources.

Satellite launchings. On Aug. 8, 1975, the European Space Agency, successor to the European Space Research Organization, launched its first satellite, a gamma-ray astronomical observatory. Seven countries contributed components to the satellite and four others provided scientific experiments.

The first satellite launched by the Japanese government, a tiny probe named Kiku, went into space on Sept. 9, 1975. Although there have been six earlier Japanese satellites, they were lofted by the University of Tokyo. The Japanese rocket that carried Kiku was built primarily from designs provided by U.S. companies.

Canada and the United States jointly developed the Communications Technology Satellite, launched by NASA in January 1976. With 200 watts of power, it is described as the world's most powerful communications satellite. It will

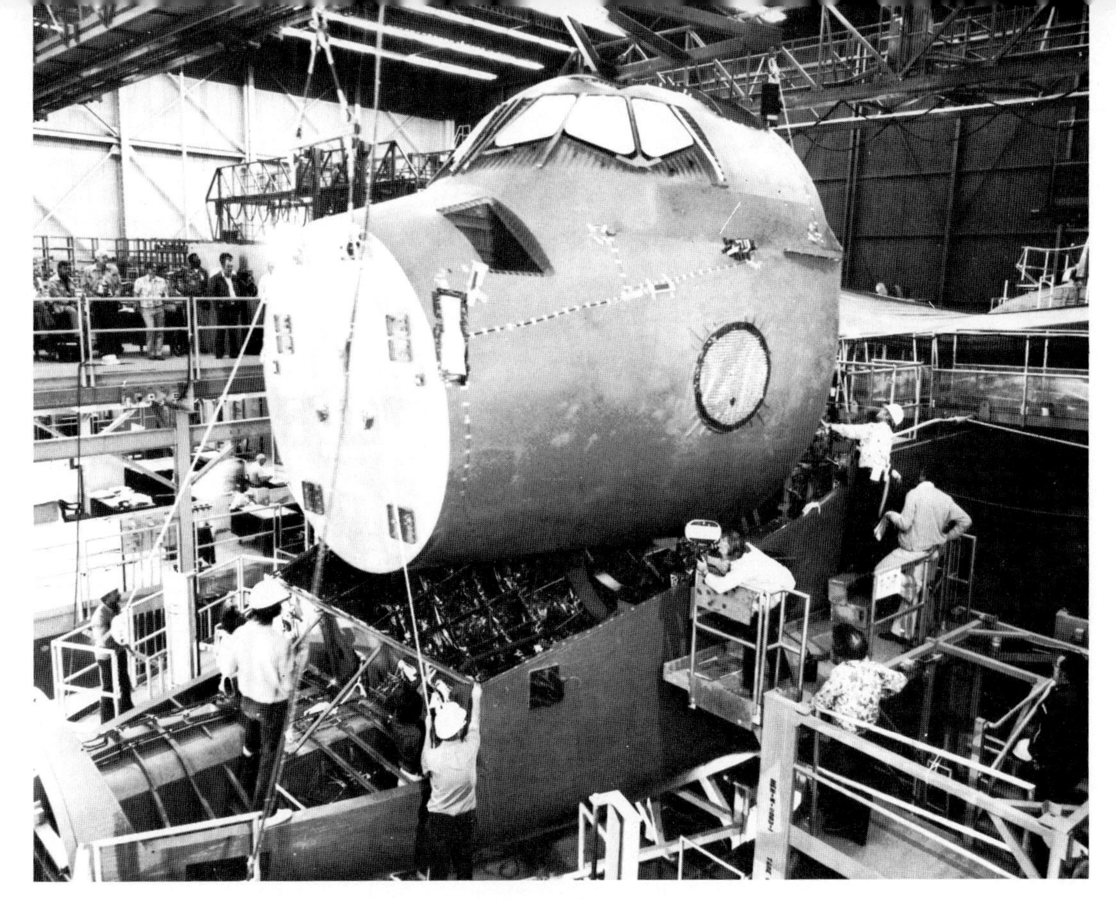

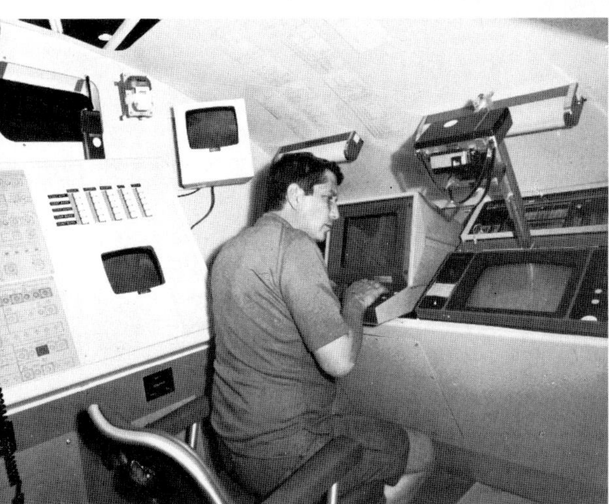

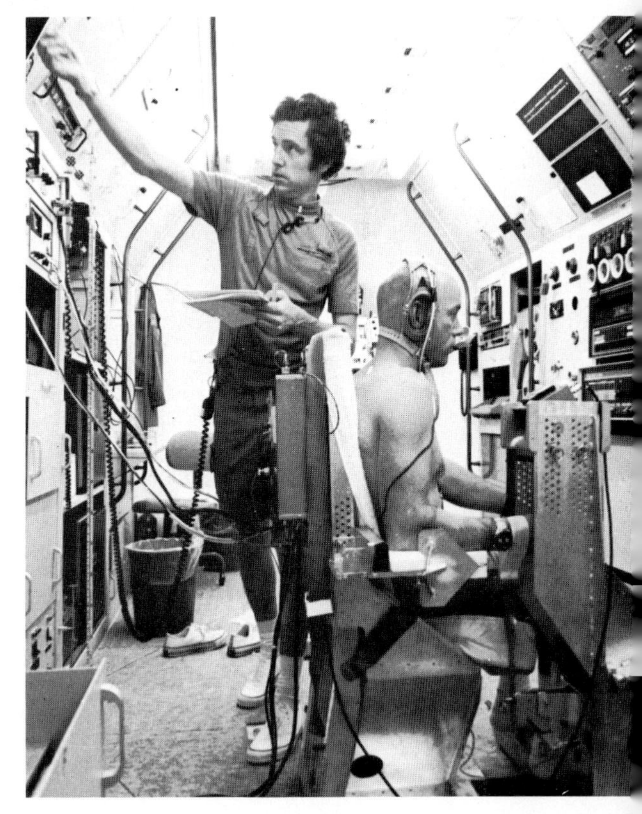

Assemblers at Rockwell International Corporation's Space Division in Palmdale, Calif., complete the forward fuselage of the reusable space shuttle, *top.* Meanwhile, nuclear chemist Robert S. Clark operates equipment, *above,* used for cosmic-ray experiments in a mock-up of the space shuttle Spacelab at the Johnson Space Center near Houston. In another part of the simulated Spacelab, *right,* physiologist Charles F. Sawin (standing) and astronaut F. Story Musgrave check equipment that will measure muscle functions during flight.

A French atmospheric and solar research satellite gets a final check before its launch on Sept. 27, 1975.

transmit data, television pictures, and two-way voice communications.

The first of a new generation of telecommunications satellites, Intelsat IV-A, owned by the International Telecommunications Satellite Organization, was launched in September 1975. It joined a network of seven Intelsat IV satellites. See COMMUNICATIONS.

Communications satellites were fast becoming the most common passengers on NASA's rockets. The agency's 1976 schedule listed 14 communications satellites among its 20 planned launches. This total included satellites for marine communications, North Atlantic Treaty Organization communications, and the Radio Corporation of America's commercial transmissions.

Nimbus 6, a U.S. satellite carrying sophisticated meteorological research instruments, was launched on June 12, 1975. It was equipped to make atmospheric measurements, trace weather balloons, and relay data from weather sensors on the ground.

NASA launched two satellites, Atmosphere Explorers D and E (Explorers 54 and 55 in orbit), on Oct. 6 and Nov. 19, 1975. They were equipped with rocket motors which could plunge them into the atmosphere to take low-altitude measurements and then power them into orbit again. They also carried sensors for studying the atmosphere's ozone layer, which some scientists believe is being harmed by the fluorocarbon propellants used in spray cans. Atmosphere Explorer E carried instruments to measure how the earth's heat and energy flow from hemisphere to hemisphere.

The simplest satellite launched during the year, a mirrored ball that is called the Laser Geodynamic Satellite (LAGEOS), had no radio transmitters or electronics of any kind. It was sent aloft on May 4, 1976, as a target for ground-based laser beams to help measure earth movements, such as continental drift, to an accuracy within as little as 2.54 centimeters (1 inch). NASA estimated that LAGEOS will have a useful lifetime of 50 years, but it will remain in orbit for more than 8-million years. [Jonathan Eberhardt]

Transportation

Major attention during 1975 and 1976 focused on progress in urban rail transit and on improving the fuel economy of the automobile.

San Francisco transit. Many of the early equipment-reliability problems encountered in the San Francisco Bay Area Rapid Transit System (BART) have been resolved. BART now carries 125,000 riders per day, 35 per cent of whom previously drove cars. Nearly 30 per cent of the people traveling across the bay to get to work now use BART, with a notable reduction in highway congestion.

BART, which started service in 1972, is a 114-kilometer (71-mile) network of automated electric trains serving communities in the San Francisco Bay area. Trains now operate at 5-minute intervals, although they were intended to operate as close as 30 seconds apart. A central computer, normally used for scheduling and dispatching, maintains the proper intervals by spacing trains one station apart.

By October 1976, a new computer control system will allow the safe spacing between trains to be reduced to two minutes. The new system uses minicomputers to check trains in and out of each section. If the trains become too close together, the computers will slow following trains.

Washington's Metro. The first link in the 158-kilometer (98-mile) Washington, D.C., subway system opened on March 29, 1976. By 1977, Metro will be extended to Washington National Airport in suburban Virginia.

Metro trains ride over tracks laid on rubber and fiberglass pads to deaden noise. As a safety measure, passenger platforms have a 1-degree slope away from the edge. A Metro study showed that the slight slope encourages waiting passengers to seek the lowest level, well back from the trains.

As in the BART System, Metro passengers use a computerized fare system to buy fare cards costing up to $20. A magnetized strip on each card records the value. A rider inserts the card in a turnstile slot when boarding the train, and inserts it again upon leaving the train. A computer calculates the distance traveled, deducts the fare from the value on the card, and prints the balance remaining for future rides.

Energy recovery from brakes. In February 1976, New York City began testing two subway cars that had been fitted with a new energy-saving device. Flywheel units beneath the cars store the mechanical energy that would otherwise be lost as heat when braking. The captured energy then is transferred back to the cars' motors, where it adds to the power needed to accelerate when the train starts up again. The equipment is expected to cut the energy needed to run the cars by one-third.

Fuel economy. Federal legislation requires automobile manufacturers to improve fuel economy to an average of at least 8 kilometers per liter (18 miles per gallon) in 1978 models and to 12 kilometers per liter (27.5 miles per gallon) by 1985. The present average is about 7 kilometers per liter (16.7 miles per gallon). A federal task force study concluded that 13 kilometers per liter (30 miles per gallon) could be achieved if drivers will use smaller cars and if manufacturers can introduce the required technology.

One suggestion for improving fuel economy is to put diesel engines into automobiles. A six-passenger, diesel-powered automobile can attain 13 kilometers per liter while meeting the 1978 emissions standards for carbon monoxide and hydrocarbons. However, it cannot yet meet the scheduled limit on nitrogen oxide emissions. General Motors (GM) is road-testing a diesel engine, and plans to introduce it commercially in 1978 if the government postpones the tighter emissions standards set to go into effect that year.

GM engineers reported in June 1975 that gains in fuel economy can be made by using grades of engine and rear axle lubricants at the lowest viscosity, or thinnest consistency, recommended by the manufacturer for a particular car model. City driving tests showed fuel economy improvements of up to 5 per cent over high-viscosity lubricants. The engineers warned, however, that extensive tests were needed to determine if viscosities lower than those recommended by the factory would increase wear on parts and oil consumption.

Safety restraints. Interest in passive restraints in automobiles continued to wane in 1976. Airbags, which suddenly inflate and act as pillows to cushion

Transportation

Continued

A keyboard, *above left,* is used to signal a driver's destination to a computerized automobile guidance system developed in West Germany. Units buried in the roadway, *above,* transmit direction instructions, plus information on road conditions, which are displayed on a receiver in the car, *above right.*

passengers during a crash, are expensive, and tests show they do not provide enough protection in side impacts and rollovers. The U.S. Department of Transportation is considering plans for a trial in one of the states to determine if the use of seat belts can be increased and to assess their effectiveness in preventing injury and death.

Meanwhile, Allied Chemical Corporation of New York City announced in June 1975 that it has combined the seat-belt idea with the airbag concept in an inflatable seat belt. The lap-and-shoulder arrangement looks like a typical seat belt, but it will expand when filled with gas under pressure.

Light stitching holds the uninflated belt in folds. A gas generator is triggered when the car is braked suddenly, the stitching rips open, and the belt inflates in a fraction of a second. When impact occurs, the wearer is already cushioned against shock. Tests with dummies indicated that the inflatable belts provide greater protection against injury to wearers than conventional seat belts can give.

Tracked levitated vehicles. Research slowed on tracked levitated vehicles (TLV) in 1976 as federal support for advanced transportation research remained at a low level. These vehicles are supported, propelled, and guided above a track by magnetic force or air pressure. TLV's are faster than high-speed trains and they provide improved safety and comfort. Maintenance costs for the guideway are lower than for conventional railways. High speed also reduces overall costs because cars can make more trips in a set period of time, reducing capital and labor costs per passenger-mile.

Experimental TLV's have run successfully in France and Germany at speeds up to 420 kilometers (260 miles) per hour. In June 1976, a demonstration run of the U.S. Prototype Tracked Air Cushion Vehicle (PTACV) in Pueblo, Colo., successfully culminated several years of development. The PTACV is propelled by a linear induction motor. It can carry 80 passengers at up to 240 kilometers (150 miles) per hour. [Herbert H. Richardson]

Zoology

Zoologists found in 1976 that, in addition to primates and some other animals, certain insects use tools. Scientists have given particular attention to the use of tools by primates, especially those considered to be related to human beings. Primates' use of tools is an indicator both of their social progress and of their reasoning abilities. With tools they can solve problems and thereby improve their efficiency in hunting, defense, or other essential activities.

Ants tool up. Social activity is also found in ants, wasps, bees, and other insects. Zoologists Joan and Gary Fellers of the University of Maryland in College Park reported in April that several species of *Aphaenogaster* ants use tools to increase their efficiency in food-gathering. Having observed that certain ants near the College Park campus used small objects to carry food, the investigators placed small portions of jelly on index cards and laid them on the ground to attract ants. The first foragers to arrive did not eat the jelly; instead, they quickly gathered small pieces of leaves and placed them on top of the jelly. Ants from a single colony spent about an hour adjusting, repositioning, and tending the leaf fragments. When they had completed the task of transferring the jelly to the bits of leaf, the ants carried the leaf pieces back to the nest.

The researchers had set up a transparent ant colony near the jelly bait so they could observe activities within the nest. The ants brought the jelly-covered leaves to a chamber where the eggs were kept and the queen and some worker ants lived. Other workers then scraped the jelly from the leaves and distributed it throughout the colony.

In addition to the leaf fragments, the ants used bits of dried mud, pine needles, pieces of dried wood, and sand grains. All the tools were from 1 to 5 millimeters (0.04 to 0.2 inch) long.

The researchers reported that the ants seemed to be selective in choosing their tools. They searched the ground near the food bait, picking up and then discarding various objects, before finally selecting a suitable tool, which they carried to the food source. One species

A rare species of peccary, thought to be extinct, was discovered still living in western Paraguay in 1975.

seemed to prefer mud chunks. The choice of tools differs, depending on the area in which an ant species lives.

Why tools? The Fellers team also tried to learn why certain ants use tools to carry food. Most ants swallow food and store it in the crop, a sort of pouch in the throat. They then transfer the food to other ants in the nest by spitting it up again.

However, this takes extra time both at the food source and in the nest, and is less efficient than using tools. The Fellerses found that the amount of jelly carried on a leaf fragment equaled the total body weight of the ant. An ant can carry only about 0.1 milligram of jelly in its crop, but 1.4 milligrams, 10 times as much, on a piece of leaf.

The *Aphaenogaster* ant's use of tools is significant in another way. Because there are many species of ants foraging in any given location, they must compete for food. Some species are highly aggressive and can prevent others from getting food by fighting and chasing them away. The less aggressive ants use tools to compete successfully.

A typical tool-using species the scientists studied was *A. rudis.* These ants quickly place leaf fragments on the food before, or even while, more aggressive ants are swallowing the food. When all the food seems to be gone and the aggressive ants leave, the *A. rudis* ants return and carry off the leaf fragments with the bits of food that their competitors overlooked.

Sleeping turtles. The green turtle, *Chelonia mydas,* which is hunted as a food delicacy, is believed to be an active species that migrates great distances through the oceans of the world. Researchers were surprised to learn in 1976 that green turtles spend the winter in a dormant, or inactive, state at the bottom of the Infiernillo Channel in the Gulf of California. Moreover, they were distressed to learn that the Seri Indians of Sonora, Mexico, and Mexican fisherman from Kino Bay were harvesting the turtles while they were in this helpless state.

R. S. Felger and K. Clifton of the Arizona-Sonora Desert Museum near Tucson and Philip J. Regal of the

"Although humans make sounds with their mouths and occasionally look at each other, there is no solid evidence that they actually communicate among themselves."

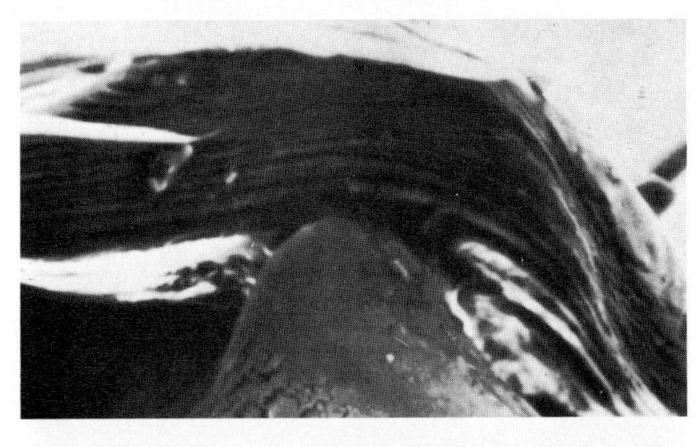

A male wolf spider, *top,* uses its palpus, a small feeler near its mouth, *second top,* to communicate. The palpus has a file and scraper at one of the joints, *second bottom,* that rub together when the spider flexes the palpus. This makes a buzzing noise. The spider also uses a series of spines on the palpus, *bottom,* to touch the ground, using it as a sounding board.

Museum of Natural History at the University of Minnesota in Minneapolis accompanied both groups of fishermen to study the turtles' habits. They found that the green turtles bury themselves in the sand or lie on submerged rocks during the cold winter months. They remain dormant until warmer ocean temperatures again stimulate them to activity in the spring. The wintering turtles are found from 4 to 15 meters (13.2 to 49.2 feet) beneath the surface. They are easy prey for the Seri Indians, who hunt them from boats with harpoons, and for the Mexican fishermen, who use scuba gear to drag the turtles into their boats.

Unfortunately, the researchers report, the turtles are now being overhunted as more modern equipment comes into use. Moreover, there may be many other areas around the world where the turtles spend the winter in this dormant and vulnerable condition. Worldwide overkill could easily endanger the species.

Trailing by tongue. Garter snakes use their tongues as a smelling organ. The flicking of the snake's forked tongue delivers odors from the environment into an internal smelling organ. John Kubie and Mimi Halpern of the Downstate Medical Center in Brooklyn, N.Y., studied this behavior in the laboratory to see if it played a role in trailing prey and reported their results in late 1975.

The researchers put garter snakes into a Y-maze that gave a choice of left or right turns where the two branching arms joined the runway. Odor trails of extracts from earthworms were painted on the floor along either the right or left arm of the maze. Most of the snakes quickly learned to follow the trail and were rewarded with a morsel of earthworm at the end. Two of the snakes learned the trick immediately, while others required 10 or more runs through the maze before they consistently chose the correct route.

Kubie and Halpern concluded that tongue-flicking is used not only in registering smells, but also is an important mechanism in trailing prey. The researchers are continuing to use the maze procedure to study other reptilian sensory functions, as well as learning, memory, and motivation.

Death trap. Zoologists often look into the past to understand how present-day animals are related to each other and what their ancestors were like. In a study that is still in progress, B. Miles Gilbert of the University of Missouri in Columbia and Larry Martin of the University of Kansas in Lawrence are investigating fossil remains in a deep, dome-shaped limestone pit called Natural Trap near Lovell, Wyo. They believe the pit trapped animals that accidentally tumbled into it over thousands of years during the Pleistocene Epoch, which ended about 10,000 years ago. Most specimens uncovered so far are estimated to be between 11,000 and 14,000 years old.

Some of the specimens that have been collected seem out of place in North America. They include the mammoth, the short-faced bear, a cheetahlike cat that resembles the modern African cheetah, and the camel. None of these animals live in North America now. Other fossil specimens are similar to animals that live in present-day Montana, such as antelope, bighorn sheep, bison, horse, and bear. However, most of the specimens are larger and all of them are quite different from their present-day counterparts. For example, the short-faced bear had long legs adapted for fast running and chasing prey.

In addition to comparing the fossil specimens with present-day animals, Gilbert and Martin expect to find evidence of the climatic changes that occurred during the Pleistocene Epoch. The pit should yield fossil specimens spanning the last great Ice Age when glaciers moved into the area and then retreated. The researchers hope to link the geologic changes that occurred then with the kinds of animals that lived there.

Singing spiders. Male wolf spiders emit noises that are especially important in attracting females and in defending territory. Until recently, scientists thought that the sounds were produced when the spider thumped its palpi – small feelers near the mouth – on the ground or on dry leaves. Scientists coined the term *drumming* to describe the sounds.

But Jerome S. Rouner of Ohio University in Athens reported in December 1975 that wolf spiders do not drum. Close examination of the palpi when the spiders were making sound revealed a file-and-scraper arrangement at one of the joints. When the palpus is flexed, the peglike scraper rubs against the ridged file, creating the characteristic buzzing or purring sounds.

Further inspection under a scanning electron microscope showed a series of long spines on the palpus that touch the ground or whatever the spider is standing on. Rouner suggests that the spines support the palpus during noisemaking. They might also act to enhance the conduction of the sounds. When the palpus is pressed against the ground, the vibrations are conducted through the ground rather than through the air. This would use the ground as a sounding board, increasing the loudness of the noise. In addition, signals transmitted through the ground would not be detected by predators that might home in on sounds emitted through the air.

Embryo transfers. Texas researchers reported in June 1976 that a full-term, normal infant male baboon was born from an embryo that had been transferred from its natural mother to a foster mother five days after conception. The experimenters were Duane C. Kraemer of Texas A&M University, Gary T. Moore of the Southwest Foundation for Research and Education, and Martin A. Kramen of the University of Texas in San Antonio.

In the experiment, the baboon mother was naturally inseminated by a male baboon. After five days, the fertilized egg was removed and implanted in a second female baboon. The embryo developed in the foster mother for 174 days, the normal gestation time for a baboon. The infant – a healthy, 875-gram (1.9 pound) male – was delivered by Caesarean section in September 1975. The implant was the only successful transfer out of 10 attempted.

Successful embryo transfers have been carried out on a commercial scale for several years in cattle and other domestic meat animals. The baboon birth is believed to be the first successful transfer in a primate. Embryo transfers in primates such as baboons, chimpanzees, and other apes could be useful in breeding specific strains of primates for research. [William J. Bell]

Rare deep-sea animals from the Gulf of Mexico went on display at the New York Aquarium in January 1976, the first time such animals have been displayed alive. Among the animals were the chain dogfish, *top,* goose barnacles, *above,* and isopod, *left.*

FROM THE SUN
HOW IT WORKS

People In Science

The rewards of scientific endeavor vary from winning a Nobel prize to successfully dealing with public officials to completing a project for a science fair. This section, which recognizes outstanding scientists, also recognizes the students who may someday follow in their footsteps.

Georges Mathé

By Alexander Dorozynski

By involving himself in politics as well as medicine, this innovative cancer researcher has become one of France's most controversial personalities

The streets of Villejuif are dark and the night shift is still on duty at the huge Paul-Brousse Hospital complex when Georges Mathé, head of the Institute of Cancerology and Immunogenetics, drives up to start his day's work. He steps from his car and heads quickly for his office. At 54, Mathé is a lithe, intense activist in medicine and politics, outspoken about the feelings he has toward his colleagues and his work. "I have the weakness that I can only work with friends," he says, "and my friendship goes to brilliant and efficient men and women."

Soon his "friends" begin arriving at the institute in the suburb south of Paris. Many of them are French physicians and biomedical researchers. But they also include scientists from Europe, the United States, South America, and Asia. Under Mathé's leadership, this team works on a treatment for cancer that involves stimulating the body's immune defense system to fight the disease. Mathé particularly concentrates on acute leukemia, a form of cancer in which white blood cells multiply wildly. Many of his patients are children.

Mathé knows no holidays. "Vacations," he says, "are just another gadget of modern civilization." His life is definitely centered around his work. He and his research team at the institute keep up a fast pace throughout the day, not even stopping for lunch. At noon, they gather, sandwiches in hand, around a conference table to discuss some aspect of the institute's many activities: the progress of cancer patients, clinical treatment procedures, cell pathology research, genetics, pharmacology, or immunology.

During the afternoon, Mathé may visit laboratories at the institute, examine patients' records, work on articles, scan publications, or see visitors. Twice a week, he teaches a course on experimental cancer research and a course on blood cell cancer to medical students at the University of Paris-South. He and other members of the institute staff also teach human biology to nonmedical scientists and courses on cancer to general practitioners.

Except on rare occasions, Mathé reserves time in the afternoon each day for visiting his patients–if only because they need reassurance. "Medicine may have become a full-fledged science," he says, "but it remains a priesthood. Every doctor should have been seriously ill at some time to qualify–to know how it feels to be ill." Mathé himself had a two-year bout with severe hepatitis in the late 1950s.

In addition to his full-time duties at the institute, Mathé finds time for what is almost a second career–a political one. "I consider it the duty of a scientist to take advantage of all platforms available to him to inform people of scientific developments and their consequences, and to suggest economic, political, and social measures that should be taken because of these consequences," he says.

Over the years, Mathé has developed considerable political muscle, and he flexed that muscle in 1976 in a spectacular showdown with Simone Veil, France's popular minister of health. Mathé was dissatisfied with the government's low cancer-research budgets, so he launched a vigorous campaign against Veil, whom he called "the gravedigger of French cancer research." The conflict spread to France's majority party, the Democratic Union for the Republic, threatening to split it. Several weeks and many harsh words later, the government allotted additional research funds, and an uneasy peace was re-established. "Politics," Mathé muses, "may well become my first career. But I'm afraid that if I left the institute, it might collapse."

His prestige helps to raise private donations, which make up about a third of the institute's operating funds. "Soliciting is part of my work," he says. Indeed, the institute's latest annual report ends with this sentence, "We live as doctors and research workers for and by leukemic children, and we are proud to beg for them."

The author:
Alexander Dorozynski is a free-lance science writer based in France.

Georges Mathé was born on July 9, 1922, to a farming couple in the rugged, mountainous region of central France. As a youth, he mistrusted physicians because he thought they were "unscientific." But his mother kept prodding him to take up medicine, and he finally agreed. When he was a student in Paris in the late 1940s, he helped make ends meet by working as a technician in a blood center. There he discovered his first scientific tool, the microscope, and his first scientific interest, the study of blood.

Georges Mathé, cancer researcher and political activist, examines a photomicrograph of a cancerous blood cell in a colleague's office at the institute. In the background hangs a portrait of Simone Veil, French minister of health and one of Mathé's political foes.

In 1947, Mathé interned at the Children's Hospital in Paris and while there, he discovered that children who were suffering from leukemia were hospitalized not so much to receive treatment, but simply to die. "Physicians, powerless in the face of the disease, turned away from it," he once wrote. "Nurses, having no prescription to deliver, could offer only care and maternal affection."

However, a year later, he watched doctors suspend the fatal course of acute leukemia by totally replacing the patient's blood. This showed him that it was possible to halt the disease at least temporarily. At about the same time, U.S. researchers reported the first successful treatment of leukemia with powerful drugs–chemotherapy. Mathé realized that a new era of cancer research had begun, and that he wanted to play a part in it.

After completing graduate studies in chemistry, biology, and the study of organic changes during the disease process, Mathé went into cancer research. His entrance into the field was a modest one, indeed. He set up his laboratory in an abandoned office in the Children's Hospital. His equipment consisted of a microscope and a few rabbits for his experiments. From there, he moved to an attic, then to a cellar in another Paris hospital.

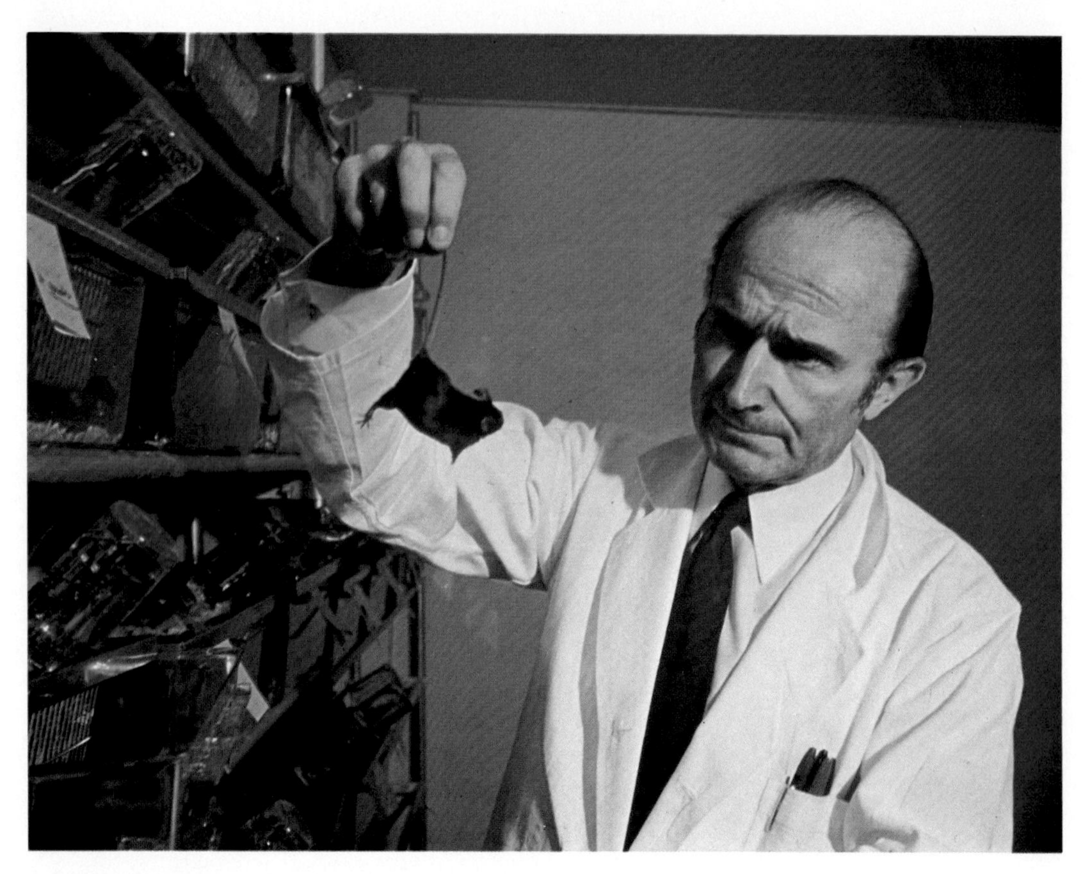

The cancer researcher gingerly checks the condition of one of his "volunteers" in an area of the institute where mice are bred specially for research on cancer treatments.

In 1955, American physician and cancer researcher Joseph H. Burchenal invited Mathé to study at the Memorial Sloan-Kettering Cancer Center in New York City. Mathé readily accepted. "With Burchenal," says Mathé, "I learned that a scientific researcher is not an abstract thinker, but a man of action, a leader, and a manager. Burchenal could wear out a full team of clinicians and technicians, yet find time to alleviate the anguish of patients and their families."

When Mathé returned to France in 1956, he practiced medicine and pursued his research at the St. Louis Hospital in Paris, experimenting with bone marrow transplants, or grafts, in mice suffering from leukemia. He was trying to determine if transplants of healthy bone marrow cells, which produce blood cells, could be used to treat leukemia. But first he had to overcome a major problem: The grafted cells reacted against the mice, producing a severe and often lethal illness. However, Mathé discovered that bone marrow transplants from one animal to another were successful when the animals receiving the marrow had first been subjected to almost-lethal doses of radiation. The radiation suppressed the animals' immune reactions and prevented rejection of the transplant. He wondered whether this could lead to a new, immunological approach to cancer treatment.

While Mathé was absorbed in this research, something occurred that suddenly thrust him into the public eye. On Oct. 15, 1958, an experimental nuclear reactor near Belgrade, Yugoslavia, accidentally released powerful bursts of gamma rays and neutrons. Six technicians were exposed to lethal doses of radiation that destroyed blood-producing cells in their bone marrow.

No treatment existed at that time, but one of the doctors on the scene knew that Mathé had been experimenting with bone marrow transplants in mice. Obviously, this technique had never been tried on human patients. But the victims of the nuclear reactor accident were doomed to die within weeks; they had nothing to lose by volunteering. Perhaps the transplants could save their lives.

Three days after the accident, the Yugoslav technicians were flown to Paris, and Mathé began searching for bone marrow donors. Donating bone marrow is not a pleasant procedure. A large needle must be inserted into the thick chest or hip bones and fresh marrow drawn out. Nevertheless, several donors volunteered.

Meanwhile, the nuclear technicians were showing the effects of lethal radiation doses. Their blood cell levels dropped rapidly, they had fevers, they began losing hair, and two of them were so weak that they could not even read.

A week after the accident, Mathé made the first human bone marrow transplant, injecting the marrow into a vein in the patient's arm. Eventually, all the technicians received the treatment.

The grafts worked precisely as Mathé expected, producing blood cells immediately after the transplant. But this was a short-term measure meant to keep the patients alive. Enough of their own bone marrow cells had escaped damage to eventually begin reproducing and thus replace the damaged marrow.

"The only object," says Mathé, "was to give them enough time, and this is what we did." Five of the patients recovered completely. But one died because, according to Mathé, he had hesitated in starting the bone marrow transplants. Nevertheless, the revolutionary treatment was hailed the world over, and Mathé was abruptly propelled into the limelight. He has never left it.

Mathé's success with the technicians opened the way for a new approach to treating leukemia patients. Acute leukemia is the most frequent form of childhood cancer. For unknown reasons, the bone marrow starts producing great quantities of cancerlike white blood cells and releases them into the bloodstream. The patient suffers such symptoms as anemia and internal bleeding.

By the time the disease is diagnosed, up to 1 trillion cancerous cells may be circulating in the body. Cancer researchers have concentrated on finding ways to destroy these cells before they overwhelm the patient's body. By the late 1950s, scientists had developed several powerful drugs that could destroy cancerous cells or stop them from multiplying. But these chemicals also destroy or damage healthy cells, and their side effects can make the patient very ill or even kill him. Because of this, patients cannot take these drugs indefinitely to keep the cancer under control. Cancer victims have remained free of the disease for up to 10 years after such treatment. But drugs can never kill off the last cancer cells, because each treatment only destroys a certain percentage of them. Similar problems exist with radiation treatment, which also damages healthy cells and probably fails to destroy all the cancerous ones. Those that are left continue to multiply.

To Mathé, bone marrow grafts represented a new approach. Transplanted marrow would produce normal white blood cells that could attack the cancer cells. First experimenting on animals, Mathé developed a transplant treatment for leukemia victims. He began by subjecting the patient's entire body to radiation to weaken the immune system. Then he injected healthy bone marrow.

But his first attempts revealed another obstacle. Radiation had weakened the patient's immune response as planned. But the immune response was so weakened that the grafted cells attacked the patient's healthy cells as well as the cancer ones. This graft-versus-host reaction can be fatal, and it remains a major obstacle to this form of therapy. Nevertheless, bone marrow grafts from compatible donors are used to treat certain marrow and blood diseases, including acute leukemia.

However, it caused Mathé to wonder whether the reverse immune reaction could be promoted. Instead of weakening the immune response so that the body would accept healthy foreign cells, could the body's entire defense system be boosted to seek out and destroy the diseased cells that escaped radiation and chemotherapy?

Mathé knew that some substances could boost immune defenses, not only against a specific disease organism, but also against other foreign invaders. For example, recent statistics indicate that children who have received the antituberculosis vaccine called Bacillus Calmette-Guérin (BCG) are less likely to be stricken with leukemia than are nonvaccinated children.

Mathé found in experiments on mice that BCG was particularly effective in boosting the immune defenses against some forms of cancer already existing in the mice. If he preceded the BCG injections with chemotherapy, or gave BCG along with a vaccine of sterilized tumor cells, he could cure some of the mice.

In collaboration with the European Organization for Research on Treatment of Cancer, Mathé tested 20 different preparations of BCG, and eventually found that a live BCG bacterial strain grown at the Pasteur Institute was by far the most effective. By 1963, he was ready for the first clinical trial.

Charles de Gaulle, then president of France, told Mathé about a 14-year-old family friend, François, who was suffering from leukemia. François received anticancer drugs and radiation treatment followed by BCG immunotherapy. That same year, Mathé started 29 other leukemia patients on a program aimed at testing the effectiveness of drug and radiation treatment followed by immunotherapy as opposed to drugs and radiation alone. All 30 patients received chemotherapy and radiation treatments, but only 20 of them also received immunotherapy with BCG plus a vaccine made from leukemic blood cells.

The results, reports Mathé, were clearly in favor of immunotherapy. In the 10 patients who did not receive immunotherapy, cancer returned within 130 days. But 7 of the 20 patients given BCG plus vaccine have remained without cancer symptoms for up to 14 years.

François, now an engineer, is married and the father of a child. Mathé examines him regularly for his "blood disorders." Like many other patients at the institute, François does not know he had leukemia. Mathé sees no purpose in discussing the details of their illness with his patients, particularly if the cancer appears to be taking a fatal course. After a lifetime of treating cancer patients, many of whom were dying, Mathé says, "Most of them want to know very little, except that there is hope for life. And there is. Physicians worthy of their profession hope to the last minute to save their patient."

Over the years, Mathé and his colleagues have developed a three-stage standard treatment. During the first stage, which lasts about one month, the patient receives anticancer drugs to reduce the number of

Mathé and staff members discuss institute business at the daily luncheon meeting, or symphagium, a word he coined from the Greek for eat together.

Mathé exchanges views with two colleagues as he hurries across the grounds of the institute.

cancerous cells from the trillion to the billion range. Then he receives drugs and radiation therapy for about four months. After this treatment, the number of cancerous cells should be reduced to 100,000 or less–the significant threshold, according to Mathé. Immunotherapy with BCG can now cope with the weakened enemy.

"It is a medieval form of treatment," he says, "but it works." He makes cuts just deep enough to cause slight bleeding in a crisscross pattern on the patient's upper arms and thighs. The total length of all these scratches is about 1 meter (3.28 feet). Then he places liquid BCG vaccine, drop by drop, on the cuts, rubs it in, lets it dry, and covers the scratches with sterile bandages. This form of treatment, called scarification, is more painful than injection. But injecting BCG directly into the bloodstream could cause the bacterial particles to clump together. A severe infection would follow that could weaken the immune defenses rather than stimulate them. Mathé generally repeats the scarification process twice a week for a month. Then he reduces the treatment to once a week for a long period of time.

The first seven patients who responded to immunotherapy received the BCG and vaccine treatment for five years. Then Mathé reduced the length of treatment to four years. So far, he has observed no relapses in the patients that were treated for the shorter time. "We really don't know how long to keep up the treatment," says Mathé. "At first, we just didn't dare stop. Even after there was no detectable disease, how could we know whether a few cancer cells hadn't survived to proliferate if given the chance?"

This voluntary infection with BCG sometimes has side effects. The patient's skin reddens and itches. Yellowish scabs form, and he has pain and fever. Sometimes the liver malfunctions and temporarily enlarges. According to Mathé, these reactions show that BCG has thoroughly penetrated the body. However, if the side effects are too severe, Mathé makes smaller cuts or temporarily halts the treatment.

The toxic cost of immunotherapy, Mathé believes, is negligible compared to that of drug treatments. At Villejuif, 300 patients have received immunotherapy and none have died of side effects. Maintenance chemotherapy, Mathé says, is less effective–patients frequently have relapses–and sometimes the side effects kill the patient.

An important factor in the success of immunotherapy appears to be the type of cancer cell involved. Mathé, who is head of the World Health Organization Reference Center for Leukemias and Related Diseases, has subdivided leukemic blood cells into four different types. One type, which he calls microlymphoblastic, appears to be particularly responsive to immunotherapy. Twenty-seven of the first 100 patients treated with BCG at the institute had cells of this type, and 22 of the 27 have survived, apparently disease-free. The other three cell types appear to be less responsive to immunotherapy.

Mathé has also used immunotherapy to treat other forms of cancer, such as cancer of the lymph nodes, skin cancer involving pigment cells, and lung and breast cancer. Preliminary results on this treatment for lung cancer are particularly encouraging, he says.

In spite of all this, Mathé's treatment is not universally followed. In fact, it has stirred a great deal of controversy among cancer researchers. There are several reasons for this. One is that Mathé appears to be carried away by enthusiasm. Although he does not ignore his failures, he nevertheless places great emphasis on his successes.

A number of researchers have tried immunotherapy without success. Mathé claims they did not follow the points that he considers essential. For example, he believes they must start immunotherapy as soon as the number of cancerous cells has been reduced by radiation and drugs to 100,000 or less. They must also use live, fresh, liquid BCG of the Pasteur strain. "An attempt at using freeze-dried BCG, for instance, has been a complete failure," he says.

There may be reluctance to accept the treatment because there is no theory to explain how it works. While Mathé and his team have concentrated on refining techniques for treating patients with immunotherapy, fundamental work aimed at explaining the mechanism of cancer immunotherapy has fallen behind. Nevertheless, Mathé points out that several cancer researchers have published results that confirm

The eloquent champion of immunotherapy presents his position at an international symposium on cancer research, *above.* Later, he chats with other cancer researchers at a reception, *right.*

the value of cancer immunotherapy as treatment for leukemia, lung cancer, and some skin cancers.

Mathé's critics within the French scientific community do not like to be quoted, because he is a powerful, influential personality. "Mathé is an imperialist," one physician says. "He has taken a powerful hold on French cancer research, and if you don't pay tribute to him, or at least agree with him, then you're against him. And this may be harmful to your career."

Mathé's approach to his work is intense and his views of medical research are unorthodox. Flair, he says, is an essential quality for a researcher. "It can be pompously called genius, although it would be better termed intuition. It can be worth all of the laboratories and all of the research teams in the world; yet it cannot replace them, because every finding becomes a discovery only from the moment it is confirmed by other scientists reproducing the results.

"It has long been believed," he continues, "that when it comes to science, logic is a good adviser. But nature is not logical, at least not in the way we mean it. How many scientists have found, at the end of their experiment, the opposite of what they were trying to demonstrate? I would say that the constant attitude of a researcher should be never to take his desires for reality. In a word, he must be modest." Mathé likes to quote French author André Gide: "A theory is good if it leads not to rest, but to greater work."

Mathé recognizes that he is more of a medical doctor than a basic researcher. Basic researchers, he says, have a disinterested objective, the advancement of knowledge. He compares them to athletes who train by running down country paths for the sheer pleasure of the effort. Medical doctors are more like marathon runners; they are trying to reach a remote but precise goal, the well-being of patients. Too often, he says, physicians are overwhelmed by a feeling of inferiority in the face of the advances of molecular biologists. But the physician's contribution to the common effort is essential. Cancer research, Mathé believes, must be a collaboration between the two.

As a politician, Mathé maintains that the high period of French research, when De Gaulle was president from 1958 to 1969, is over. "Today, research is in a low period," he says. "Technocrats have taken power. Research does not interest them."

During De Gaulle's presidency, Mathé was an adviser to the ministry of health, and he prompted the creation of the National Institute for Medical Research. Mathé is still a member of its board. In 1972, he chaired the government's science and technology advisory committee and also headed the subcommittee on health, society, and environment of the superagency that oversees the organizations involved in government science policies.

Mathé has learned to cultivate politicians, administrators, and journalists, and to play the political game. His controversial stand on the issue of tightened research budgets in 1976 may have earned him the

Surrounded by some of the art he has acquired on his travels, Mathé browses through journals in the living room of his home near Paris, *above.* In a rare moment of relaxation, *right,* he lounges in the backyard with his wife, daughter, and two grandchildren.

label of maverick–a man to be handled with care. But the public conflict with the minister of health confirmed that he can exert great political leverage. "I was surprised, and pleasantly so," he says. "Why, even the Communist Party supported me."

Mathé believes that European science must struggle to avoid domination by United States science. His major argument in the funding controversy was that U.S. cancer-research budgets are 66 times larger than comparable French budgets.

"In the United States," he says, "public opinion is aroused, for instance, by revealing the diagnosis and prognosis to all cancer patients, so that a real phobia about cancer has taken hold there. This adversely affects the morale of the patients, but it fills the cashboxes of laboratories. In America, wealthy people willingly contribute to research, if only because of favorable tax legislation."

Mathé believes this publicity and funding has contributed to the power of U.S. science. From his point of view, this makes it difficult for European scientists. "Americans impose upon us their language, nomenclatures, classifications, and conclusions. They drown us in innumerable journals. The French, it is true, attempt to resist. But they shouldn't believe they can do it only by speaking French." Mathé refers to the first years of De Gaulle's presidency, when the government urged French scientists to speak French at international meetings. Mathé and many others resisted this measure as unrealistic. "Anglo-Saxons will make no effort to understand this language. The only good weapon of our country is the originality of its research."

Between medicine and politics, Mathé has little time for anything else. His wife, Marie-Louise, and Catherine, their only daughter, share the remaining hours of the day. But one could hardly call him a family man. "When I sit at home, I may say to myself, 'I could as well go to the laboratory.' And I don't need to travel for the dubious pleasure of tourism. When I travel to international meetings, I discover other countries, other cities. In world capitals, the best motion pictures, theaters, and concert halls are available to me. During my business travels, I've collected a few smiling pre-Columbian figurines, Persian miniatures, and Chinese glass paintings. This is the extent of my hobbies–unless politics can be called a hobby."

Mathé has many friends and acquaintances from varied walks of life. "I once became friends with a young man dying of leukemia at a time when we were unarmed against this disease. He was about to become a doctor. We became friends because he was conscious that, as he fought for his life, he was helping me in my research. He left life without tears, knowing he had helped fight a terrible disease."

In addition to scientists, political figures, businessmen, and publishers, Mathé even counts a gangster among his acquaintances. "It's true," says Mathé, "I've treated him for cancer. And he offered, one day, to organize a holdup of the Bank of France for the benefit of cancer research." Mathé pauses. "How I regret that I didn't accept."

FLEXING THE HEXAFLEXAGON
DESCRIPTION
BACKGROUND

DENVER CONVENTION COMPLEX
CURRIGAN EXHIBITION HALL
INTERNATIONAL SCIENCE
AND ENGINEERING FAIR
MAY 12-13-14 1976
THANKS FOR COMING TO DENVER

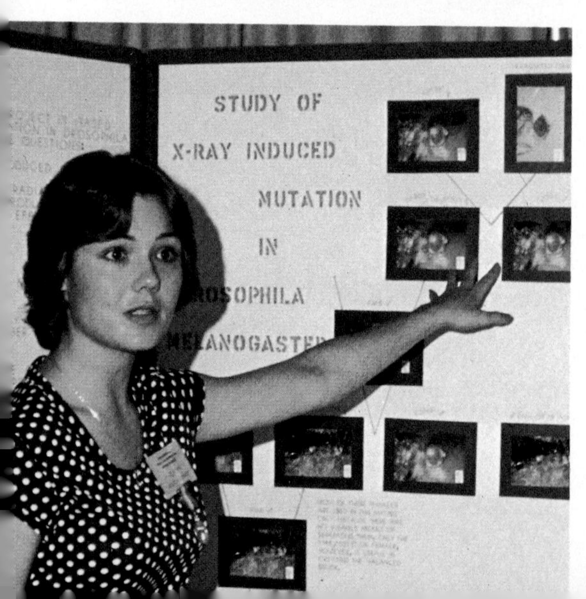
STUDY OF
X-RAY INDUCED
MUTATION
IN

The International Science And Engineering Fair

By William J. Cromie

Aspiring student scientists dazzle Denver with displays of their research expertise

Richard H. Ebright, 16, brought his butterflies from South Temple, Pa. Jan Erik Hult, 19, brought a gas chromatograph flame ionization detector and recorder from Göteborg, Sweden. Wanda Colón, 15, brought lichens from Puerto Rico. They and 404 other young people met at Currigan Hall in Denver in May 1976 for the 27th International Science and Engineering Fair (ISEF). All the exhibitors were finalists from 227 regional fairs in the United States or national fairs in other countries. Now they would compete with each other at the World Series of science fairs.

Currigan Hall, which is the size of a large airplane hangar, became jammed on Monday and Tuesday, May 10 and 11, with students and adults who swarmed around unpacking wooden crates and cardboard boxes. They had come to Denver by airplane, pickup truck, trailer, and family car. Clayton D. Farnham, 17, flew from Alaska with two stuffed grouse nestled in his lap. They were part of his project on trapping activities in Alaska's Copper River Basin. Robin J. Smith, 15, and her parents unpacked water hyacinths, still fresh after a 3½-day drive from Ormond Beach, Fla. Her project demonstrated how the hyacinths might be used to control water pollution. Collin J. McKinney, 18, and his parents drove from El Paso, Tex., in a truck

Facts About The Fair

Every spring, students who are interested in science show their expertise by exhibiting their work in science fairs. Their projects range from insect collections to research on ways to detect cancer.

Elementary school students show their exhibits at school and local fairs. High school students also exhibit their work first at school and local fairs, but the winners go on to larger city, regional, and state fairs. The top of the pyramid is the International Science and Engineering Fair (ISEF), held during the second week in May.

The ISEF, sponsored by Science Service of Washington, D.C., grew out of smaller science fairs that began about 1930. The first ISEF was held in 1950. It featured the work of 30 finalists chosen from some 15,000 students who competed at 13 regional fairs. The ISEF grew rapidly until the competition now involves more than 1 million students from high schools in the United States and other nations.

The quality and the complexity of the projects have increased with the number of participants. Many of the entries at the first ISEF were simple collections and models. Typical projects in recent years have included "A Holographic Study of the Sporangiophore of Phycomyces" and "A Myo-Electric Prosthetic Terminal Device."

Only students who have not reached their 21st birthday by May 1 are eligible to participate. Each finalist must be accompanied by an adult. Expenses are paid by the sponsoring local or regional fair.

The ISEF is one of the few contests in which the judges outnumber the contestants. Some 500 scientists, engineers, and physicians judge about 400 exhibits.

Science Service awards first, second, third, and fourth prizes in 11 categories that range from behavioral science to zoology. About 30 private, professional, and government organizations also present awards.

Students who want to compete in science fairs can obtain information from their science teachers, science fair directors, or by writing to Science Service, 1719 N Street NW, Washington, D.C. 20036. [W.J.C.]

loaded with the electronic equipment McKinney used to measure electrical resistance in plants.

The participants took different competitive routes to the fair. Linda Lee Terry, 18, of Flint, Mich., was a senior at Kearsley High School, which sends its students directly to the Flint Area Science Fair. Terry's project on changing the attitudes of young people toward the deaf won at the area fair, then she went on to the ISEF in Denver. Richard M. Bryant, Jr., 14, of Tulsa, Okla., took a longer route. He won first prize at the Byrd Junior High School fair in Tulsa with a project on how water pollution affects plants. From there, he went on to the Greater Tulsa Science Fair, then to Denver.

In the lobby of Currigan Hall, young people crowded around the registration desk. Dorothy Schriver, ISEF coordinator, handed medals and certificates to all who had made it this far in the international competition. The medal, a silver medallion with an embossed science emblem and the student's name, hangs from a rainbow-colored ribbon. The certificate bears a color reproduction of the medal, together with the finalist's name, project title, and high school. Similar certificates of commendation would also be given to each participant's sponsoring teacher and school.

The author:
William J. Cromie is a free-lance science writer and executive director of the Council for the Advancement of Science Writing. He wrote "Earthquake Early Warning" for the 1976 edition of *Science Year.*

Students gather for the ISEF at Currigan Hall in Denver, *top.* Fair coordinator Dorothy Schriver (at left) waits to register them, *above.* Later, a student shows off her medal and certificate, *left.*

Boxes and crates litter the hall as students set up their exhibits, *right.* Robin Smith arranges water hyacinths, while Peter Deegan puts up an electronic seismometer, *below.* Jeff Marks tries out his experiment on mouthwash, *below right.*

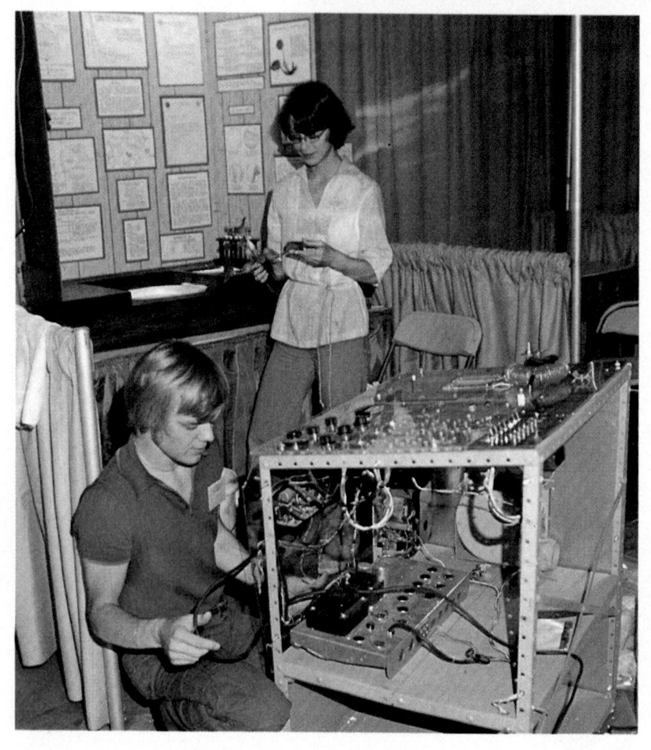

Dorothy Schriver has directed the ISEF program since it began 27 years ago. She works as assistant director of Science Service, a non-profit corporation in Washington, D.C., which organizes and is the principal continuing sponsor of the fair. Science Service was established in 1921 to popularize and encourage public interest in science. In 1950, Science Service representatives and people who held science fairs in various parts of the country agreed to sponsor a national science fair. One has been held every year since.

The loudspeaker at Currigan Hall blared an announcement that all exhibits had to be set up and approved by inspectors before 6 P.M. on Tuesday. The inspection process went smoothly, except for one case: A 17-year-old student who had used mice in his project had failed to follow strict procedures set up to ensure that all animal experiments are properly supervised. Medical experts must certify that the experiment is relevant and carried out without subjecting the animals to unnecessary pain. Reluctantly, members of the rules committee disqualified the project.

Promptly at 9 A.M. on Wednesday, 509 scientists, engineers, college professors, and other professional men and women entered Currigan Hall, and the grueling, nail-biting, 12-hour judging routine began. Some 230 judges from the Denver area had the task of selecting students who would receive the ISEF monetary awards in 11 categories of science, such as biochemistry and physics. Another 279 experts considered projects for the special awards presented by professional societies, federal organizations, and corporations, such as the American Astronautical Society, the U.S. Patent Office, and the Eastman Kodak Company. These prizes range from magazine subscriptions to trips to other countries.

Judges checked the unmanned exhibits in the morning, tentatively selecting and eliminating projects in their fields of expertise. Then they conferred with other experts to get opinions on work outside their areas of competence. "It's a battle of attrition," one judge remarked. Another explained, "I make a list of projects I think deserve further consideration. Then I put these together with the selections of 29 other judges in my team. In the process, I may change my choices."

The judges returned in the afternoon to interview all participants whether they were considered for an award or not. Each student talks with at least three judges about his or her project. Judges are instructed to talk to students individually, not as a group, and to spend at least 10 minutes on each interview.

Judging can be as challenging for the judges as it is for the students. Alexander Cruz, a University of Colorado biologist, had difficulty understanding 18-year-old Masami Hasegawa's explanation of his study of the growth and behavior of a certain species of lizard in Japan. A Japanese-English phrasebook was of little help, so an interpreter had to be found. An Army major approached one exhibit and exclaimed, "Good grief, this kid knows more about the subject than I

Judges carefully review each exhibit, preparing for individual interviews with the participants.

do!" An Air Force general who toured the projects called it "a very humbling experience." "Judging serves as an educational experience," pointed out Gary McClelland, a University of Colorado psychologist. "Students receive new information, feedback on their work, and encouragement. They can ask questions and discuss their field of interest with professionals."

The judges carried scoring cards on which they rated each project as fair, good, or excellent for creative ability, scientific thought, thoroughness, skill, and clarity. "When I look at a project," McClelland explained, "I ask: 'Is it designed to answer a reasonable question, one that can be handled with resources a student has at his or her disposal? Did the student know how to properly collect and handle data? Were good, clear results obtained? Do the results have significance for the student, or the world at large?' In the interview, I determine how much of the work was done by the student and how much by others. All the judges have enough experience to know if a student is trying to snow us. I participated in science fairs as a student, so I know most of the tricks that might be tried."

Finalists say that being interviewed by the judges is the most enjoyable–and sometimes the scariest–part of the fair. "I enjoyed meeting

Army officers listen intently as Karen Mikkelson explains how she developed a new way to diagnose disorders of optic nerves and parts of the brain.

Jan Erik Hult, one of the Swedish exhibitors, tells a judge about the chemical applications of his gas chromatograph flame ionization experiments.

With the aid of an interpreter, Masami Hasegawa of Japan answers questions about his exhibit on the growth and behavior of a species of Japanese lizard.

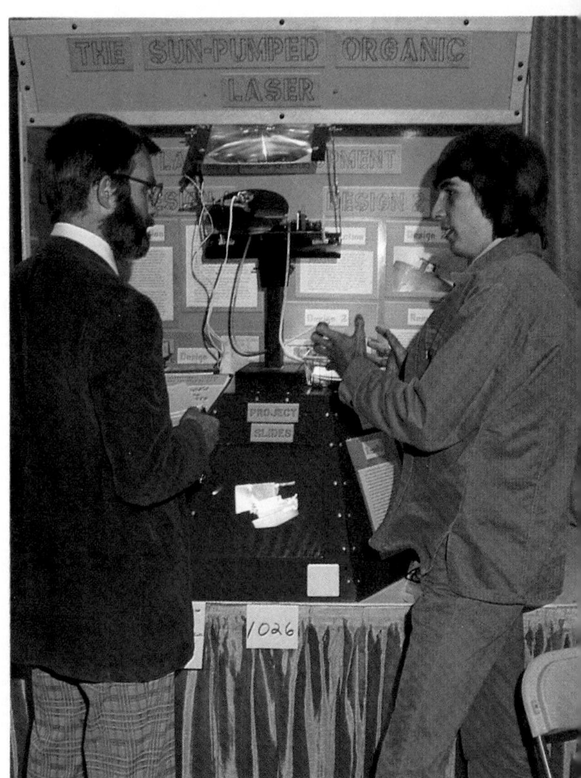

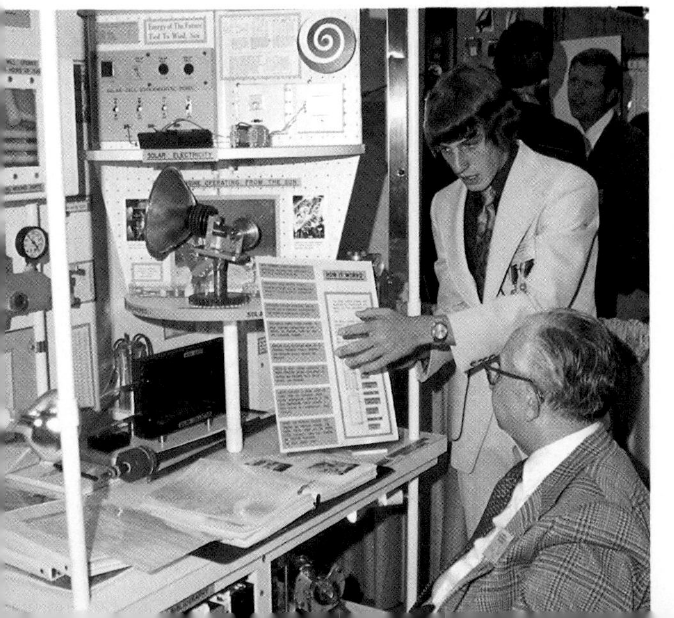

A judge listens attentively as Greg Weeks describes how he designed a laser that could be powered by sunlight instead of electricity, *above.* Randy Elliott, *left,* tells how he also used sunlight as an energy source for an air-conditioning system he developed.

and talking with the judges," said Deborah Joy Nelson, 17, of Bagley, Minn. "But I don't know if they bought what I told them." Her project, on using dry poultry waste as a feed supplement for sheep, later won a fourth-place zoology award.

Richard Bryant said he tried to "keep cool" during the interviews, but "it was kind of hard when they asked questions and I didn't have the answers. Like the teacher always calls on you when you don't know the answer. They asked me about things I hadn't figured out."

Hasegawa remarked that competition and judging were tougher in Japan than in the United States. Hult said competition was tougher in the United States than in Sweden. "Only 45 finalists competed in the Swedish national fair," he said. "There, it is more like an exhibition; here, there is competition with hundreds who have beaten thousands of others. I don't know how I came out." Both Hasegawa and Hult were to take second prizes for their work, Hasegawa in the zoology category and Hult in chemistry.

The fair was opened to the public on Thursday morning. Students with odd-numbered exhibits manned their booths to explain their work to the visitors, many of them students and teachers. In the afternoon, the odd-numbered exhibitors had a free half day, while their even-numbered neighbors manned the exhibit hall.

Most students took advantage of the half day off to enjoy one of the free tours available to finalists and their escorts. The Denver area is second only to Washington, D.C., as a center of federal activity, and students could visit the laboratories of the National Bureau of Standards, the U.S. Geological Survey, and the National Center of Atmospheric Research. Other tours included half-day trips to colorful, historic mining towns in the mountains west of Denver; exploratory hikes over the alpine tundra near Loveland Pass; and visits to Denver's zoo, botanic gardens, planetarium, and museum of natural history. Students could also talk to researchers at the University of Colorado's Cancer Research Center, School of Engineering, and departments of biology, biophysics, and genetics. Or they could tour Lowry Air Force Base, United Airlines' pilot-training center, the Colorado School of Mines, and several local medical centers and industrial laboratories.

Tension built up on Thursday as the students awaited the first awards banquet that night. Some students attended in long gowns or their best suits; others came in jeans and tennis shoes. Twenty-one certificates and plaques were presented by the American Dental Association (ADA), American Medical Association (AMA), and American Veterinary Medical Association (AVMA) for projects dealing with aspects of human or animal health. Winners of ADA and AMA superior-achievement awards were invited to the annual meetings of these associations as special guests. ADA merit-achievement winners received $100 to buy scientific equipment.

Michael Lomont, 16, of Fort Wayne, Ind., and his sponsor, Dianne Dunfree, were especially proud of the AVMA animal health award

Many students toured the Denver area in their free time. Some visited the old mining town of Georgetown, where they saw the jail, *top,* the hotel kitchen, *left,* and the wine cellar, *above.*

Canadian Michael Demerling, who participated in the 1975 fair, won an award from Motorola. He explains his laser communications link to Motorola research engineers who are working on similar projects.

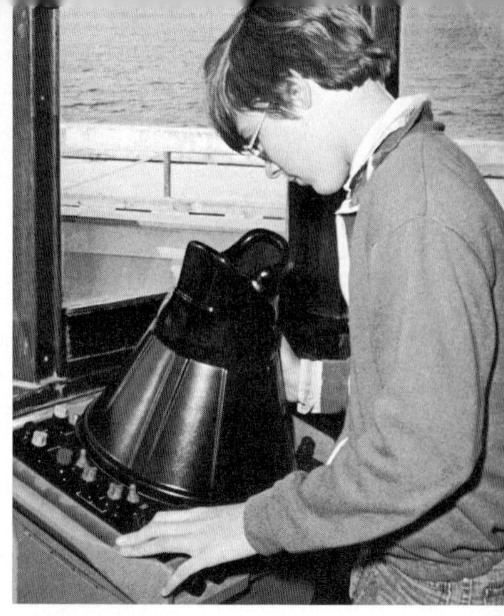

Kenneth Egan checks the radar aboard the Navy research ship he sailed on as a 1975 winner.

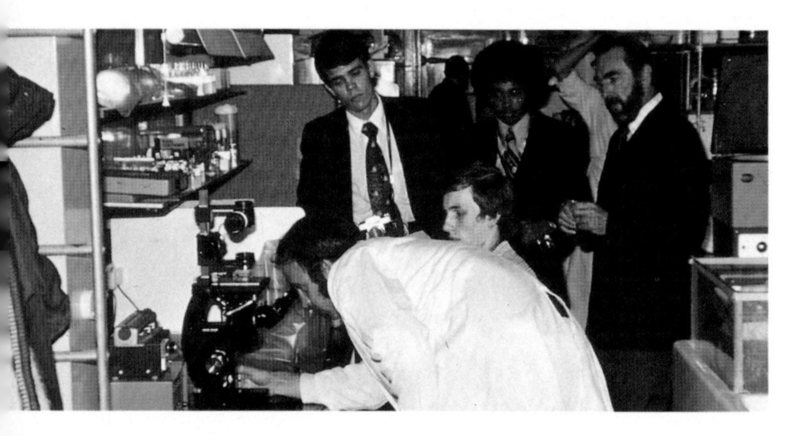

David Eslinger, Reginald Jenkins (both standing), and Jon Huppenthal (sitting), visited a London laboratory, *above,* en route to the 1975 Nobel ceremonies in Sweden. Darcy McGinn and Holly Ann Barrett spent four days in Tokyo, *below.* The five trips were awards given at the 1975 ISEF.

Cindy Johnson won a 1975 summer job in a national park where she helped measure forest regrowth.

that he won. Most high schools do what they can to help a student with a science project. But when Lomont found that his school could not help, Mrs. Dunfree, a medical technologist and family friend, took him to the hospital where she works. The staff there taught him how to use the electrophoresis equipment he needed to experiment with antiserum as a treatment for bacterial blood poisoning. Lomont worked there on his project for 18 months. When he went to the podium to accept his award, Mrs. Dunfree had tears in her eyes.

On Friday morning, 275 additional special awards were presented at the Auditorium Theatre across the street from Currigan Hall. Federal, professional, and private organizations awarded trips abroad, summer jobs, oceanographic cruises, cash prizes, magazine subscriptions, books, medals, plaques, and certificates.

The U.S. Army, Navy, and Air Force each awarded a student an expense-paid trip to Japan to represent the United States at the 20th Annual Japan Student Science Awards Fair in Tokyo in January 1977. For this trip, called "Operation Cherry Blossom," Army judges chose Karen S. Mikkelson, 18, of San Diego, who developed a new way to diagnose disorders of the optic nerves and parts of the brain. The Navy selected Robert J. Partyka, 18, of Columbus, Ohio, who designed and built three linear accelerators, then tested the effects of their radiation beams on human cancer cells. The Air Force picked John H. Runnels, 15, of Baton Rouge, La., for his mathematics project "A Combinatorial Problem: Paving with Integral Squares."

Each branch of the armed services also chose a winner and an alternate for a December 1976 trip to the Nobel prize ceremonies in Stockholm, Sweden. The Air Force winner, David T. Leighton, Jr., 17, of Arlington, Va., developed a new radiocarbon-dating method that does not require complex, bulky equipment. It can be used by geologists and archaeologists in the field. Explaining why only one of his teachers was able to help him with the project, Leighton said, "Most of them really did not understand what I was doing."

Richard Ebright was the Army's choice for the Stockholm trip. Participants who had been at the 1975 ISEF remembered Ebright as the young man who got up at 5 A.M. every day to catch butterflies. By studying the endocrine glands of these insects, he discovered a previously unknown hormone that is essential for proper growth of butterflies and moths. His research may lead to a new biological pesticide that will destroy harmful insects without producing any undesirable environmental side effects.

The Navy winner was Diane H. Wooden, 17, of Rockville, Md. She worked out a new method of counting stars to determine the detailed structure of dark nebulae, the clouds of dust and gas out of which new stars form. In the process, she taught herself FORTRAN (FORmula TRANslation), a computer language, in one weekend, and wrote the computer programs for processing her data in another. She also did all her own photographic work.

Koshare Indian dancers, from Explorer Boy Scout Post 2230, entertained at the closing banquet.

On Friday evening, ISEF officials announced the winners of fourth-place ($25), third-place ($50), second-place ($75), and first-place ($100) ISEF awards in 11 categories. Each category had one or two first-place awards, two to four second-place awards, two to five third-place awards, and up to seven fourth-place awards. Many students who had won special awards took additional ISEF awards.

Edward G. Sherburne, Jr., director of Science Service, also announced two winners of a new award. The students who won it will represent the United States at the Second Science Fair of the Americas in São Paulo, Brazil, in September 1976. Sarah E. Dennis, 15, of San Antonio, Tex., was one winner. Her project involved 1,400 experiments and tests to analyze dye fixatives and methods used to bond dyes to protein fibers such as wool. Philip A. Anfinrud, 17, of Aneta, N. Dak., won for experiments supporting the theory that freon released from aerosol cans can destroy the atmospheric ozone that protects the earth from harmful ultraviolet radiation.

Students who won the special trips and summer jobs can look forward to unforgettable and invaluable experiences, according to the 1975 winners. Their experiences dealt not only with science, but also touched on all aspects of their lives.

Three of the Puerto Rican student exhibitors exchange enthusiastic comments about their experiences at the fair during a banquet before the Friday-night awards ceremony.

Michael Lomont smiles modestly as a friend congratulates him on the animal health award he has won.

Karen Mikkelson excitedly embraces her father upon hearing her science project is one of the top special-award winners, as her teacher looks on. The U.S. Army gave her an expense-paid trip to the Japan Student Science Awards Fair in Tokyo in 1977.

For a change of pace after a day of manning their science exhibits, ISEF participants enjoy themselves at an evening mixer for students.

Glenn Seaborg, Science Service president and winner of a Nobel prize, congratulates Diane Wooden, David Leighton, and Richard Ebright, who won trips to the Nobel prize ceremonies in Stockholm, Sweden.

John M. Huppenthal, 19, of Michigan City, Ind., who is now attending Purdue University, went to the Nobel award ceremonies in December 1975. He described the world's most honored scientists as being pretty much like everyone else.

"That surprised me," Huppenthal confessed. "I expected they would be completely different from other people. They acted normal and friendly, and even offered to help me. Once you see they are the same kind of people as you are, it makes you try harder because you feel there is a chance of winning a Nobel prize yourself. Something you once thought was impossible becomes possible."

Holly Ann Barrett, 19, of Essex, Iowa, discovered a whole new world in Japan when she visited Tokyo as a representative at the Japan Student Science Awards Fair. "I lived in a small town in Iowa all my life," she said. "I was totally unprepared to find an entire nation with a completely different way of looking at life. The people everywhere I went impressed me. Students are dedicated, and adults have their minds set on a real purpose in life. They seem less flighty and preoccupied with recreation than people in the United States."

Barrett's science fair project "greatly influenced" her choice of a major in college. The project involved testing the effects of induced anxiety and frustration on elementary and high school students. Through the project, she found that "a great deal of research needs to be done in psychology, and it seems like a field where I can apply my talents to make a contribution."

Christopher K. York, 19, of Annandale, Va., won a 1975 summer job with the National Park Service as a result of his study of birds of prey. He worked at Isle Royale National Park in Michigan, searching for old nesting sites and cliffs where peregrine falcons could build nests. The information will be used when Cornell University carries out a plan to re-establish the endangered bird in locations where it was once abundant. The job gave York an opportunity to do what he wants most in life–"to study animals in their natural environment."

Steven P. Houtchens, 16, of Lancaster, Tex., won an expense-paid week at the Naval Postgraduate School in Monterey, Calif. By experimenting with models he carved and sailed in a homemade tank, he developed a new way to measure the resistance that ships of different shapes encounter in moving through water. Houtchens was considering a career in ocean engineering, but a day on the rolling, pitching research ship *Acania* in Monterey Bay dampened his plans. "I don't think I could get used to all that moving up real quick, then falling down again," he admitted. Kenneth W. Egan, 16, of Taylors, S.C., took the same cruise. However, it impressed him so much that he applied for the United States Naval Academy, something he never thought of doing before the trip.

Of the 407 participants, 221 won awards at the 1976 ISEF. But, in a way, all were winners. All finalists had the opportunity to discuss their projects and interests with scientists, engineers, and other professionals who attended the fair as judges and observers. One visitor to the exhibits was Glenn T. Seaborg, a co-winner of the 1951 Nobel prize for chemistry, one of the developers of the atomic bomb, discoverer of six chemical elements, and now president of Science Service. Seaborg talked to many of the students during the week. Friendships grew among students who shared common interests, and they made valuable contacts that may lead to jobs, scholarships, new ideas, and better designed projects in later years.

Being chosen as an ISEF finalist demonstrates the type of achievement that college admissions officers seek in addition to grades. The experience and responsibility of getting an exhibit to the fair, setting it up, being judged, and explaining it to scientists and visitors is a valuable by-product. In sum, the ISEF gives students a chance to broaden their horizons both in science and in life.

As Karen Mikkelson put it: "I worked on my project for 18 months to 2 years. In the summers, I worked 10 to 15 hours every day at a local hospital. At times, I got very discouraged and wondered if it was all worthwhile. But the ISEF proved it was. I've always wanted to do research and to help people. I wasn't always sure I could do it. But now I feel much more confident."

First-Place ISEF Winners

Behavioral and Social Sciences:
Kenneth J. Lohmann, 16,
West Lafayette, Ind.

Biochemistry:
Brenda I. Troche, 16,
Mayagüez, Puerto Rico

Botany:
Maggie P. Murray, 17,
Melbourne, Fla.

Chemistry:
Sarah E. Dennis, 15,
San Antonio, Tex.;
Diane L. Medved, 18,
Kansas City, Mo.

Earth and Space Sciences:
Diane H. Wooden, 17,
Rockville, Md.;
Philip A. Anfinrud, 17,
Aneta, N. Dak.

Engineering:
Peter A. Sandborn, 16,
Fort Collins, Colo.;
Jeffrey L. Kegarise, 16,
Newburgh, Ind.

Mathematics and Computers:
John H. Runnels, 15,
Baton Rouge, La.

Medicine and Health:
Sue Vorderbruggen, 18,
Bagley, Minn.;
Lise A. Desquenne, 18,
Harrisville, R.I.

Microbiology:
Jerry Jackson, 16,
Rockledge, Fla.

Physics:
Peter J. Weaver, 17,
Dallastown, Pa.

Zoology:
Ardellie Rivera, 17,
San Germán, Puerto Rico;
Richard H. Ebright, 16,
South Temple, Pa.

Two Danish physicists, Aage N. Bohr, left, and Ben Mottelson, shared the 1975 Nobel prize in physics with James Rainwater of Columbia University.

Awards And Prizes

A listing and description of the year's major awards and prizes in science, and the men and women who received them

Earth and Physical Sciences

Chemistry. Major awards in the field of chemistry included:

Nobel Prize. An Australian and a Swiss chemist shared the 1975 Nobel prize in chemistry. They are John W. Cornforth, 58, a professor at the University of Sussex in England, and Vladimir Prelog, 69, professor at the Federal Technical University in Zurich, Switzerland.

Cornforth and Prelog shared the $143,000 prize for their work in stereochemistry, which deals with the position of atoms in substances with differing optical and chemical properties.

Cornforth, who has been deaf since childhood, left his native Australia to study at the University of Oxford and has spent most of his life in England. During World War II, he worked with Sir Robert Robinson to determine the structure of penicillin.

Prelog was born in Sarajevo, Yugoslavia, but is now a citizen of Switzerland. He developed the system for determining D- and L-isomerism in molecules. This is the right- or left-handedness of certain compounds that have the same chemical formulas.

Perkin Medal. Lewis H. Sarett, president of Merck Sharp & Dohme Research Laboratories in Rahway, N.J., received the 1976 Perkin Medal. The Perkin award is presented in the United States by the American section of the Society of Chemical Industry.

Sarett, a leading authority on steroid chemistry, was the first to synthesize cortisone, which is now used in controlling rheumatoid arthritis and other inflammatory diseases. In addition, he played a part in the synthesis of dexamethasone, a corticosteroid that is 35 times more potent than cortisone.

Sarett also headed Merck teams that discovered amprolium, which combats coccidiosis in poultry, and thaibendazole, which controls gastrointestinal parasites in other livestock.

Priestley Medal, the American Chemical Society's highest award, was presented to George S. Hammond in 1976. Hammond is professor of chemistry at the University of California, Santa Cruz, and foreign secretary of the National Academy of Sciences (NAS).

Hammond has published more than 250 scientific papers and four books during 27 years as a teacher and researcher. He is an international authority on photochemistry, the branch of chemistry dealing with chemical changes caused by radiant energy. He is considered an extraordinary teacher and a noted lecturer on photochemistry and the role of science and chemistry.

A native of Auburn, Me., Hammond received a B.S. degree *magna cum laude* from Bates College in 1943 and a Ph.D. in chemistry from Harvard University in 1947. He taught at Iowa State University and California Institute of Technology before joining the Santa Cruz faculty in 1972.

Physics. Awards recognizing major work in physics included:

Nobel Prize. The 1975 Nobel prize in physics was awarded to an American and two Danish physicists: James Rainwater, 58, of Columbia University; Aage N. Bohr, 53, of the Institute for Theoretical Physics in Copenhagen, Denmark; and Ben Mottelson, 49, of the Nordic Institute of Technology in Copenhagen. They shared the $143,000 prize "for the discovery of the connection between collective motion and particle motion in atomic nuclei and the development of the theory of the structure of the atomic nucleus based on this connection."

Bohr is the son of Niels Bohr, who won the 1922 Nobel prize in physics for his studies of the structure of atoms and their radiations. Aage Bohr and Mottelson have worked together for several years in Copenhagen. With Rainwater, they developed the collective model of the nucleus.

Buckley Prize. George Feher, professor of physics at the University of California, San Diego, since 1960, won the 1976 Oliver E. Buckley Solid State Physics Prize. He was honored for his "development of electron nuclear double resonance and the application of spin resonance to a wide range of problems in the physics of condensed matter."

Feher, a native of Czechoslovakia, received his B.S., M.S., and Ph.D. degrees from the University of California, Berkeley. He was elected to the National Academy of Sciences in 1975.

Franklin Medal. John Bardeen, professor of physics and electrical engi-

Physicist Samuel C. C. Ting received an E. O. Lawrence Memorial Award for his contributions to the field of atomic energy, including the discovery of new particles of matter.

neering at the University of Illinois in Urbana, won the 1975 Franklin Medal, highest award of the Franklin Institute of Philadelphia. Bardeen was honored for his work on the electrical conductivity of solids, semiconductor devices, and superconductivity.

Bardeen was the first person to win a Nobel prize twice for work in the same field. He shared the physics prize in 1956 for inventing the transistor effect, and again in 1972 for his work on superconductivity.

John Bardeen

E. O. Lawrence Memorial Award. Five scientists received the Ernest O. Lawrence Memorial Award in 1975 for outstanding contributions in the field of atomic energy. The award honors the late physicist Ernest O. Lawrence, the inventor of the cyclotron. The winners:

Evan H. Appelman, of Argonne National Laboratory, Argonne, Ill.; Charles E. Elderkin, Battelle Pacific Northwest Laboratories, Richland, Wash.; William A. Lokke, Lawrence Livermore Laboratory, Livermore, Calif.; Burton Richter, Stanford Linear Accelerator Center, Palo Alto, Calif.; and Samuel C. C. Ting, the Massachusetts Institute of Technology (M.I.T.) in Cambridge.

Geosciences. Awards for important work in the geosciences include:

Agassiz Medal. Walter Heinrich Munk, professor of geophysics at the Institute of Geophysics and Planetary Physics, University of California, San Diego, received the 1976 Agassiz Medal in oceanography. The award was "for outstanding experimental and theoretical research on the spectrum of motion in the oceans and the earth."

Penrose Medal. Preston E. Cloud, Jr., professor of biogeology at the University of California, Santa Barbara, received the 1976 Penrose Medal from the Geological Society of America.

Thompson Gold Medal. James M. Schopf of the Coal Geology Laboratory at Ohio State University received the $1,000 Mary Clark Thompson Medal in 1976. The award honored his "remarkable high level of scientific and scholarly work in both systematic paleontology and stratigraphy."

Life Sciences

Biology. Among the awards presented in biology were the following:

Carski Award. Elizabeth R. Hall, professor of bacteriology and public health at Washington State University in Pullman, won the Carski Foundation Distinguished Teaching Award for 1976. The American Society of Microbiology makes the $1,000 award annually to recognize distinguished teaching of microbiology to undergraduates.

Horwitz Prize. Two Swedish chemists shared the 1975 Louisa Gross Horwitz Prize. They are Sune Bergstrom, professor of chemistry and rector at the Karolinska Institute in Stockholm, and Bengt Samuelsson, Karolinska professor of medical and physiological chemistry.

Bergstrom and Samuelsson were honored for their work with prostaglandins, the hormonelike regulators of many body functions. The $25,000 Horwitz Prize is awarded annually for outstanding research in biology.

The two Swedish scientists determined the structures of the prostaglandins and synthesized them chemically. Their work has potential application in the treatment of stomach ulcers, fevers, allergies, and many other maladies.

Lilly Award. Ronald W. Davis, assistant professor of biochemistry at the Stanford University School of Medicine, received the 1976 Eli Lilly and Company Award in Microbiology and Immunology. The $1,000 award is for his work on gene manipulation.

A native of Maroa, Ill., Davis earned his B.S. degree at Eastern Illinois and did graduate studies at the California Institute of Technology. He also studied at Harvard University.

NAS Award for Environmental Quality. David M. Evans, a consulting engineer-geologist in Denver, Colo., received the 1976 National Academy of Sciences Award for Environmental Quality. Evans was honored for "his key discovery that injection of liquids deep underground can generate earthquakes, and for his subsequent studies and their environmental implications." The $5,000 award is given to recognize outstanding contributions in science or technology to improve the environment.

Physician Hans Selye shared the Kittay Award for work demonstrating the role that stress plays in disease.

Life Sciences

Continued

Biologist Renato Dulbecco shared the 1975 Nobel prize in medicine with two former pupils, David Baltimore and Howard M. Temin, for research that linked cancer and viruses.

Tyler Ecology Award. Three scientists noted for environmental work shared the $150,000 John and Alice Tyler Ecology Award in 1976. Zoologist Charles Elton of Oxford University in England, microbiologist René J. Dubos of Rockefeller University in New York City, and sanitary engineer Abel Wolman of Johns Hopkins University in Baltimore were honored for their accomplishments in ecology.

Elton, considered by many as the founder of modern ecology, has written important books on the subject and founded the Bureau of Animal Populations. Dubos, who demonstrated the feasibility of obtaining germ-fighting drugs from microbes more than 30 years ago, helped to prepare the United Nations Stockholm Conference on the Human Environment in 1972. Wolman developed methods by which city water is chlorinated. He also pioneered in wastewater-purification methods, established the U.S. Atomic Energy Commission's (AEC) sanitary engineering standards, and helped to establish the World Health Organization.

Medicine. Major awards in medical sciences included the following:

Nobel Prize. Renato Dulbecco, 61, and two of his former pupils shared the 1975 Nobel prize in physiology or medicine for discovering how viruses can cause malignant tumors in human beings. Dulbecco, an Italian-born American now at the Imperial Cancer Research Fund Laboratory in London, shared the prize with David Baltimore, 37, professor of microbiology at the M.I.T. Center for Cancer Research in Boston, and Howard M. Temin, 41, professor of oncology, or tumor research, at the University of Wisconsin's McArdle Memorial Laboratory.

Baltimore was a graduate student under Dulbecco at the Salk Institute for Biological Studies in San Diego, and Temin studied under Dulbecco at the California Institute of Technology.

Dulbecco laid the groundwork for most of the recent advances in tumor virology. He developed the plaque technique, the first laboratory method of rapidly counting deoxyribonucleic acid (DNA) viruses. Temin and Balti-

Life Sciences
Continued

Abel Wolman

more independently discovered reverse transcriptase in 1970. Reverse transcriptase is an enzyme that allows a ribonucleic acid (RNA) virus to make a DNA copy of itself. The process may play some role in initiating the cancers that some RNA viruses can cause.

Gairdner Awards. Six medical scientists received Gairdner Foundation International Awards of $10,000 each in 1975. The awards are made by the Gairdner Foundation of Toronto, Canada, for outstanding research. They went to:

Dr. Ernest Beutler of the City of Hope National Medical Center in Duarte, Calif., for his contributions to the field of hematology.

Dr. Baruch S. Blumberg, epidemiologist at the Institute for Cancer Research in Philadelphia, who discovered the Australia antigen and opened new avenues of research in liver disease and blood transfusion.

Biochemist Henri G. Hers of the Physiological Chemistry Laboratory in Louvain, Belgium, for his work in carbohydrate and glycogen metabolism.

Physiologist Hugh E. Huxley of the Laboratory of Molecular Biology in England, who has increased understanding of the skeletal muscles.

Dr. John D. Keith of the University of Toronto for research on congenital heart disease.

Dr. William T. Mustard of the Hospital for Sick Children in Toronto, for work in cardiovascular surgery.

Kittay Award. Two behavioral scientists, James Olds and Hans Selye, shared the 1976 International Kittay Award. The $25,000 award is given annually by the Kittay Scientific Foundation for work on mental health.

Olds, professor of behavioral biology at the California Institute of Technology in Pasadena, identified specific areas in the brain called pleasure centers. His work indicated the physiological basis for hedonism, devotion to pleasure.

Selye, director of the Institute of Experimental Medicine and Surgery at the University of Montreal, has demonstrated the role stress plays in several diseases and the body's reaction to the wear and tear of living.

Space Sciences

Glynn S. Lunney

Aerospace. The highest awards granted in the aerospace sciences included:

Collier Trophy. David S. Lewis, chairman of the board of General Dynamics Corporation and the U.S. Air Force-Industry team that produced the F-16 fighter aircraft, received the Robert J. Collier Trophy for 1975. The award was given for significant advancements in aviation technology leading to innovative fighter aircraft effectiveness.

The F-16, incorporating outstanding design and technological innovations, set significant new performance standards. The aircraft was selected in 1975 by the U.S. Air Force and a group of four North Atlantic Treaty Organization nations.

Goddard Astronautics Award. Edward Price, visiting professor at the Georgia Institute of Technology in Atlanta, received the Goddard Astronautics Award in 1976. The award is presented for outstanding contributions in the science of propulsion or energy conversion.

Price was honored "for original contributions and leadership in the development of solid rocket propulsion with emphasis on combustion technology."

Hill Space Transportation Award. Glynn Lunney received the 1975 Louis W. Hill Space Transportation Award, the last to be given. Lunney is manager of the shuttle performance, integration, and development program for the National Aeronautics and Space Administration. He was honored "for pioneering efforts . . . that have contributed significantly to the success of every U.S. manned space flight program."

Astronomy. Among the top honors awarded for important contributions in astronomy was the following:

Bruce Medal. Estonian astronomer Ernst J. Öpik was awarded the Catherine Wolfe Bruce Gold Medal in 1976. The award is presented by the Astronomical Society of the Pacific.

Öpik is a theoretical astronomer known for his contributions to the study of meteors. He also invented the rocking mirror telescope.

General Awards

Kenneth S. Pitzer

Science and Humanity Awards for outstanding contributions to science and humanity during the past year included the following:

Founders Medal. Manson Benedict, a pioneer in the development of atomic energy technology, was awarded the 1976 Founders Medal. The award is given by the National Academy of Engineering (NAE).

Benedict is institute professor emeritus at M.I.T. He was awarded the medal for "his outstanding engineering accomplishments . . . his contributions to the development of atomic energy technology and his leadership in nuclear engineering education."

Benedict is known for his outstanding work in chemical and nuclear engineering, and especially for his contributions to the gaseous diffusion process for the separation of uranium isotopes. During World War II, he was in charge of the process design of the K-25 gaseous diffusion plant for separating uranium-235 from natural uranium at Oak Ridge, Tenn. Benedict was also responsible for operations analysis for the Atomic Energy Commission during 1951 and 1952.

National Medal of Science. The United States highest award in science, mathematics, and engineering went to 13 Americans in 1976. President Gerald R. Ford, in announcing the winners, said that examination of their accomplishments "demonstrates the importance of science and engineering to the nation."

The winners were selected from 204 nominations by the NAS, various professional societies, and a number of colleges and universities. Those honored are:

Nicholaas Bloembergen, applied physics, Harvard University

Britton Chance, biophysics, University of Pennsylvania

Erwin Chargaff, biochemistry, Columbia University

Paul J. Flory, chemistry, Stanford University, Palo Alto, Calif.

William A. Fowler, physics, California Institute of Technology

Kurt Gödel, mathematics, Institute for Advanced Study, Princeton University

Rudolph Kompfner, electronics, Stanford University

James Van Gundia Neel, genetics, University of Michigan, Ann Arbor

Linus Pauling, chemistry, Stanford University

Ralph B. Peck, civil engineering, Albuquerque, N. Mex.

Kenneth S. Pitzer, chemistry, University of California, Berkeley

James A. Shannon, biomedicine, National Institutes of Health, Silver Springs, Md.

Abel Wolman, sanitary engineering, Johns Hopkins University, Baltimore

Oersted Medal. Victor F. Weisskopf, institute professor of physics at the Massachusetts Institute of Technology in Cambridge, was awarded the 1976 Oersted Medal. The medal is awarded annually by the American Association of Physics Teachers for outstanding contributions to the teaching of physics.

Weisskopf came to the United States from Vienna, Austria, in 1937. He has taught at M.I.T., the University of Rochester in New York, and in Zurich, Switzerland. He has also written several books and articles. His book *Knowledge and Wonder, The Natural World as Man Knows It* (1962) was chosen by the Thomas Alva Edison Foundation as the best science book of the year for youth. Weisskopf, who believes that students should not concentrate on science to the exclusion of art, music, and literature, was honored as an expositor of physics, a dedicated humanist, and a teacher of teachers.

Zworykin Award. C. Kumar N. Patel, director of the Bell Laboratories Electronics Research Laboratory in Murray Hill, N.J., received the Vladimir K. Zworykin Award for 1976. The $5,000 award is presented each year by the National Academy of Engineering.

Patel was honored for outstanding achievement in electronics, particularly his original contributions in laser technology. He invented a flowing gas laser in 1965. As a result of his discoveries, lasers are now being used in metal and fabric cutting, for welding, and for military uses.

Patel is a native of Baramati, India, and was graduated from Poona University in India. He also holds graduate degrees from Stanford University in Palo Alto, Calif. [Joseph P. Spohn]

Major Awards and Prizes

Award winners treated more fully in the first portion of this section are indicated by an asterisk (*)

Adler Prize: Edward A. Boyse
*Agassiz Medal (oceanography): Walter H. Munk
Alan T. Waterman Award (mathematics): Charles L. Fefferman
AAAS-Rosenstiel Award (oceanography): Kenneth O. Emery
AAAS-Socio-Psychological Prize: Gregory B. Markus, R. B. Zajonc
ACS Medicinal Chemistry Award: Bernard Belleau
American Institute of Chemists Gold Medal Award: Kenneth S. Pitzer
American Physical Society High Polymer Physics Prize: Richard S. Stein
Arthur L. Day Medal (geophysics): Hans Ramberg
Arthur Weber Prize (heart research): George Rona
Bonner Prize (nuclear physics): John Schiffer
Bowie Medal (geophysics): Jule G. Charney
*Bruce Medal (astronomy): Ernst J. Öpik
*Buckley Solid State Physics Prize: George Feher
*Carski Foundation Award (teaching): Elizabeth R. Hall
*Collier Trophy (astronautics): David S. Lewis
Daniel Elliot Medal (zoology): Howard E. Evans
Davisson-Germer Prize (optics): Ugo Fano
Debye Award (physical chemistry): Robert Zwanzig
*Founders Medal (engineering): Manson Benedict
*Franklin Medal (physics): John Bardeen
*Gairdner Awards (medicine): Ernest Beutler, Baruch S. Blumberg, Henri G. Hers, Hugh E. Huxley, John D. Keith, William T. Mustard
Garvan Medal (chemistry): Isabella L. Karle
Gibbs Medal (chemistry): John C. Neidermair
*Goddard Astronautics Award: Edward Price
Guggenheim Medal (aeronautics): Duane Wallace
Haley Astronautics Award: Gerald Carr, Edward Gibson, William Pogue
Heineman Prize (American Physical Society): S. W. Hawking
Heineman Prize (Göttingen [West Germany] Academy of Sciences): Philip W. Anderson
*Hill Space Transportation Award: Glynn Lunney
*Horwitz Prize (biology): Sune Bergstrom, Bengt Samuelsson
Ives Medal (optics): Ali Javan
*Kittay Award (psychiatry): James Olds, Hans Selye
Klumpke-Roberts Prize (astronomy): Chesley Bonestell
Krupp-Halbach Energy Research Award: Erwin Gärtner, Hans Heublein, Alwin E. Petzold, Hans-Joachim Leuschner, Ludger Dilla, Hermann Pieper
Langmuir Prize (chemical physics): John S. Waugh
Lasker Awards (medical research): John E. Baer, Karl H. Beyer, Jr., Frank J. Dixon, Roger C. L. Guillemine, Godfrey N. Hounsfield, Henry G. Kunkel, Frederick Novelle, William H. Oldendorf, Andrew V. Schally, James M. Sprague, Jules Stein
*Lawrence Memorial Award (atomic energy): Evan H. Appelman, Charles E. Elderkin, William A. Lokke, Burton Richter, Samuel C. C. Ting
*Lilly Award (microbiology): Ronald W. Davis
Meggers Award (spectroscopy): Jean Blaise
*NAS Award for Environmental Quality: David M. Evans
*National Medal of Science: Nicholaas Bloembergen, Britton Chance, Erwin Chargaff, Paul J. Flory, William A. Fowler, Kurt Gödel, Rudolph Kompfner, James Van Gundia Neel, Linus Pauling, Ralph B. Peck, Kenneth S. Pitzer, James A. Shannon, Abel Wolman
Niels Bohr Gold Medal (atomic energy): Hans A. Bethe
*Nobel Prize: chemistry, John W. Cornforth, Vladimir Prelog; physics, Aage N. Bohr, Ben Mottelson, James Rainwater; physiology or medicine, David Baltimore, Renato Dulbecco, Howard M. Temin
*Oersted Medal (teaching): Victor F. Weisskopf
Oppenheimer Memorial Prize (physics): Yoichiro Nambu
*Penrose Medal (geology): Preston E. Cloud, Jr.
*Perkin Medal (chemistry): Lewis H. Sarett
Pfizer Award (enzyme chemistry): M. L. Gefter
*Priestley Medal (chemistry): George S. Hammond
*Thompson Gold Medal (geology): James M. Schopf
Trumpler Award (astronomy): Robert B. Hanson
*Tyler Ecology Award: René J. Dubos, Charles Elton, Abel Wolman
U.S. Steel Foundation Award (molecular biology): Daniel Nathans
Waksman Award (microbiology): Wallace P. Rowe
*Zworykin Award (engineering): C. Kumar N. Patel

Deaths of Notable Scientists

Detlev W. Bronk

Theodosius Dobzhansky

John R. Dunning

Notable scientists who died between June 1, 1975, and June 1, 1976, include those listed below. An asterisk (*) indicates that a biography appears in *The World Book Encyclopedia.*

*__Aalto, Alvar__ (1898-May 11, 1976), Finnish architect, pioneer in the interplay of light, acoustics, and form in such buildings as Baker House at the Massachusetts Institute of Technology.

Ballard, Bristow G. (1902-Sept. 22, 1975), Canadian electrical engineer and president of the National Research Council of Canada from 1963 to 1967.

Bjerknes, Jacob (1897-July 7, 1975), Swedish-born meteorologist, developed the air-mass analysis method of weather forecasting with his father, Vilhelm.

Bronk, Detlev W. (1897-Nov. 17, 1975), biophysicist, president emeritus of Rockefeller University and former president of the National Academy of Sciences.

Dam, Henrik (1895-April 17, 1976), Danish biochemist, co-winner of the 1943 Nobel prize for physiology or medicine who discovered the role of vitamin K in coagulating blood.

Dennison, David M. (1900-April 3, 1976), theoretical physicist who discovered the spin of the proton, important in understanding molecular structure and atomic and nuclear structure.

Dobzhansky, Theodosius (1900-Dec. 18, 1975), Russian-born zoologist and biologist. His book *Genetics and the Origin of Species* (1937) was one of the most important of its kind since Charles Darwin's classic study.

Dragstedt, Lester R. (1893-July 15, 1975), physiologist and surgeon who performed the first successful separation of Siamese twins in 1955.

*__Dunning, John R.__ (1907-Aug. 25, 1975), physicist who helped develop the method of isolating the uranium-235 used in nuclear weapons and pioneered in research on the discharge of neutrons from uranium fission.

Egtvedt, Claire L. (1892-Oct. 19, 1975), aeronautical engineer and former president of the Boeing Company, developed the B-17 Flying Fortress.

Evans, Alice (1881-Sept. 5, 1975), pioneer bacteriologist, helped identify raw milk as a source of undulant fever.

Fairley, Gordon H. (1930-Oct. 23, 1975), British oncologist and a leading authority on the drug treatment of cancer. He was killed by a bomb explosion in London.

Ferri, Antonio (1912-Dec. 28, 1975), Italian-born aeronautical engineer who led research in the development of wind tunnels for testing supersonic aircraft.

György, Paul (1893-March 1, 1976), Hungarian-born nutritionist and pediatrician who discovered the vitamins riboflavin, pyridoxine (B6), and biotin.

Haddow, Sir Alexander (1907-Jan. 21, 1976), British pathologist, one of the first researchers to realize the possibility of treating cancer with drugs and by stimulating the body's own defenses.

*__Heisenberg, Werner K.__ (1901-Feb. 1, 1976), German physicist who won the 1932 Nobel prize for physics for founding quantum mechanics, which led to a more precise theory about atoms. He also developed the uncertainty principle, which states that the position and velocity of an electron in motion cannot be simultaneously measured precisely.

Heremans, J. F. (1927-Oct. 29, 1975), Belgian immunologist who isolated and characterized immunoglobulin A (IgA) and coined the word immunoglobulin.

*__Hertz, Gustav__ (1887-Oct. 30, 1975), German nuclear physicist who shared the 1925 Nobel prize for physics for proving the validity of Niels Bohr's theory of the atom.

Hogben, Lancelot (1895-Aug. 22, 1975), British physiologist and one of the great popularizers of science whose *Mathematics for the Million* (1936) has been translated into 30 languages.

Josi, Enrico (1885-Sept. 1, 1975), Italian archaeologist who discovered the tomb of Saint Peter under the Vatican Basilica in 1949.

Kay, Marshall (1904-Sept. 3, 1975), Canadian-born geologist and winner of the 1971 Penrose Medal for his early model of continental drift, which paved the way for global tectonics.

Lippisch, Alexander M. (1894-Feb. 11, 1976), German-born aeronautical engineer who designed the first delta-winged aircraft and the first airplane to fly faster than the speed of sound.

May, Jacques M. (1896-June 30, 1975), French-born physician, nutritionist, and director of the American Geographical Society's Department of

Deaths of Notable Scientists

Continued

Werner K. Heisenberg

Jacques Monod

Edward L. Tatum

Medical Geography whose maps of the global distribution of infectious disease became tools in teaching medicine throughout the world.

McCauley, George V. (1882-April 19, 1976), physicist who supervised the design and casting of such giant mirror disks as the 508-centimeter (200-inch) mirror of the Hale telescope located at Palomar Observatory.

Minkowsky, Rudolph L. (1895-Jan. 4, 1976), German-born astronomer who supervised the National Geographic Society-Palomar Observatory Sky Survey for many years, providing photographs used by astronomers throughout the world.

***Monod, Jacques** (1910-May 31, 1976), French biochemist who shared the 1965 Nobel prize for physiology or medicine with André Lwoff and François Jacob for discoveries concerning how genes are controlled.

Nyquist, Harry (1889-April 4, 1976), engineer who developed the Nyquist criterion, a set of conditions necessary to keep feedback circuits stable that is also used in the study of such human reaction processes as the responses of automobile drivers.

Penfield, Wilder G. (1891-April 5, 1976), neurologist who founded the Montreal Neurological Institute in Canada in 1934 with a $1.2-million grant from the Rockefeller Foundation, and pioneered surgical techniques in the treatment of epilepsy.

Polanyi, Michael (1891-Feb. 22, 1976), Hungarian-born British physical chemist noted for his work on the theory of reaction rates and the X-ray diffraction pattern of natural plant fibers, which led to the rotating crystal method of X-ray analysis.

Pool, Judith G. (1919-July 13, 1975), research physiologist who developed a simple method for extracting from normal blood the protein needed for hemophilia patients, thus enabling hemophiliacs to treat themselves for their disease at home.

Rogers, William M. (1900-Sept. 2, 1975), heart researcher who headed a team in 1962 that developed a tiny stethoscope that can be ingested by the patient, allowing the physician to record the sound of blood flow in the aorta and hear otherwise inaudible heart sounds.

Sauer, Carl O. (1889-July 18, 1975), geographer, winner of the 1957 Vega Medal. His book *Northern Mists* (1968) contended that America was discovered by Irish monks several centuries before Leif Ericson landed about A.D. 1000.

Stillman, Irwin M. (1895-Aug. 26, 1975), physician and author of best-selling diet books advocating the drinking of large quantities of water.

Stong, Clair L. (1902-Dec. 9, 1975), electrical engineer whose wide-ranging column "The Amateur Scientist" in *Scientific American* covered everything from how to build an atomic particle accelerator in your own garage to the study of color vision in pigeons.

***Tatum, Edward L.** (1909-Nov. 5, 1975), biochemist, shared the 1958 Nobel prize for physiology or medicine for discovering that genes regulate specific chemical processes. He and George W. Beadle found that mutations caused by X rays, ultraviolet light, and biochemical processes are passed on to successive generations.

Thomas, Sir James Tudor (1893-Jan. 23, 1976), British eye surgeon whose research in the 1930s laid the foundation for the world's first eye bank in 1944 in New York City.

Thomson, Sir George Paget (1892-Sept. 10, 1975), British physicist, co-winner of the 1937 Nobel prize for physics with American Clinton J. H. Davisson for discovery of the diffraction of electrons by crystals.

Tomkins, Gordon M. (1926-July 22, 1975), biochemist who provided evidence for a low molecular weight compound such as a steroid hormone specifically changing the conformation of a protein molecule.

Whipple, George H. (1878-Feb. 1, 1976), pathologist, co-winner of the 1934 Nobel prize for physiology or medicine for discovering that pernicious anemia could be controlled by a liver diet.

Wildt, Rupert (1905-Jan. 9, 1976), German-born astrophysicist, won the 1966 Eddington Gold Medal for work on solar and planetary atmospheres.

Wrinch, Dorothy (1894-Feb. 11, 1976), Argentine-born biochemist of the University of Oxford in England who devised the cyclo theory of peptide structure and later extended the theory to polypeptide chains. [Irene B. Keller]

The Choice To Die

By Robert M. Veatch

The new medical technology, together with an increase in the number of people demanding a hand in the decision, raises hard questions about the ethics of ending a life

A baby I shall call Jimmy was born recently with a severe malformation of the spinal column, a condition called spina bifida with myelomeningocele. His vertebrae were not completely developed. The spinal cord, covered only by a thin membrane, protruded from an opening in his back. From the position of the opening we knew that Jimmy would never control his leg muscles, bladder, or bowels.

In addition, Jimmy's head was abnormally large. Fluid that forms in the brain normally passes into the spinal cord. When this fluid is blocked because of the spinal deformity, pressure enlarges the brain and the skull, and retardation often results. To avoid this, surgeons must immediately implant a tube from the brain under the skin into a vein near the heart. The tube will probably have to be replaced occasionally. The severe curvature of the spine will require surgery, and

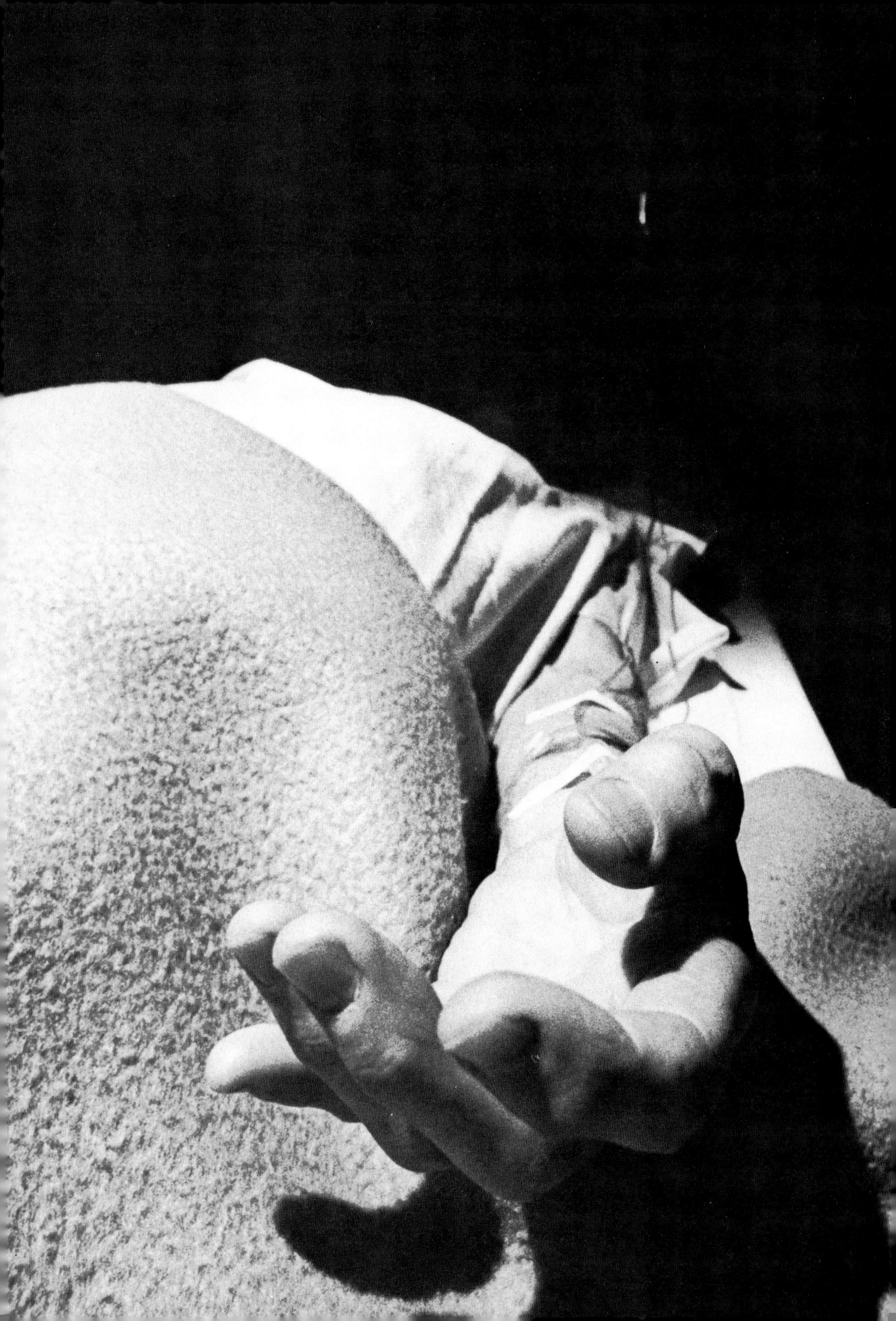

other surgery will be needed to close the back and provide artificial control of the bowels.

Such cases–there are 2 in every 1,000 births–are posing serious ethical and legal dilemmas for parents and physicians, and for the agencies that must judge whether the state can permit parents or physicians to decide either not to operate or to pay for a lifetime of expensive care if the operations are performed. Today, a number of physicians are using criteria such as severe paralysis, spinal curvature, and an enlarged head to decide to forego surgery. Can society tolerate decisions not to treat a human being? Can decisions be made on the basis of scientific measures, or is the issue fundamentally ethical? Moreover, if we may let a person die, is it permissible to actively hasten his death? How can we determine which treatments can be refused? What policies should we use to make these decisions?

A century or even a decade ago, these questions would not have arisen; Jimmy would have died from infection, of the spinal opening or of his bladder. Now infection can be halted with antibiotics, the spinal opening can be surgically closed, and mental retardation controlled, at least partially. But in such conditions, biomedical intervention cannot be totally successful. Technology permits us to save life–or at least prolong it–but does not ensure that the life will be worth living.

The author:
Robert M. Veatch is Staff Director, Research Group on Death and Dying, the Institute of Society, Ethics and the Life Sciences. He is the author of *Death, Dying and the Biological Revolution.*

We can remove a human heart and keep a patient alive mechanically for many days waiting for a heart transplant, but we cannot completely control the rejection of the new heart when it is put into the body. For someone whose brain is destroyed through an accident or illness, we can control breathing, blood pressure, and other functions with respirators, drugs, and special diet. But we can neither restore the brain's capacity to control the functions nor, more important, to think or feel or respond. We can prolong the life of a victim of advanced cancer, but cannot ultimately stop the disease process. Nor can we avoid the effects of the disease or all the side effects of the treatment–the nausea of radiation sickness, the loss of tissues and limbs, and the psychological trauma.

The new medical technologies–cardiac pacemakers, kidney machines, intravenous feedings–and drugs that control virtually every body function, save some lives and prolong others at a level of quality that everyone would agree is high. The problem is, however, that the same technologies, when applied to terminally ill patients, prolong the process of dying.

The new technologies are not alone to blame. In the 1970s, there are more points in the progress of disease where decisions can be made, and there are more decision makers. Not only the doctor, but also relatives, hospital administrators, the clergy, and the law are involved in the decisions. Most important, the patient is more and more becoming central. In the social revolution of the last 10 years, equality, freedom, and rights are being demanded by racial minorities, women, students, and the elderly. Not surprisingly, a patients' rights movement has also emerged, with patients demanding to participate in any decisions affecting their bodies.

One reason for this is the neglected fact that medical choices require not only technical expertise but also ethical values. We used to believe without question that the "doctor knows best." If you had pneumonia, the choice was whether or not to take penicillin. Unless you had a religious objection to the drug, penicillin was preferable to the alternative...death. Since the physician knew the technical aspects of the treatment, and there was no disagreement on the values, it would not occur to you that you had a role in deciding whether or not to take the penicillin. Today, however, when the choice in a terminal disease is between two weeks at home in peace with your family, or two months, or two years, in a hospital with repeated radical surgery and tubes to control diet and breathing and elimination, the decision is not so obvious. This is why some patients insist on a role in making decisions on their treatment.

People who are going to be central decision makers about their own medical care, or want to participate in the public policies controlling death and dying in our society, must be aware of several crucial distinctions. One of the most basic is the difference between deciding what death means and what is acceptable care for someone who is dying. It used to be perfectly obvious when a person was dead. Within a few seconds or minutes the heart, respiration, and brain all stopped functioning. If these processes did not reverse spontaneously, the person was considered dead. Now, however, the technological ability to continue the functions of the heart and lungs, even in a patient whose brain has been completely destroyed, has given rise to a debate on the "definition of death."

One side says that we should pronounce a person dead whose brain is destroyed, even if he is breathing on a respirator and his heart is beating because the respirator is providing oxygen. An Ad Hoc Committee of the Harvard Medical School to Examine the Definition of Brain Death developed clear scientific measures for predicting that a person's brain has been completely destroyed. The committee, made up of distinguished scholars including a lawyer, a historian of science, and a theologian as well as several physicians, submitted a report in 1968, in which it outlined four measures. There must be: (1) total unawareness of external stimulation; (2) no movements or breathing; (3) no reflexes; and (4) a confirming flat electroencephalogram, indicating the absence of any brain activity. The tests must be repeated at least 24 hours later with no change. Conditions resulting from drugs that depress the central nervous system and from abnormally low body temperature must be excluded. These criteria have been confirmed without exception in thousands of cases. By mid-1976, 11 states had adopted legislation sanctioning the use of such criteria.

People on the other side of the debate raise philosophical questions. They argue that it is not the brain that dies, but the person as a whole.

We should be careful to use the criteria, these people say, only as measures of when the person as a whole is dead. Also, some people are concerned that the brain-oriented criteria are favored primarily because it makes it easier to obtain organs for transplantation. These two objections suggest more cautious phrasing in any legislation. The law should say clearly that a *person,* not simply his *brain,* is dead. And the person ought to be considered dead when his brain is destroyed, regardless of whether he is a potential organ donor.

More recently, two other problems have emerged. First, once it has been determined that a person's brain is irreversibly destroyed, we still must decide whether he should be considered dead. The first question is technical and must be resolved by medical experts. But the second is a policy question calling for value judgments. Defining death depends not only on medical facts, but also on our religious and philosophical values about what is essential to human nature. This second question must be answered by citizens like you and me.

The second problem is even more difficult. Should we call a person dead if he loses his higher brain functions, but still retains reflexes controlled by the lower brain centers? We need both a technical and a policy answer. The technical question is: Can medical experts be positive that the patient has irreversibly lost higher brain functions when he still retains some lower brain activity? The policy question is: Is the essential part of a person his capacity to think and feel, or simply to carry out the functions of breathing and pumping blood controlled by the lower centers? It is a complex issue, for if we use the capacity to think and feel as the essential element of what it means to be human, we are in danger of defining the catatonic schizophrenic mental patient or the mentally retarded as "dead" in the sense that they do not have the essential human characteristics.

Nevertheless, I think we can safely decide that a person is dead if he has lost all higher brain function while he retains lower brain reflexes, and still consider him alive if he retains minimal capacity to use his higher functions. But that decision cannot be left to technical experts. In current practice, different physicians–sometimes in the same hospital–are using different concepts of death. If we are not convinced that we now can be sure when higher brain functions are irreversibly lost, perhaps we should maintain the simpler policy of calling people dead only when they lose the function of their entire brain.

But what of the person we all decide is still minimally alive? Here we must distinguish between cases where further medical intervention would simply prolong an inevitable dying process and those where treatment could maintain the life of a patient indefinitely, even though he would never be "cured." Consider a baby I once saw born with a genetic disease called Trisomy-18. He had severe breathing difficulties as well as physical abnormalities. The physician was certain that treatment would only prolong the inevitable; the child would die within one or two weeks. The situation is radically different in the case of Jimmy, the baby with the spinal deformity. Although he will almost certainly die without treatment, he may have a relatively normal lifespan if we intervene. Whatever our moral feelings, the two cases seem different. It is hard to see what can be gained by requiring short-term treatment simply to prolong an inevitable dying process for a couple of weeks.

If it is acceptable to let a baby die, would it not be better to actively hasten death and cut short the suffering? Is simply letting die ethically and legally the same as active killing? Motives may lead to complete exoneration of the mercy killer in countries such as Germany, Italy, Sweden, Switzerland, and Uruguay. A tolerant attitude toward euthanasia is also found in Denmark, Holland, Yugoslavia, and Spain. But in the United States, it is clearly illegal to actively kill, even for reasons of mercy.

There are several reasons offered for the distinction between active killing and simply letting die. One is that psychologically we feel different about actively killing a person than we do about letting him die. However, this may be simply because we have been taught that active killing is "more wrong." It is not very logical to use that as a reason for distinguishing between the two acts.

Some people point out that the active killing of a dying, suffering patient is at variance with the traditional role of the medical profession–to preserve life. This is not necessarily true, however, if we can accept the case where the physician does not use all measures available to save his patient. It also does not pertain if we accept that another individual may do the killing. Some people argue that the intention is different in letting die than in active killing, but this is not always the case either. Death may intentionally be brought about by omission–by not starting a respirator, for example. Or, a doctor may actively kill a patient unintentionally, when, for example, he lets the knife slip during surgery or gives a drug to relieve pain that causes an allergic reaction that stops breathing.

There are more plausible arguments against active killing. If we accept it in an exceptional case for reasons of mercy, it may lead to active killing for less altruistic reasons. For example, during the 1930s and 1940s the Nazis practiced euthanasia to "purify the race."

Another consequence might be a decision to kill that is a mistake. Mercy killing is often done in passion, before the medical condition is really clear. In 1973, for instance, 26-year-old George Zygmaniak broke his neck in a motorcycle accident. Convinced that he would be paralyzed from his neck down for the rest of his life and that he could never adjust to such a life, he pleaded with his brother, Lester, to put him out of his misery. Three days after the accident, his brother killed him with the family shotgun. Many people would argue that even if the merciful act was ethically acceptable, there had not been enough time to make sure that the paralysis was permanent or to explore the possibilities for George to adjust to a new life.

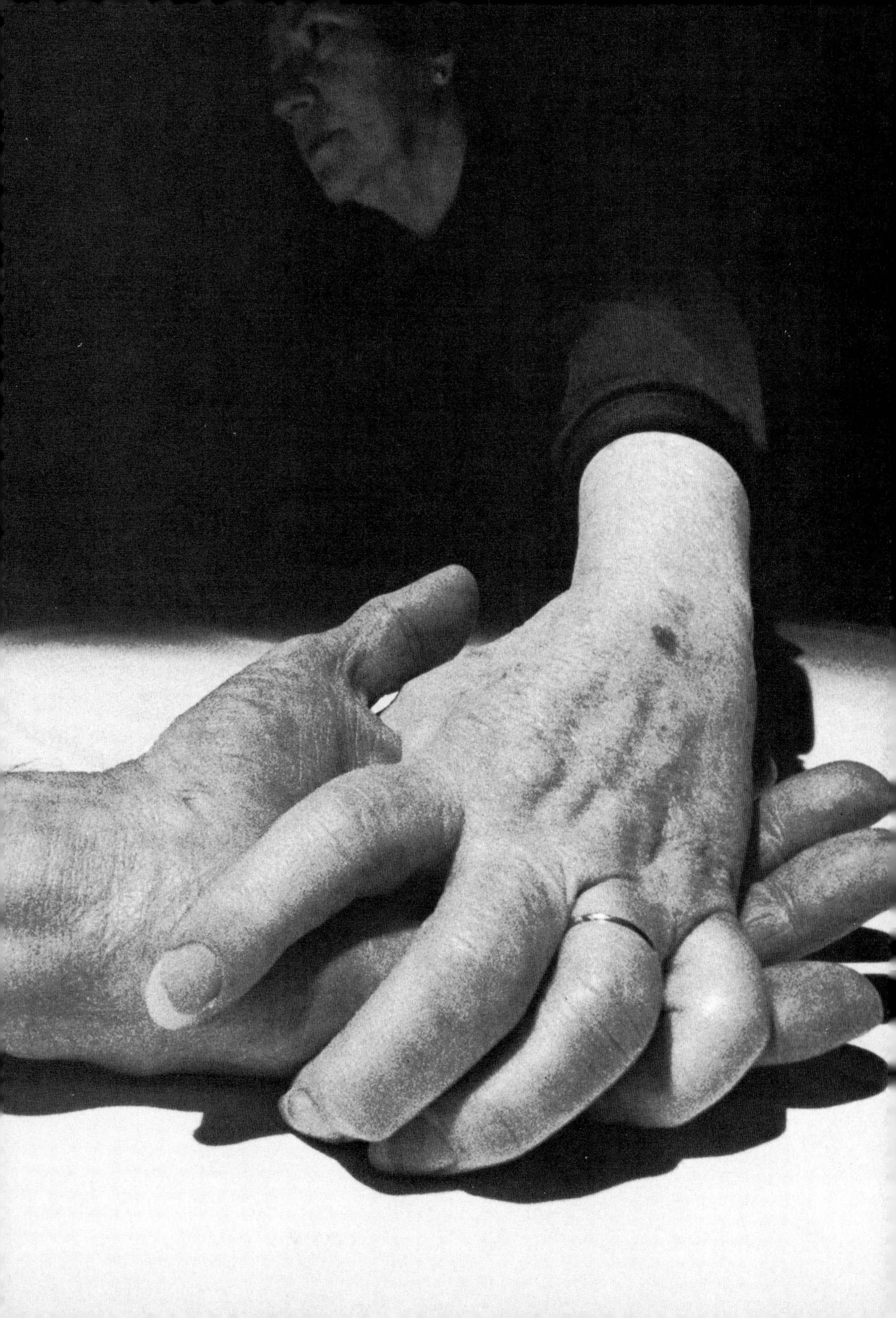

There is no question that the cause of death is different when one actively kills than when one simply steps aside to "allow nature to take its course." Is this morally relevant, though, if the expected consequences are the same? It seems to me an oversimplification either to maintain that there is no difference or to argue for an absolute moral separation. Consider the case of a patient in a coma who feels no pain, and will probably die within 24 hours. He needs a drug injected every hour to regulate his blood pressure. You can either kill him actively by injecting an air bubble into a vein, or indirectly by omitting the blood pressure drug. Which course would you choose?

Determining whether there is indeed a moral difference between active killing and letting die does not always resolve practical dilemmas, however. It may be ethical to kill someone who is inevitably dying and in unrelievable agony, for instance. On the other hand, many physicians claim that there is no pain that cannot be relieved with painkillers. If this is true then, even if active killing may be condoned as theoretically moral, there may be no cases where it is necessary. And even if it is considered moral, perhaps it should still be illegal because of the danger of errors or abuse.

The real issue, then, seems to be not in justifying occasional active killing for mercy, but deciding when it is ethically and legally acceptable to stop treatment so that the patient can do his own dying. But we also have to deal with the question of what treatments may be reasonably refused. Treatments thought to be morally required are sometimes called "ordinary" while those thought expendable are called "extraordinary." An intravenous feeding is called "ordinary," but a heart transplant is not. Yet, defining the reasonableness of a treatment by how common or ordinary it is implies that one should logically assume that a treatment given often, even if foolishly or uselessly, should be morally required.

Theologians and philosophers have never defined "ordinary" simply as the common treatment. They consider the condition of the specific patient. A complicated, experimental treatment such as a transplant might be reasonable for a young patient with a good chance for a long, healthy life. On the other hand, an intravenous feeding might be "extraordinary" for a senile patient with terminal cancer and a bad heart who would otherwise actively resist treatment.

The next problem is how to distinguish useful treatments from useless ones. If the treatment will do no good, it does not seem reasonable to give it. But this judgment depends on personal values. Is prolongation of life in an irreversible coma desirable? It may be if one believes in the value of biological life per se. And even if it is established that the treatment is indeed useful, it does not necessarily follow that the treatment is required. The courts have consistently and correctly allowed competent adult patients who are members of the Jehovah's Witnesses movement to refuse blood transfusions, even though the treatment was judged necessary to save their lives. The Roman Catho-

lic Church considers several obviously lifesaving treatments to be "extraordinary"–for example, continued use of a kidney machine for a patient who cannot cope with the psychological burden of having his blood dialyzed three times a week for the rest of his life.

Pope Pius XII, in a 1957 address to the International Congress of Anesthesiologists in Rome, said, "Normally, one is held to use only ordinary means–according to circumstances of persons, places, times, and culture–that is to say, means that do not involve any grave burden for oneself or another."

The key term is "grave burden," which I think needs to be analyzed. Burden to others (inconvenience to hospital staff, relatives, or taxpayers) should not be as weighty a consideration as burden for the patient himself–whether directly, in terms of pain, suffering, and degradation, or indirectly, if he decides that his continued treatment is consuming family resources. Taking this "patient perspective" also means that it is not sufficient for the physician alone to decide if a treatment is a grave burden for the patient. It must be the judgment of the patient himself. Our right to privacy gives us the fundamental right to control of our bodies, even if death may be the result.

Where the patient is not competent to judge, however, matters become more complex. A patient who is a minor, or is judged mentally incompetent by a court, has someone–usually a parent–presumed or designated to be his guardian. This guardian has the right and the duty to do what he believes is in the best interests of the patient. The incompetent patient has as much right to be spared the burden or degradation as a competent patient. The difference in the case of the guardian refusing to let treatment continue, however, is that society requires that the guardian be "reasonable."

The case of Karen Ann Quinlan raises almost every issue I have discussed. On the night of April 15, 1975, a police rescue squad brought 21-year-old Quinlan to a hospital emergency room in Newton, N.J. She had apparently stopped breathing for two periods of at least 15 minutes. Urine and blood tests revealed the presence of quinine, aspirin, barbiturates, and traces of the tranquilizers Valium and Librium. While the exact nature of her brain damage was not determined precisely, additional tests led to the conclusion that she was in what was termed a "persistent vegetative state."

By July, Karen Ann Quinlan's parents had authorized treatment to be stopped. When the physicians refused, Joseph Quinlan brought the case to court, seeking to be clearly designated as his daughter's guardian for the purpose of authorizing the discontinuance of all extraordinary treatments. In the court hearings, it was at first argued that the woman might be dead according to brain criteria for pronouncing death. It is not clear whether New Jersey law permits death to be pronounced on the basis of brain criteria, however. In any event, it later became clear that although she was in a "persistent vegetative state," she did not meet the Harvard Committee criteria. She could breathe without the respirator for periods of time and electroencephalograms showed evidence of brain activity. She was not dead by any commonly debated definition, including complete loss of higher brain activity. Nor was it clear whether she was "dying." Life such as hers may be prolonged indefinitely. Yet she is not like the patient on the kidney machine nor the baby with the deformed spine.

There has been much disagreement about whether the Quinlans' refusal of further treatment for their daughter was sufficiently reasonable. Was turning off a respirator more like active killing (a view held by many physicians) or like letting die (the position of most philosophers and theologians)?

In such cases, once we decide what is ethically acceptable and legal, we are still left with the unpleasant question of what ought to be done when we face a death in our own family or when the state considers laws to change public policy. Some may opt for the simple solution that we ought not to decide anything because to do so would be to play God. Yet, if decisions about medical intervention were trespassing in territory not appropriate for humans, there could be no medical practice or hospitals. The Bible tells us that when God created human beings, He gave them dominion over the earth. To fail to make difficult decisions when they are called for is to fail in our task as human beings. It is an inevitable result of our complicated society that we have to make a choice about which interventions are appropriate and which are not.

There are some basic policies to consider if our fundamental rights and responsibilities that are now in effect are to be protected. The most common policy is to let the individual physician–who may or may not consult others–decide at the critical moment what action is to be taken. Now, those who still believe that the doctor knows best may be trying to escape the burden of making decisions or, more likely, they are confusing two basic issues. The technical question of what the possible treatment alternatives are and the likely outcome is the appropriate domain of the physician. But the policy question we must determine ourselves: What *ought* to be done in this case, given the possible outcome of the treatment alternatives? On that, the physician is no more an expert than you and I. He must draw on his own personal value system to reach an answer. It would be wrong to let him inflict his values on us. That is why it was so important that the law in Quinlan's case ultimately recognized her right of privacy.

New Jersey Superior Court Judge Robert Muir, Jr., ruled on Nov. 10, 1975, that "the determination whether or not Karen Ann Quinlan be removed from the respirator is to be left to the treating physician." Apparently, Judge Muir confused the expertise the physician has in determining Karen Quinlan's physical condition with the special skill or authority required to make the ethical judgment about what ought to be done with a person in her condition. Fortunately, on March 31, 1976, the New Jersey Supreme Court unanimously decided, in effect,

to reverse that judgment. They held that, for a patient in this particular condition, the decision is one to be made by the father acting as her guardian. He was given the authority to select a physician who could determine whether there is any reasonable possibility that his daughter would return to a cognitive sapient state. This physician must decide if treatment should be terminated. The provision makes very little sense, because the court gave the father the right to privacy in making the choice and the right to select physicians for his daughter's care. If the physician does decide that termination of treatment is called for, the matter will be referred to an ethics committee. If that committee confirms that there is no reasonable hope for recovery to a cognitive sapient state, then treatment may be stopped.

While I agree with the court's decision, I think there are dangerous confusions in the way it specified that the decision be made. The committee was poorly defined. We are not told whether it is to be made up of lay people, medical professionals, or a combination of both. We are not told whether the committee must rule by majority vote or be unanimous. Furthermore, the committee is given only one task–to determine if "there is no reasonable possibility of Karen ever emerging from her present comatose condition to a cognitive, sapient state." The question is purely technical–no ethical decision is to be made by the "ethics committee."

Nevertheless, I think the court was right in asking the committee to perform only the technical task. While the committee should guard against any extreme position taken by the physician, its members would not necessarily hold the same religious and ethical values as the father acting on behalf of his daughter. To ask the committee to make an ethical judgment would be a breach of the patient's privacy. If the patient is to be deprived of the right to privacy by having his guardian's decision overridden on grounds of unreasonableness, it should only be done by a duly constituted public body such as the court.

There are options that consider the incompetent patient's perspective. Their virtue is that they try to make the decision that the patient would have made. One such option is the informal letter that you or I could write while competent, giving instructions about terminal care. An example is the Living Will. It is a good informal device that can serve as the basis of discussions with your family or physician about your terminal care should you become incompetent. However, since it is written while you are healthy and often years before it is used, it is difficult to be specific. Your opinions, or technology, or both may change. Furthermore, it is not legally binding, so a physician cannot be compelled to carry out your instructions.

Another option would involve a more specific set of instructions to the legal apparatus. I have been experimenting with a draft of a power of attorney document that would simply transfer the decision-making authority to some trusted individual. It is meant to be legally binding, although this has not yet been tested in court.

Still another option is a public one. Most people who end up in a coma do so without having put anything in writing. There have been at least 20 attempts in 16 states to introduce legislation to clarify what can be done in such cases. None had passed by mid-1976, but they provide interesting examples of what a more public policy might be. Some states, such as Montana and Oregon, have considered bills which would make active killing legal on request. These bills have found little support, in part because of moral objections, and in part because they do not get at the real problem of decision making for the incompetent. You cannot ask to be killed if you are a baby, or if you are an adult in a coma.

A second kind of bill would legalize the Living Will type of letter. Such bills have been considered in Florida and Massachusetts. They would be helpful only if patients put their wishes in writing while they were still competent.

A third kind of legislation makes the most sense to me. It follows many of the principles of the Quinlan decision. It would first confirm the patient's right to refuse treatment while competent. A bill introduced in the West Virginia legislature, for example, presumes that in the case of an incompetent patient the next of kin would become the guardian or, if there were no next of kin, the court would appoint a guardian. I favor such a bill, but feel that, in addition, the individual should have the right to legally designate a guardian ahead of time. This would resolve the problem if for some reason you did not want your next of kin to make such decisions, or if you had two relatives of equal degree of kinship. A bill of this sort should also make clear that a physician terminating treatment under such instructions would not be guilty of a crime. And it should protect the physician whose conscience would not permit him to discontinue treatment, provided he found some other physician who would.

Medical technologies have become so complex and treatment alternatives are so controversial that policy questions–which depend on religious and other ethical values–may no longer be left out of any medical decision. It is essential to develop a decision-making method that will protect the rights of the patient to receive or refuse treatment according to his own values.

Difficult decisions will always be with us. We might overcome tissue rejection in organ transplants, control the spread of malignancy, and eliminate heart disease through diet and drugs, but there will always be new medical problems. Whatever we do, disease and death will never be completely conquered.

While we will probably never find the choices easy, we need to know what our options are–our rights and responsibilities as individuals and as family members to decide what treatments are so useless or burdensome from the patient's perspective that they may be refused. People have a right to the best possible medical care. They also have a right to refuse that care, even if it means death.

Index

This index covers the contents of the 1975, 1976, and 1977 editions of *Science Year,* The World Book Science Annual.

Each index entry is followed by the edition year in *italics* and the page numbers:

Muons, *77*-325, *76*-323

This means that information about Muons begins on the pages indicated for each of the editions.

An index entry that is the title of an article appearing in *Science Year* is printed in boldface italic letters: ***Archaeology.*** An entry that is not an article title, but a subject discussed in an article of some other title, is printed: **Plutonium.**

The various "See" and "See also" cross references in the index are to other entries within the index. Clue words or phrases are used when the entry needs further definition or when two or more references to the same subject appear in *Science Year.* These make it easy to locate the material on the page.

Neurology, *77*-312, *76*-312, *75*-321; *Special Report,* *76*-54. See also **Brain; Nervous system; Neuroanatomy.**

The indication *"il."* means that the reference is to an illustration only, as:

Alberta Fireball, *il., 77*-216

Index

A

B

Index

C

Index

D

E

F

Index

G

H

Index

I

J

K

L

M

Index

N

Index

Index

S

T

Index

U

V

Acknowledgments

The publishers of *Science Year* gratefully acknowledge the courtesy of the following artists, photographers, publishers, institutions, agencies, and corporations for the illustrations in this volume. Credits should be read from left to right, top to bottom, on their respective pages. All entries marked with an asterisk (*) denote illustrations created exclusively for *Science Year*. All maps were created by the *World Book* Cartographic Staff.

Cover
Ed Hoppe*

Advisory Board
7 Lee Boltin*; Joseph A. Erhardt*; Joseph A. Erhardt*; Roland Patry*; Dennis Galloway*; John Swanberg*; Theodore Polumbaum*

Special Reports
10 Ernest Norcia*; Lee Balterman*; Intel Corp.; D. M. Yermanos, University of California; Jean Helmer*
12-14 Western Geophysical Division, Litton Industries
15 Seiscom Delta Inc.
16-24 Western Geophysical Division, Litton Industries
26 Joseph A. Erhardt*
30-33 George Suyeoka*
36 Christopher Morrow
37 Don Richards, The Menninger Foundation
40 Ed Hoppe*
43 Ed Hoppe*; V. Mueller (Joseph A. Erhardt*)
44 Marge Moran*
46-47 Ed Hoppe*
48 José Luis Salazar, M.D., University of Illinois; Marge Moran*
50-51 Alex Ebel*
52 Joseph A. Erhardt*
54 Robert T. Bakker; David Cunningham*
55 Robert T. Bakker
57 Alex Ebel*
58 David Cunningham*
60-61 Alex Ebel*
63 James Teason*
64 Jack J. Kunz*
67 Rudolph Schmid, University of California
68 Anne G. Vietmeyer
70-71 Noel D. Vietmeyer
73 D. M. Yermanos, University of California
74 D. M. Yermanos, University of California; Office of Arid Land Studies, University of Arizona
75 Noel D. Vietmeyer
77 T. N. Khan, University of Papua New Guinea; International Institute of Tropical Agriculture, Ibadan, Nigeria; Noel D. Vietmeyer
78 Larry Bagnall and James Hentges, University of Florida; NASA
80-81 Nemo Warr
82 John Levis
83 John Levis; John Levis; John Levis; Jeanne Riddle, M.D.; Jeanne Riddle, M.D.
84-85 Outer Coffin of the Chantress of Amun, Henttowy. The Metropolitan Museum of Art, New York City. Museum Excavations, 1923-1924, Rogers Fund 1925; John Levis
86 Nemo Warr; John Levis; John Levis; John Levis
89 George Lynn; John Levis; John Levis; John Levis
90 George Lynn
91-92 John Levis
94-95 Ernest Norcia*
96 Kathryn Sederberg*
98-100 Ernest Norcia*
101 Halton Arp and Francesco Bertola
103-105 Ernest Norcia*
108 U.S. Forest Service
110 Marge Moran
111 Jim Dunn, National Ski Patrol
112-113 U.S. Forest Service
114 Marge Moran*
116 U.S. Forest Service
117 Marge Moran*; Marge Moran*; U.S. Forest Service
118 U.S. Forest Service; Marge Moran
119 Marge Moran
120 Jim Dunn, National Ski Patrol
122-123 Alexander Low, Woodfin Camp, Inc.
124 Jerry Elliott, San Diego State College
125 Bob Day from Nancy Palmer; Hans Reinhard, Bruce Coleman Inc.; Adam Woolfitt, Woodfin Camp, Inc.
126-127 Jean Helmer*
129 Cyril Ponnamperuma, University of Maryland
130-131 Jean Helmer*
132-133 Woods Hole Oceanographic Institution; Ernest Braun; Shelly Grossman, Woodfin Camp, Inc.; Kal Muller, Woodfin Camp, Inc.
135 Jean Helmer*
136 S. Brooke from Sidney W. Fox, University of Miami, Coral Gables, Fla.; L. A. Nagy from *Science*. Copyright 1974 by the American Association for the Advancement of Science
138-139 Centurion Industries Inc.
141 United Press Int.; Hewlett Packard; Mitchell Funk, The Image Bank; Hewlett Packard
142-143 Intel Corp.; IBM Corp.; IBM Corp.; IBM Corp.; IBM Corp.
144-145 Henry Groskinsky; Intel Corp.
147 Calvin Campbell, Massachusetts Institute of Technology; Leeds & Northrup Company; The Perkin-Elmer Corporation
148 Sears, Roebuck and Company (Joseph A. Erhardt*); Joseph A. Erhardt*; Joseph A. Erhardt*
149 Bennett Pump Company; Amana Refrigeration, Inc.; Chrysler Corporation (Joseph A. Erhardt*); Bell Laboratories
152-153 Bobbye Cochran*
154 Robert Hamburger
155 E. Bulba, Department of Environmental Sciences, Harvard University
156-157 Bobbye Cochran*
158 Henry Metzger, National Institutes of Health
160 Bobbye Cochran*
165 Robert H. Glaze, Artstreet
166 Morbark Industries, Inc.
167 U.S. Forest Service
168-169 Donald Meighan*; Henry Koval*
170 Union Electric Company; Union Electric Company; Occidental Research Corporation
173 Donald Meighan*; U.S. Army Natick Laboratories
174 Tate & Lyle Limited
176 Ernest Norcia*
178 Kathryn Sederberg*
181 Copyright 1975 by the Association of Universities for Research in Astronomy, Inc. John Luthes, Kitt Peak National Observatory
182 Jet Propulsion Laboratory
184-185 Ernest Norcia*
186 Sovfoto
187 Jet Propulsion Laboratory
190-191 Derek Grinnell*
193 Brown Bros.; Bettmann Archive; Bettmann Archive
194 Brown Bros.
195 Bettmann Archive
196-198 Product Illustration
200 Lee Balterman*
202 Joseph A. Erhardt*
203 Kinuko Craft*

204-206 Lee Balterman*
207 Kinuko Craft*
208 Lee Balterman*
210-213 The Center for Short-Lived Phenomena
214 Ricardo Ferro, Black Star
215-222 The Center for Short-Lived Phenomena

Science File

224 U.S. Department of Agriculture; © National Geographic Society; Walter Kale, *Chicago Tribune;* Cerro Tololo Inter-American Observatory; Jim Wallace, Duke University
226 Walter Kale, *Chicago Tribune*
227 Smallford Planters Ltd.
228 U.S. Department of Agriculture
231 David Brill, © National Geographic Society; © National Geographic Society; Cleveland Museum of Natural History; © National Geographic Society; © National Geographic Society
233-234 Wide World
235 Anthony F. Aveni, Colgate University; Mas Nakagawa*
236 Garbage Project, University of Arizona
237 Donald W. Lathrop
238 Copyright © 1975 by permission of *Saturday Review* and Nick Hobart
239 Jet Propulsion Laboratory
241 U.S. Naval Research Laboratory
242 R. M. Williamon, Fernbank Science Center Observatory
244 Cerro Tololo Inter-American Observatory; Mas Nakagawa*
246 H. Spinrad and H. E. Smith, Kitt Peak National Observatory
247 D. Berg, J. Davies, B. Allett and J.-D. Rochaix, University of Geneva; Mas Nakagawa*
248 Henry Martin, *American Scientist,* Journal of Sigma XI, the Scientific Research Society of North America, Inc.
249 P. M. Andrews and C. R. Hackenbrock, University of Texas Health Science Center
253 Andrew H. Knoll and Elso S. Barghoorn, Harvard University. © 1975 by the American Association for the Advancement of Science
255 C-E Tyler Industrial Products, subsidiary of Combustion Engineering, Inc.
256 U. S. Department of Agriculture
259 Argonne National Laboratory
261 R. Henderson and P. N. T. Unwin, Medical Research Council Laboratory of Molecular Biology, Cambridge
263 Communications Satellite Corporation
265 Carl Iwasaki, *Newsweek*
267 Lawrence E. Gilbert
268 *The New York Times*
270 Copyright © 1976 by permission of *Saturday Review* and James Estes
272 The British Post Office
273 General Electric Research and Development Center
275 Westinghouse Electric Corporation; Solarex Corporation; Manley Commercial Photography, Inc.
278 General Electric Research and Development Center; Mas Nakagawa*
279 Cornell University
280 Industrie Pirelli SPA
281 Pacific Gas and Electric Co.
282 Jack D. Griffith, Stanford University School of Medicine
283 Jim Wallace, Duke University
284 NASA
285 Smithsonian Institution; Mas Nakagawa*
287 U.S. Geological Survey
289 NASA
292 Sidney Harris, *American Scientist,* Journal of Sigma XI, The Scientific Research Society of North America, Inc.; David L. Dilcher, Department of Plant Sciences, Indiana University, Bloomington.
294 Paul A. Farber, Steven Specter, and Herman Friedman, Temple University and Albert Einstein Medical Center
295 Erling Johansen, D. M.D., School of Medicine and Dentistry, University of Rochester
296 Drawing by Ross; © 1975 The New Yorker Magazine, Inc.
297 Wide World
300 F. N. G. Clarke Limited, Varian Associates
302 Mas Nakagawa*
303 Mort Bond, The Children's Hospital of Philadelphia
304 Mas Nakagawa
305 National Severe Storms Laboratory, NOAA; Rick Bellati, *Stillwater News-Press*
306-307 NOAA
308 Dana M. Fowlkes, Gerald B. Dooher, and William M. O'Leary, Cornell University Medical College
309 Richard Blakemore, Woods Hole Oceanographic Institution, from *Science.* Copyright 1975 by the American Association for the Advancement of Science
312 Martha Constantine-Paton and Robert R. Capranica, Cornell University from *Science.* Copyright 1975 by the American Association for the Advancement of Science
314 Los Alamos Scientific Laboratory, University of California
316 California Polytechnic State University; Carol Miller
318 Vahan Shirvanian, reprinted from *Audubon,* The Magazine of the National Audubon Society
320 Mas Nakagawa*; Hawker Siddeley Dynamics Ltd.
321 Stanford Research Institute
322 M. Isaacson, J. Cangmore, and A. V. Crewe, University of Chicago
324 Mas Nakagawa*
325 Fermi National Accelerator Laboratory
327 Los Alamos Scientific Laboratory, University of California
328 Lawrence Berkeley Laboratory, University of California
329 Sovfoto
331 Peter Williams, Charles Evans, Martin Grossbeck, and Howard Birnbaum, Materials Research Laboratory, University of Illinois
333 Paul J. Weldon and Daniel L. Hoffman, Bucknell University
334 Alan Copeland, NYT Pictures
337 Calvin Grondahl, *Deseret News*
340 Wide World
341 Sidney Harris, *American Scientist,* Journal of Sigma XI, The Scientific Research Society of North America, Inc.
343-345 NASA
346 S. A. Engins Matra
348 Blaupunkt
349 Ralph M. Wetzel, University of Connecticut
350 Sidney Harris, *American Scientist,* Journal of Sigma XI, The Scientific Research Society of North America, Inc.
351 Jerome S. Rovner, Ohio University
353 Bill Meng, New York Zoological Society

People in Science

354 Bruno Barbey, Magnum*; Joseph A. Erhardt*; Joseph A. Erhardt*; Joseph A. Erhardt*; Bruno Barbey, Magnum*
356 Bruno Barbey, Magnum*
359-360 Garafalo, *Paris Match*
363 Bruno Barbey, Magnum*
364 Simon and Ginfray, Gamma
366-368 Bruno Barbey, Magnum*
370-379 Joseph A. Erhardt*
380 Motorola Inc.; U.S. Navy; Albert M. Leahy, Office of Naval Research; SP5 M. L. Frommel, U.S. Army; A. C. Caprio, Yellowstone National Park
382-384 Joseph A. Erhardt*
386 Pictorial Parade
388 CERN; University of Illinois
389 © Karsh, Ottawa from Kittay Scientific Foundation
390 Pictorial Parade
391 Pepperdine University; NASA
392 University of California
394 American Association for the Advancement of Science; Franklin Institute; Columbia University
395 Pictorial Parade; Kamera Bild; Keystone
397-408 Joseph A. Erhardt*

Typography
Display—Univers
Total Typography, Inc., Chicago
Text—Baskerville Linofilm
Total Typography, Inc., Chicago
Text—Baskerville Linotron
Black Dot Computer Typesetting
Corporation, Chicago

Offset Positives
Collins, Miller, & Hutchings, Chicago
Schawkgraphics, Inc., Chicago

Printing
Kingsport Press, Inc., Kingsport, Tenn.

Binding
Kingsport Press, Inc., Kingsport, Tenn.

Paper
Text
Childcraft Text, Web Offset (basis 60 pound)
Mead, Escanaba, Mich.

Cover Material
Oyster White Lexotone
Holliston Mills, Inc., Kingsport, Tenn.
Riverside Lithox
Columbia Mills, Inc., Minetto, N.Y.